1992

This book evaluates the increasing wealth of knowledge which has accumulated concerning the regulation of synthesis and assembly of structural components of the bacterial cell. The state of this discipline is now such that it is possible, in many cases, to trace the exact sequence of events triggered by a change in the physical or chemical environment of a bacterial cell: signalling, gene expression, transport of the gene product to its correct location and assembly into a functional structure. The scope of this volume is broad, ranging from the organisation of the nuclear material itself to the sequence of events leading to differentiation and development; from the synthesis of intracellular storage material to the assembly of specialised photosynthetic membranes, periplasmic electron transfer chains and heat resistant spores. It will therefore provide an authoritative teaching text for students as well as a comprehensive series of reviews of structural aspects of microbial physiology and biochemistry.

PROKARYOTIC STRUCTURE AND FUNCTION

A New Perspective

SYMPOSIA OF THE SOCIETY FOR GENERAL MICROBIOLOGY*

Series editor (1991–1996): Dr Martin Collins, Department of Food Microbiology, The Queen's University of Belfast

1 THE NATURE OF THE BACTERIAL SURFACE
2 THE NATURE OF VIRUS MULTIPLICATION
3 ADAPTATION IN MICRO-ORGANISMS
4 AUTOTROPHIC MICRO-ORGANISMS
5 MECHANISMS OF MICROBIAL PATHOGENICITY
6 BACTERIAL ANATOMY
7 MICROBIAL ECOLOGY
8 THE STRATEGY OF CHEMOTHERAPY
9 VIRUS GROWTH AND VARIATION
10 MICROBIAL GENETICS
11 MICROBIAL REACTION TO ENVIRONMENT
12 MICROBIAL CLASSIFICATION
13 SYMBIOTIC ASSOCIATIONS
14 MICROBIAL BEHAVIOUR 'IN VIVO' AND 'IN VITRO'
15 FUNCTION AND STRUCTURE IN MICRO-ORGANISMS
16 BIOCHEMICAL STUDIES OF ANTIMICROBIAL DRUGS
17 AIRBORNE MICROBES
18 THE MOLECULAR BIOLOGY OF VIRUSES
19 MICROBIAL GROWTH
20 ORGANIZATION AND CONTROL IN PROKARYOTIC AND EUKARYOTIC CELLS
21 MICROBES AND BIOLOGICAL PRODUCTIVITY
22 MICROBIAL PATHOGENICITY IN MAN AND ANIMALS
23 MICROBIAL DIFFERENTIATION
24 EVOLUTION IN THE MICROBIAL WORLD
25 CONTROL PROCESSES IN VIRUS MULTIPLICATION
26 THE SURVIVAL OF VEGETATIVE MICROBES
27 MICROBIAL ENERGETICS
28 RELATIONS BETWEEN STRUCTURE AND FUNCTION IN THE PROKARYOTIC CELL
29 MICROBIAL TECHNOLOGY: CURRENT STATE, FUTURE PROSPECTS
30 THE EUKARYOTIC MICROBIAL CELL
31 GENETICS AS A TOOL IN MICROBIOLOGY
32 MOLECULAR AND CELLULAR ASPECTS OF MICROBIAL EVOLUTION
33 VIRUS PERSISTENCE
34 MICROBES IN THEIR NATURAL ENVIRONMENTS
35 INTERFERONS FROM MOLECULAR BIOLOGY TO CLINICAL APPLICATIONS
36 THE MICROBE 1984 PART I: VIRUSES
PART II: PROKARYOTES AND EUKARYOTES
37 VIRUSES AND CANCER
38 THE SCIENTIFIC BASIS OF ANTIMICROBIAL CHEMOTHERAPY
39 REGULATION OF GENE EXPRESSION – 25 YEARS ON
40 MOLECULAR BASIS OF VIRUSES DISEASE
41 ECOLOGY OF MICROBIAL COMMUNITIES
42 THE NITROGEN AND SULPHUR CYCLES
43 TRANSPOSITION
44 MICROBIAL PRODUCTS: NEW APPROACHES
45 CONTROL OF VIRUS DISEASES
46 THE BIOLOGY OF THE CHEMOTACTIC RESPONSE

*Published by the Cambridge University Press, except for the first Symposium, which was published by Blackwell's Scientific Publications Limited.

PROKARYOTIC STRUCTURE AND FUNCTION

A New Perspective

EDITED BY

S. MOHAN, C. DOW

AND J. A. COLES

FORTY-SEVENTH SYMPOSIUM OF THE SOCIETY FOR GENERAL MICROBIOLOGY HELD AT THE UNIVERSITY OF EDINBURGH APRIL 1991

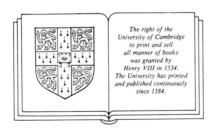

The right of the University of Cambridge to print and sell all manner of books was granted by Henry VIII in 1534. The University has printed and published continuously since 1584.

Published for the Society of General Microbiology

CAMBRIDGE UNIVERSITY PRESS

CAMBRIDGE

NEW YORK PORT CHESTER

MELBOURNE SYDNEY

Published by the Press Syndicate of the University of Cambridge
The Pitt Building, Trumpington Street, Cambridge CB2 1RP
40 West 20th Street, New York, NY 10011-4211, USA
10 Stamford Road, Oakleigh, Victoria 3166, Australia

First published 1992

Printed in Great Britain at The Bath Press, Bath, Avon

A catalogue record of this book is available from the British Library

Library Congress Cataloguing in Publication Data

Society for General Microbiology. Symposium (47th: 1991: University of Edinburgh)
Prokaryotic structure and function: a new perspective: Forty-seventh Symposium of the Society for General Microbiology held at the University of Edinburgh, April 1991/edited by S. Mohan, C. Dow, and J. A. Cole.
p. cm.
"Published for the Society of General Microbiology."
ISBN 0-521-41570-5
1. Prokaryotes–Ultrastructure–Congresses. 2. Prokaryotes–Physiology–Congresses. I. Mohan, S. (Sudesh), 1944–
II. Dow, Crawford, S. III, Cole, J. A. IV. Society for General Microbiology. V. Title.
QR75.S63 1992
589.9–dc20 91-26377 CIP

ISBN 0 521 41570 5 hardback

CONTRIBUTORS

THOMAS, C. M., School of Biological Sciences, University of Birmingham, PO Box 363, Birmingham B15 2TT, UK

JAGURA-BURDZY, G., School of Biological Sciences, University of Birmingham, PO Box 363, Birmingham B15 2TT, UK

ADAMS, D. G., Department of Microbiology, University of Leeds, Leeds LS2 9JT, UK

DAWES, E. A., Department of Applied Biology, University of Hull, Hull HU6 7RX, UK

DREWS, G., Institute of Biology II, Microbiology, Albert-Ludwigs University, 7800 Freiburg, Germany

SHIMKETS, L. J., Department of Microbiology, University of Georgia, Athens, Georgia 30602, USA

HODGSON, D. A., Department of Biological Sciences, University of Warwick, Coventry CV4 7AL, UK

DOOLITTLE, W. FORD, Department of Biochemistry, Dalhousie University, Halifax, Nova Scotia B3H 4H7, Canada

LAM, W., Department of Biochemistry, Dalhousie University, Halifax, Nova Scotia B3H 4H7, Canada

SCHALKWYK, L., Department of Biochemistry, Dalhousie University, Halifax, Nova Scotia B3H 4H7, Canada

STRAGIER, P., Institut de Biologie Physico-Chimique, 13 rue Pierre et Marie Curie, 75005 Paris, France

NANNINGA, N., Section of Molecular Cytology, Department of Molecular Cell Biology, University of Amsterdam, The Netherlands

WIENTJES, F. B., Section of Molecular Cytology, Department of Molecular Cell Biology, University of Amsterdam, The Netherlands

MULDER, E., Section of Molecular Cytology, Department of Molecular Cell Biology, University of Amsterdam, The Netherlands

WOLDRINGH, C. L., Section of Molecular Cytology, Department of Molecular Cell Biology, University of Amsterdam, The Netherlands

WHEALS, A. E., Microbiology Group, School of Biological Sciences, University of Bath, Bath BA2 7AY, UK

MOIR, A., Department of Molecular Biology and Biotechnology, University of Sheffield, Sheffield S10 2UH, UK

PUGSLEY, A. P., Unité de Génétique Moléculaire, Institut Pasteur, 25 rue du Dr Roux, Paris 75724 Cedex 15, France

FERGUSON, S. J., Department of Biochemistry, University of Oxford, South Parks Road, Oxford OX1 3QU, UK

BI, E., Department of Microbiology, Molecular Genetics and Immunology, University of Kansas Medical Center, Kansas City K56 6103, USA

LUTKENHAUS, J., Department of Microbiology, Molecular Genetics and Immunology, University of Kansas Medical Center, Kansas City K56 6103, USA

ROUVIÈRE-YANIV, J., Laboratoire de Physiologie Bactérienne, Institut de Biologie Physico-Chimique, 13 rue Pierre et Marie Curie, 75005 Paris, France

KISELEVA, E., Institute of Cytology and Genetics, Academy of Sciences of the USSR Siberian Brance, 630090 Novosibirsk, USSR. Present address: Department of Molecular Genetics, Medical Nobel Karolinska Institute, Stockholm, Sweden

BENSAID, A., Laboratoire de Physiologie Bactérienne, Institut de Biologie Physico-Chimique, 13 rue Pierre et Marie Curie, 75005 Paris, France

ALMEIDA, A., Laboratoire de Physiologie Bactérienne, Institut de Biologie Physico-Chimique, 13 rue Pierre et Marie Curie, 75005 Paris, France

DRLICA, K., Public Health Research Institute, 455 First Avenue, New York, NY 10016, USA

CONTENTS

EDITORS' PREFACE

The preface to the previous Symposium Volume concerned with structure and function in the prokaryote cell began with two observations. First, 13 years had elapsed since the previous SGM Symposium on the subject. Secondly, the 1978 volume, unlike that of 1965, was not only restricted to prokaryotes but also unable to include all of the rapidly expanding aspects that were prominent in the research literature. In view of the explosive expansion of molecular genetic data during the subsequent 13 years, it is inevitable that the editors of the current 1991 volume have had to be even more selective. Nevertheless, it is hoped that the reader will benefit from our approach which has been to re-examine structure–function–regulation relationships starting from the centre of the bacterium and working outwards. Superimposed on this theme is a new dimension: for the first time it is possible for each author to describe in reasonable detail not only the various structural components and their biochemical composition, but also the organization and DNA sequence of the genes which direct their synthesis, as well as the various control circuits which regulate their expression.

The volume begins by considering the evolution and basic features of gene and genome structure, tracing the roots of current progress to the seminal ideas presented at a 1971 Symposium of the Society by R. Y. Stanier and C. R. Woese. This is followed by an analysis of the possible roles of proteins such as HU in maintaining the regular, but elaborate, structure of the bacterial nucleoid. This topic was avoided in the 1978 volume, presumably because the ideas of Kellenberger, Pettijohn and Worcel were still too controversial to be presented with conviction. A critical review of the replicon hypothesis, which was formulated in 1963 by Jacob, Brenner and Cuzin, again traces the development of current ideas to those presented by W. H. Hayes in the 1965 symposium. And so the volume progresses from internal storage polymers to the genetics of bacterial cell division and a comparison of the prokaryote and eukaryote cell cycle to events at the cell surface – envelope growth, uptake and secretion across bacterial membranes, the synthesis and organization of intracytoplasmic membranes and the importance of the bacterial periplasm. The final four chapters are concerned with bacterial differentiation and multicellularity – sporulation, spore germination, cell–cell interactions controlling development of myxobacteria and differentiation in the Actinomycetes.

Why was it considered necessary to re-examine so many topics that had been presented in the earlier Society Symposia? The answer lies in the structure of many of the chapters which elegantly illustrate that it is now possible, perhaps for the first time, to summarize how environmental changes are sensed by bacteria; how the information is transmitted to the genome to stimulate a different pattern of gene expression; how the resulting new gene products are directed to their correct cellular locations; and, finally, how they are assembled into functional complexes. No longer must we be content to record the structure of a particular component: rather, it is increasingly possible to deduce evolutionary relationships between components, and even predict the consequences of their malfunction. The major regret is that many recent developments in our understanding of structure–function relationships had to be omitted for lack of space.

S. B. MOHAN, C. DOW and J. A. COLE, September 1991

EVOLUTION AND BASIC FEATURES OF GENE AND GENOME STRUCTURE

W. FORD DOOLITTLE, WAN LAM and LEO SCHALKWYK

Department of Biochemistry, Dalhousie University, Halifax, Nova Scotia B3H 4H7, Canada

INTRODUCTION

Our ability to discern evolutionary *pattern* – whether genealogical relationships between species, geographical distributions of populations within species or frequencies of alleles within populations – has sharpened because genes and their primary products can be looked at directly. At the highest level, Zuckerkandl and Pauling's (1965) promise that comparing gene and protein sequences directly or indirectly would allow phylogeny to be reconstructed has been realized many times over. People may still argue about whether humans are closer to chimpanzees or gorillas (Holmquist, Miyamoto & Goodman, 1988), or whether archaea are holophyletic or paraphyletic (Lake, 1991), but such arguments cannot obscure the fact that molecular data have provided many very robust trees not obtainable in other ways, and that it is possible, for instance, to conclude confidently that chimpanzees are truly remote from archaea *purely* on the basis of molecular data.

Molecular biologists can also focus on evolutionary *process*. Indeed, Darwin's premise that patterns of species similarity and difference result from the process of natural selection could not be fully understood without an intimate understanding of how genotype is coupled to phenotype. The range and randomness of mutations exposed to selection could not be appreciated until there was knowledge of mutational mechanisms. And the view that much of what happens in genome evolution (pseudogene formation, concerted evolution, parasitic DNA behaviour) is effectively uncoupled from the evolution of phenotype emerged only from molecular data, collected for other reasons.

Life's history seems most readable, however, when *both* pattern and process can be looked at. Evolutionary inferences about the development of a specific complex trait, say a behaviour, a biochemical pathway, or a molecular organelle like the ribosome, should account ideally for the distribution of components of those behaviours, pathways or organelles among species of known relationship. Primitive features, present at the beginning of development of the trait, should be spread broadly throughout the tree, while features perfected more recently might be of more restricted distribution.

One of evolutionary molecular biology's central goals, surely, is to use this and other analytical principles in reconstructing events leading not only to organelles like the ribosome but to the entire apparatus of gene replication and expression. To the extent that all cells are alike in features of replication, transcription and translation, pattern gives us no clues; guesses about process can be evaluated only for physical and chemical feasibility and for the indirect plausibility that analogy with other known biological mechanisms might provide.

To the extent that cells differ, however, cladistic inferences *can* be made about primitive and derived states of development, and thus reconstruction can be hoped for. When the archaea were discovered, it appeared that they might serve as an outgroup which would allow us to determine, for those features of molecular biology in which eubacteria and eukaryotes were known to differ, which were primitive and which derived. Failing that, it seemed that they should at least provide a third perspective on early cell and molecular biology. This chapter is about why the archaea held such promise, the extent to which they have fulfilled it, and what should be done next.

THE DEVELOPMENT OF THE THREE KINGDOM CONCEPT

Before Woese

At the Twentieth Symposium of the Society of General Microbiology, Roger Stanier (1970) described the two then prevailing hypotheses about evolutionary links between prokaryotes and eukaryotes. The older school held for '*direct filiation*' – the belief that simple prokaryotes gave rise to complex prokaryotes from which simple and then complex eukaryotes descended (Allsopp, 1969; Bogorad, 1975; Cavalier-Smith, 1975). Because cyanobacteria and eukaryotic algae engage in oxygen-evolving photosynthetic processes too similar in detail to have arisen independently, the logical place to site the prokaryote-to-eukaryote transition was between them, and the transition then could be an easy, gradual one. By this view, chloroplasts and mitochondria arose through compartmentalization of the once-unsegregated genes for photosynthesis and respiration, and all eukaryotic life descended from some early 'Uralga'.

The newer school, with younger and more vigorous (if, at that time, no more numerous) adherents, joined Lynn Margulis (1970) in her belief that the eukaryotic cell was an evolutionary chimaera, put together from at least three quite different, previously prokaryotic, lineages. In many ways a revival of earlier nineteenth-century German notions, this '*serial endosymbiosis hypothesis*' (Taylor, 1974) claimed that chloroplasts descended from once-free-living cyanobacteria which had taken up residence as symbionts within the cytoplasm of some proto-eukaryotic host cell, and then had lost autonomy through the transfer of some genes to the nucleus and the outright deletion of others. Similarly, mitochondria were seen as descendants of some respiring bacterium – physiological, biochemical and direct sequence comparisons implicated a relative of *Paracoccus denitrificans* (John & Whatley, 1975).

The host for these endosymbioses, the previously prokaryotic cell whose newly enveloped genome became the eukaryotic nucleus, was less easily identified. Stanier and Margulis both suggested a eubacterium able to engulf other cells because it was wall-less and had developed some endocytotic activity, but there was not agreement about the place of such a cell within known prokaryotic lineages. Margulis, following Searcy, nominated *Thermoplasma*, a wall-less thermophile then described as an unusual mycoplasma producing actin-like and histone-like proteins (Searcy, Stein & Green, 1978). *Thermoplasma* is now known as one of the archaea, but molecular evidence fails to place it especially close to the eukaryotic nuclear lineage.

Either hypothesis, direct filiation or endosymbiosis, needs to account somehow for the additional substantial differences between prokaryotic and eukaryotic cells which was still just coming to be appreciated in the 1960s and 1970s. Stanier and van Niel had summarized differences at the cellular and ultrastructural level in 1962 (Stanier & van Niel, 1962). During the next 10 to 15 years we began to understand how the prokaryote : eukaryote dichotomy was manifested at the level of gene structure and function, the relevant contrast being that between prokaryotic and nuclear genomes. Promoters, RNA polymerases, transcription and translation signals, ribosome : mRNA recognition systems, RNA modification systems, and the organizations of genes for related functions seemed to differ between prokaryotes (or at least *E. coli*) and the nuclear components of eukaryotes (or at least animals and yeast). Introns, discovered in eukaryotic genes in 1977, represented perhaps the most flamboyant example (Gilbert, 1978).

Neither hypothesis offered a particularly compelling explanation for these differences. Both would have eukaryotes, or more precisely the nuclear evolutionary lineage, derived from some branch well within an already quite bushy prokaryotic phylogenetic tree. If prokaryotic gene structure had been established before that branching, nearer the root of the tree – and the similarities between gene structure and function across the then known prokaryotes (all eubacteria) suggested that it had – then substantial changes were wrought in all the genes of the developing nuclear lineage after its emergence, to give rise to the distinctive features alluded to above (and many others). Although the need to differentiate and co-ordinate the activities of three different compartmentalized genomes might be a driving force for change (no matter how this situation came about), it is still difficult to see why, or how, such basic features as polymerase-promoter recognition systems or transcriptional linkage into operons should, or could have been, altered for so many genes.

After Woese

Believers in either direct filiation or endosymbiosis bolstered their beliefs by reference to comparative anatomy, biochemistry, physiology and the fossil record, data sets permitting much argument over weighting and relevance. Sequence data, although not unarguable, at least put issues in quantitative, rigorously analysable terms. Unfortunately, the largest collections of sequence data available before the late 1970s were inappropriate or inadequate for looking at deep branchings. Cytochromes and ferredoxins, the most frequently sequenced proteins, track mitochondrial and chloroplast lineages. Their study did identify likely prokaryotic ancestors of these organelles, but told nothing of the history of the nucleus. 5S ribosomal RNA (rRNA), which does mark both nuclear and organellar history, and for which there was a large data base, is simply too short and too conservative to provide a reliable clock. Schwartz and Dayhoff (1978) had used it to root the eukaryotic nuclear lineage into a bacterial phylogeny – between *E. coli* and *Bacillus subtilis* – where no one would now place eukaryotes.

Carl Woese, who had begun phylogenetic studies with 5S rRNA, switched his attention to the much larger (1500 nucleotide rather than 120 nucleotide) 16S rRNA in the early 1970s. He argued that 16S (and 23S) rRNAs make the best molecular chronometers, since they: (i) have long, highly conserved regions useful for looking at distant phylogenetic relationships, interspersed with variable regions valuable for close relationships; (ii) are of universal conservative function, not prone to spates of rapid sequence change due to selection; (iii) are coded for by genomes of all prokaryotes, eukaryotes and organelles; and (iv) are not likely candidates for horizontal gene transfer.

Because large RNAs could not then be sequenced, and DNAs could not be cloned, Woese and his colleagues began by accumulating 'T1 oligonucleotide catalogs' – lists of the enzymically determined sequences of all G-terminated oligonucleotides released by digestion of *in vivo*-[^{32}P]-labelled 16S rRNA (Pechman & Woese, 1972). By the end of the decade, they had accumulated catalogues from several hundred prokaryotic 16S and eukaryotic 18S rRNAs, had tabulated pairwise distance measures ('S_{AB} values', or numbers of coincident oligonucleotides, weighted by length) and constructed, from these distance matrices, cladograms which allowed them to see the overall phylogenetic structure of the living world (Woese & Fox, 1977*a*,*b*), and pick out many details within prokaryotic genealogy (Fox *et al.*, 1980).

Woese's data unequivocally divided the living world into three 'primary kingdoms', which he and Fox then called eubacteria, archaebacteria and urkaryotes (the nuclear-cytoplasmic lineage of eukaryotic cells). All the bacteria about which anything was known before then turned out to be eubacteria, and *none of them was any closer to the eukaryotic nuclear lineage than any other*. The importance of this simple fact has been seriously underappreciated, although Woese and Fox made themselves quite clear in 1977 (Woese & Fox, 1977*b*), in a discussion of the prokaryote : eukaryote dichotomy as definitive as Stanier and van Niel's, 15 years before. Most molecular biologists believed (and still believe) that 'eukaryotes evolved from prokaryotes'. As an articulation of the faith that simpler cells appeared before complex ones, this seems likely to be true, if unprovable. However, it is nonsensical to suggest that eukaryotes somehow evolved from all prokaryotes together, and clear, on the basis of Woese's analyses, that the nuclear lineage came from no particular prokaryotic lineage identifiable at the time. Instead, the eukaryote : prokaryote division was as deep in phylogeny as it was in ultrastructure. There is neither an *a priori* reason to believe that features which now distinguish nuclear genes from those of prokaryotes are recent additions to an earlier *E. coli*-like molecular biology, nor, indeed, any way to decide whether or not contemporary eukaryotic or prokaryotic genes more closely resemble those of the first cells.

Several investigators have used protein molecular clocks or combinations of clock and paleontological data to assign an age of between 1.6 and 3 billion years to the 'prokaryote:eukaryote divergence'. Doolittle, Anderson and Feng (1989) come up with 1.8 billion years for what must really be the eubacteria : eukaryote divergence, using a collection of ten protein data sets. It is now accepted that there are fossil prokaryotes 3.8 billion years old, and all the rRNA and protein sequence data, including those of R. Doolittle, say that the eubacteria : eukaryote divergence predates any divergence within the surviving eubacteria. Thus, we must assume that all but one of the (presumably very many) lineages stemming from the cells which left the first fossils became extinct some 1.8 billion years ago, and that all modern life was re-founded from the single survivor. Furthermore, if the stromatolite-like appearance of the earliest fossils bespeaks their cyanobacterial nature, the lucky lineage was cyanobacterial. So, the root of the universal tree must lie within the cyanobacteria, and grossly un-clocklike behaviour must be invoked to account for the current shape of that tree. This dating discrepancy might best be resolved by admitting that there remain problems in calibrating molecular clocks, and that 1.8 billion is really *remarkably close* to 3.8, all things considered!

Woese (Woese & Fox, 1977*b*; Woese, 1987) instead favoured the notion that the last common eubacterial : eukaryote ancestor was, in fact, very early and primitive – a '*progenote*' still in the throes of completing the evolution (refinement) of the links between genotype and phenotype (Woese, 1982). He argued that, at early stages in life's history, the machineries of replication, transcription and translation were fashioned imperfectly, and that cellular evolution differed both in tempo (because of high error rates) and mode (because much selection was directed at improving the genotype : phenotype coupling, not the environmentally defined fitness of phenotype) from evolution today. He claimed that difference in gene and cell structure and function between eubacteria, eukaryotes and archaebacteria are so profound that:

> the only solution to the problem is for the universal ancestor to be a progenote. Since the progenote is far simpler and more rudimentary than extant organisms, the significant *differences* in basic molecular structures and processes that distinguish the three major types of organisms would be attributes that the universal ancestor never possessed. In other words, the more rudimentary versions of a function present in the progenote would become refined and augmented independently, and so uniquely, in each of the progeny lineages. This independent *refinement* (and augmentation) of a more rudimentary function – not the *replacement* of one complex function by a different complex version thereof (the beginning stages of which would be strongly selected against) – is why remarkable differences in detail have evolved for the basic functions in each of the urkingdoms (Woese, 1987).

Position, monophyletic nature and renaming of the archaebacteria

During the first large international gathering of molecular archaebacteriologists in Munich in 1981 (published as Kandler, 1982), notions of how the three primary kingdoms might be related to each other were vague and ill formed. Fox, Luehrsen and Woese (1982) entertained the possibility that lateral gene transfer or its precellular equivalent were so rampant at the time of the last common ancestor as to make it meaningless to try to define the branching order of complete genomic-cellular lineages. By the time of the second gathering in Munich in 1985 (published as Kandler & Zillig, 1986), accumulating ribosomal RNA catalogue and sequence data had affirmed the coherence (monophyletic nature) of the archaebacteria. More limited protein sequence data supported this notion (archaebacterial proteins being more like each other than like their eubacterial or eukaryotic homologs), and gave hope that gene transfer had not muddied evolution's tracks hopelessly. However, no tree could be rooted, because no data set could provide an 'outgroup' for all life. Both kinds of data did show a deep division *within* the archaebacteria, separating (roughly speaking) the extreme thermophiles from methanogens and halophiles (Woese, 1987).

A variant position has been consistently taken by James Lake (summarized in Lake, 1989, 1991), although the arguments he has advanced in support of it have changed over the years. Initially, from differences in ribosome morphology detectable in the electron microscope, Lake claimed that thermophilic archaebacteria (his 'eocytes') are closer to eukaryotes than to the rest of the eubacteria, that halophiles are a sister group of eubacteria (together 'photocytes'), and that only methanogens could comprise a third kingdom meaningfully. In other words, archaebacteria are polyphyletic (derive from separate ancestors, at least one of which we would not call archaebacterial) or paraphyletic (have given rise to one or more groups not considered archaebacteria) or both. Lake's interpretations of ribosome morphology were sharply questioned (Stoffler & Stoffler-Meilicke, 1986), and some of the key traits he considered evidence for sisterhood of the halo- and eu-bacteria (e.g. light-driven proton pumps) were surely homoplasies. He has argued since that the apparent monophyly of the archaebacteria is an artefact of the unequal rate effect described by Felsenstein (1978; slowly evolving lineages seem to be sister groups because they both resemble the universal ancestor more closely) and devised his own method, evolutionary parsimony, to deal with a subset of the data. In his hands, this yields once again the 'eocyte tree' (Lake, 1991).

In what seems to be the most methodologically comprehensive and ideologically even-handed analysis to date, Gouy and Li (1989) showed that two further techniques, neighbour-joining and maximum parsimony, support a monophyletic, holophyletic (one ancestor, no non-archaebacterial descendants) archaebacterial assemblage, and that evolutionary parsimony is a poor method of tree reconstruction with simulated data. Moreover, when 23S ribosomal RNA sequence data was considered, all three methods supported a monophyletic holophyletic archaebacterial 'kingdom'.

Rooting the universal tree seemed an exercise in metabiology until Iwabe and coworkers (1989) used new protein sequence data and a technique employed more than a decade earlier by Schwartz and Dayhoff (1978), addressing in fact the same issue. The approach focuses on ancient gene duplications, that is, duplications which predate the earliest splitting within the tree relating all contemporary living things. Potentially, all living things can contain one gene descended from each of the two ancient duplication products (call them A and B). A tree can be constructed for all organisms using A sequences, and rooted using any B sequence, or vice versa. There is as yet very little of the appropriate sort of data for looking at the root of the universal tree, but Iwabe *et al* (1989) successfully employed the gene pairs EF-Tu/EF-G and F_1 ATPase α and β, both of the former having been sequenced in *Methanococcus*, and both of the latter in *Sulfolobus*. In all four cases, the archaebacterial sequences cluster with their eukaryotic homologs, and all four possible rootings support a tree in which archaebacteria and the nuclear lineage of eukaryotic cells share a common ancestor more recent than either shares with eubacteria, the root lying then between the eubacterial and the archaebacteria/eukaryote branches (Fig. 1). This is an enormously important conclusion for all of evolutionary biology, and we will rest easier when there are more cases of duplicated genes on which to base it.

This same conclusion *is* consistent with primary sequence data for most (non-duplicated) protein-coding genes, and especially those determining the components of the transcription and translation apparatus. For RNA polymerase subunit genes, and for many ribosomal protein and translation factor genes (Zillig *et al.*, 1989: for summary see Dennis & Matheson, 1989), archaebacterial coding sequences more closely resemble their eukaryotic than their

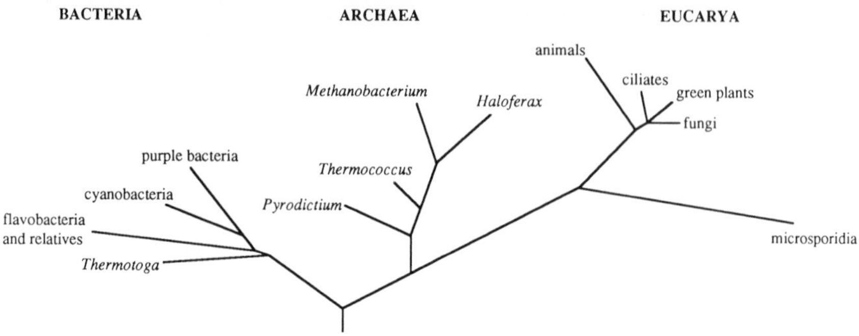

Fig. 1. The rooted phylogenetic tree proposed by Woese, Kandler and Wheelis (1990), on the basis of Woese's ribosomal RNA sequence data, and the rooting procedure of Iwabe *et al.* (1989).

eubacterial homologs. Without a rooting, however, this could be explained by assuming that archaebacterial and eukaryotic protein-coding sequences evolve more slowly than eubacterial sequences, and the universal tree could potentially have any of three possible rooted topologies.

Woese, Kandler & Wheelis (1990) have welcomed warmly the rooting of Iwabe and coworkers, using it as the basis for a new description, and a renaming, of the basic divisions of the living world. Some of their suggestions may seem simply semantic, but the concern with how names influence the way organisms are thought about, or even how experiments are designed, is not misplaced. The term archae*bacteria*, for instance, encourages us to think of these organisms as, when all is finally said and done, just another kind of bacteria. Yet the rooted tree indicates that their closest relatives are eukaryotes.

The eukaryote:prokaryote distinction has been misleading, Woese *et al.* (1990) continue. What unites the archaebacteria and eubacteria as prokaryotes are negative features, principally, the lack of a nucleus. Stanier and van Niel had come to terms with this in 1962, when the only prokaryotes known were eubacteria. These authors defined several features which they felt united what were then seen as the most disparate prokaryotes – true bacteria and cyanobacteria ('blue-green algae'). These were (i) absence of membranes separating the genetic material or the enzymes of photosynthesis and respiration from the rest of the cytoplasm; (ii) nuclear division by fission; and (iii) presence of a cell wall containing a specific mucopeptide (peptidoglycan) as its strengthening agent. Some of these (but not iii) may be characteristic of archaebacteria, too, but it is not right to assume this, or that any other characteristics which have since come to be expected for eubacteria (Pribnow box-type promoters, Shine/Dalgarno ribosome: mRNA recognition signals, 70S ribosomes, absence of splicesomal introns) will also be found in archaea, if the true tree is as shown in Fig. 1.

In recognition of this, Woese *et al.* (1990) renamed the primary kingdoms (now 'domains') as Eukarya, Archaea and Bacteria. The new term for archaebacteria (informal singular form is archaeon) indeed does discourage the otherwise natural assumption that archaea will prove to be just odd *E. coli*. It preserves, perhaps regrettably, the notion that this group of organisms is somehow more primitive, older, or more *archae-* than the others. With this quibble registered, the new names are used henceforth.

WHAT *ARE* THE ARCHAEA LIKE?

Early molecular and genetic work on archaea had focused on the gene for bacteriorhodopsin in halophiles, since this membrane protein has been chosen as a model light-driven proton pump by biophysicists, and on frequent mutations in this locus, soon shown to be the result of the activities of insertion sequence elements (for review, see Charlebois & Doolittle, 1989). Some molecular studies on ribosomes and ribosomal proteins had been motivated by interest in molecular adaptations to high ionic strength.

In the dozen or so years since the first claims for the unique evolutionary importance of the archaea, many additional genes have been cloned (through the use of heterologous or oligonucleotide probes, immunological screening or complementation of *E. coli* auxotrophs) and sequenced (see collections of papers in Kandler, 1982; Kandler & Zillig, 1986; Dennis & Matheson, 1989). Much can be deduced by simple scrutiny of these sequences, and their comparison to each other and to eubacterial and eukaryotic gene sequences (for example, see Zillig *et al.*, 1988; Brown, Daniels & Reeve, 1989). Some salient conclusions from such comparisons are listed below.

1. Archaea almost certainly use the 'universal genetic code'.
2. Archael protein-coding genes sequences so far lack introns, but two thermophile 23S ribosomal RNA genes and many tRNA genes boast introns, which seem to be processed through common (proteinaceaous) enzymatic cleavage and ligation steps (Thompson & Daniels, 1989; Kjems & Garrett, 1991).
3. Although many archaeal messenger RNAs (especially those from methanogens) do exhibit purine-rich stretches complementary to the 3′-end of 16S rRNA, in the right position to serve the function ascribed to the Shine/Dalgarno sequence in bacteria, some show too short a 5′ leader (zero or two nucleotides, for instance; Betlach, Shand & Leong, 1989).
4. Alignments of 5′-flanking regions reveal conserved A and B 'boxes' more similar in location and sequence to motifs conserved in the promoters of eukaryotic RNA polymerase-II transcribed genes than to eubacterial Pribnow box-like regions.
5. Many archaeal genes are tightly linked, indeed overlapping (Brown *et al.*, 1989; Shimmin *et al.*, 1989; Lam *et al.*, 1990). Although polar muta-

tions and other genetic evidence for transcriptional linkage and coordinate regulation is scant, it seems obvious that archaea have operons.

Archaea *do* show many unique features at the molecular level, and one can in fact easily tell, with a stretch of rRNA sequence, or a pattern of modification of a tRNA species, or a promoter region, whether it is of archeal, eukaryal (eukaryotic) or bacterial origin. Nevertheless, there is a feeling of disappointment. Nothing archaea have so far revealed seems as surprising as introns did when discovered in 1977, or as baffling as the vast bestiary of repetitive sequences uncovered in eukaryotic DNA by renaturation experiments in the 50s, or the byzantine phenomenon of RNA editing more recently stumbled across. What is surprising, and perhaps provides a more abiding mystery than any such molecular curiosities, is an unexpected combination of eukaryote-like features at the level of gene sequence and expression and bacteria-like features at the level of gene and chromosome organization.

Archaeal protein-coding genes, in addition to resembling their eukaryotic homologs in primary sequence, appear to be transcribed by eukaryote-like RNA polymerases, from eukaryote-like promoters. Consensus sequence determinations (Zillig *et al.*, 1988, see Fig. 2) – suggesting that RNA polymerase interacts with DNA at a region about 25 base pairs upstream from the transcription start site (box A) and again but less strongly around the start site (box B) – have been confirmed by DNA footprinting analyses (e.g. Thomm *et al.*, 1989), and more dramatically recently by *in vitro* transcription experiments (Reiter, Hudepohl & Zillig, 1990). In addition, archaeal RNA polymerases are like their eukaryotic counterparts in subunit composition and in the sequences of subunits, so much so that the archaeal enzymes group with eukaryotic RNA polymerases II and III, to the exclusion of RNA polymerase I and bacterial polymerases, in phylogenetic trees constructed from their sequences (Zillig *et al.*, 1989).

Such observation are, of course, consistent with the rooted tree shown in Fig. 1. Inconsistent with these observations, and with that tree, is most of what we know about gene organization in archaea. For example, genes for B (B″, B′), A and C subunits of RNA polymerase are found in operons comparable in organization to those for β and β' in *E. coli*. For ribosomal protein genes, the resemblance is even more striking. Clusters comparable to the streptomycin and spectinomycin operons of *E. coli* have been sequenced from *Methanococcus* (Auer, Lechner & Bock, 1989). These show 5 and 15 genes, respectively, homologous to those of *E. coli*, in the same order as in *E. coli*. (There are 'extra' genes in the archaeal operons – these correspond to ribosomal proteins found in eukaryotic, but not bacterial, ribosomes.) This cannot be the result of convergence!

In the authors' laboratory, efforts have focused around the development and systematic exploitation of tools for proper genetic analysis in archaea, with *Haloferax volcanii* as model. Procedures for transformation with phage,

Eubacteria

```
-35                              -10       +1

TTGACA-----------------TATAAT-------
```

Archaebacteria

```
          -28                               +1
```

$\mathrm{TTTA^{A}{}_{T}A}$----------------------$\mathrm{{}^{A}{}_{T}TG^{A}{}_{C}}$

Eukaryotes (RNA poymerase II)

```
          -28                               +1

        TATAAA------------------------CA
```

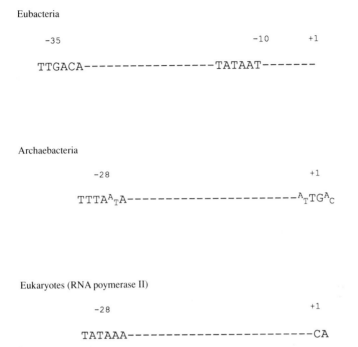

Fig. 2. Promoter consensus sequences for the three kingdoms. From Bucher and Trifonov (1986) and Reiter *et al.* (1990).

plasmid and chromosomal DNAs have been discovered and optimized (Cline, Schalkwyk & Doolittle, 1989), a variety of shuttle vectors prepared (Lam & Doolittle, 1989), transformation of auxotrophs has been used as a means for finding (and then sequencing) amino acid biosynthetic genes within an ordered cosmid bank (Lam *et al.*, 1990), and a physical (contig) map has been constructed for the entire genome (R. L. Charlebois, L. C. Schalkwyk, J. D. Hofman & W. F. Doolittle, unpublished data). The latter exercise shows *Haloferax* to have a 2900 kbp chromosome, and plasmids of 690, 442, 86 and 6 kbp, and the mapping of genes to the chromosome by a combination of physical and genetic techniques has begun. One of the genetic systems concentrated on comprises the genes for tryptophan biosynthesis (Lam *et al.*, 1990; W. Lam, A. Cohen & W. F. Doolittle, unpublished observations). All these map to one of two widely separated chromosomal locations. In each cluster, genes overlap and seem very likely to be cotranscribed, and the expression of the two clusters is coordinately repressed by tryptophan. Comparing the order of the *trp* genes in *Haloferax* to the extensive set of gene order data for eubacteria and eukaryotes (Fig. 3) allows two conclusions. First, *Haloferax* is unique – no eubacterium or eukaryote has its *trp* genes organized in just this way. Second, *Haloferax* nevertheless

EUBACTERIA										
Purple bacteria:										
α subdivision										
Rhizobium and relatives			E-----G		D	C	F	B	A	
Zymomonas mobilis					D	C	F	B	A	
Caulobacter crescentus			E		D	C	F	B	A	
β subdivision										
Pseudomonas acidovorans			E	G	D	C	F	B	A	
γ subdivision										
Escherichia coli and relatives			E-----G		D	C-----F		B	A	
Serratia and relatives			E	G	D	C-----F		B	A	
Pseudomonas aeruginosa and relatives			E	G	D	C	F	B	A	
Acinetobacter calcoaceticus			E	G	D	C	F	B	A	
Gram-positive eubacteria:										
low (G+C) subdivision										
Bacillus subtilis and relatives			E		D	C	F	B	A	G
Lactobacillus casei					D	C	F	B	A	
high (G+C) subdivision										
Brevibacterium lactofermentum			E	G	D	C-----F		B	A	
Spirochetes:										
Spirochaeta aurantia			E							
Leptospira biflexa			E	G						
ARCHAEBACTERIA										
Methanococcus voltae							F	B	A	
Haloferax volcanii	D	F	E	G		C		B	A	
EUKARYOTES										
Saccharomyces cerevisiae			E	G-----------C			F	A-----B		D
Neurospora crassa			E	G-----------C-----F				A-----B		D
Aspergillus nidulans				G-----------C-----F				A-----B		
Coprinus cinereus								A-----B		
Penicillium chrysogenum				G-----------C-----F						
Arabidopsis thaliana									B	

Fig. 3. Organization of *trp* genes in members of the three kingdoms. From Crawford and Milkman (1991) and references therein, and W. Lam, A. Cohen & W. F. Doolittle, unpublished data. Dotted lines indicate coding regions which are fused into single genes. Underlining indicates genes which are contranscribed from common promoters (in the left-to-right direction).

does retain the linkage between B and A genes which characterizes all eubacteria, as well as the simple fact of clustering of *trp* genes. There seems no escaping the conclusion that the last common ancestor of eubacteria and archaea (and thus by the new tree, the last common ancestor of *all* life) did have a *trp* operon of some sort, variously disrupted in the several prokaryotic lineages and, like all operons, replaced by a set of transcriptionally and genetically unlinked genes (or fused reading frames) in eukaryotic nuclear genomes.

This, and the much more extensive evidence for homology in the structures of clusters of genes encoding components of the transcription and translation apparatus, must mean that these and probably other basic features of the organization of genes in genomes were set in place at the root of the universal tree (Fig. 1). The fact that eukaryotes are different now means that they have changed radically (destroyed operons, in this case) somewhere between the point of their divergence from archaea and the deepest point at which we have any information on eukaryotic nuclear gene organization (trypanosomes – above the microsporidia in Fig. 1). This, in a sense, casts us back to the position in the mid-1970s, before Woese began to publish the results of his grand phylogenetic survey. There is, in fact, a 'prokaryotic' lineage which is closer to – which could be said to be ancestral to – the eukaryotic nuclear lineage. It is, again, sensible to suggest that there *was* a 'prokaryote' to eukaryote transition, and that, as a cause or consequence of this transition, some rather drastic changes were wrought in the nuclear genome. The ironic difference here is that the ancestral 'prokaryotic' lineage – the archaeal lineage – is not one we knew in the mid-1970s, and it may be 'prokaryotic' only in so far as this term describes a structural condition, not a phylogenetic position.

This new (and certainly still highly speculative) view also calls into question the claim, endorsed by many, that the last common ancestor of all extant life was in fact 'the progenote'. The existence of such a primitive entity remains almost a logical certainty, but *perhaps* much of the necessary refinement of genes and genomes had already occurred before the earliest bifurcation of the tree. That is, the last common ancestor may have already become a proper 'genote'.

It is clearly worth learning more about the archaea, but we may soon find that we are learning mostly about specific gene systems underlying specific adaptations of archaea to the extreme environments they prefer, and not about major early changes in gene structure and function. For the 1990s, it might be better to focus attention on the lowest *eukaryotic* branches of the tree of life.

W. F. Doolittle is a Fellow of The Canadian Institute for Advanced Research.

REFERENCES

Auer, J., Lechner, K. & Bock, A. (1989). Gene organization and structure of two transcriptional units from *Methanococcus* coding for ribosomal proteins and elongation factors. *Canadian Journal of Microbiology*, **35**, 200–4.

Allsopp, A. (1969). Phylogenetic relationships of the prokaryota and the origin of the eukaryotic cell. *New Phytologist*, **68**, 591–612.

Betlach, M. C., Shand, R. F. & Leong, D. M. (1989). Regulation of the bacterio-opsin gene of a halophilic archaebacterium. *Canadian Journal of Microbiology*, **35**, 134–40.

Bogorad, L. (1975). Evolution of organelles and eukaryotic genomes. *Science*, **188**, 891–8.

Brown, J. W., Daniels, C. J. & Reeve, J. N. (1989). Gene structure, organization and expression in archaebacteria. *CRC Critical Reviews in Microbiology*, **16**, 287–338.

Bucher, P. & Trifonov, E. N. (1986). Compilation and analysis of eukaryotic POL II promoter sequences. *Nucleic Acids Research*, **13**, 1009–26.

Cavalier-Smith, T. (1975). The origin of nuclei and of eukaryotic cells. *Nature*, London, **256**, 463–8.

Charlebois, R. L. & Doolittle, W. F. (1989). Transposable elements and genome structure in halobacteria. In *Mobile DNA*, eds. M. Howe & D. Berg, pp. 297–307. American Society for Microbiology, Washington, DC.

Cline, S. W., Schalkwyk, L. C. & Doolittle, W. F. (1989). Transformation of the archaebacterium *Haloferax volcanii* with genomic DNA. *Journal of Bacteriology*, **171**, 4987–91.

Crawford, I. P. & Milkman, R. (1991). In *Evolution at the Molecular Level*. Eds. R. K. Selander, A. G. Clark & T. S. Whittam, pp. 77–95, Sinauer Associates, Inc, Sunderland, Massachusetts.

Dennis, P. P. & Matheson, A. T., eds. (1989). Proceedings of the Third International Conference on Molecular Biology of the Archaebacteria – 1988. *Canadian Journal of Microbiology*, **35**, 1–244 (special conference edition).

Doolittle, R. F., Anderson, K. L. & Feng, D.-F. (1989). Estimating the prokaryote–eukaryote divergence time from protein sequences. In *The Hierarchy of Life*, eds B. Fernholm, K. Bremer & H. Jornvall, pp. 73–86. Excerpta Medica, Amsterdam.

Felsenstein, J. (1978). Cases in which parsimony or compatibility methods will be positively misleading. *Systematic Zoology*, **27**, 401–10.

Fox, G. E., Luehrsen, K. R. & Woese, C. R. (1982). Archaebacterial 5S Ribosomal RNA. In *Archaebacteria*. Ed. O. Kandler, pp. 330–45. Gustav Fischer Verlag, Stuttgart.

Fox, G. E., Stackebrandt, E., Hespell, R. B., Gibson, J., Maniloff, J., Dyer, T. A., Wolfe, R. S., Balch, W. E., Tanner, R., Magrum, L. J., Zablen, L. B., Blakemore, R., Gupta, R., Bonen, L., Lewis, B. J., Stahl, D. A., Luehrsen, K. R., Chen, K. N. & Woese, C. R. (1980). The phylogeny of prokaryotes. *Science*, **209**, 457–63.

Gilbert, W. (1978). Why genes-in-pieces? *Nature*, London, **271**, 501.

Gouy, M. & Li, W. H. (1989). Phylogenetic analysis based on rRNA sequences supports the archaebacterial rather than the eocyte tree. *Nature*, London, **339**, 145–7.

Holmquist, R., Miyamoto, M. M. & Goodman, M. (1988). Analysis of higher-primate phylogeny from transversion differences in nuclear and mitochondrial DNA from Lake's method of evolutionary parsimony and operator metrics. *Molecular Biological Evolution*, **5**, 217–36.

Iwabe, N., Kuma, K.-I., Hasegawa, M., Osawa, S. & Miyata, T. (1989). Evolutionary relationship of archaebacteria, eubacteria, and eukaryotes inferred from phylogenetic trees of duplicated genes. *Proceedings of the National Academy of Sciences, USA*, **86**, 9355–9.

John, P. & Whatley, F. R. (1975). *Paracoccus denitrificans* and the evolutionary origin of mitochondria. *Nature*, London, **254**, 495–8.

Kandler, O. (1982). *Archaebacteria*. Gustav Fischer Verlag, Stuttgart.

Kandler, O. & Zillig, W. (1986). *Archaebacteria '85*. Gustav Fischer Verlag, Stuttgart.

Kjems, J. & Garrett, R. A. (1991). Ribosomal RNA introns in archaea and evidence for RNA conformational changes associated with splicing. *Proceedings of the National Academy of Sciences, USA.*, **88**, 439–43.

Lake, J. A. (1989). Origin of the eukaryotic nucleus: eukaryotes and eocytes are genotypically related. *Canadian Journal of Microbiology*, **35**, 109–18.

Lake, J. A. (1991). Tracing origins with molecular sequences: metazoan and eukaryotic beginnings. *Trends in Biochemical Sciences*, **16**, 46–50.

Lam, W. & Doolittle, W. F. (1989). Shuttle vectors for the archaebacterium *Haloferax volcanii*. *Proceedings of the National Academy of Sciences, USA*, **86**, 5478–82.

Lam, W. L., Cohen, A., Tsouluhas, D. & Doolittle, W. F. (1990). Genes for tryptophan biosynthesis in the archaebacterium *Haloferax volcanii*. *Proceedings of the National Academy of Sciences, USA*, **87**, 6614–18.

Margulis, L. (1970). *Origin of Eukaryotic Cells*. Yale University Press, New Haven.

Pechman, K. J. & Woese, C. R. (1972). Characterization of the primary structural homology between the 16S ribosomal RNAs of *Escherichia coli* and *Bacillus megaterium* by oliogomer cataloging. *Journal of Molecular Evolution*, **1**, 230–40.

Reiter, W. D., Hudepohl, U. & Zillig, W. (1990). Mutational analysis of an archaebacterial promoter: essential role of a 'TATA box' for transcription efficiency and start site selection *in vitro*. *Proceedings of the National Academy of Sciences, USA.*, **87**, 9509–13.

Schwartz, R. M. & Dayhoff, M. O. (1978). Origins of prokaryotes, eukaryotes, mitochondria and chloroplasts. *Science*, **199**, 395–403.

Searcy, D. G., Stein, D. B. & Green, G. R. (1978). Phylogenetic affinities between eukaryotic cells and a thermoplasmic mycoplasm. *BioSystems*, **10**, 19–28.

Shimmin, L. C., Newton, C. H., Ramirez, C., Yee, J., Downing, W. L., Louie, A., Matheson, A. T. & Dennis, P. P. (1989). Organization of genes encoding the L11, L1, L10, and L12 equivalent ribosomal proteins in eubacteria, archaebacteria, and eucaryotes. *Canadian Journal of Microbiology*, **35**, 164–70.

Stanier, R. Y. (1970). Some aspects of the biology of cells and their possible evolutionary significance. *Symposium of the Society of General Microbiology*, **20**, 1–38.

Stanier, R. Y. & van Niel, C. B. (1962). The concept of a bacterium. *Archiv für Mikrobiologie*, **42**, 17–35.

Stoffler, G. & Stoffler-Meilicke, M. (1986). Electron microscopy of archaebacterial ribosomes. In *Archaebacteria '85*. Eds. O. Kandler & W. Zillig, pp. 123–30. Gustav Fischer Verlag, Stuttgart.

Taylor, F. J. R. (1974). Implications and extensions of the serial endosymbiosis theory for the origin of eukaryotes. *Taxon*, **23**, 229–58.

Thomm, M., Wich, G., Brown, J. W., Frey, G., Sherf, B. A. & Beckler, G. S. (1989). An archaebacterial promoter sequence assigned by RNA polymerase binding experiments. *Canadian Journal of Microbiology*, **35**, 30–5.

Thompson, L. D. & Daniels, C. J. (1989). A tRNA(Trp) intron endonuclease from *Halobacterium volcanii*, unique substrate recognition properties. *Journal of Biological Chemistry*, **263**, 17951–9.

Woese, C. R. (1982). Archaebacteria and cellular origins: an overview. In *Archaebacteria*. Ed. O. Kandler, pp. 1–17. Gustav Fischer Verlag, Stuttgart.

Woese, C. R. (1987). Bacterial evolution. *Microbiological Reviews*, **51**, 221–71.

Woese, C. R. & Fox, G. E. (1977*a*). The phylogenetic structure of the prokaryotic domain: the primary kingdoms. *Proceedings of the National Academy of Sciences, USA*, **74**, 5088–90.

Woese, C. R. & Fox, G. E. (1977*b*). The concept of cellular evolution. *Journal of Molecular Evolution*, **10**, 1–6.

Woese, C. R., Kandler, O. & Wheelis, M. L. (1990). Towards a natural system of organisms: proposal for the domains Archaea, Bacteria, and Eucarya. *Proceedings of the National Academy of Sciences, USA*, **87**, 4576–9.

Zillig, W., Palm, P., Reiter, W. D., Gropp, F., Puhler, G. & Klenk, H. P. (1988). Comparative evaluation of gene expression in archaebacteria. *European Journal of Biochemistry*, **173**, 473–82.

Zillig, W., Klenk, H. P., Palm, P., Puhler, G., Gropp, F., Garrett, R. A. & Leffers, H. (1989). The phylogenetic relations of DNA-dependent RNA polymerases of archaebacteria, eukaryotes, and eubacteria. *Canadian Journal of Microbiology*, **35**, 73–80.

Zuckerkandl, E. & Pauling, L. (1965). Molecules as documents of evolutionary history. *Journal of Theoretical Biology*, **8**, 357–66.

PROTEIN HU AND BACTERIAL DNA SUPERCOILING

J. ROUVIÈRE-YANIV,*[1] E. KISELEVA,[2] A. BENSAID,[1] A. ALMEIDA,[1] and K. DRLICA[3]

[1]*Laboratoire de Physiologie Bactérienne, Institut de Biologie Physico-chimique*
13 rue Pierre et Marie Curie, 75005 Paris, France
[2]*Institute of Cytology and Genetics*
Academy of Sciences of the USSR Siberian Branch,
630090 Novosibirsk, USSR.
[3]*Public Health Research Institute, 455 First Avenue,*
New York, NY 10016, USA

To the memory of Abe Worcel, whose intense fascination with biology was an inspiration to us

INTRODUCTION

The DNA contained in the nucleus of eukaryotic cells is condensed into a nucleoprotein structure called chromatin. In chromatin, the DNA is organized in a structural unit, the nucleosome, which contains 174 bp of DNA wrapped roughly twice around an octamer composed of pairs of the four core histones. Prokaryotes lack a nucleus *sensu stricto*, and DNA was considered for a long time as protein free, non-condensed and non-structured. Eventually, Kellenberger, Ryter and Séchaud (1958) showed that, in fixed cells, the DNA is found condensed in a 'nuclear-like' core linked at certain points to the cell membrane. The absence of a nuclear membrane delimiting such a chromosome rendered its isolation and its study difficult. Taking advantage of the observation that counter-ions stabilize this condensed nucleus-like structure, the groups of Pettijohn and Worcel isolated and studied this structure called 'the bacterial nucleoid' (Stonington & Pettijohn, 1971; Worcel & Burgi, 1972). Using physico-chemical techniques and electron microscopic observations, they showed that bacterial DNA is organized into topologically independent domains. Later, the electron microscopic studies of Griffith (1976) suggested that the bacterial chromosome is found in the cell in a condensed but very unstable structure composed of repetitive units which resembled eukaryotic nucleosomes, albeit of larger dimensions.

*Corresponding author.

These results prompted us to enquire whether bacteria contain chromosomal proteins that can play the same role as histones in compacting DNA.

The protein HU that we have isolated from *E. coli* is the most abundant of the bacterial DNA-binding proteins (Rouvière-Yaniv & Gros, 1975). HU is found associated with the nucleoid and is present in the cell mainly as a heterodimer $\alpha\beta$ (Rouvière-Yaniv & Gros, 1975). The two subunits, which are 70% homologous, are encoded by the *hupA* and *hupB* genes (Kano *et al.*, 1986, 1987). As with histones, HU is well conserved in bacteria and has homologues in eukaryotic organelles (mitochondria and chloroplasts). However, the property that is most suggestive of the potential role of HU in maintaining the structure of the bacterial chromosome is its capacity to introduce negative supercoiling into a relaxed, circular DNA molecule in presence of topoisomerase I and to condense DNA into nucleosome-like structures (Rouvière-Yaniv, Yaniv & Germond, 1979). The isolation of *hup* mutants should help in understanding the exact role of HU *in vivo*. From these mutants it is also hoped to gain a better view of the mechanisms which affect the structure of DNA and regulate essential biological functions such as DNA replication and transcription.

Bacterial chromosome structure has also been investigated from another point of view. Shortly after Pettijohn isolated intact nucleoids, Worcel showed that the chromosome is under negative tension (Stonington & Pettijohn, 1971; Worcel & Burgi, 1972). The discovery of topoisomerases by Wang and Gellert (Wang, 1971; Gellert *et al.*, 1976*a,b*) suggested that chromosomal supercoiling might have biological importance. That prompted us to ask how perturbations of topoisomerases affect supercoiling and the activities of bacterial chromosomes.

Since purified HU can wrap DNA and alter DNA supercoiling, it is important to consider the topics of histone-like proteins and DNA supercoiling together. A summary of DNA supercoiling is presented here, followed by an outline of the properties of HU. At the end we speculate how the two topics may be tied together.

DNA SUPERCOILING

When closed circular DNA molecules are extracted from bacterial cells, they have a deficiency of duplex turns relative to linear DNAs of the same length. This deficiency places strain on the DNA, strain that causes the molecule as a whole to coil. Coiling arising from a deficiency of duplex turns is referred to loosely as negative supercoiling (an excess of duplex turns would give rise to positive supercoiling). The strain, and thus supercoiling, is spontaneously relieved (relaxed) by nicks or breaks in the DNA that allow strand rotation; consequently, supercoiling is found only in DNA molecules that are circular or are otherwise constrained so the strands cannot rotate.

Since processes that separate DNA strands relieve negative superhelical strain, such processes will tend to occur more readily in supercoiled than in relaxed DNA. Processes necessitating DNA strand separation include replication and transcription. Negative supercoiling also affects the three-dimensional configuration of DNA, facilitating loop formation and wrapping around proteins (DNA wrapped into a left-handed toroidal coil is topologically equivalent to DNA containing a negative supercoil). In a sense, negatively supercoiled DNA is energetically activated.

DNA topoisomerases and the control of supercoiling

Topoisomerases, enzymes that change supercoiling, were discovered in the 1970s. They alter the number of turns in DNA by a strand-breaking and rejoining process. Such a reaction mechanism also allows knots in DNA to be tied or untied and DNA circles to be catenated or decatenated. Mutations and inhibitors became available for perturbing topoisomerase activities in living cells, and these perturbations were seen to affect many chromosomal activities (for review see Drlica, 1984).

Gyrase and topoisomerase

Two bacterial topoisomerases are involved in the control of supercoiling. One is DNA gyrase, the product of the *gyrA* and *gyrB* genes (Gellert, 1981). *In vitro*, this enzyme hydrolyses ATP and introduces negative supercoils by a double-strand passage mechanism. Soon after gyrase was discovered, specific inhibitors of gyrase were found to block the introduction of supercoils into bacteriophage lambda DNA upon superinfection of a lysogen (Gellert *et al.*, 1976*a*,*b*). This clearly established gyrase as a source of negative supercoiling.

The second enzyme is topoisomerase I, which is encoded by *topA*. *In vitro*, this enzyme relaxes negative supercoils using a strand-passage mechanism in which only one strand of the DNA is broken (Wang, 1985). A point mutation in *topA* leads to abnormally high levels of supercoiling, consistent with the idea that topoisomerase I modulates the supercoiling effect of gyrase. In *E. coli*, deletion of *topA* causes cells to grow poorly. Normal cell growth is restored by compensatory mutations, many of which map in the gyrase genes and lower supercoiling (Dinardo *et al.*, 1982; Pruss, Manes & Drlica, 1982). Thus, there is little doubt that supercoiling is physiologically important.

Two aspects of the topoisomerases tend to reduce variation in supercoiling under stable growth conditions. One is the response of the enzymes to the topological state of the DNA. Gyrase is more active on a relaxed DNA substrate and topoisomerase I on a more negatively supercoiled one. Another is the homeostatic effect of supercoiling on expression of the *gyrA*, *gyrB*, and *topA* genes (Menzel & Gellert, 1983; Tse-Dinh, 1985; Tse-Dinh & Beran, 1988). Lowering negative supercoiling raises gyrase expression and lowers topoisomerase I expression. Raising supercoiling raises levels of topoisomerase I expression.

Cellular energetics and DNA supercoiling

Since ATP is required for gyrase to introduce negative supercoils into DNA and since in the absence of ATP gyrase removes supercoils, it is likely that cellular energetics also affects supercoiling. *In vitro* the ratio of [ATP] to [ADP] strongly influences the level of supercoiling reached in the presence of purified gyrase, regardless of whether gyrase is introducing or removing supercoils (Westerhoff *et al.*, 1988). Higher values for the ratio generate greater negative superhelical density. Thus ADP appears to interfere with the supercoiling interaction of gyrase with DNA while allowing a competing relaxing reaction to occur. Intracellular DNA supercoiling might also depend on [ATP]/[ADP], probably through its effect on the balance between the supercoiling and relaxing activities of gyrase. Parallels between [ATP]/[ADP] and supercoiling have been observed following a shift to anaerobiosis (Hsieh, Burger & Drlica, unpublished observations) and following exposure to high-concentrations of sodium chloride (Hsieh, Rouvière-Yaniv & Drlica, 1991).

Transcription and supercoiling

The interaction between supercoiling and transcription is complex, and each influences the other. Negative supercoiling is expected to facilitate the DNA strand separation associated with initiation of transcription, and supercoiling should enhance expression from some genes. Many biochemical studies support this idea. However, moderate and high levels of supercoiling frequently inhibit initiation of transcription, presumably by interfering with proper promoter recognition by RNA polymerase. Thus there are optimal levels of supercoiling for transcription, and they vary from one gene to another (Brahms *et al.*, 1985). This means that perturbations of gyrase that relax DNA inside cells will cause expression to increase, decrease, or remain unchanged depending on the gene examined.

Transcription also affects supercoiling. This was first observed with plasmid DNA isolated from *topA* mutants: the supercoiling of pBR322 DNA was much more negative than that of a closely related plasmid unable to transcribe the *tet* gene (Pruss & Drlica, 1986). It was subsequently proposed that tracking of the transcription complex along DNA generates positive supercoils ahead of the complex and negative supercoils behind (Liu & Wang, 1987; Wu *et al.*, 1988). Topoisomerase I would normally remove the negative supercoils and gyrase the positive ones. Thus, negative supercoils will accumulate in a *top A* mutant, and positive supercoils will accumulate when gyrase is inhibited.

If the topoisomerases are balanced and suitably active, transcription and similar translocation processes would have only a very transient and local effect on supercoiling. However, cases have been found with wild-type cells where induction of very high levels of transcription from a plasmid results in abnormally high levels of negative supercoiling (Figueroa & Bossi, 1988). Thus the corrective action of gyrase and topoisomerase I can be overwhelmed.

Extracellular environment and supercoiling

Since DNA supercoiling can have a profound effect on the expression of large numbers of genes, it has been attractive to imagine that DNA supercoiling might be sensitive to the environment in a way that would facilitate bacterial adaptation to changing environments. The first hint that this might be the case came from a report that some mutations that prevent anaerobic growth map at or near the gyrase genes. Subsequently, it was shown that supercoiling differs in plasmids extracted from cells grown under aerobic or anaerobic conditions and that supercoiling is more negative in chromosomes extracted from cells growing anaerobically (Higgins *et al.*, 1988; Hsieh *et al.*, unpublished observations). Plasmid supercoiling also becomes more negative when the growth medium osmolarity is increased. Thus certain environmental conditions appear to affect the regulation of DNA supercoiling. Whether these changes in supercoiling actually promote adaptation has not been demonstrated clearly.

HISTONE-LIKE PROTEIN HU

Properties

The HU protein which has been isolated from *E. coli* is a 9 kD DNA-binding protein which is thermostable and acid soluble; its amino-acid composition, rich in lysine and alanine, is quite unusual for bacterial proteins but resembles those of the lysine-rich histones, H2b or H1 (Rouvière-Yaniv & Gros, 1975). This protein, which migrates during electrophoresis in SDS gels and in urea-acidic gels as a single band, is in fact composed of two polypeptide chains as revealed by sequence analysis (Laine *et al.*, 1978) and by the separation into two bands (HU1 and HU2, Fig. 1(*a*)) in urea–Triton gels (Rouvière-Yaniv & Kjeldgaard, 1979). Furthermore, it was shown that HU is present in the cell mainly (90–95%), as an $\alpha\beta$ heterodimer (Fig. 1(*b*)), composed of two polypeptides of 90 amino acids each which share 70% amino acid similarity. Reassociation studies with the separated subunits after denaturation in urea revealed that, at equilibrium, the formation of heterodimeric HU is highly favoured and only very low amounts of homodimers are present (Rouvière-Yaniv, unpublished observations). In *Bacillus stearothermophilus* there is only one HU species, so why does *E. coli* have two different HU subunits, and are the three forms of HU equivalent?

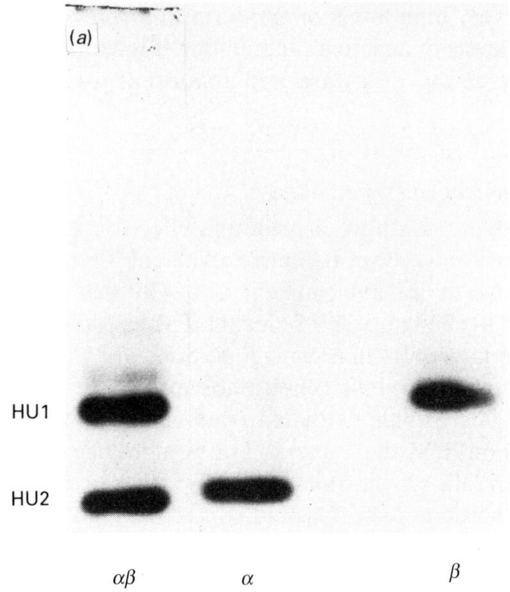
(a)
HU1
HU2
αβ
α
β

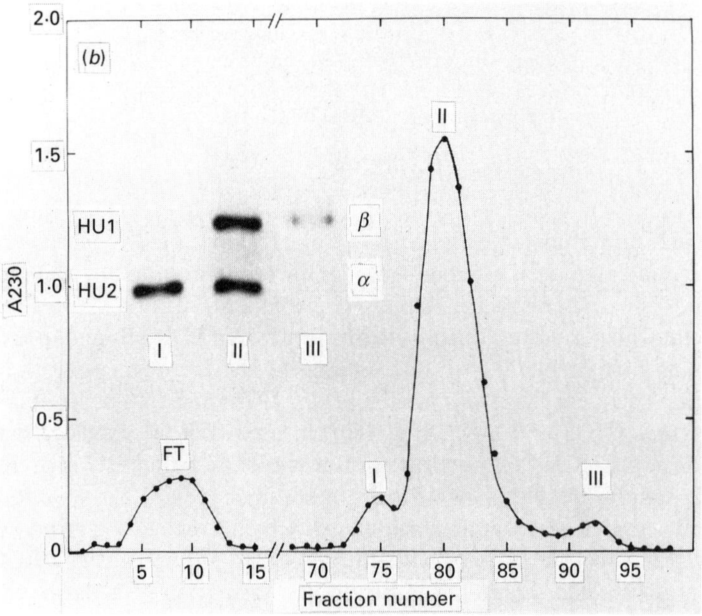
2·0
(b)
1·5
1·0
0·5
0
A230
HU1
HU2
β
α
I
II
III
FT
I
II
III
5
10
15
70
75
80
85
90
95
Fraction number

In vitro *studies*

HU is a histone-like protein

Contrary to unwinding proteins which destabilize the DNA double helix, HU increases the T_m of ds-DNA (Rouvière-Yaniv *et al.*, 1977). Interestingly, it was also shown that HU binds ds-DNA, ss-DNA and RNA (Rouvière-Yaniv & Gros, 1975). In addition, like histones which have remained quasi-invariant throughout evolution, HU is well conserved (for review see Drlica & Rouvière-Yaniv, 1987). It was found that cyanobacteria, which are separated from enterobacteria by about 3×10^9 years, possess an HU protein which cross-reacts immunologically with serum prepared against *E. coli* HU (Haselkorn & Rouvière-Yaniv, 1976). Analysis of amino acid sequences showed that the evolution rate is of the order of about 1% difference in 5×10^7 years compared with the 1% per 6×10^7 years for histones H2A and H2B (Aitken & Rouvière-Yaniv, 1979). Furthermore, proteins resembling HU have also been isolated from yeast mitochondria and from plant chloroplasts (Caron, Jacq & Rouvière-Yaniv, 1979; Briat *et al.*, 1984).

To examine the hypothesis that HU has functions similar to histones, it was asked whether HU was able to form nucleosome-like structures with DNA. First, a protein–DNA assembly protocol had to be developed that would work for both HU and histones since the high-salt procedure in use with histones would not allow HU, which dissociates from DNA at 0.4 M NaCl, to bind to DNA. This difference in binding strength between HU and histones, which are dissociated from DNA only in the presence of 2.0 M salt, could explain in part the instability of the bacterial nucleoid. It was found that SV40 chromatin can be reconstituted normally at low salt (Germond *et al.*, 1979) and that HU can be assembled into a condensed structure with circular DNA molecules (Rouvière-Yaniv *et al.*, 1979). As with histones, beaded structures were formed. However, these were less stable than nucleosomes (cross-linking of HU and DNA was necessary for EM studies), and the size and the number of beads were different (Fig. 2). Nevertheless, the final compaction ratio was identical with that found for nucleosomes. The assembly of HU and circular DNA also fulfilled another criterion: introduction of negative superhelical turns in the presence of topoisomerase I activity

Fig. 1. HU is composed of two subunits that form both hetero- and homo-dimers. (*a*) Purified HU protein was fractionated under denaturing conditions on a CM-cellulose column in the presence of 7 M urea. Aliquots from the input material ($\alpha\beta$) and from the two peaks obtained (α and β) were analysed on a urea–Triton–X100 polyacrylamide gel. According to current nomenclature, HU1 corresponds to the β subunit (gene *hup*B) and HU2 to the α subunit (gene *hup*A), respectively. (*b*) HU purified by DNA-cellulose column affinity chromatography followed by a Sephadex sizing column was absorbed to a phosphocellulose column and eluted with a salt gradient. Peaks I, II, and III correspond to $\alpha 2$, $\alpha\beta$ and β, respectively. The presence of homo- and hetero-dimers was confirmed by gel filtration and dimethylsuberimidate chemical cross-linking.

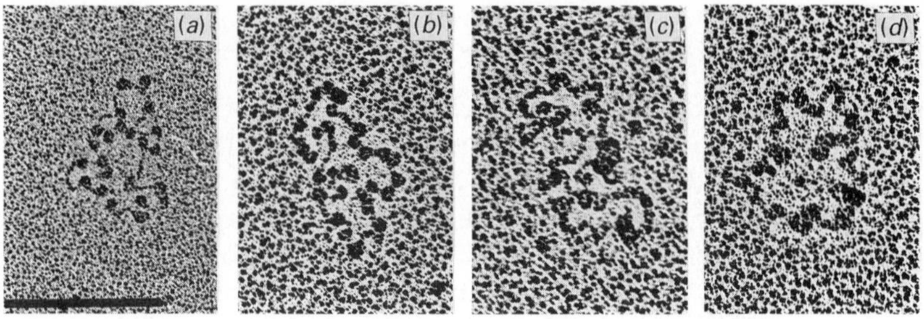

Fig. 2. Electron microscopic observations of complexes formed between SV40 circular DNA and histones or HU protein. Protein–DNA complexes were formed in the presence of a chromatin salt wash enriched in topoisomerase I activity. (*a*) SV40 minichromosomes isolated from infected monkey CV1 cells; (*b*) SV40 minichromosomes assembled from superhelical DNA and the four core histones (H2a, H2b, H3 and H4) prepared from calf thymus. (*c*) Complex formed between HU and relaxed (flr) SV40 DNA; (*d*) complex formed between HU and superhelical (F1) SV40 circular DNA. The HU–DNA complexes were fixed with 15 mM glutaraldehyde. All samples were absorbed on to carbon grids and rotary shadowed as described by Dubochet *et al.* (1971). The four images are at the same magnification. The bar represents 200 nm.

(Fig. 3). The numbers of superturns was in agreement with the number of beads observed by electron microscopy. It was concluded that there are about 275 bp of DNA per bead and that 8–10 dimers of HU were present in each bead (Rouvière-Yaniv *et al.*, 1979). These structures could represent a sort of primitive nucleosome with the heterodimer HU furnishing a primitive dyad axis of organization. These data were later confirmed by Broyles and Pettijohn (1986).

Two features are not yet resolved. One is the lability of the prokaryotic 'nucleosome'. This could generate a dynamic state in the nucleoid in which compacted chromatin could be in equilibrium with expanded structures. In higher organisms, such a flexibility is not observed for most of the DNA in the nucleus. Another problem is the quantity of HU present in the cell. By radioimmunoassays the amount of HU has been estimated to be of the order of 30,000 dimers per cell (Rouvière-Yaniv, 1978). Assuming that the stoichiometry necessary in the *in vitro* assays has to be maintained *in vivo*, it was calculated that 5–10 times more HU would be needed to cover the entire *E. coli* chromosome with nucleosome-like structures. Perhaps these two points can be rationalized if HU, which is not strongly bound to DNA, is continually binding to, and dissociating from, DNA. Fewer molecules of HU might then be required to maintain a partially condensed structure.

Role of HU in replication

In the reconstituted *dnaA* and *oriC*-dependent system for initiation of DNA replication developed by Kornberg and co-workers, HU was shown to stimulate one of the early steps (Dixon & Kornberg, 1984). The formation of

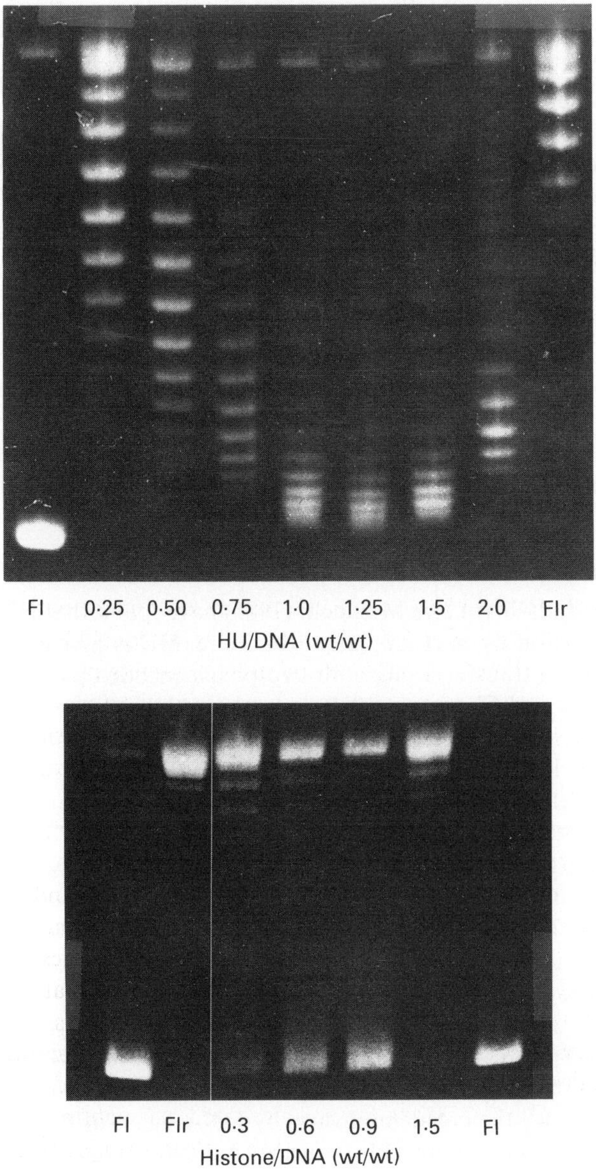

Fig. 3. Both HU (top) and histones (bottom) introduce negative superhelical turns in relaxed circular DNA in the presence of topoisomerase I. Circular relaxed SV40 DNA (FIr) was incubated with increasing amounts of HU (top) or histones (bottom) in the presence of a chromatin salt wash enriched in topoisomerase I for 45 min at 37°C in 50 mM NaCl (HU) or for 90 min in 150 mM NaCl at 37°C (histones). The reactions were stopped by addition of SDS (0.5% final concentration). After 15 min incubation at 55°C, the deproteinized DNA was analysed on 1% agarose gels in Tris–phosphate–EDTA buffer at pH 7.7. The gels were stained with ethidium bromide and photographed under UV light.

the open complex is greatly stimulated by a low concentration of HU (5 molecules/*oriC*), probably by facilitating DNA bending (Bramhill & Kornberg, 1988). High levels of HU (136 molecules/*oriC*) inhibit DNA replication when RNA polymerase is absent (Baker & Kornberg, 1988). This inhibition could be due to the titration by HU of negative superhelical density, which is required for initiation. In recent work, Skarstadt, Baker and Kornberg (1990) showed that opening of the DNA duplex required for initiation can be greatly facilitated by an RNA–DNA hybrid in the absence of the stimulatory effect of HU. This was effective even when the RNA hybridized well away from the origin of replication.

RNA polymerase and HU also interfere with the origin-dependent replication of bacteriophage λ in an *in vitro* system. When RNA polymerase is absent, high levels of HU inhibit the reaction while, in its presence, the inhibition by HU is abolished. Moreover, HU seems to interfere positively with the initiation complex (Mensa-Wilmot, Carroll & McMacken, 1989). In this system HU restores the physiological linkage between transcription and the initiation of λ DNA replication.

Role of HU in site-specific recombination

Craigie, Ardndt-Jovin and Mizuuchi (1985) have found that HU is required for transposition by bacteriophage Mu. Here, HU is involved in the early stage of strand transfer along with two phage-encoded proteins, MuA and MuB. Surette and Chaconas (1989) have shown by immunoelectron microscopy that HU is required for the formation of Type-1 transposomes (DNA–complex A). In fact, HU has much more affinity for DNA–complex-A than for DNA itself (Lavoie & Chaconas, 1990). Whether or not there is a direct interaction between protein A and HU is not clear. Similarly HU (and IHF) stimulate Tn10 transposition in an *in vitro* assay (Morisato & Kleckner, 1987). Still another case is seen with the site-specific inversion associated with flagellar phase variation in *Salmonella* spp. This reaction, which requires the *hin* recombinase and factor II, is stimulated about 10-fold by HU (Johnson, Bruist & Simon, 1986).

HU facilitates the cyclization of short linear DNA molecules by ligase

Hodges-Garcia, Hagerman and Pettijohn (1989) have shown that HU greatly increases ($\times 10^5$) the covalent closure by DNA ligase of linear DNA molecules smaller than 150 bp. This is further indication that HU can condense DNA into structures that may resemble nucleosomes.

In vivo *studies*

Role in transposition

In vivo as *in vitro*, HU is required for the transposition of bacteriophage Mu in *E. coli*. Using null HU mutants, it has been found that Mu growth and lysogeny are unaffected by the absence of either one of the HU subunits. In contrast, in bacteria mutated in both HU genes Mu is unable to grow, and it lysogenizes very poorly (at least 100-fold reduced). Similarly, intermolecular transposition of Mu and mini-Mu is not altered in single *hup* mutants but is less frequent in the double mutant (Huisman *et al.*, 1989). Similar results were obtained by Imamoto and Kano (1990). These observations establish that HU plays a key role in Mu transposition and that, for this process, the three forms of HU are equivalent: the heterodimer can be replaced by either of the homodimers.

The situation in *S. typhimurium* is somewhat different. Hillyard and collaborators (1990) found that in *S. typhimurium* LT2, which Mu does not infect, the transposition efficiencies of Mu d1 introduced on bacteriophage P22 is affected only weakly (4-fold) by the absence of HU. More experiments are necessary to understand the differences found between *E. coli* and *S. typhimurium*; however, the authors suggest that one of the differences could be the synthesis in the *hupAB* mutant of a short polypeptide not present in the *hup*+. This has not yet been observed in *E. coli*. On the contrary, the frequency of DNA inversion at the *fljB* locus of *S. typhimurium* is strongly reduced in the double *hupAB* mutant as expected from the *in vitro* role of HU in the phase variation reaction.

Very recent experiments by Mendelson, Gottesman and Oppenheim (1991) show that HU introduces topological changes in the cohesive ends of bacteriophage λ. The DNA–protein structures obtained at the *cos*-sites with HU can compensate for the lack of IHF in promoting terminase action and phage DNA maturation. These authors therefore suggest that either HU or IHF are required to establish the higher order DNA–protein structure at *cos* that is the substrate for λ terminase.

Role in replication

It was shown that the *hupAB* double mutation is not lethal for the cell although they grow rather poorly (Wada *et al.*, 1988; Huisman *et al.*, 1989). More recently, Ogawa *et al.* (1989) found that, even though deleterious effects are caused by the absence of HU on the replication of *oriC*-containing minichromosomes, replication of the bacterial chromosome seems to be normal. The origin and the direction of DNA replication were examined by marker frequency analysis and no difference was found between the *hup* double mutant and the wild type. The authors stressed that these data indicate that it is unlikely that another mechanism is responsible for replication in the *hupAB* mutant. The fact that *oriC* plasmids cannot be maintained in *hupAB* cells unless the *mioC* promoter is present reinforces the idea that transcription by RNA polymerase is required at an early step of replication. Further experiments are needed to clarify the implied roles for both RNA polymerase and HU in this process. Indeed, HU was originally isolated as a factor stimulating transcription of bacteriophage λ DNA by *E. coli*

RNA polymerase. Certainly, it will be of interest to investigate further this property of HU.

The above observations concerning replication in *hupAB* mutants imply that either HU is not absolutely required for *E. coli* replication *in vivo* or that, in its absence, HU can be replaced by another protein already present or synthesized *de novo*. It is the second hypothesis that is favoured by the authors. In the absence of HU, survival and growth rate depends on the genetic background and the bacteria appear to accumulate compensatory mutations (Huisman *et al.*, 1989).

Role in segregation

HU double mutants grow poorly and form colonies of heterogeneous size on plates. This heterogeneity varies with the genetic background. When analysed by microscopy these cells exhibit a high proportion of anucleate cells. It is not yet understood at what level HU is acting to affect cell partitioning. However, it is one of the first functions visibly compensated (see below).

Three different types of function seem to affect the stability of F′ episomes: replication, cell division, and partition. Since HU seems to affect the segregation of the chromosome, it was of interest to see if plasmid segregation is also affected. Ogura *et al.* (1990) recently showed that the *hupB* gene is required for the stable maintenance of mini-F plasmids since the *hopD* mutation can be complemented by the *hupB* gene. Maintenance and stability of various plasmids are affected in *hup* mutants to different levels; in particular, HU is required for the replication of mini-F and mini-PI plasmids. In *S. typhimurium*, the rates of F′ segregation was measured in *hup* mutants. It is affected little by *hupA*, but, in a *hupB* mutant, segregation is only 3% of the wild-type strain and, in a double *hupAB* mutant, it is down 120-fold.

WHAT CAN COMPENSATE FOR THE ABSENCE OF HU?

Compensatory mutations

As has been seen, growth and cell division are greatly perturbed in the double *hup* mutants with up to 10–15% of the cells becoming anucleate. Moreover, these mutant cells are extremely fragile; they plate very poorly, and they are extra-sensitive to cold, to salt and to antibiotics (Huisman *et al.*, 1989). The authors found that these cells acquire additional mutations which suppress these phenotypes, even though they still lack HU and are refractory to Mu growth. These pseudo-revertants grow better and most no longer form anucleate cells. Their occurrence might reflect the acquisition of compensatory mutations which can control the topology of DNA in the absence of HU.

3-D structure of HU

HU from *B. stearothermophilus*, HBs, has been crystallized. Analysis of X-ray diffraction patterns leads to a model for the three-dimensional structure of HU. It is different from the classical model of bacterial DNA binding proteins derived from repressors. Two identical HBs monomers interlock in their N-terminal parts. The dimer has a hydrophobic interior, and two long arms extend to give this protein a lobster-shaped structure (Tanaka *et al.*, 1984). Most of the charged residues are on the two flexible arms. These are also the most highly conserved regions. These flexible arms could surround the double helix by fitting either in the major or the minor groove (Yang & Nash, 1989). Recent models, built on refined data analysis carried to a resolution of about 2.0 Å (White *et al.*, 1989), underline not only the obvious homologies of the HU proteins from different bacteria but also homologies with two other histone-like proteins, integration host factor, IHF (Nash, 1981; for review see Friedman, 1988) and TF1, a protein isolated from *B. subtilis* after bacteriophage SPO1 infection (Greene & Geiduscheck, 1985; Schneider & Geiduscheck, 1990). These three proteins should interact with DNA in the same way if the DNA-binding domain of the protein is the only source of interaction. Thus IHF was a good candidate to substitute for the absence of HU.

Comparative studies of the interactions of HU and IHF

IHF, in addition to its homologies with HU, is also a dimeric protein composed of two homologous subunits which plays a role in several site-specific DNA interactions. IHF was discovered as an essential protein for the integration of bacteriophage λ (Miller & Nash, 1981) and was shown to bind consensus sequences necessary for this specific recombination event (Craig & Nash, 1984; Leong *et al.*, 1985). This binding sequence is found in a large number of locations including on IS1 (Prentki, Chandler & Galas, 1987) and in *oriC* (Polaczeck, 1990; Filutowicz & Roll, 1990). In contrast, HU, probably like the histones, does not seem to recognize specific DNA sequences.

To ascertain whether IHF can, and does, replace HU in its absence, two types of studies were carried out *in vitro* and *in vivo*. To avoid topological problems, short, synthetic DNA fragments (double or single-stranded; ds or ss) of different sizes and sequences were used for this work. The complexes obtained with the different forms of HU ($\alpha\beta$, $\alpha2$, $\beta2$) and IHF were studied by a gel retardation technique. As the quantity of HU was increased with 21 bp or longer fragments of ds DNA, complexes of DNA-HU heterodimers appeared (Fig. 4 represents an experiment with an 87 bp fragment). Each dimer occupied about 9 bp and was spaced regularly on the DNA fragment. It was found that the homodimers form similar structures, which explains why the single HU mutants have nearly normal phenotypes. It should be

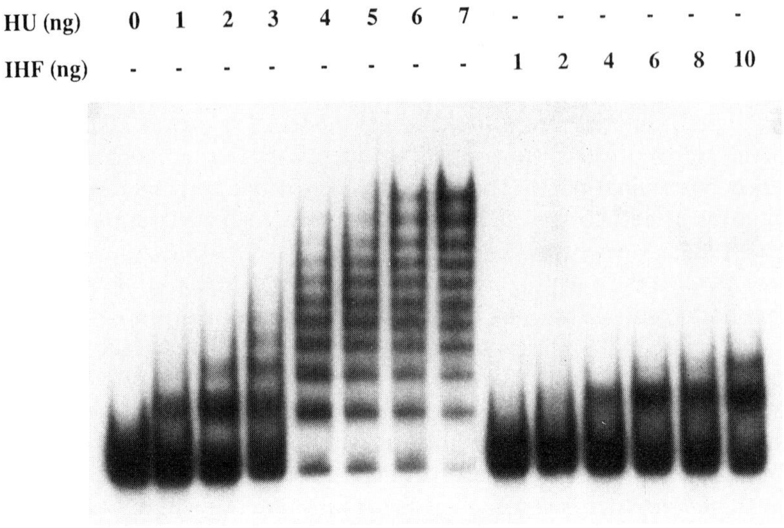

Fig. 4. Complexes formed with HU or IHF and a 87 bp fragment containing an IHF binding site. An end-labelled DNA fragment of 87 bp containing the left end of transposable element IS1 including a high affinity IHF binding site (Prentky *et al.*, 1987) was incubated with the indicated amounts of HU (left) or IHF (right) at room temperature in 10 mM Tris–HCl (pH 7.6), 15 mM KCl, 2 mM spermidine, 15% glycerol, 0.1 mM EDTA and 0.1 mg/ml BSA. The samples were then loaded onto a 5% polyacrylamide gel in 0.25 × TBE (1 × TBE is 89 mM Tris base, 89 mM boric acid and 2.5 mM EDTA). The gel was fixed, dried and autoradiographed. Note the appearance of ten retarded bands with HU and only a major one with IHF.

noted, however, that $\beta 2$ has a lower affinity for this random DNA than the two other forms ($\alpha\beta$, $\alpha 2$) (Bonnefoy & Rouvière-Yaniv, 1991). In contrast, IHF forms a different complex with these DNAs, whether or not it carries its consensus sequence. A maximum of two bands with a 42 bp DNA could be observed at protein saturation and predominantly a single complex is observed with the 87 bp fragment which carries one consensus IHF sequence (Fig. 4). Furthermore, IHF does not bind at all to ss-DNA whereas HU forms a strong complex. Interestingly, the HU homodimers bind much more avidly than the heterodimer to ss-DNA. Finally, it was found that IHF greatly prefers to bind to a curved sequence rather than to an uncurved fragment; HU does not show such preference (Bonnefoy & Rouvière-Yaniv, 1991). These data suggest that IHF is unlikely to compensate for HU.

If IHF were to replace HU for at least some essential function, it would seem reasonable that the synthesis of these two proteins might be somehow coordinated. Using promotor fusions integrated in IHF mutants and *hup* mutants no significant cross-talk in the expression of the IHF and HU genes has been observed (Boubrik, Bonnefoy & Rouvière-Yaniv, 1991). Moreover, the defects observed in the growth of HU double mutants were not suppressed by an overproduction of IHF (Bonnefoy & Rouvière-Yaniv, 1991). This

Table 1. *Over-expression of IHF in* hup *mutants.*

	Vector	pIHF	pIHF + IPTG
Generation time (min) during exponential growth			
wt	36	48	78
hup AB	120	120	120
Optical density (A600) at the stationary phase			
wt	1.8	1.6	0.9
hup AB	0.6	0.6	0.6

can be seen in Table 1 where the two parameters, generation time and stationary phase density, are unaffected.

HU AND DNA STRUCTURE

Localization of HU in vitro and *in vivo*

HU is localized on the isolated E. coli *nucleoid*

For these studies the *E. coli* nucleoid was isolated by two techniques. In the first one, which permits the isolation and characterization of membrane-free structures as described by Worcel and Pettijohn, the lysis was performed in the presence of 1.0 M Nacl (Stonington & Pettijohn, 1971; Worcel & Burgi, 1972). The second involved use of low salt (0.1 M NaCl) combined with spermidine (Kornberg, Lockwood & Worcel, 1974). With the second technique, the nucleoid is associated with the cellular membrane. Since it had been shown previously that HU is acid soluble and heat stable, HU could be purified and characterized with the help of specific immuno-serum from the bulk of nucleoid-associated proteins (mainly membrane proteins). It was found that most of the HU present in the cell is associated with the nucleoid (Rouvière-Yaniv, 1978). A small fraction (about 5%) was recovered from the 0.3 M salt, ribosomal wash. Logically enough, the protein was not found on the nucleoid prepared with 1.0 M NaCl, since HU is eluted from DNA cellulose by 0.4 M salt.

These experiments showing that the majority of HU is associated with the nucleoid could be compared with the results of experiments performed at the same time by Griffith (1976). Griffith's elegant work showed that a condensed minichromosome of bacteriophage λ, having visible nucleosome-like structures, could be observed by electron microscopy when a very gentle procedure of lysis included buffers of low ionic strength. Just as the authors failed to find HU associated with the high-salt nucleoid, Griffith failed to detect nucleosome-like structures with lysates prepared at

1.0 mNaCl. These results were the first demonstration that the λ chromosome could be structured in a similar fashion to the eukaryotic chromosome.

Using a different isolation procedure, Varshavsky *et al.* (1977) obtained *E. coli* DNA covered regularly by two proteins of 9 kD and 17 kD. HU was easily characterized with a specific serum as the 9 kD protein. An ambiguity still persists regarding the 17 kD protein. It was argued for a long time that it was the *firA*-encoded protein described by Lathe *et al.* (1980) but it may also have been the bacterial H1 protein (Cukier-Kahn, Jacquet & Gros, 1972; Higgins *et al.*, 1990). Neither the Griffith work nor the Varshavsky study has been extended by subsequent work.

*HU on the edge of the nucleoid**

Kellenberger, using an elaborate thin-section electron microscopic technique, studied the nucleoid and localized on it some of the potential nucleoid-associated proteins (Durrenberger *et al.*, 1988). His group found that the *E. coli* nucleoid, which is not separated from cytoplasm by a membrane, occupied all the space free of ribosomes. This nucleoid is therefore profoundly and irregularly built with a deep 'fjord-like' appearance and intimately mixed with the cytoplasm. The border between the matrix of DNA and the transcriptional–translational apparatus is therefore very extensive. Interestingly, it had been found long ago by Ryter and Chang (1975) that nascent mRNA, visualized by silver grains, is localized preferentially at the border of the nucleoid. Using serum prepared against HU protein or against topoisomerase I coupled with gold particles to mark the antigen–antibody complexes, Kellenberger and coworkers, more recently, visualized HU and topoisomerase I at the same place where the messenger RNA had been localized previously (Durrenberger *et al.*, 1988). They concluded that HU is primarily located outside the nucleoid. Additional experiments are obviously needed to solve this discrepancy between the different points of view, especially since even the most gentle fixation procedure can introduce changes from the living situation.

*Finally HU on the DNA, by two other techniques**

Two independent groups using different approaches seem to agree about the localization of HU. Recently, Pettijohn and his group have developed methods for permeabilizing bacterial cells to study the distribution of fluorescein-labelled HU directly in the bacteria (personal communication). They found that the added HU binds to the DNA and segregates with the nucleoid during cell division. The distribution of the labelled HU seems to be indistinguishable from that of the fluorescent probe DAPI which uniformly stains the DNA of the nucleoid.

Another approach was taken by a group from Novosibirsk who used either the Kleinschmidt or the Miller technique of spreading DNA (Klein-

*See note added in Proof.

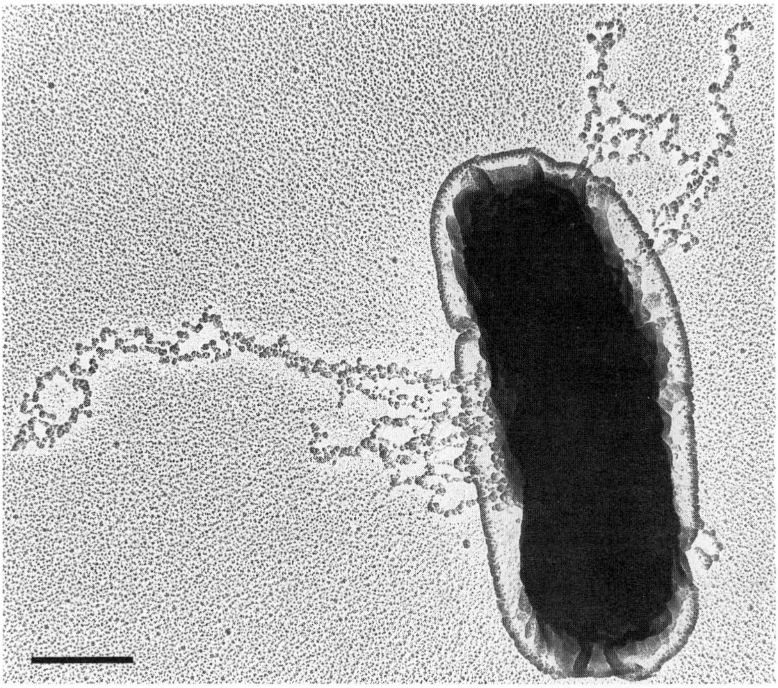

Fig. 5. Supercoiled loops and fibrils observed in the *E. coli* nucleoid which had been released partially from spheroplasts subjected to mild osmotic shock. The material was spread by the technique of Kleinschmidt (1968) and rotary shadowed. Note the closely packed particles with the diameter of 30–35 nm. The bar represents 500 nm.

schmidt, 1968; Miller & Beatty, 1969) and a very gentle way of lysing the bacteria (Kiseleva *et al.*, 1986, 1988). The organization of the *E. coli* nucleoid, after a mild osmotic shock and a quick preparation of the samples, can be seen in Fig. 5. Coming out of the lysed spheroplast are particles having a size of about 30–35 nm. These particles are very fragile, since, when the lysis was performed on whole cells instead of spheroplasts, they were not visible. If the time between lysis, spreading, and fixation is increased, decondensation of the nucleoid occurred (Fig. 6; Kiseleva *et al.*, 1988). This expanded nucleoid is like Kavenoff's classical image of the *E. coli* nucleoid spread with extended loops (Kavenoff & Ryder, 1976). Decondensation could arise from loss of proteins such as HU from the compact particules. Since the half-life of HU–DNA complexes is very short (of the order of one minute; Broyles & Pettijohn, 1986) one attractive possibility is that HU is popping on and off the chromosome. This would explain the instability observed for the condensed structures.

Finally, Fig. 7 shows that, in spores of Streptomyces which have been rapidly and gently lysed and spread, DNA-associated HU protein can be detected by specific antibodies labelled subsequently with protein A coupled

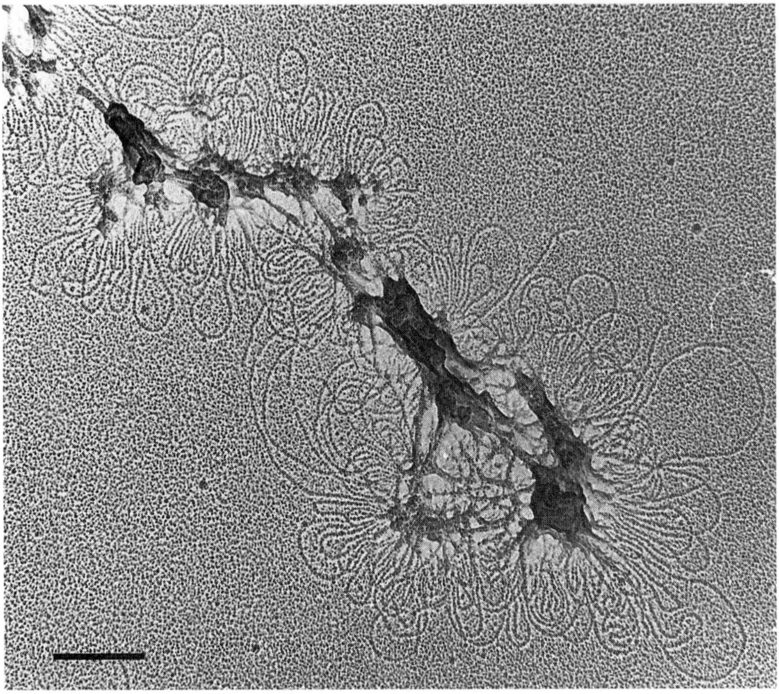

Fig. 6. *E. coli* nucleoids spread under low ionic strength conditions. Spheroplasts were prepared from *E. coli* grown to stationary phase. Nucleoids were released and incubated for 30 min in low ionic strength conditions before being spread by the Kleinschmidt technique followed by rotary shadowing. The bar represents 500 nm.

to gold (Kiseleva *et al.*, 1988). Using these techniques, the nucleoid isolated from *E. coli hup* mutants is being studied by the authors. Preliminary studies reveal HU in single mutants and no gold grains in *hupAB* cells. Furthermore, preliminary results suggest that the nucleoid isolated from the *hup* double mutant decondenses much faster than usual (E.K. and JRY unpublished observations).

HU and supercoiling

Constrained and unconstrained supercoiling

As pointed out above, DNA extracted from cells has a deficiency of duplex turns which can be detected as negative superhelical tension by a number of assays. Inside cells, however, the deficiency of turns can be divided into at least two components. About 40% of the total deficiency is free to exert superhelical tension that can be detected by supercoiling assays. The rest of the deficiency is not detected by an effect on supercoiling presumably because the DNA is either denatured by proteins (denaturation removes negative supercoils) and/or wrapped around proteins (wrapped

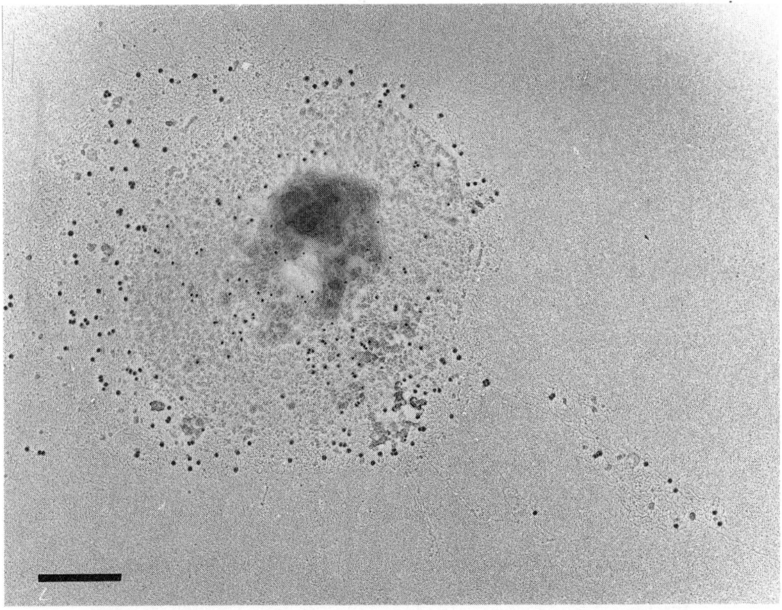

Fig. 7. Binding of anti-HU antibodies to a nucleoid partially released from a Streptomyces spore. The material was spread by the mild Miller's method (Miller & Beatty, 1969) and reacted with anti-HU antibodies. Protein A gold complex (dark dots) was used to detect the HU–antibodies complex. Note that HU is abundant on DNA loops extruding from the central mass. The bar represents 500 nm.

DNA constrained by proteins could be unresponsive to probes of supercoiling) (Pettijohn & Pfenninger, 1980; Bliska & Cozzarelli, 1987). Since HU wraps DNA *in vitro*, it is a candidate for constraining some of the supercoils *in vivo*.

If HU constrains DNA *in vivo*, *hup* mutations would be expected to affect DNA supercoiling. The general idea is that gyrase and topoisomerase I would maintain a constant level of superhelical tension, but, if HU were absent, the level of restrained supercoils would be lower than in wild-type cells. After extraction of DNA, the supercoiling would be less in the mutant. Small effects on the level of supercoiling have been seen in *hupAB* mutants in *S. typhimurium* (Hillyard *et al.*, 1990). Figure 8 shows the effect in *E. coli*. Plasmid pBR322, isolated from wild-type and *hup* mutants, was separated into topoisomers by electrophoresis in the presence of the intercalating dye chloroquine. At the dye concentration shown, the more highly supercoiled topoisomers migrate more rapidly. Plasmid topoisomers obtained from wild-type cells migrate more rapidly than those from *hupAB* double mutants (compare lanes a–c and k–m). This represents a difference of about 10–15%, a number much lower than expected if HU wrapped half of the DNA and if the double mutant simply lacked constrained supercoils. Plasmid

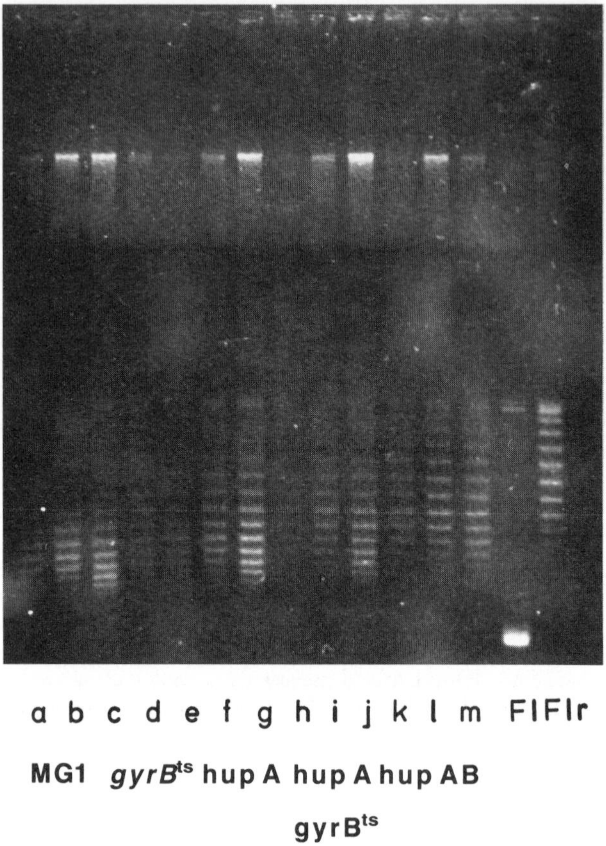

Fig. 8. Superhelical density of plasmids obtained from *hup* mutants. Strain MG1 (a,b,c) or its $gyrB^{ts}$ (d), *hup*A (e,f,g), *hup*A *gyr*B^{ts} (h,i,j) or *hup*AB (k,l,m) derivatives were transformed with pBR322 plasmid DNA and were grown at 40 °C (a,e,h,k) 90 min before being harvested. Aliquots were transferred to 30 °C (b,f,i,l) or 42 °C (c,g,j,m). Plasmid DNA was purified as previously described (Pruss, 1985) and run on an agarose gel in the presence of 15 μg/ml chloroquine. Fl and Flr correspond to supercoiled and relaxed pBR322 DNA.

obtained from the *hupB* mutant are similar to wild-type (not shown); plasmid isolated from the *hupA* mutant falls between wild type and double mutant while an *hupA* $gyrB^{ts}$ double mutant resembles the *hupAB* double mutant.

HU and gyrase

Although the supercoiling data are consistent with HU being a wrapping protein *in vivo*, it could be causing the observed effect by an interaction with gyrase. Yang and Ames (1988) had found that gyrase shows some preference for binding to clusters of *rep*, the repeated sequences which are scattered around the chromosome. HU enhances the ability of gyrase to bind these sequences (Yang & Ames, 1990). Thus, gyrase may utilize HU while introducing supercoils, and the absence of HU would make gyrase less effective. The authors would like to point out that this explanation for lower supercoiling in *hup* mutants does not exclude the explanation given in the previous paragraph.

CONCLUSION

Although the first reconstitution studies in which HU altered DNA topology made it seem obvious that HU and DNA supercoiling are related, it has been difficult to obtain experimental support for this being true in living cells. A potential interaction with gyrase has added a new dimension, and now a serious effort needs to be made to define that interaction both *in vivo* and *in vitro*. One approach is to assume that, if the products of the genes interact, it should be possible to find suppressor mutations that will help uncover the mechanism. Suppressors of HU should be easy to obtain, since it is likely that most strains carrying the *hup* double mutation are already suppressed. The authors speculate that some of these may map in gyrase, and they are currently exploring that possibility.

It is clear from the above that the authors tend to believe that the main function of HU is through its effect on superhelicity. Could all of its effects observed *in vitro* and the physiological defects caused by its absence *in vivo* be primarily due to structural effects on the bacterial DNA? Since their current idea is that the absence of HU is balanced by the establishment of a novel equilibrium in the activity of the cellular topoisomerases, it could be that the majority of the known targets for HU action are all compensated in the same way. This is certainly what happens for the replication process where no severe defect seems to be caused by the absence of HU. It is likely that, for essential processes, like replication, alternative pathways exist which are only employed under specific conditions, for example, the absence of HU. In addition, HU could have more specific roles which are not yet understood and which might be partially fulfilled by IHF. The authors have clearly shown that IHF does not usually form the same complexes with DNA, but some preliminary data suggest that on special occasions they can replace each other.

However, another function of HU seems to emerge from the studies of several laboratories, including the authors'. HU could be an enhancer or a silencer of the activity of some genes. Flashner and Gralla (1988) have shown that HU stimulates the binding of the *lac* repressor and of CRP to their specific sites on the *lac* operon whereas it inhibits the specific binding of the *trp* repressor. The authors suggest that HU induces flexible structures on the DNA which will favour or disfavour the specific binding of certain proteins to their sites. If this is the case, HU will certainly play an important role in the regulation of the expression of some genes. Moreover, this regulation could be positive or negative, as the two examples cited above have shown. It has been related in a previous section how HU similarly enhances the binding of gyrase to its sites (Yang & Ames, 1990) and, finally, the authors will cite results, where they have shown that HU increases by a factor of 10 the binding of IHF to DNA which does not carry the IHF consensus binding sequence (Bonnefoy & Rouvière-Yaniv, 1991). This mechanism could play a role in regulating the positioning of IHF at certain sites.

To conclude, we would like to recall that a still unresolved question is why *E. coli*, contrary to most other bacteria, has a heterotypic HU and, in addition, why the HU1 subunit, encoded by the *hupB* gene, is degraded by a *lon*-dependent mechanism in the absence of HU2 (Bonnefoy, Almeida & Rouvière-Yaniv, 1989; Rouvière-Yaniv *et al.*, 1990). Do the two subunits have the same role? A start has been made to answer this question by showing differences in their potential structure with DNA and their potential enhancing activity, but certainly other studies are necessary to decipher all the facets of this histone-like protein.

ACKNOWLEDGEMENTS

The authors would like to thank Elisabeth Mouray, Eliette Bonnefoy and Jacqueline Plumbridge for their help during the preparation of this manuscript. The work in Paris was supported by the Centre de la Recherche Scientifique (URA1139), the Association de la Recherche sur le Cancer and la Fondation pour la Recherche Médicale; and in New York by a grant from the National Science Foundation (PMB 8718115).

Note added in proof

After this paper was delivered at the SGM symposium (8–12 April 1991), two papers have appeared (May 1991) discussing the problem of the *E.coli* nucleoid: one by the group of Kellemberger (Bohrmann *et al.*, 1991) which shows the coralline shape of the bacterial nucleoid very nicely using a new procedure of cryofixation, the other one by Shellman and Pettijohn (1991) which shows that HU binds primarily to DNA and is distributed equally throughout the nucleoid.

Bohrmann, B., Villiger, R., Johansen, R. & Kellemberger, E. (1991). Coralline shape of the bacterial nucloid after cryofixation. *Journal of Bacteriology*, **173**, 3149–58.

Shellman, V. L. & Pettijohn, D. E. (1991). Introduction of proteins into living bacterial cells: distribution of labelled HU protein in *Escherichia coli. Journal of Bacteriology*, **173**, 3047–59.

REFERENCES

Aitken, A. & Rouvière-Yaniv, J. (1979). Amino and carboxy terminal sequences of the DNA-binding protein HU from the cyanobacterium *Synechocystis* PCC. *Biochemical and Biophysical Research Communications*, **91**, 461–7.

Baker, T. A. & Kornberg, A. (1988). Transcriptional activation of initiation of replication from the *Escherichia coli* chromosomal origin: an RNA–DNA hybrid near oriC. *Cell*, **55**, 113–23.

Bliska, J. & Cozzarelli, N. (1987). Use of site-specific recombination as a probe of DNA structure and metabolism *in vivo*. *Journal of Molecular Biology*, **94**, 205–18.

Bonnefoy, E., Almeida, A. & Rouvière-Yaniv, J. (1989). Lon-dependent regulation of the DNA binding protein HU in *Escherichia coli. Proceedings of the National Academy of Sciences, USA*, **86**, 7691–5.

Bonnefoy, E. & Rouvière-Yaniv, J. (1991). HU and IHF, two homologous histone-like proteins of *Escherichia coli*, form different protein–DNA complexes with short DNA fragments. *EMBO Journal*, **10**, 687–96.

Boubrik, F., Bonnefoy, E. & Rouvière-Yaniv, J. (1991). HU and IHF: similarities and differences in *Escherichia coli*, the lack of HU is not compensated for by IHF. *Research in Microbiology*, in press.

Brahms, J., Dargouge, O., Brahms, S., Ohara, Y. & Vagner, V. (1985). Activation and inhibition of transcription by supercoiling. *Journal of Molecular Biology*, **181**, 455–65.

Bramhill, D. & Kornberg, A. (1988). A model for initiation at origins of DNA replication. *Cell*, **54**, 915–18.

Briat, J.-F., Letoffe, A., Mashe, R. & Rouvière-Yaniv, J. (1984). Similarity between the bacterial histone-like protein HU and a protein from spinach chloroplasts. *FEBS Letters*, **172**, 75–9.

Broyles, S. & Pettijohn, D. E. (1986). Interaction of the *Escherichia coli* HU protein with DNA: evidence for formation of nucleosome-like structures with altered DNA helical pitch. *Journal of Molecular Biology*, **187**, 47–60.

Caron, F., Jacq, C. & Rouvière-Yaniv, J. (1979). Characterization of a histone-like protein extracted from yeast mitochondria. *Proceedings of the National Academy of Sciences, USA*, **76**, 4265–9.

Craig, N. L. & Nash, H. A. (1984). *Escherichia coli* integration host factor binds to specific sites in DNA. *Cell*, **39**, 707–16.

Craigie, R., Ardndt-Jovin, D. & Mizuuchi, D. (1985). A defined system for the DNA strand-transfer reaction at the initiation of bacteriophage Mu transposition: protein and DNA substrate requirements. *Proceedings of the National Academy of Sciences, USA*, **82**, 7570–4.

Cukier-Kahn, R., Jaquet, M. & Gros, F. (1972). Two heat-resistant, low molecular weight proteins from *Escherichia coli* that stimulate Dna-directed RNA synthesis. *Proceedings of the National Academy of Sciences, USA*, **69**, 3643–7.

Dinardo, S., Voelkel, K. A., Sternglanz, R., Reynolds, A. E. & Wright, A. (1982). *Escherichia coli* DNA topoisomerase I mutants have a compensatory mutation in DNA gyrase genes. *Cell*, **31**, 43–51.

Dixon, N. & Kornberg, A. (1984). Protein HU in the enzymatic replication of the chromosomal origin of *Escherichia coli*. *Proceedings of the National Academy of Sciences, USA*, **81**, 424–8.

Drlica, K. (1984). Biology of bacterial deoxyribonucleic acid topoisomerases. *Microbiological Reviews*, **48**, 273–89.

Drlica, K. & Rouvière-Yaniv, J. (1987). Histone-like proteins of bacteria. *Microbiological Reviews*, **51**, 301–19.

Dubochet, J., Ducommun, M., Zollinger, M. & Kellemberger, E. (1971). A new preparation method for dark-field electron microscopy of biomacromolecules. *Journal of Ultrastructure Research*, **35**, 147–67.

Durrenberger, M., Bjornsti, M. A., Uetz, T., Hobot, J. A. & Kellemberger, E. (1988). Intracellular location of the histone-like protein HU in *Escherichia coli*. *Journal of Bacteriology*, **170**, 4757–68.

Figueroa, N. & Bossi, L. (1988). Transcription induces gyration of the DNA template in *Escherichia coli*. *Proceedings of the National Academy of Sciences, USA*, **85**, 9416–20.

Filutowicz, M. & Roll, J. (1990). The requirement of IHF protein for extrachromosomal replication of the *Escherichia coli oriC* in a mutant deficient in DNA polymerase activity. *The New Biologist*, **2**, 818–27.

Flashner, Y. & Gralla, J. D. (1988). DNA dynamic flexibility and protein recognition:

differential stimulation by bacterial histone-like protein HU. *Cell*, **54**, 713–21.
Friedman, D. I. (1988). Integration host factor: a protein for all reasons. *Cell*, **55**, 545–54.
Gellert, M. (1981). DNA topoisomerases. *Annual Review of Biochemistry*, **50**, 879–910.
Gellert, M., O'Dea, M. H., Itoh, T. & Tomizawa, J. (1976*a*). Novobiocin and coumermycin inhibit DNA supercoiling catalyzed by DNA gyrase. *Proceedings of the National Academy of Sciences, USA*, **73**, 4474–8.
Gellert, M., O'Dea, M. H., Mizuuchi, K. & Nash, H. (1976*b*). DNA gyrase an enzyme that introduces superhelical turns into DNA. *Proceedings of the National Academy of Sciences, USA*, **73**, 3872–6.
Germond, J. E., Rouvière-Yaniv, J., Yaniv, M. & Brutlag, D. (1979). The nicking-closing enzyme assembles nucleosome-like structures *in vitro*. *Proceedings of the National Academy of Sciences, USA*, **76**, 3779–83.
Greene, J. & Geiduschek, E. P. (1985). Site-specific DNA binding by the bacteriophage SPO1-encoded type II DNA-binding protein. *EMBO Journal*, **4**, 1345–9.
Griffith, J. D. (1976). Visualization of procaryotic DNA in regulatory condensed chromatin-like fiber. *Proceedings of the National Academy of Sciences, USA*, **73**, 563–7.
Haselkorn, R. & Rouvière-Yaniv, J. (1976). Cyanobacterial DNA-binding protein related to *Escherichia coli* HU. *Proceedings of the National Academy of Sciences, USA*, **73**, 1917–20.
Higgins, C. F., Dorman, J., Stirling, D. A., Waddell, L., Booth, R., May, M. & Bremer, E. (1988). A physiological role for DNA supercoiling in the osmotic regulation of gene expression in *Salmonella typhimurium* and *Escherichia coli*. *Cell*, **52**, 569–84.
Higgins, C. F., Hinton, J. C., Hulton, C. S. J., Owen-Hughes, T., Pavitt, G. D. & Seirafi, A. (1990). Protein H1: a role for chromatin structure in the regulation of bacterial gene expression and virulence. *Molecular Microbiology*, **4 (12)**, 2007–12.
Hillyard, D. R., Edlund, M., Hughes, K., Marsh, M. & Higgins, N. P. (1990). Subunit-specific phenotypes of *Salmonella typhimurium* HU mutants. *Journal of Bacteriology*, **172**, 5402–7.
Hodges-Garcia, Y., Hagerman, P. J. & Pettijohn, D. E. (1989). DNA ring closure mediated by protein HU. *Journal of Biological Chemistry*, **264**, 14621–3.
Hsieh, L., Rouvière-Yaniv, J. & Drlica, K. (1991). Bacterial DNA supercoiling and ATP/ADP: changes associated with salt shock. *Journal of Bacteriology*, in press.
Huisman, O., Faelen, M., Girard, D., Jaffé, A., Toussaint, A. & Rouvière-Yaniv, J. (1989). Multiple defects in *Escherichia coli* mutants lacking HU protein. *Journal of Bacteriology*, **171**, 3704–12.
Imamoto, F. & Kano, Y. (1990). Physiological characterization of deletion mutants of the *hupA* and *hupB* genes in *Escherichia coli*. In *The Bacterial Chromosome*. Drlica, K. & Riley, M. eds. American Society of Microbiology, pp. 259–66.
Johnson, R. C., Bruist, M. F. & Simon, M. I. (1986). Host protein requirements for *in vitro* site-specific DNA inversion. *Cell*, **46**, 531–9.
Kano, Y., Wada, M., Nagaso, T. & Imamoto, F. (1986). Genetic characterisation of the gene *hupB* encoding the HU1 protein of *Escherichia coli*. *Gene*, **45**, 37–44.
Kano, Y., Osato, K., Wada, M. & Imamoto, F. (1987). Cloning and sequencing of the HU2 gene of *Escherichia coli*. *Molecular and General Genetics*, **209**, 408–10.
Kavenoff, R. & Ryder, O. A. (1976). Electron microscopy of membrane-associated folded chromosomes of *Escherichia coli*. *Chromosome*, **55**, 13.
Kellenberger, E., Ryter, A. & Séchaud, J. (1958). Electron microscopy study of

DNA-containing plasmall. Vegetative and mature phage DNA as compared with normal bacterial nucleoids in different physiological states. *Journal of Biophysical and Biochemical Cytology*, **4**, 671–8.

Kiseleva, E. V., Kuliba, N. P., Bayborodin, S. I., Panfilova, Z. I., Khristolyubova, N. B., Koslav, A. V. & Salganic, R. I. (1988). Electromicroscopic study of DNA structural organization in nucleoids of spores and mycelium of streptomycetes. *Proceedings of the National Academy of Sciences, USSR*, **229**, 1486–8.

Kiseleva, E. V., Likhoshway, E. V., Khristolyubova, N. B. & Serdukova, N. B. (1986). An electromicroscopic analysis of the levels of the structural organization of *E. coli* chromosome. *Proceedings of the National Academy of Sciences, USSR*, **289**, 1235–7.

Kleinschmidt, A. K. (1968). Monolayer technique in electron microscopy of nucleic acid molecular. In Grossman, L. & Moldave, K., eds., *Methods in Enzymology*, vol. **12b**, pp. 361–77, Academic Press, New York & London.

Kornberg, T., Lockwood, A. & Worcel, A. (1974). Replication of chromosome with a soluble enzyme system. *Proceedings of the National Academy of Sciences, USA*, **71**, 3189–93.

Laine, B., Sautiere, P., Biserte, G., Cohen-Solal, M., Gros, F. & Rouvière-Yaniv, J. (1978). The amino and carboxy-terminal amino acid sequences of protein HU from *Escherichia coli*. *FEBS Letters*, **89**, 116–20.

Lathe, R., Buc, H., Lecocq, J. & Bautz, E. (1980). Prokaryotic histone-like protein interacting with RNA polymerase. *Proceedings of the National Academy of Sciences, USA*, **77**, 3548–52.

Lavoie, B. D. & Chaconas, G. (1990). Immunoelectron microscopic analysis of the A, B, and HU protein content of bacteriophage Mu transposomes. *Journal of Biological Chemistry*, **265**, 1623–7.

Leong, J., Nunes-Duby, S., Lesser, C., Youderian, P., Susskind, M. & Landy, A. (1985). The φ 80 and P22 attachment sites. Primary structure and interaction with *Escherichia coli* integration host factor. *Journal of Biological Chemistry*, **260**, 4468–77.

Liu, L. & Wang, J. (1987). Supercoiling of the DNA template during transcription. *Proceedings of the National Academy of Sciences, USA*, **84**, 7024–7.

Mendelson, I., Gottesman, M. & Oppenheim, A. B. (1991). HU and integration host factor function as auxiliary proteins in cleavage of phage lambda cohesive ends by terminase. *Journal of Bacteriology*, **173**, 1670–6.

Mensa-Wilmot, K., Carroll, K. & McMacken, R. (1989). Transcriptional activation of bacteriophage λ DNA replication *in vitro*: regulatory role of histone-like protein HU of *Escherichia coli*. *EMBO Journal*, **8**, 2393–402.

Menzel, R. & Gellert, M. (1983). Regulations of the genes for *Escherichia coli* DNA gyrase: homeostatic control of DNA supercoiling. *Cell*, **34**, 105–13.

Miller, H. I. & Nash, H. (1981). Direct role of the *himA* gene product in phage lambda integration. *Nature*, London, **290**, 523–6.

Miller, J. R. & Beatty, B. R. (1969). Visualization of nucleolar genes. *Science*, **164**, 955–7.

Morisato, D. & Kleckner, N. (1987). Tn10 transposition and circle formation *in vitro*. *Cell*, **51**, 101–11.

Nash, H. A. (1981). Integration and excision of bacteriophage I: the mechanism of conservative site-specific recombination. *Annual Review of Genetics*, **15**, 143–67.

Ogawa, T., Wada, M., Kano, Y., Imamoto, F. & Okasaki, T. (1989). Dna replication in *Escherichia coli* mutants that lack protein HU. *Journal of Bacteriology*, **171**, 5672–9.

Ogura, T., Niki, H., Kano, Y., Imamoto, F. & Hiraga, S. (1990). Maintenance of plasmids in HU and IHF mutants of *Escherichia coli. Molecular and General Genetics*, **220**, 197–203.

Pettijohn, D. & Pfenninger, O. (1980). Supercoils in prokaryotics DNA restrained *in vivo. Proceedings of the National Academy of Sciences, USA*, **77**, 1331–5.

Polaczeck, P. (1990). Bending of the origin of replication of *Escherichia coli* by bending of IHF at a specific site. *The New Biologist*, **2**, 265–71.

Prentki, P., Chandler, M. & Galas, D. (1987). *Escherichia coli* host factor bends the DNA at the ends of IS1 and in an insertion hotspot with multiple IHF binding sites. *EMBO Journal*, **6**, 2479–87.

Pruss, G. J. (1985). DNA topoisomerase mutants: increased heterogeneity in linking number and other replicon-dependent changes in DNA supercoiling. *Journal of Molecular Biology*, **185**, 51–63.

Pruss, G., Manes, S. H. & Drlica, K. (1982). *Escherichia coli* DNA topoisomerase mutants: increased supercoiling is corrected by mutations near gyrase gene. *Cell*, **31**, 35–42.

Pruss, G. & Drlica, K. (1986). Topoisomerase I mutants: the gene of pBR322 that encodes resistance to tetracycline affects plasmid DNA supercoiling. *Proceedings of the National Academy of Sciences, USA*, **83**, 8952–6.

Rouvière-Yaniv, J. & Gros, F. (1975). Characterization of a novel, low molecular weight DNA-binding protein from *Escherichia coli. Proceedings of the National Academy of Sciences, USA*, **72**, 3428–32.

Rouvière-Yaniv, J., Gros, F., Haselkorn, R. & Reiss, C. (1977). Histone-like proteins in prokaryotic organisms and their interaction with DNA. In *The Organization and Expression of the Eukaryotic Genome*. Bradbury, E. M. & Javaherian, K., eds, pp. 211–31, Academic Press, New York.

Rouvière-Yaniv, J. (1978). Localization of the HU protein on the *Escherichia coli* nucleoid. *Cold Spring Harbor Symposia on Quantitative Biology*, **42**, 439–47.

Rouvière-Yaniv, J., Yaniv, M. & Germond, J. (1979). *Escherichia coli* DNA-binding protein HU forms nucleosome-like structure with circular double-stranded DNA. *Cell*, **17**, 265–74.

Rouvière-Yaniv, J. & Kjeldgaard, N. (1979). Native *Escherichia coli* HU protein is a heterotypic dimer. *FEBS Letters*, **106**, 297–300.

Rouvière-Yaniv, J., Bonnefoy, E., Huisman, O. & Almeida, A. (1990). Regulation of HU protein synthesis in *Escherichia coli*. In *The Bacterial Chromosome*. Drlica, K. & Riley, M., eds, pp. 247–57, American Society of Microbiology, Washington DC.

Ryter, A. & Chang, A. (1975). Localization of transcribing genes in the bacterial cell by means of high resolution autoradiography. *Journal of Molecular Biology*, **98**, 797–810.

Schneider, G. J. & Geiduscheck, P. E. (1990). Stoichiometry of DNA binding by the bacteriophage SPO1-encoded type II DNA-binding protein TF1. *Journal of Biological Chemistry*, **265**, 10198–200.

Skarstadt, K. T., Baker, A. & Kornberg, A. (1990). Strand separation required for initiation of replication at the chromosomal origin of *Escherichia coli* is facilitated by a distant RNA–DNA hybrid. *EMBO Journal*, **9**, 2341–8.

Stonington, O. & Pettijohn, D. (1971). The folded genome of *Escherichia coli* isolated in protein–DNA–RNA complex. *Proceedings of the National Academy of Sciences, USA*. **68**, 6–9.

Surette, M. G. & Chaconas, G. (1989). A protein factor which reduces the negative supercoiling requirement in the Mu DNA strand transfer reaction is *Escherichia coli* integration host factor. *Journal of Biological Chemistry*, **264**, 3028–34.

Tanaka, I. K., Appelt, K., Dijk, J., White, S. W. & Wilson, K. S. (1984). 3-Å resolution structure of a protein with histone-like properties in prokaryotes. *Nature*, London, **310**, 376–81.

Tse-Dinh, Y. (1985). Regulation of the *Escherichia coli* topoisomerase I gene by DNA supercoiling. *Nucleic Acids Research*, **13**, 4751–63.

Tse-Dinh, Y. & Beran, R. (1988). Multiples promoters for transcription of the *Escherichia coli* DNA topoisomerase I gene and their regulation by DNA supercoiling. *Journal of Molecular Biology*, **202**, 735–42.

Varshavsky, A. J., Nedospasov, S. A., Bakayev, V. V., Bakeyeva, T. G. & Georgiev, G. (1977). Histone-like proteins in the purified *Escherichia coli* deoxyribonucleoprotein. *Nucleic Acids Research*, **4**, 2725–45.

Wada, M., Kano, Y., Ogawa, T., Okasaki, T. & Imamoto, F. (1988). Construction and characterization of the deletion mutant of *hupA* and *hupB* genes in *Escherichia coli*. *Journal of Molecular Biology*, **204**, 581–91.

Wang, J. C. (1971). Interaction between DNA and an *Escherichia coli* protein. *Journal of Molecular Biology*, **53**, 523–33.

Wang, J. C. (1985). DNA topoisomerases. *Annual Review of Biochemistry*, **54**, 665–97.

Westerhoff, H., O'Dea, M., Maxwell, A. & Gellert, M. (1988). DNA supercoiling by DNA gyrase. Astatic head analysis. *Cell Biophysics*, **12**, 157–81.

White, S. P., Appelt, K., Wilson, K. S. & Tanaka, I. (1989). A protein structural motif that bends DNA. *Proteins: Structure, Function, and Genetics*, **5**, 281–8.

Worcel, A. & Burgi, E. (1972). On the structure of the folded chromosome of *Escherichia coli*. *Journal of Molecular Biology*, **71**, 127–47.

Wu, H., Shy, S., Wang, J. C. & Liu, L. F. (1988). Transcription generates positively and negatively supercoiled domains in the template. *Cell*, **53**, 433–40.

Yang, Y. & Ames, G. F. L. (1990). The family of repetitive extragenic palindromic sequences: interaction with DNA gyrase and histone-like protein HU. In *The Bacterial Chromosome*, Drlica, K. & Riley, M., eds, pp. 211–25, American Society of Microbiology.

Yang, Y. & Ames, G. F. L. (1988). DNA gyrase binds to the family of prokaryotic repetitive extragenic palindromic sequences. *Proceedings of the National Academy of Sciences, USA*, **85**, 8850–4.

Yang, C. C. & Nash, H. A. (1989). The interaction of *Escherichia coli* IHF protein with its specific binding sites. *Cell*, **57**, 869–80.

REPLICATION AND SEGREGATION: THE REPLICON HYPOTHESIS REVISITED

CHRISTOPHER M. THOMAS and GRAZYNA JAGURA-BURDZY

School of Biological Sciences, University of Birmingham, PO Box 363, Birmingham B15 2TT, UK

INTRODUCTION

The model of the 'replicon' or 'self-replicating genetic unit' proposed by Jacob, Brenner and Cuzin (1963) in a paper entitled 'On the regulation of DNA replication in bacteria' has influenced our thinking about replication and its control for almost 30 years. Their original paper contains many ideas which appear generally to be lumped together under the rather ill-defined 'replicon hypothesis'. However, it offered to the scientific world not one but two hypotheses (Fig. 1). One hypothesis, the so-called 'replicon model', defined 'replicon' as a unit of replication and proposed its minimal requirements. The second hypothesis, the 'surface attachment model', offered an explanation for the apparent coregulation or coupling of replicon duplication and equal distribution of DNA molecules between daughter cells. The aim of this chapter is to examine how the data collected from experiments with bacteria during the last three decades have changed our perception of 'replicon' and the mechanisms controlling replication and segregation of replicons.

THE REPLICON MODEL

When Jacob *et al.*, (1963) first proposed the replicon model, little was known about the mechanisms controlling DNA replication. The model was based on the prevailing knowledge of regulatory elements involved in protein synthesis at that time. They proposed that a unit capable of independent replication (a 'replicon') would be composed of: (i) a gene encoding a specific initiator; and (ii) an operator of replication, i.e. a site of action for the corresponding initiator (Fig. 1a). Since then, DNA replication and its regulation have been the subject of much investigation and it appears that this simple replicon structure is basically valid for all systems which have been studied. All encode a specific initiator which may be a protein or an RNA molecule and in all cases these initiators act at a specific site, which is now known as the origin of replication (*ori*). Although Jacob *et al.*, (1963) called the whole plasmid or chromosome a replicon, according to their definition a 'replicon' consists of the minimal elements essential for autonomous repli-

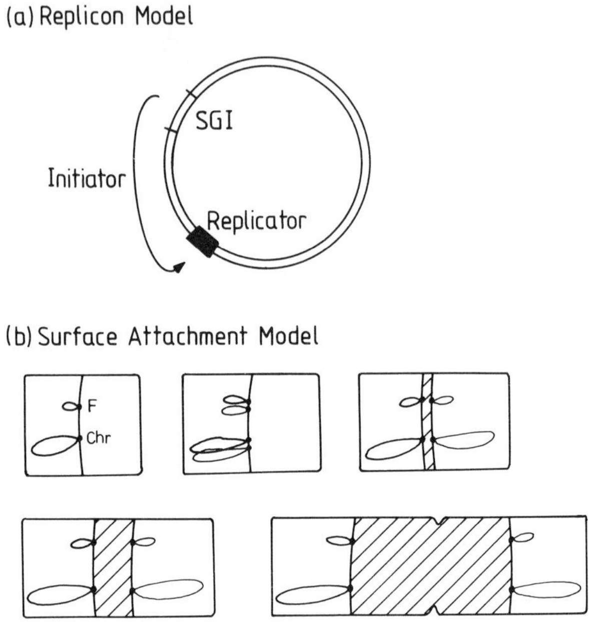

Fig. 1. The replicon hypothesis: (a) replicon model; (b) surface attachment model. SGI is the structural gene for the initiator.

cation and it is that definition which we will use in this chapter. In this section, we review current knowledge about the initiation of replication, concentrating on the nature of known initiator elements and their corresponding origins of replication in bacterial chromosomal and plasmid systems.

Most bacterial replicons are clustered

Recombinant DNA techniques have allowed the dissection of bacterial genomes to define the segments which are capable of autonomous replication. This approach has been particularly successful for bacterial plasmids (Fig. 2; Thomas, 1987). Such analyses have shown that all essential *cis*- and *trans*-acting sequences are generally clustered in a 1 to 3 kb region (mini-replicon). Some plasmids such as the classical sex factor F appear to have arisen by multiple recombination events and contain more than one replicon (F contains at least three; Womble & Rownd, 1988). Even for plasmids such as those of the IncP group from which it initially proved impossible to isolate a compact mini-replicon in this way, molecular analysis has shown that the ancestral replicon was confined to a small segment which has subsequently been disrupted by insertion of transposable elements or other DNA (Smith & Thomas, 1989).

The *cis*-acting sequences that allow replication when supplied with *trans*-

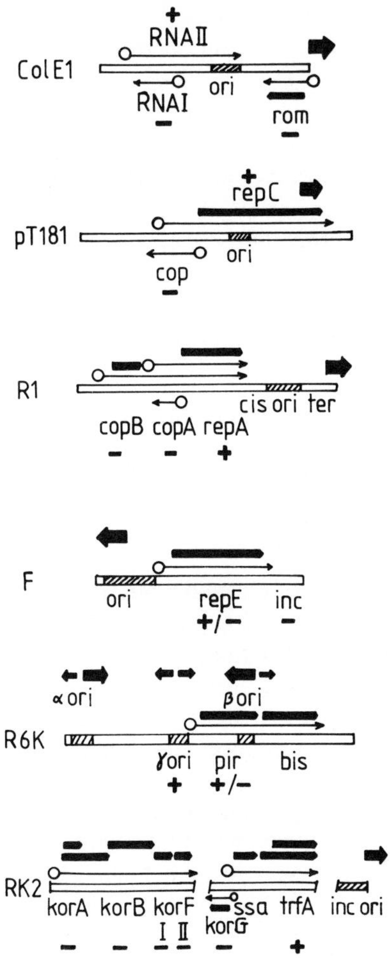

Fig. 2. Examples of the structures of plasmid replicons: ColE1 (Polisky, 1988); pT181 (Novick, 1989); R1/NR1 (Womble & Rownd, 1988); F (Murotsu *et al.*, 1981); R6K (Kolter, Inuzuka & Helinski, 1978; Crosa, 1980; Mukhopadhyay, Filutowicz & Helinski, 1986; Filutowicz *et al.*, 1986); RK2 (Thomas, Mayer & Helinski, 1980; Thomas & Smith, 1987).

acting functions encoded by the parent replicon can be defined by joining random fragments to a selectable marker on a DNA segment which is not able to replicate. The *trans*-acting function required to activate the *ori* can be located by cloning random fragments into a vector and determining which hybrid can support replication of *ori* joined to the selectable marker.

In the case of the bacterial chromosome similar analyses have permitted the isolation of the segment containing the replication origin (*oriC*) (Yasuda & Hirota, 1977; Messer *et al.*, 1978). The *trans*-acting initiators for the chromosome are much more difficult to define because one cannot get rid of the resident chromosome which provides all of the host genes. However, analysis of replication *in vivo* using conditional lethal mutations (Hirota, Ryter & Jacob, 1968) and *in vitro* using purified host proteins (Fuller & Kornberg, 1983; Kaguni & Kornberg, 1984) allows the activator proteins and the genes that encode them to be identified. The *dnaA* gene encodes what is regarded as the specific initiator (Hansen & Rasmussen, 1977). In *Bacillus subtilis* (Ogasawara *et al.*, 1985) as well as *Pseudomonas aeruginosa* and *P.putida* (Yee & Smith, 1990) *dnaA* is located directly adjacent to *oriC*. *Escherichia coli* and related enteric bacteria appear to be the exception in having a gap of 40 kb between *oriC* and *dnaA* which has probably arisen by inversion of a chromosomal segment (Ogasawara *et al.*, 1985).

The nature of replication initiators

To date, all replicons have been found to encode specific initiators. Initiators can be either RNA or protein and can act in a number of different ways.

1. RNA as an activator

In the well-studied example of plasmid ColE1, the initiator is an RNA molecule. The lack of a positively acting plasmid-encoded protein was originally demonstrated by showing that the plasmid could replicate after injection into a bacterial cell which had been treated with chloramphenicol to stop protein synthesis (Donoghue & Sharp, 1978; Kahn & Helinski, 1978). The role of RNA as a specific initiator was demonstrated by analysis of the replication process *in vitro* (Itoh & Tomizawa, 1978). A 550-base transcript, RNA II, produced by RNA polymerase passes through a series of conformations until it adopts a structure which allows it to remain associated with the plasmid DNA near *ori* (Masukata & Tomizawa, 1984; Wong & Polisky, 1985). It is then cleaved by RNaseH, thus producing a primer from which DNA PolI extends the leading strand for replication (Itoh & Tomizawa, 1980) (Fig. 3).

2. Initiator proteins which nick the origin

Specific initiator proteins can act in one of two ways. The simpler and better understood mechanism is by nicking one strand of the replication origin and thus initiating rolling circle replication. This is employed by a variety of both high and low copy number plasmids from Gram-positive bacteria such as *Bacillus subtilis*, *Staphylococcus aureus*, *Streptomyces* sp and *Streptococcus* sp (Novick, 1989). Analysis of the Rep protein from pT181 and related plasmids has shown that their C-terminal domains are responsible for DNA binding specificity while the nicking activity is conferred by a domain with a tyrosine residue in the middle of the protein (Thomas, Balson & Shaw, 1988; Fig. 4). Mutations which alter binding but not nicking and vice versa can be isolated. After initiation, the Rep protein is covalently attached to

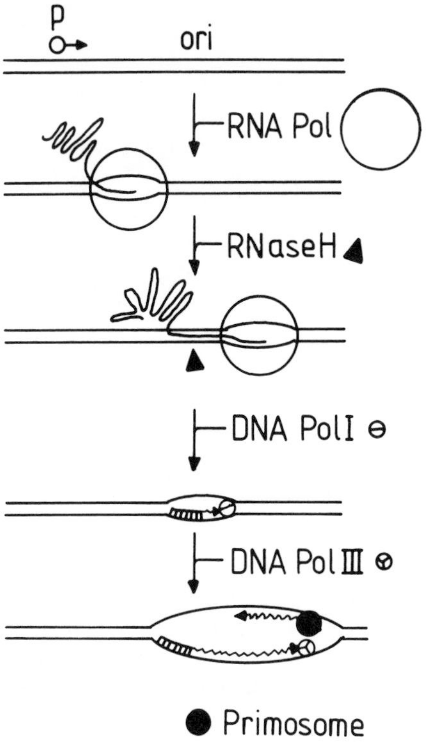

Fig. 3. Initiation of ColE1 replication at *ori* by RnaseH processing of a transcript from an upstream promoter.

the 5′ end generated at the nick, the energy of the protein–DNA bond being conserved to allow closure of the DNA circle after completion of replication (Koepsel *et al.*, 1985; Thomas *et al.*, 1988). The nicking–closing (topoisomerase) activity can be separated from the nicking activity both by the ionic conditions *in vitro* and by mutational changes. Intragenic complementation indicates that RepC functions at least as a dimer (quoted in Novick, 1989).

3. *Initiator proteins which cause unwinding*

All initiator proteins of this type so far studied are site-specific DNA binding proteins. DnaA is probably the most ubiquitous protein that activates an origin from which replication proceeds after strand separation (Fujita, Yoshikawa & Ogasawara, 1990). It binds to numerous specific sites on the bacterial chromosome as well as on many plasmids (Fuller, Funnell & Kornberg, 1984). It not only activates the origins of replication of the chromosome and several plasmids but is also involved in regulation of many genes including *dnaA* itself (Hansen & Rasmussen, 1977). The *dnaA* gene is thus an autogenously regulated gene, although whether this repression is mediated

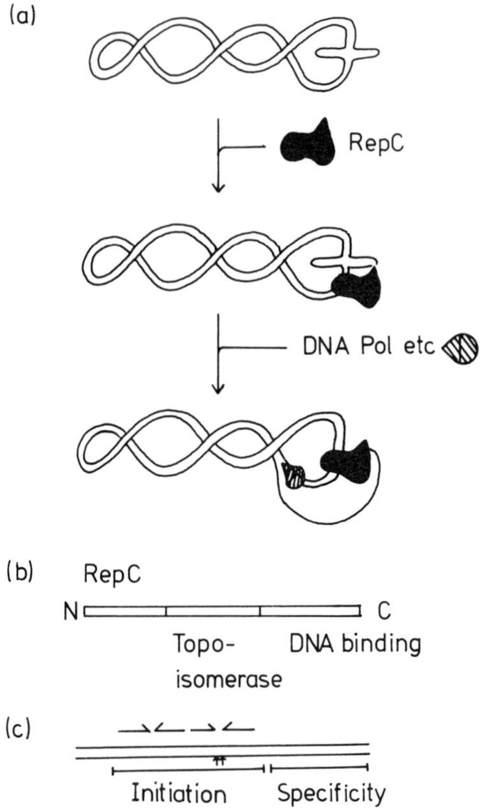

Fig. 4. Initiation of rolling circle replication by nicking a supercoiled template: (a) the general process on pT181; (b) the identified domains of pT181 RepC protein; (c) features of the replication origin of pT181 (for review see Novick, 1989).

at the level of transcription initiation (Kücherer *et al.*, 1986; Polaczek & Wright, 1990), termination (DnaA can block transcription when bound to DNA in one orientation within a gene; Schaefer & Messer, 1989) or at a post-transcriptional stage is not clear.

DnaA can exist in both active and inactive forms. It is active when bound to ATP (Sekimizu, Bramhill & Kornberg, 1987) (a type I nucleotide-binding domain can be identified in the primary sequence of DnaA) and is inactive in the absence of ATP or when bound to ADP. Conversion back to the active form can be brought about by cAMP (Hughes, Landoulsi & Kohiyama, 1988) or certain acidic phospholipids (Sekimizu & Kornberg, 1988). This latter effect is consistent with DnaA occupying a membrane location (Yung & Kornberg, 1988).

Analysis of the initiation process catalysed by DnaA suggests that the protein probably contains at least two domains which interact with DNA.

One is the region which binds the primary 9 bp repeats (Fuller *et al.*, 1984) and the second is the region which interacts with and melts the 13 bp repeats (Bramhill & Kornberg, 1988a; Yung & Kornberg, 1989). DnaA probably also interacts with other proteins such as DnaB (Frey, Chandler & Caro, 1984) and RNA polymerase (Atlung, 1984).

The *rep* genes of bacterial plasmids also encode proteins with multiple functions. These functions include activation of replication, repression of replication and autogenous repression of *rep* transcription. A variety of recent observations suggest that these different functions may be mediated by different forms of the Rep proteins, which may correspond to different oligomers. For some plasmids one class of copy number mutations maps in the *rep* gene (Filutowicz, McEachern & Helinski, 1986; Durland *et al.*, 1990; Muraiso, Mukhodadhyay & Chattoraj, 1990). The negative roles (both transcriptional repression and origin occlusion by 'looping' or 'handcuffing'; see page 61) of Rep proteins appear to be associated with multimeric forms. With R6K, mutations in the *pir* gene which reduce autogenous repression also reduce dimerization suggesting that it is the dimer which mediates autogenous regulation by the Rep protein (M. Filutowicz, personal communication). In P1 the heat shock proteins are required for replication and biochemical analysis suggests that their role is to convert an inactive or inhibitory dimer to an active monomer (Tilly & Yarmolinsky, 1989; Wickner, 1990; S. Wickner, personal communication). For RK2 also, fractions of purified TrfA in which the protein is largely dimeric are relatively inactive *in vitro*, while the most active fractions contain largely monomeric TrfA (D. Porter & C. M. Thomas, unpublished observations). That Rep protein forms different complexes at the origin repeat sequences called 'iterons' and at the *repE* gene operator has also been established for F plasmids (Masson & Ray, 1986).

At present we do not know the structure of any Rep protein sufficiently well to understand the differences in the ways in which monomers and dimers may interact with DNA or whether tetramers must be formed for looping or handcuffing. However, since dimerization of Rep proteins appears to be important in modulation of their activity, it is interesting that a short leucine zipper motif can be found in some but not all Rep proteins which dimerize (Giraldo *et al.*, 1989). Apart from this motif and the identification of α helix–turn–α helix motifs, relatively little is yet known about the structure of these proteins.

The nature of DNA replication origins

1. Origins activated by RNA

In the case of ColE1, recognition of the origin is specified by the sequence of the RNA which remains associated to form an RNA-DNA duplex which is cleaved by RNaseH (Itoh & Tomizawa, 1980) (Fig. 3). If cleavage does not occur, the position at which DNA Polymerase I initiates leading strand synthesis from the RNA primer can vary (Ohmori, Murakami & Nagata, 1987). This suggests that the DNA at the origin may not possess unique properties which distinguish it from the surrounding DNA.

2. Origins which are nicked

For those plasmids whose replication is initiated by a nick, the origin serves both to initiate and terminate replication. A dimeric molecule containing two origins will generate monomeric products due to termination of replication i.e. circularization of the nascent product at the second origin (Gros, Te Riele & Ehrlich, 1987; Iordanescu & Projan, 1988). Two families of such plasmids have been studied, the archetypes for which are pT181 and pC194 (Novick, 1989). The main conserved features of their origins are the segment I, in which nicking occurs, and segment II, a downstream (3′) site recognized by the specific initiator protein (Fig. 4). Segment I of pT181 *ori* consists of an inverted repeat which may be extruded to form a cruciform structure in supercoiled DNA (quoted in Novick, 1989). This greatly increases the efficiency with which the cognate Rep protein acts upon it. Related Rep proteins which bind poorly to the adjacent segment II can nick with low efficiency *in vitro* when the template is supercoiled but not when it is relaxed. The *cmp* locus of pT181 appears to control the supercoiling density of the plasmid and hence the efficiency of initiation (Gennaro & Novick, 1988).

3. Origins which initiate theta-form replication

Perhaps the most intriguing replication origins are those in which protein–DNA interactions lead to the unwinding of the DNA duplex, allowing entry of primase and thus initiation of leading strand synthesis which is generally closely followed by primosome assembly and initiation of lagging strand synthesis. Such origins can initiate replication unidirectionally or bidirectionally.

The most common feature of these origins is the presence of repeated sequences (iterons) to which the initiator protein binds. these can be present in direct or inverted orientation. In *oriC* of *E.coli* and a variety of other Gram-negative bacteria (*Micrococcus luteus*, *Proteus mirabilis*, *Salmonella typhimurium* and *Serratia marcescens*) there are four DnaA binding sites (a sequence of nine bases with the consensus 5′ TTATCCACA 3′) in alternating orientation within a *cis*-acting region of 245 bp which is sufficient for autonomous replication when supplied with *trans*-acting factors (Fig. 5c) (Meijer *et al.*, 1979; Zyskind *et al.*, 1983). In contrast, the *oriC* of *Bacillus subtilis* requires two regions which are separated by the *dnaA* gene (Moriya, Ogasawara & Yoshikawa, 1985). Both these regions contain copies of the DnaA binding site; one contains eight and the other three (Fig. 5c). The *oriC*s of *Pseudomonas aeruginosa* and *P. putida* have a similar chromosomal

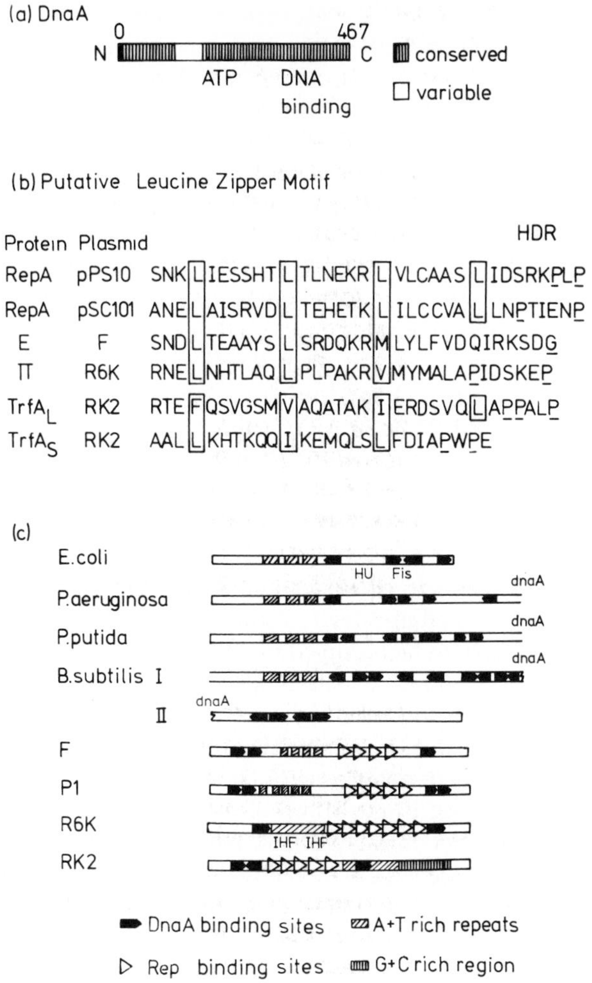

Fig. 5. (a) General structure of the DnaA proteins; (b) leucine zipper motif identified in a variety of Rep proteins; (c) general structure of some replication origins from which theta-form replication is initiated.

location to the *B. subtilis oriC* but an internal organization more like *E. coli* although the spacing of the DnaA binding sites is different (Yee & Smith, 1990).

The repeats in plasmid replication origins tend to be much more evenly spaced, occurring in tandem arrays with a repeat unit of 22 to 23 bp (Fig. 5c). The number of repeats varies: for example, there are four in *oriS* of F (Murotsu, Tsutsui & Matsubara, 1984); five in P1 (Abeles, Snyder & Chattoraj, 1984) and RK2 (Stalker, Thomas & Helinski, 1981; Cross, Warne

& Thomas, 1986); and seven in R6K gamma origin (Stalker, Kolter & Helinski, 1979). Finally, in R6K additional repeats placed at a distance are essential for activity of *oriα* and *oriβ* (Shafferman & Helinski, 1983; Shafferman *et al.*, 1987; Shon, Germino & Bastia, 1982).

In addition, origins generally contain: an A + T-rich region with additional repeats where strand separation is likely to occur (Bramhill & Kornberg, 1988b); a G + C-rich region which might help to focus melting in the A + T-rich region (this is not present so frequently; Stalker *et al.*, 1981); binding sites for additional accessory proteins such as IHF (Drlica & Rouviere-Yaniv, 1987; Filutowicz & Appelt, 1988; Gamas *et al.*, 1986; Stenzel, Patel & Bastia, 1987), HU (a histone-like protein; Drlica & Rouviere-Yaniv, 1987; Dixon & Kornberg, 1984; Ogawa *et al.*, 1989) and Fis (a protein also involved in inversion of Mu). Many plasmid origins contain DnaA boxes which in some cases are essential (Hansen & Yarmolinsky, 1986; Gaylo, Turjman & Bastia, 1987) while in others they just play a stimulatory role (Tang, Womble & Rownd, 1989). Finally, in some origins Dam methylation sites are present and serve to cause a delay in reinitiation of replication (see below).

DnaA, HU and Fis at *oriC* create a complex in which the DNA is tightly folded (Fig. 6). DnaA binding sites 1, 2 and 4 at *oriC* have a high affinity for DnaA and are likely to be occupied by DnaA under most physiological conditions. The lower affinity of site 3 for DnaA means that only when most DnaA-binding sites on the chromosome are occupied will this site be filled (W. Messer, personal communication). When it is filled, DnaA causes melting of the A + T-rich region of the origin (Bramhill & Kornberg, 1988*a*). This might be due to migration of a melted region brought about by distortion of the DNA through DnaA binding (Bramhill & Kornberg, 1988*b*). Alternatively, a second domain of DnaA might interact directly with the 13-base repeats in this region and thus cause melting, possibly by electrostatic repulsion (Yung & Kornberg, 1989). This unwinding allows entry of the DnaB/DnaC complex which unwinds the origin further and allows entry of priming enzymes for leading and lagging strands. Transcription from outside the origin is required for efficient initiation (Baker & Kornberg, 1988; Skarstadt, Baker & Kornberg, 1990). At plasmid replication origins a related activation occurs through a combination of plasmid Rep protein and host proteins, most commonly DnaA.

THE SURFACE ATTACHMENT MODEL

Jacob *et al.*, (1963) defined not only the minimal requirements for replicon structure but also offered explanations for: (i) the control of replication; (ii) coordination of key processes in cell growth and division; and (iii) the equal distribution of DNA molecules between daughter cells. Self-replicating units were assumed to be bound through the initiator sites on the DNA

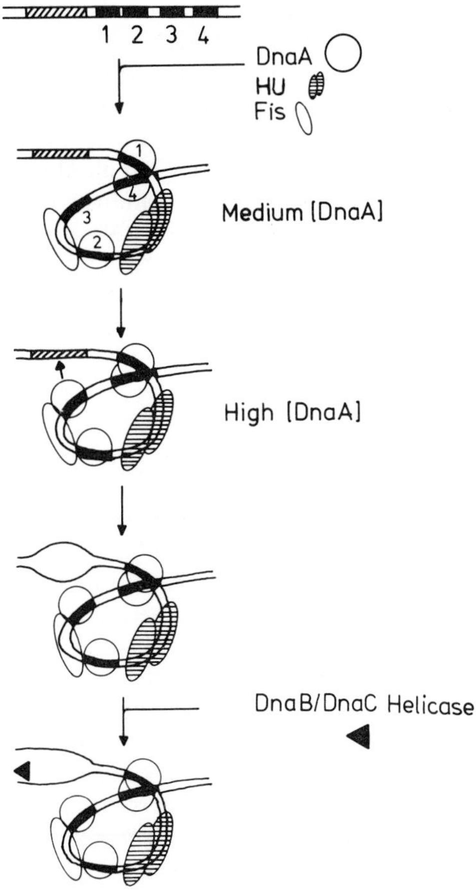

Fig. 6. Replisome and initiation of replication at *oriC* based on a combination of models from the laboratories of A. Kornberg and W. Messer.

to defined sites at the cell equator (Fig. 1b). At a specific stage during the cell cycle the attachment sites would be duplicated, promoting initiation of replication. This would allow control over the frequency and timing of replication. It was also proposed that membrane growth and septum formation would assure the spatial separation of the newly formed replicons to the opposite halves of the dividing cell. For the model to be feasible the surface attachment would have to involve the cell wall and possibly the outer membrane since lateral diffusion in the cytoplasmic membrane is now known to be too rapid to provide the physical strength to pull the chromosomes apart (Quinn, 1981). Therefore in the following sections 'membrane attachment' refers generally to structures that involve cell wall and/or outer membranes. This could occur at sites such as Bayers bridges or periplasmic annuli where inner and outer membranes are thought to fuse at gaps in the cell wall (Bayer *et al.*, 1982; MacAlister *et al.*, 1983). A second requirement is that the cell surface growth occurs primarily from the centre so that attached chromosomes will move outwards. While some growth might occur in this way, the present consensus appears to be that it is a much more disperse process (Green & Schaechter, 1972; Lin, Hirota & Jacob, 1971). Nevertheless, it is impossible to rule out some active role for membrane growth on these grounds. Therefore we still consider it worthwhile discussing the proposal of Jacob *et al.* (1963) that there is a tight coupling between cell division, replicon duplication and partitioning, with the site-specific DNA-membrane complex playing an essential regulatory function in all of these processes.

Does replication occur at membrane sites?

A central idea of the surface attachment model is that replication occurs at membrane sites. The ability to develop soluble *in vitro* replication systems for plasmids and chromosomes suggested that membranes were not involved, but accumulating experimental evidence suggests that membranes do play an important role in the initiation of chromosome replication. Numerous DNA/membrane complexes isolated from different bacterial species and for various replicons have the potential to synthesize DNA *in vitro* without the addition of enzymes or regulatory factors (for a review see Firshein, 1989). In *E.coli* a membrane complex which contains the replication origin of the chromosome (*oriC*) was isolated from cells at the time of initiation of chromosome replication (Nagai *et al.*, 1980; Gayama *et al.*, 1990) and was further purified (Hendricson *et al.*, 1982; Jacq *et al.*, 1983). Similarly, early replicative intermediates from the origin region were isolated in an outer membrane fraction of *E.coli* (Yoshimoto *et al.*, 1986). These observations support the idea that there is a direct connection between the initiation event at *oriC* and membrane binding. Temperature-sensitive initiation mutants of *E.coli* (*dnaA*, *dnaB*) and *B.subtilis* show abnormalities in membrane protein patterns (Inouye & Pardee, 1970; Siccardi *et al.*, 1971; Harmon & Taber, 1977; Imada, Carroll & Sueoka, 1976). Similarly, inhibition of initiation by a variety of drugs or antibiotics leads to the degradation of DNA–membrane complexes (Yoshikawa, Ogasawara & Seiki, 1980; Ogasawara, Seiki & Yoshikawa, 1981). Studies to determine which replication protein(s) may interact with membranes lead to conclusive results with the DnaA initiator protein. Especially interesting is data suggesting that phospholipids play a role in activation of DnaA by ATP (Sekimizu, Yung & Kornberg, 1988). Analysis of DNA–membrane complexes indicated the specificity of binding between *oriC* and certain outer membrane proteins (Wolf-Watz & Masters, 1979). It has been shown that a 463 bp region of *oriC* specifically interacts with the outer membrane fraction through more than one binding site (Hendricson *et al.*, 1982, Kusano *et al.*, 1984).

Coordination of DNA replication and cell division

Although there is considerable evidence that replication takes place at a membrane site, is it the appearance of such a site at a specific stage during the cell cycle that triggers initiation? By varying the growth rate it was possible to show that chromosome replication is initiated not at a constant stage during the cycle but rather at a constant time before cell division so that at fast growth rates the next round of replication is initiated before the cell cycle is completed (Cooper & Helmstetter, 1968). Replication appears to be linked directly to growth, initiation of replication occurring at a constant DNA to mass ratio (Donachie, 1968). Experiments with the 'baby machine' of Cooper and Helmstetter (Helmstetter & Pierucci, 1976; Fig. 7a) and by flow cytometry (Skarstadt, Boye & Steen, 1986) also showed that initiation at *oriC* occurs synchronously. What mechanisms are responsible for this synchrony?

Recent experiments in *E.coli* indicate that synchrony is due at least in part to the *dam* methylation system (Fig. 7b,c). In wild-type bacteria DNA becomes hemimethylated during replication. The 245 bp minimal *oriC* region has 11 GATC sequences which are the targets for Dam methylation whereas, on average, only a single site would be expected in a fragment of this size (Meijer *et al.*, 1979). The importance of the methylation state was indicated by the observation that *dam* mutants showed asynchronous initiation of DNA replication (Bakker & Smith, 1989) but synchrony could be restored by induction of a cloned *dam* gene to give normal levels of Dam enzyme activity (Fig. 7b) (Boye & Lobner-Olesen, 1990). Over-expression of the *dam* gene results in too rapid remethylation and loss of synchrony once again (Messer *et al.*, 1985; Boye & Lobner-Olesen, 1990). The fact that *dam* mutants are transformed very poorly with methylated *oriC* plasmids (Messer *et al.*, 1985; Smith *et al.*, 1985; Russell & Zinder, 1987) but are easily transformed with unmethylated DNA suggested that, once *oriC* DNA has become hemimethylated due to replication, it is inhibited from further initiations until it has been remethylated. There is now evidence that hemimethylated DNA becomes attached to the bacterial membrane which inhibits further initiation for an eclipse period (Ogden, Pratt & Schaechter, 1988; Landoulsi *et al.*, 1990) (Fig. 7c). *In vivo* it has been found that remethylation of *oriC* is delayed for up to 10 min (Ogden *et al.*, 1988; Campbell & Kleckner, 1990) whereas remethylation of other sites tested was completed in 1.5 to 3 min (Campbell & Kleckner, 1990). Another factor which might delay reinitiation is the state of supercoiling of the nucleoid domain containing *oriC* (Leonard, Hucul & Helmstetter, 1982). By the time the DNA has become supercoiled again, remethylated, and has been released from the membrane,

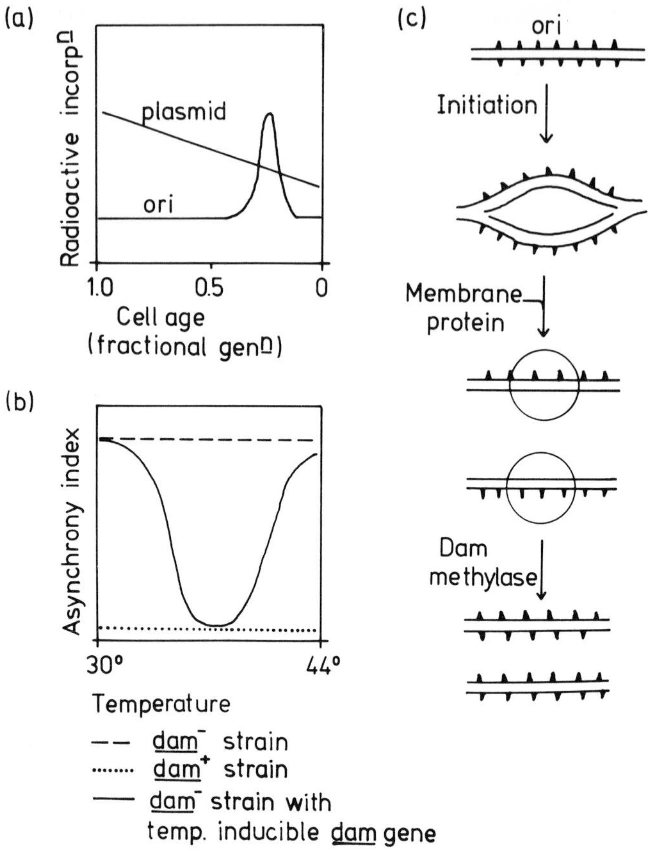

Fig. 7. Coordination of DNA replication and cell division. (a) The 'baby machine' of Cooper & Helmstetter revealed synchronous initiation of *oriC* at a specific stage in the cell cycle but exponential replication of plasmids. (b) Typical results from flow cytometry showing the influence of the Dam methylase of *E.coli* on synchrony of initiation of replication. (c) Cartoon to illustrate binding of hemimethylated *oriC* to a membrane protein to inhibit replication. (See text for references).

the potential for replication has decayed.

What creates this peak in the potential for initiation? From the sequence of the *E.coli* chromosome that is available at present it appears that many DnaA binding sites are concentrated in the chromosome segment that is replicated soon after initiation. It is proposed that replication first displaces DnaA, increasing the concentration of free DnaA and then, due to duplication of the DnaA binding sites, results in the titration of free DnaA to a level well below that required for initiation of replication (F. Hansen & T. Atlung, personal communication; Mahaffy & Zyskind, 1989). The first initiation event therefore leads to a short-lived burst of initiation potential.

A different interpretation of the observations cited above has been proposed recently (Norris, 1990). The 'Membrane Detachment Model' proposes that transient changes in the membrane composition at a specific stage during the cell cycle lead to dissociation of the chromosome from the membrane, allowing initiation to occur. Reassociation with the membrane occurs due to the hemimethylated state of the newly replicated *oriC* DNA.

Although control over the initiation process is important in regulating DNA synthesis, it is not the only step which is linked to bacterial growth. Although the frequency of chromosome initiation can increase when the level of DnaA is raised by induction of a strain containing the *dnaA* gene under the control of an inducible promoter (Atlung *et al.*, 1985) the DNA to mass ratio still remains constant suggesting that there are other factors which limit chromosome replication after initiation has occurred (T. Atlung, personal communication). Even when the initiation mass is changed by replacing *oriC* with a plasmid R1 origin of replication and inducing more frequent initiations, chromosome replication does not occur in proportion to the number of initiations – the rate of elongation fork movement may become a limiting factor (Nordstrom, Bernander & Dasgupta, 1990). Similarly, over-initiation at *oriC* in *B. subtilis* does not lead to uncontrolled replication – elongation stops at a point approximately 200 kb from the replication origin and only resumes in the next division cycle (Henckes *et al.*, 1989). Therefore it appears that a number of factors may couple DNA replication to growth but not specifically to the cell cycle.

The surface attachment model proposed that there should be tight control over the number of initiation events. That this is clearly not the case in *E. coli* is indicated by the fact that *oriC* plasmids do not compete with the chromosome, i.e. they do not express incompatibility (Messer *et al.*, 1978; Yasuda & Hirota, 1977). Indeed, they continue to replicate in synchrony with the chromosome even when there are multiple copies in the cell (Helmstetter & Leonard, 1987; Leonard & Helmstetter, 1986). Thus in *E. coli*, control appears to consist of providing excess replication potential for a short period but limiting each origin to a single initiation event by the inability of hemimethylated DNA to initiate. On the other hand, *B. subtilis* lacks a Dam methylation system and yet chromosome replication is still synchronous (E. Boye, personal communication). Introduction of *dnaA* boxes on a plasmid inhibits chromosome replication, probably by titration of DnaA, suggesting that there is only a limited potential for replication (Moriya *et al.*, 1988). This is consistent with recent experiments which indicate that the level of DnaA protein is a major factor, if not the major factor, in determining the number of initiation events which can occur simultaneously in *B. subtilis* (Moriya *et al.*, 1990). The DnaB protein of *B. subtilis* is also required for initiation and is a membrane protein (Hoshino *et al.*, 1987; Sueoka, Hoshino & McKenzie, 1988). These two elements, limited replication potential and membrane attachment, suggest that *B. subtilis* may turn out to fit the surface attachment model quite well.

Since plasmids such as F and R1 are present at one to two copies per chromosome, it was of interest to determine whether they are also replicated at a specific stage during the cell cycle. The results of density shift experiments indicated clearly that plasmid molecules are replicated randomly both in time and with respect to the choice of the individual molecule that is selected for replication (Gustafsson, Nordstrom & Perram, 1978). Recent experiments with the 'baby machine' of Cooper and Helmstetter have confirmed that almost all plasmids replicate exponentially throughout the cell cycle (Leonard & Helmstetter, 1988; Keasling, Palsson & Cooper, 1991). The only plasmid over which there seems to be some doubt is F for which the results from different laboratories lead to contradictory conclusions (Keasling *et al.*, 1991; C. Helmstetter, personal communication).

Control of DNA replication

The observed relationships of replication to growth indicate that many factors, both positive and negative, can be involved. Although elongation rate may provide some control when over-initiation has occurred, this is atypical. Normally initiation appears to be the control point. So what is it that controls the triggering of replication?

At an SGM symposium in 1969, Pritchard, Barth and Collins proposed the 'Inhibitor Dilution Model'. In this model, replication is controlled by the concentration of an inhibitor which is synthesized constitutively from the genome. As the genome is replicated, the number of genes producing the repressor rises so that the inhibitor concentration rises and eventually shuts off replication. As growth of the cell dilutes the inhibitor its concentration falls allowing further replication. Until recently there were no examples of inhibitors acting directly at the origin. However, the purification of the IciA protein (33 kD; Hwang & Kornberg, 1990) which can bind to the DnaA 13 bp A + T-rich repeats of *oriC* and thus block the initiation process, has provided the first direct evidence that repressor molecules may be involved in the regulation of chromosome replication of *E.coli.* The 40 kD membrane protein that inhibits initiation of hemimethylated *oriC* may also count as such (Landoulsi *et al.*, 1990). An inhibitor of chromosome replication in *B. subtilis* has also been identified by J. Laffan and W. Firshein (personal communication).

In systems where there is no evidence for direct inhibition of initiation, how is replication controlled? Many experiments to answer this question have been conducted with plasmids since they all appear to possess some mechanism for inhibiting their replication when the copy number rises above a certain level. This was first demonstrated with hybrids between the high copy number plasmid ColE1 (15 copies per chromosome) and the lower copy number plasmid pSC101 (5 copies per chromosome; Cabello, Timmis & Cohen, 1976). Elevation of the pSC101 copy number three-fold above its normal level resulted in complete inhibition of replication from its origin. Related behaviour is illustrated by experiments carried out by Highlander and Novick (1987) with plasmid pT181 of *Staphylococcus aureus*. They studied the kinetics with which pT181 or its copy mutant repopulate bacteria after their replication has been inhibited to dilute them to one copy per cell. The plasmid with normal replication control (specified by an antisense RNA) over-shoots its steady state copy number but then is inhibited from further replication until its copy number has fallen below the steady state level. The copy number mutant replicates rapidly at first but then slowly approaches a steady level as it runs out of replication potential. The wild type therefore appears to control its replication by an inhibitory activity while the mutant is controlled by the availability of positively required factors.

The ability to shut off replication at high copy number has been linked to a plasmid-encoded inhibitor only in specific cases. The best studied examples are ColE1, pT181 and R1/NR1 all of which encode small antisense RNA molecules which form RNA duplexes with transcripts which are required either directly or indirectly for replication. In ColE1, RNAI binding prevents RNAII from adopting the conformation required for productive binding at the origin (Polisky, Zhang & Fitzwater, 1990; Tomizawa *et al.*, 1981; Tomizawa, 1990) (Fig. 8). In pT181 the Cop RNA interaction favours the formation of a structure which causes premature termination of *rep* transcription (Novick *et al.*, 1989; Fig. 8). In R1/NR1 CopA RNA interferes with RepA translation from its mRNA (Nordstrom *et al.*, 1988). It is not yet clear how the repression occurs but suggestions still being considered include indirect occlusion of the *repA* ribosome binding site (Dong, Womble & Rownd, 1987) and RNase attack on the CopA.CopT RNA duplex to initiate degradation of the mRNA (Blomberg, Wagner & Nordstrom, 1990). Interaction between CopA and CopT starts with 'kissing' followed by the effectively irreversible formation of an RNA duplex (Persson, Wagner & Nordstrom, 1988). The effectiveness of CopA depends on its ability to form the 'kissing' complex rather than on the stability of the complex once it has formed (Persson, Wagner & Nordstrom, 1990). R1/NR1 also encodes a transcriptional repressor protein, CopB, the concentration of which falls in proportion to copy number thus derepressing *repA* transcription (Riise & Molin, 1986). In all these cases, replication is controlled by the level of an initiator which, in turn, is regulated by a repressor, whose concentration rises directly in proportion to plasmid copy number. In other types of plasmids, such repressors have not been identified.

In the absence of such an inhibitor, one explanation for the inhibition of plasmid replication when the copy number rises above a certain level is offered by a recent hypothesis, the 'handcuffing' model. This suggests that the plasmid DNA itself can act as the inhibitor. As the plasmid copy number rises, origin–Rep–Rep–origin complexes are formed and block the

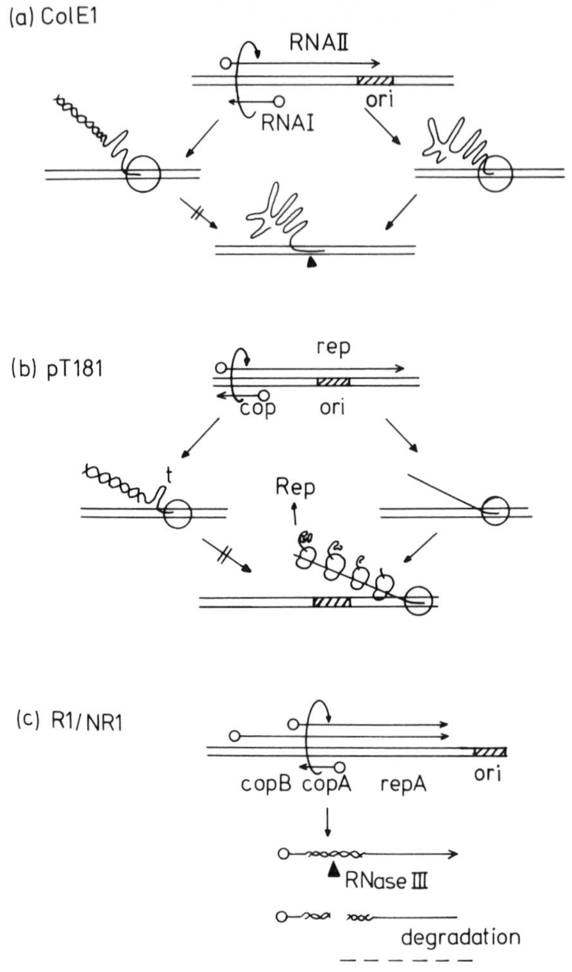

Fig. 8. Control of plasmid replication by antisense RNA. For reviews see Polisky (1988), Novick (1989) and Nordstrom *et al.* (1990).

entry of proteins essential for initiation (Fig. 9). This model was originally proposed for R6K where the π protein was shown to mediate interaction between the DNA molecules containing π binding sites (Mukherjee, Erickson & Bastia, 1988; McEachern *et al.*, 1989) and is related to ideas developed to explain the role of iterons in copy number control of P1 (Pal & Chattoraj, 1988). Further evidence for 'handcuffing' has been provided by showing that *in vitro* replication of RK2 is inhibited by copies of the TrfA-binding iterons on a supercoiled plasmid. Since the inhibition is not relieved by addition of excess TrfA it was argued that it is not due to titration of positively

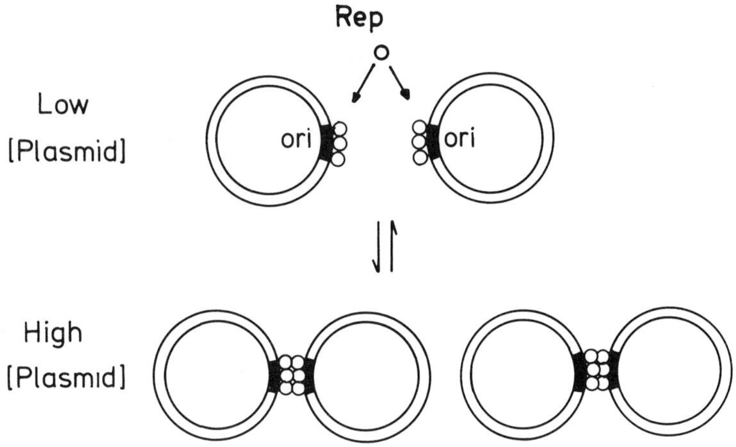

Fig. 9. Handcuffing model for control of plasmid replication based on the proposals of Helinski and coworkers.

acting TrfA. The repeats must therefore be having some sort of direct negative effect (Kittel & Helinski, 1991).

Alternative explanations have also been offered. Sompayrac & Maaloe (1973) proposed that the initiator, or a protein synthesized in a coordinated way with it, would control the rate of initiator synthesis (Fig. 10). It has been found that all plasmid systems, as well as the bacterial chromosome, control the expression of the initiator gene either by autogenous regulation (chromosomal *dnaA*, Hansen & Rasmussen, 1977; *repE* of F, Masson & Ray, 1986, Tokino, Murotsu & Matsubara, 1986; *repA* of P1 prophage, Swack *et al.*, 1987; *repA* of pSC101, Linder *et al.*, 1985, Vocke & Bastia, 1985, Yamaguchi & Masamune, 1985; *pir* of R6K, Filutowicz *et al.*, 1985) or by independent repressor molecules whose expression are autogenously regulated (IncP, Thomas & Helinski, 1989; and IncQ plasmids, Haring & Scherzinger, 1989). The regulation of *trfA* gene expression on RK2 is especially interesting. To date at least five genes, whose products repress *trfA* transcription, have been identified (Thomas & Helinski, 1989; Jagura-Burdzy, Ibbotson & Thomas, 1991). The small increases in the concentration of each repressor as the plasmid copy number increases may have a cumulative effect and result in severe repression of *trfA* expression consistent with the apparent ability of these circuits to provide the overriding copy number control in RK2 (Thomas & Hussain, 1984).

In some plasmid replication systems these regulatory circuits result in Rep proteins being present at or close to limiting levels (RepA of P1 exists at 40 monomers per unit copy of plasmid DNA which contains 14 RepA binding sites, Swack *et al.*, 1987; RepE of F is present at about 100 copies per cell, Kline, 1988; while for TrfA of RK2 the combined total of P_{285}

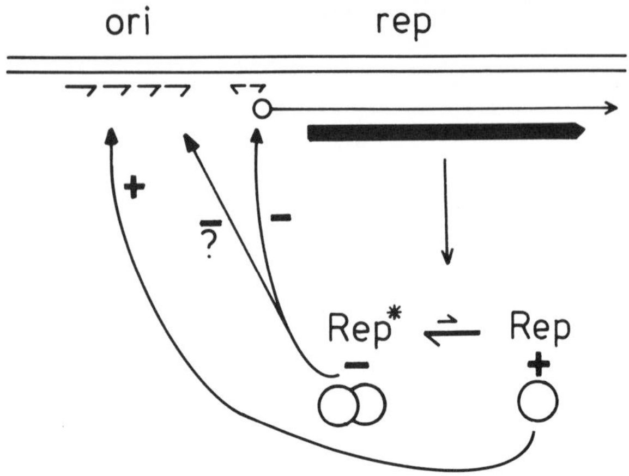

Fig. 10. Model for autogenous control of plasmid copy number based on ideas from Sompayrac and Maaloe (1973), Trawick & Kline (1985) and M. Filutowicz (personal communication).

and P_{382} is 300 monomers compared to 135 binding sites per cell, Durland & Helinski, 1990). There are others in which the concentration of initiator is 100-fold higher than required for replication (3500 to 10 000 dimers of π protein per cell for R6K replicons; Filutowicz *et al.*, 1986). Obviously, the total concentration of Rep protein cannot be the sole determinant controlling initiation. The existence of the active form of DnaA protein in complex with ATP versus inactive forms (DnaA–ADP, DnaA bound to acidic phospholipids or DnaA-free; Sekimizu *et al.*, 1987) raises the possibility of modifying the pool size of active initiator without changing its concentration in the cell. The idea that there are two Rep protein forms, one active and the other inhibitory, was proposed by Trawick & Kline (1985). The dimeric forms of RepA of P1 and TrfA of RK2 seem to be relatively inactive in replication (S. Wickner, personal communication; D. Porter & C. M. Thomas, unpublished observations) while the dimeric form of π protein of R6K appears to interfere with replication both by repressing *pir* gene transcription and possibly also by blocking the positive action of the monomer at the origin (M. Filutowicz, personal communication).

Clearly, there are many ways in which replication initiation can be inhibited but for many systems it remains to be determined which effects are the most important under normal growth conditions and at normal copy numbers.

Replication and partitioning functions can be separated

The surface attachment model proposed that a single attachment site between DNA and membranes would allow both replication and active partitioning. This proposal is easily tested by examining the properties of the DNA seg-

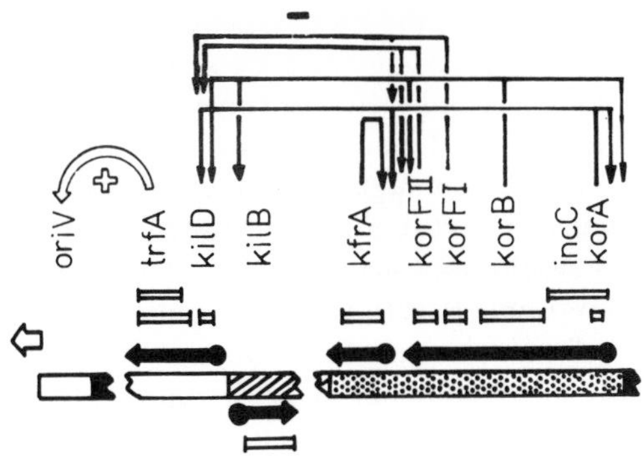

Fig. 11. Coordinate expression of replication, partitioning and conjugative transfer functions of broad host range plasmid RK2 (reviewed in Thomas & Helinski, 1989). The segments encoding these functions are shown as follows: replication functions, open blocks; partitioning functions, stippled blocks; conjugation functions, hatched blocks.

ments isolated by the genetic dissection of genomes described above. For example, the minichromosomes which contain *oriC* are very unstable despite relatively high copy numbers, i.e. possession of the site which allows replication does not ensure efficient segregation (Messer *et al.*, 1978; Yasuda & Hirota, 1977). Although it has not yet proven possible to define the chromosomal segment (if there is one) which provides the ability to segregate efficiently to daughter cells, dissection of plasmids has revealed the existence of specific partitioning functions which are distinct from the replicon and which can be used as cassettes to stabilize unrelated replicons (Austin, 1988). For example, the partitioning functions from F can stabilize *oriC* plasmids (Ogura & Hiraga, 1983).

Although it appears that plasmid replication and partitioning functions are found as separate cassettes, the plasmid RK2 shows intriguing coordination of the replication and partitioning functions it carries (Fig. 11). The central regulatory operon of RK2 encodes several proteins, four of which (KorA, KorB, KorFI and KorFII) act as repressors for the replication initiation gene *trfA* as well as other RK2 operons (Thomas & Helinski, 1989; Jagura-Burdzy *et al.*, 1991). Two products of the *korABF* operon, IncC and KorB, show homology to ParA/B proteins of P1 and SopA/B proteins of F, which have been established to play a role in active partitioning. It has been shown that *incC* is part of an RK2-encoded stability locus which will also stabilize heterologous plasmids (Motallebi-Veshareh *et al.*, 1990). Because of this tight coupling of replication and partitioning functions on this plasmid, it is difficult to separate the two systems genetically.

Does membrane growth drive active partitioning?

If membrane growth is the force which drives partitioning, then the copies of the bacterial chromosome should move apart slowly as the cell grows.

However, when the location of the bacterial nucleoids is visualized by staining with fluorescent dyes, they can be seen to remain at the central pole until replication is complete and then suddenly jump to the quarter positions which represent the equatorial plane of daughter cells (Donachie & Begg, 1989; Hiraga *et al.*, 1989). This suggests that, even if membrane attachment is an important aspect of partitioning, something else may be responsible for the movement of the chromosome from one location to another. Attempts to analyse the chromosome partitioning mechanism have concentrated on isolating mutants of *E. coli* which are defective in partitioning. The phenotype initially chosen was the production of cells which were elongated due to arrest of cell division following the failure to partition the daughter chromosomes (Hirota *et al.*, 1968, 1971). The analysis of these *par* mutants proved that *parA* and *parD* are at least double mutants with defects in *gyrB* and *gyrA* (DNA gyrase subunits B and A) (Kato, Nishimura & Suzuki, 1989; Hussain *et al.*, 1987*a,b*); *parC* was found to code for a topoisomerase (Kato *et al.*, 1988, 1990); while *parB* was found to be an allele of *dnaG* coding for primase activity which is essential for DNA replication (Norris *et al.*, 1986). Moreover, the same 'Par' phenotype was identified for a *gyrB* mutant (DNA gyrase subunit B; Orr *et al.*, 1979; Steck & Drlica, 1984). Thus the partitioning defect of these mutants is due to impaired replication or decatenation of chromosomes.

More recently, Hiraga and coworkers have chosen a different phenotype to isolate 'Par' mutants. These show no defect in cell division. The daughter chromosomes are replicated and decatenated normally but fail to be positioned at the cell quarters. This leads to the formation of normal sized anucleate cells and others with two chromosomes (Hiraga *et al.*, 1989). The *mucA1* mutation of this class is located in *tolC* which is known to code for an outer membrane protein (Hiraga *et al.*, 1989). This confirms the likely importance of membrane binding in partitioning. The other mutation of this sort is *mukB106* which lies in a gene coding for a 180 kD protein with three domains: an amino-terminal globular domain with a nucleotide binding sequence; a central region containing two α-helical coiled-coil domains; and a large carboxyl terminal globular domain (Niki *et al.*, 1991; Fig. 12). This structure is typical of force-generating enzymes in eukaryotic cells like myosin and kinesin heavy chains. MukB and HMP protein (a second 180 kDa polypeptide recently identified in *E. coli* cells by use of antibodies against yeast myosin; Casaregola *et al.*, 1990) seem to be good candidates for providing the active force to position the daughter chromosomes at the cell quarters. We also identified recently a gene of plasmid RK2, *kfrA*, which encodes a protein resembling myosin and kinesin domains with high α-helical content and a repetitive heptad sequence typical for coiled-coil proteins (Thomas *et al.*, 1990; Jagura-Burdzy & Thomas, unpublished observations; Fig. 12). The structure of the protein and the location of the gene encoding it adjacent to an operon already implicated in partitioning suggest that it might play

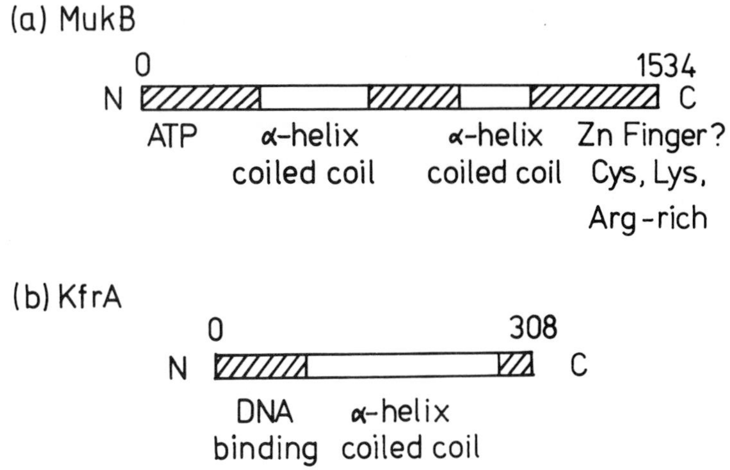

Fig. 12. Structural features of (a) the MukB protein (Niki *et al.*, 1991), and (b) the krfA protein (Thomas *et al.*, 1990; G. Jagura-Burdzy & C. M. Thomas, unpublished observations).

a role in plasmid partitioning related to that of MukB in chromosome partitioning.

Bacterial plasmids provide useful model systems for studying the partitioning process. Whereas high copy number plasmids seem to be randomly distributed between daughter cells, low copy number plasmids have evolved different mechanisms to ensure stable inheritance in the population of dividing host cells. Among these, active partitioning systems are the most interesting in the present context. The related partitioning functions of F and P1 each consist of two *trans*-acting products (SopA/SopB and ParA/ParB) and a specific *cis*-acting sequence (*sopC*/*parS*) which is thought to be analogous to the centromere of eukaryotic chromosomes (for review see Austin, 1988; Fig. 13). Studies on the purified SopB protein of F which binds to *sopC* shows that it is membrane associated in the presence of Mg^{2+} ions, suggesting that it might directly provide a bridge between plasmid DNA and the membranes (Watanabe *et al.*, 1989). The second protein, SopA, is related to ParA of P1 which has recently been shown to be an ATPase which is stimulated by ParB (M. Davis & S. Austin, personal communication). Dissection of *parS* of P1 has shown that the full region consists of both ParB and IHF binding sites (Davis & Austin, 1987; Funnell, 1988). Plasmids with this *parS* compete with each other. A minimal system lacking the IHF binding site can also function and again plasmids carrying it compete with each other. However, plasmids with the full *parS* do not compete with plasmids with the minimal *parS*. This has led to the suggestion that pairing is brought about directly by ParB rather than by binding at host (membrane) sites (reviewed in: Austin, 1988; Austin & Nordstrom, 1990; Nordstrom & Austin,

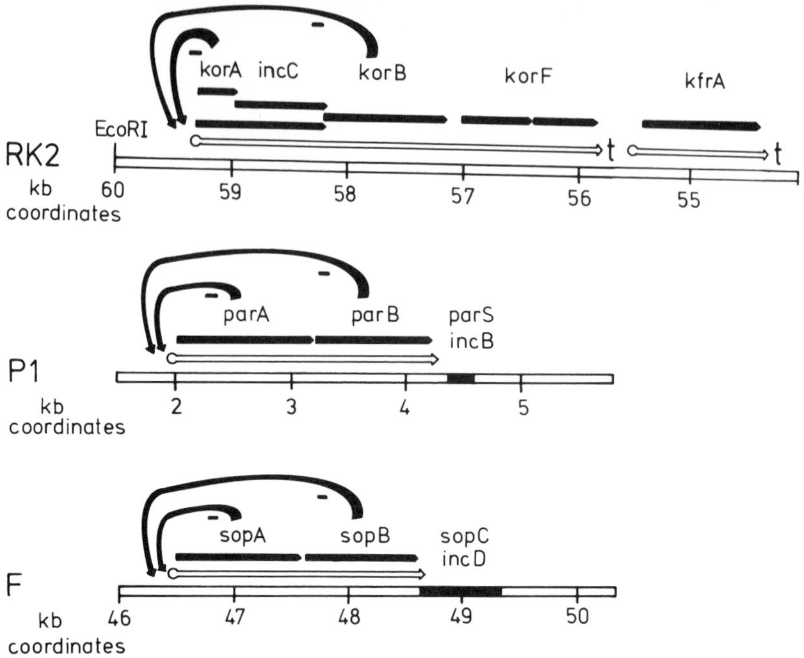

Fig. 13. Partitioning functions of P1, F and RK2 (see Motallebi-Veshareh *et al.*, 1990).

1989). Removal of the IHF-binding site, and therefore loss of IHF participation, is proposed to cause a change in specificity of such pairing. This is easier to envisage than that the two types of complex (with and without IHF) should bind completely different host sites. The model requires that once pairing has taken place, one plasmid copy should be segregated to each half of the cell. Phosphorylation of ParB by the ATPase activity of ParA might facilitate the separation of paired plasmids (Motallebi-Veshareh, Rouch & Thomas, 1990). Additional features of the equivalent region of RK2 may provide further clues (Motallebi-Veshareh *et al.*, 1990). Not only does this region also encode membrane-associated histone-like proteins (Jagura-Burdzy, Ibbotson & Thomas, 1991) which may substitute for IHF in a complex at the RK2 centromere-like sequence, but also the KfrA protein (Thomas *et al.*, 1990) may be involved by allowing attachment of the plasmids to host migratory filaments built of the dynamin- or myosin-like proteins described above (Fig. 14).

Such a model could also be applicable to chromosome partitioning. Centromere-like sequences could be located near the terminus. Chromosome replication may trigger the formation of a paired complex at these sequences that could then associate with the filaments. This would result in the translocation of the chromosomes toward the poles once chromosome replication

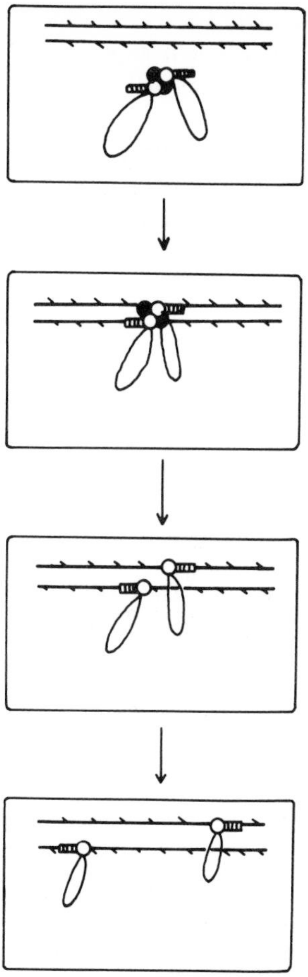

Fig. 14. Partitioning model involving pairing, attachment to fibrils, phosphorylation to separate pairs and movement towards opposite cell poles.

was complete. Although the cell envelope would play a role by providing the framework relative to which the chromosome moved, it would not be membrane growth which provided the active movement.

CONCLUSIONS

Despite the fact that many specific details of the replicon hypothesis turn out to be incorrect, it has been immensely important in the development of ideas about replication and segregation. The 'replicon model' appears to be generally applicable to most plasmids for which it has proven possible to define a single initiator molecule which activates the replication origin. Although membranes might play a crucial role in replication and partitioning it appears that the single mechanism envisaged by Jacob *et al.* (1963) is too simple and mechanical. First, it suggests the existence of single sites for both replication and partitioning and second, it suggests that the limiting number of these sites provides the replication control mechanism. Neither of these hypotheses appear to be correct. Replication does not appear to be universally regulated by the number of sites at which it can occur and membrane growth does not appear to be the driving force behind partitioning.

ACKNOWLEDGEMENTS

We thank Deepan Shah for helpful discussion and all colleagues who have supplied information before publication. Current work cited from this laboratory is supported by MRC Project Grants G8710910CB, G8807218CB and G8919550CB.

REFERENCES

Abeles, A., Snyder, K. M. & Chattoraj, D. K. (1984). P1 plasmid replication: replicon structure. *Journal of Molecular Biology*, **173**, 307–24.

Atlung, T. (1984). Allele-specific suppression of *dnaA*(Ts) mutations by *rpoB* mutations in *Escherichia coli. Molecular and General Genetics*, **200**, 442–50.

Atlung, T., Rasmussen, K. V., Clausen, E. S. & Hansen, F. G. (1985). Role of the dnaA protein in control of replication. In *Molecular Biology of Bacterial Growth*, eds M. Schaechter, F. C. Neidhart, J. Ingraham, & N. O. Kjeldgard, pp. 282–97, Jones and Bartlett, Boston.

Austin, S. (1988). Plasmid partition. *Plasmid*, **20**, 1–9.

Austin, S. & Nordstrom, K. (1990). Partition-mediated incompatibility of bacterial plasmids. *Cell*, **60**, 351–4.

Autissier, F., Jaffe, A. & Kepes, A. (1971). Segregation of galactoside permease, a membrane marker during growth and cell division of *Escherichia coli. Molecular and General Genetics*, **112**, 276–88.

Baker, T. A. & Kornberg, A. (1988). Transcriptional activation of initiation of replication from the *E. coli* chromosomal origin: An RNA–DNA hybrid near *oriC. Cell*, **55**, 113–23.

Bakker, A. & Smith, D. W. (1989). Methylation of GATC sites is required for precise timing between rounds of DNA replication in *Escherichia coli. Journal of Bacteriology*, **171**, 5738–42.

Bayer, M. H., Costello, G. P. & Bayer, M. E. (1982). Isolation and partial characterization of membrane vesicles carrying markers of membrane adhesion sites. *Journal of Bacteriology*, **149**, 758–676.

Blomberg, P., Wagner, G. E. H. & Nordstrom, K. (1990). Control of replication of plasmid R1: the duplex between the antisense RNA, CopA, and its target CopT, is processed specifically *in vivo* and *in vitro* by RNase III. *EMBO Journal*, **9**, 2331–40.

Boye, E. & Lobner-Olesen, A. (1990). The role of *dam* methyltransferase in the control of DNA replication in *E. coli. Cell*, **62**, 981–9.

Bramhill, D. & Kornberg, A. (1988*a*). Duplex opening by dnaA protein at novel sequences in initiation of replication at the origin of the *E. coli* chromosome. *Cell*, **52**, 743–55.

Bramhill, D. & Kornberg, A. (1988*b*). A model for initiation at origin of DNA replication. *Cell*, **54**, 915–18.

Cabello, F., Timmis, K. N. & Cohen, S. N. (1976). Replication control in a composite plasmid constructed by *in vitro* linkage of two distinct replicons. *Nature*, London, **259**, 285–90.

Campbell, J. & Kleckner, N. C. (1990). *E. coli oriC* and the *dnaA* gene promoter are sequestered from *dam* methyltransferase following the passage of the chromosomal replication fork. *Cell*, **62**, 967–79.

Casaregola, S., Norris, V., Goldberg, M. & Holland, I. B. (1990). Identification of a 180 kD protein in *Escherichia coli* related to a yeast heavy-chain myosin. *Molecular Microbiology*, **4**, 505–11.

Cooper, S. & Helmstetter, C. E. (1968). Chromosome replication and the division cycle of *Escherichia coli* B/r. *Journal of Molecular Biology*, **31**, 578–90.

Crosa, J. H. (1980). Three origins of replication are active *in vivo* in the R-plasmid RSF1040. *Journal of Biological Chemistry*, **255**, 11075–7.

Cross, M. A., Warne, S. R. & Thomas, C. M. (1986). Analysis of the vegetative origin of replication of broad-host-range plasmid RK2 by transposon mutagenesis. *Plasmid*, **15**, 132–46.

Davis, M. A. & Austin, S. J. (1987). Recognition of the P1 plasmid centromere analog by the ParB protein is modified by a specific host factor. *EMBO Journal*, **7**, 1881–8.

Dixon, N. E. & Kornberg, A. (1984). Protein HU in the enzymatic replication of the chromosomal origin of *Escherichia coli*. *Proceedings of the National Academy of Sciences, USA*, **81**, 424–8.

Donachie, W. D. (1968). Relationship between cell size and time of initiation of DNA replication. *Nature*, London, **219**, 1077–9.

Donachie, W. D. & Begg, K. J. (1989). Chromosome partition in *Escherichia coli* requires postreplication protein synthesis. *Journal of Bacteriology*, **171**, 5405–9.

Dong, X., Womble, D. D. & Rownd, R. H. (1987). Transcriptional pausing for plasmid NR1 replication control. *Journal of Bacteriology*, **169**, 5353–63.

Donoghue, D. J. & Sharp, P. A. (1978). Replication of colicin E1 plasmid DNA *in vivo* requires no plasmid-encoded proteins. *Journal of Bacteriology*, **133**, 1287–94.

Drlica, K. & Rouviere-Yaniv, J. (1987). Histonelike proteins of bacteria. *Microbiological Reviews*, **51**, 301–19.

Durland, R. H. & Helinski, D. R. (1990). Replication of the broad-host-range plasmid RK2: direct measurement of intracellular concentrations of the essential TrfA replication proteins and their effect on plasmid copy number. *Journal of Bacteriology*, **172**, 3849–58.

Durland, R. H. Toukdarian, A., Fang, F. & Helinski, D. R. (1990). Mutations in the *trfA* replication gene of the broad-host-range plasmid RK2 result in elevated plasmid copy numbers. *Journal of Bacteriology*, **172**, 3859–67.

Filutowicz, M. & Appelt, K. (1988). The integration host factor of *Escherichia coli* binds to multiple sites at plasmid R6K gamma origin and is essential for replication. *Nucleic Acids Research*, **16**, 3829–41.

Filutowicz, M., Davis, G., Greener, A., Helinski, D. R. (1985). Autorepressor properties of the π-initiation protein encoded by plasmid R6K. *Nucleic Acids Research*, **13**, 103–14.

Filutowicz, M., McEachern, M. J. & Helinski, D. R. (1986). Positive and negative roles of an initiator protein at an origin of replication. *Proceedings of the National Academy of Sciences, USA*, **83**, 9645–9.

Firshein, W. (1989). Role of the DNA/membrane complex in prokaryotic DNA replication. *Annual Reviews of Microbiology*, **43**, 89–120.

Frey, J., Chandler, M. & Caro, L. (1984). Over-initiation of chromosome and plasmid replication in a *dnaAcos* mutant of *Escherichia coli* K12: evidence for *dnaA-dnaB* interactions. *Journal of Molecular Biology*, **179**, 171–83.

Fujita, M. Q., Yoshikawa, H. & Ogasawara, N. (1990). Structure of the *dnaA* region of *Micrococcus luteus*: conservation and variations among eubacteria. *Gene*, **93**, 73–8.

Fuller, R. S., Funnell, B. E. & Kornberg, A. (1984). The dnaA protein complex with the *E. coli* chromosomal origin of replication (*oriC*) and other DNA sites. *Cell*, **38**, 889–900.

Fuller, R. S. & Kornberg, A. (1983). Purified DnaA protein in initiation of replication at the *Escherichia coli* chromosomal origin of replication. *Proceedings of the National Academy of Sciences, USA*, **80**, 5817–21.

Funnell, B. E. (1988). Participation of *Escherichia coli* integration host factor in the P1 plasmid partition system. *Proceedings of the National Academy of Sciences, USA*, **85**, 6657–61.

Gamas, P., Burger, A. C., Churchward, G., Caro, L., Galas, D. & Chandler, M. (1986). Replication of pSC101: effects of mutations in the *E. coli* DNA binding protein IHF. *Molecular and General Genetics*, **204**, 85–9.

Gayama, S., Kataoka, T., Wachi, M., Tamura, G. & Nagai, K. (1990). Periodic formation of the *oriC* complex of *Escherichia coli*. *EMBO Journal*, **9**, 3761–5.

Gaylo, P. J., Turjman, N., Bastia, D. (1987). DnaA protein is required for replication of the minimal replicon of the broad-host-range plasmid RK2 in *Escherichia coli*. *Journal of Bacteriology*, **169**, 4703–9.

Gennaro, M. L. & Novick, R. P. (1988). An enhancer of DNA replication. *Journal of Bacteriology*, **170**, 5709–19.

Giraldo, R., Nieto, C., Fernandez-Tresquerres, M. E. & Diaz, R. (1989). Bacterial zipper. *Nature*, London, **342**, 866.

Green, E. W. & Schaechter, M. (1972). The mode of segregation of the bacterial cell membrane. *Proceedings of the National Academy of Sciences, USA*, **69**, 2312–16.

Gros, M., Te Riele, H. & Ehrlich, S. D. (1987). Rolling circle replication of the single-stranded plasmid pC194. *EMBO Journal*, **6**, 3863–9.

Gustafsson, P., Nordstrom, K. & Perram, J. W. (1978). Selection and timing of replication of plasmids R1drd-19 and F'lac in *Escherichia coli*. *Plasmid*, **1**, 187–202.

Hansen, F. G. & Rasmussen, K. V. (1977). Regulation of the *dnaA* product in *Escherichia coli*. *Molecular and General Genetics*, **183**, 463–72.

Hansen, E. B. & Yarmolinsky, M. B. (1986). Host participation in plasmid maintenance: Dependence upon *dnaA* of replicons derived from P1 and F. *Proceedings of the National Academy of Sciences, USA*, **83**, 4423–7.

Haring, V. & Scherzinger, E. (1989). Replication proteins of the IncQ plasmid RSF1010. In *Promiscuous Plasmids of Gram-negative Bacteria*, ed. C. M. Thomas, pp. 95–124, Academic Press, New York.

Harmon, J. M. & Taber, H. W. (1977). Altered accumulation of a membrane protein unique to a membrane-deoxyribonucleic acid complex in a *dna* initiation mutant of *Bacillus subtilis*. *Journal of Bacteriology*, **130**, 1224–37.

Helmstetter, C. E. & Leonard, A. C. (1987). Coordinate initiation of chromosome and minichromosome replication in *Escherichia coli*. *Journal of Bacteriology*, **169**, 3489–94.

Helmstetter, C. E. & Pierucci, O. (1976). DNA Synthesis during the division cycle of three substrains of *Escherichia coli* B/r. *Journal of Molecular Biology*, **102**, 477–86.

Henckes, G., Harper, F., Levine, A., Vannier, F. & Seror, S. J. (1989). Overreplication of the origin region in the *dnaB37* mutant of *Bacillus subtilis*: Postinitiation control of chromosomal replication. *Proceedings of the National Academy of Sciences, USA*, **86**, 8660–3.

Hendricson, W. G., Kusano, T., Yamaki, H., Balakrishan, R., King, M., Murchie, J. & Schaechter, M. (1982). Binding of the origin of replication of *Escherichia coli* to the outer membrane. *Cell*, **30**, 915–23.

Highlander, S. K. & Novick, R. P. (1987). Plasmid repopulation kinetics in *Staphylococcus aureus*. *Plasmid*, **17**, 210–21.

Hiraga, S., Niki, H., Ogura, T., Ichinose, C., Mori, H., Ezaki, B. & Jaffe, A. (1989). Chromosome partitioning in *Escherichia coli*: Novel mutants producing anucleate cells. *Journal of Bacteriology*, **171**, 1496–505.

Hirota, Y., Ryter, A. & Jacob, F. (1968). Thermosensitive mutants of *E.coli* affected in the processes of DNA synthesis and cellular division. *Cold Spring Harbor Symposium of Quantitative Biology*, **33**, 677–93.

Hirota, Y., Ricard, M. & Shapiro, B. (1971). The use of thermosensitive mutants of *E.coli* in the analysis of cell division. In *Biomembranes*, ed. L. A. Manson, pp. 13–31, Plenum Publ. Co., New York.

Hoshino, T., McKenzie, T., Schmidt, S., Tanaka, T. & Sueoka, N. (1987). Nucleotide sequence of *Bacillus subtilis dnaB*: a gene essential for DNA replication initiation and membrane attachment. *Proceedings of the National Academy of Sciences, USA*, **84**, 653–7.

Hughes, P., Landoulsi, A. & Kohiyama, M. (1988). A novel role for cAMP in the control of the activity of the *E. coli* chromosome replication initiator protein, DnaA. *Cell*, **55**, 343–50.

Hussain, K., Begg, K. J., Salmond, G. P. C. & Donachie, W. D. (1987*a*). ParD: A new gene coding for a protein required for chromosome partitioning and septum localization in *Escherichia coli*. *Molecular Microbiology*, **1**, 73–81.

Hussain, K., Elliott, E. J. & Salmond, G. P. C. (1987*b*). The *parD* mutant of *Escherichia coli* also carries a *gyrA* mutation. The complete sequence of *gyrA*. *Molecular Microbiology*, **1**, 259–73.

Hwang, D. S. & Kornberg, A. (1990). A novel protein binds a key origin sequence to block replication of an *E. coli* minichromosome. *Cell*, **63**, 325–31.

Imada, S., Carroll, L. E. & Sueoka, N. (1976). DNA/membrane complex in *Bacillus subtilis*. In *Microbiology*, ed. D. Schlessinger, pp. 116–27.

Inouye, M. & Pardee, A. B. (1970). Changes of membrane proteins and their relationship to deoxyribonucleic acid synthesis and cell division of *Escherichia coli*. *Journal of Biological Chemistry*, **245**, 5813–19.

Iordanescu, S. & Projan, S. J. (1988). Replication termination for staphylococcal plasmids: plasmids pT181 and pC221 cross react in the termination process. *Journal of Bacteriology*, **170**, 3427–34.

Itoh, T. & Tomizawa, J. (1978). Initiation of replication of plasmid ColE1 DNA by RNA polymerase, ribonuclease H and DNA polymerase I. *Cold Spring Harbour Symposium of Quantitative Biology*, **43**, 409–18.

Itoh, T. & Tomizawa, J. (1980). Formation of an RNA primer for initiation of replication of ColE1 DNA by ribonuclease H. *Proceedings of the National Academy of Sciences, USA*, **77**, 2450–4.

Jacq, A., Kohiyama, M., Lother, H. & Messer, W. (1983). Recognition sites for a membrane-derived DNA binding protein preparation in the *E. coli* replication origin. *Molecular and General Genetics*, **191**, 460–5.

Jacob, F., Brenner, S. & Cuzin, F. (1963). On the regulation of DNA replication in bacteria. *Cold Spring Harbour Symposium of Quantitative Biology*, **28**, 329–48.

Jagura-Burdzy, G., Ibbotson, J. P. & Thomas, C. M. (1991). The *korF* region of broad host range plasmid RK2 encodes two polypeptide products capable of repressing transcription. *Journal of Bacteriology*, **173**, 826–33.

Kaguni, J. M. & Kornberg, A. (1984). Replication initiated at the origin (*oriC*) of the *E. coli* chromosome reconstituted with purified enzymes. *Cell*, **38**, 183–90.

Kahn, M. & Helinski, D. R. (1978). Construction of a novel plasmid-phage hybrid: use of the hybrid to demonstrate ColE1 DNA replication *in vivo* in the absence of a ColE1-specified protein. *Proceedings of the National Academy of Sciences, USA*, **75**, 2200–4.

Kato, J.-I., Nishimura, Y., Yamada, M., Suzuki, H. & Hirota, Y. (1988). Gene organization in the region containing a new gene involved in chromosome partition in *Escherichia coli*. *Journal of Bacteriology*, **170**, 3967–77.

Kato, J., Nishimura, Y. & Suzuki, H. (1989). *Escherichia coli parA* is an allele of the *gyrB* gene. *Molecular and General Genetics*, **217**, 178–81.

Kato, J., Nishimura, Y., Imamamura, R., Niki, H., Hiraga, S. & Suzuki, H. (1990). New topoisomerase essential for chromosome segregation in *E.coli*. *Cell*, **63**, 393–404.

Keasling, J. D., Palsson, B. O. & Cooper, S. (1991). Cycle-specific F plasmid replication: regulation by cell size control of initiation. *Journal of Bacteriology*, **173**, 2673–80.

Kittell, B. L. & Helinski, D. R. (1991). Iteron inhibition of plasmid RK2 replication *in vitro*: Evidence for intermolecular coupling of replication origins as a mechanism for RK2 replication control. *Proceedings of the National Academy of Sciences, USA*, **88**, 1389–93.

Kline, B. C. (1988). Aspects of plasmid F maintenance in *Escherichia coli*. *Canadian Journal of Microbiology*, **34**, 526–35.

Koepsel, R., Murray, R. W., Rosenblum, W. D. & Khan, S. A. (1985). The replication initiator protein of plasmid pT181 has sequence-specific endonuclease and topoismomerase-like activities. *Proceedings of the National Academy of Sciences, USA*, **82**, 6845–9.

Kolter, R., Inuzuka, M. & Helinski, D. R. (1978). Trans-complementation-dependent replication of a low molecular weight origin fragment from plasmid R6K. *Cell*, **15**, 1199–208.

Kücherer, C., Lother, H., Kolling, R., Schauzu, M. -A. & Messer, W. (1986). Regulajstion of transcription of the chromosomal *dnaA* gene of *Escherichia coli*. *Molecular*
and General Genetics, **205**, 115–21.

Kusano, T., Steinmetz, D., Hendricson, W. G., Murchie, J., King, M. *et al.* (1984). Direct evidence for specific binding of the replicative origin of the *Escherichia coli* chromosome to the membrane. *Journal of Bacteriology*, **158**, 313–16.

Landoulsi, A., Malki, R., Kohiyama, M. & Hughes, P. (1990). The *E. coli* cell surface specifically prevents the initiation of DNA replication at *oriC* on hemimethylated DNA templates. *Cell*, **63**, 1053–60.

Leonard, A. C. & Helmstetter, C. E. (1986). Cell cycle-specific replication of *Escherichia coli* minichromosomes. *Proceedings of the National Academy of Sciences, USA*, **83**, 5101–5.

Leonard, A. C. & Helmstetter, A. C. (1988). Replication patterns of multiple plasmids in *Escherichia coli*. *Journal of Bacteriology*, **170**, 1380–3.

Leonard, A. C. Hucul, J. A. & Helmstetter, C. E. (1982). Kinetics of minichromosome replication in *Escherichia coli* B/r. *Journal of Bacteriology*, **149**, 499–507.

Lin, E. C. C., Hirota, Y. & Jacob, F. (1971). On the process of cellular division in *Escherichia coli*. VI. Use of a methyl cellulose autoradiographic method for the study of cellular division in *Escherichia coli*. *Journal of Bacteriology*, **108**, 375–85.

Linder, P., Churchward, G., Guixan, X., Yi-Yi, Y. & Caro, L. (1985). An essential replication gene, *repA*, of plasmid pSC101 is autoregulated. *Journal of Molecular Biology*, **181**, 383–93.

MacAlister, T. J., MacDonald, B. & Rothfield, L. I. (1983). The periseptal annulus: an organelle associated with cell division in Gram-negative bacteria. *Proceedings of the National Academy of Sciences, USA*, **80**, 1372–6.

Mahaffy, J. & Zyskind, J. (1989). A model for the initiation of replication in *Escherichia coli*. *Journal of Theoretical Biology*, **140**, 453–77.

Masson, L. & Ray, D. S. (1986). Mechanism of autonomous control of *Escherichia coli* F plasmid: different complexes of the initiator/repressor protein bound to its operator and to an F plasmid replication origin. *Nucleic Acids Research*, **14**, 5693–711.

Masukata, H. & Tomizawa, J. (1984). Effects of point mutations on formation and structure of the RNA primer for ColE1 replication. *Cell*, **36**, 513–22.

McEachern, M. B., Bott, M. A., Tooker, P. A. & Helinski, D. R. (1989). Negative control of plasmid RbK replication: possible role of intermolecular coupling of replication origins. *Proceedings of the National Academy of Sciences, USA*, **86**, 7942–6.

Meijer, M., Beck, E., Hansen, F. G., Bergmans, H. E. N., Messer, W., Von Meyenburg, K. & Schaller, H. (1979). Nucleotide sequence of the origin of replication of the *Escherichia coli* K12 chromosome. *Proceedings of the National Academy of Sciences, USA*, **76**, 580–4.

Messer, W., Bellekes, U. & Lother, H. (1985). Effect of *dam* methylation on the activity of the *E. coli* replication origin, *oriC*. *EMBO Journal*, **4**, 1327–32.

Messer, W., Bergmans, H. E. N., Meijer, M., Womack, J. E., Hansen, F. G. & Von Meyenburg, K. (1978). Minichromosomes: plasmids which carry the *E. coli* chromosome replication origin. *Molecular and General Genetics*, **162**, 269–75.

Messer, W., Seufert, W., Schaefer, C., Gielow, A., Hartmann, H. & Wende, M. (1988). Functions of the DnaA protein of *Escherichia coli* in replication and transcription. *Biochimica et Biophysica Acta*, **951**, 351–8.

Moriya, S., Fukuoka, T., Ogasawara, N. & Yoshikawa, H. (1988). Regulation of initiation of chromosomal replication by DnaA-boxes in the origin region of the *Bacillus subtilis* chromosome. *EMBO Journal*, **7**, 2911–17.

Moriya, S., Kato, K., Yoshikawa, H. & Ogasawara, N. (1990). Isolation of a *dnaA* mutant of *Bacillus subtilis* defective in initiation of replication: amount of DnaA protein determines cells-initiation potential. *EMBO Journal*, **9**, 2905–10.

Moriya, S., Ogasawara, N. & Yoshikawa, H. (1985). Structure and function of the region of the replication origin of the *Bacillus subtilis* chromosome. III Nucleotide sequence of some 10 000 base pairs in the origin region. *Nucleic Acids Research*, **13**, 2251–65.

Motallebi-Veshareh, M., Rouch, D. A. & Thomas, C. M. (1990). A family of ATPases involved in active partitioning of diverse bacterial plasmids. *Molecular Microbiology*, **4**, 1455–63.

Mukherjee, S., Erickson, H. & Bastia, D. (1988). Enhancer–origin interaction in plasmid R6K involves a DNA loop mediated by initiator-protein. *Cell*, **52**, 375–83.

Mukhopadhyay, P., Filutowicz, M. & Helinski, D. R. (1986). Replication from one of the three origins of the plasmid R6K requires coupled expression of two plasmid-encoded proteins. *Journal of Biological Chemistry*, **261**, 9534–9.

Muraiso, K., Mukhopadhyay, G. & Chattoraj, D. K. (1990). Location of a P1 plasmid replication inhibitor determinant within the initiator gene. *Journal of Bacteriology*, **172**, 4441–7.

Murotsu, T., Matsubara, K., Sugisaki, H. & Takanami, M. (1981). Nine unique repeating sequences in the a region essential for replication and incompatibility of the mini-F plasmid. *Gene*, **15**, 257–71.

Murotsu, T., Tsutsui, M. & Matsubara, K. (1984). Identification of the minimal essential region for the replication origin of mini-F plasmid. *Molecular and General Genetics*, **186**, 373–8.

Nagai, K., Hendricson, W., Balakrishnan, R., Yamaki, H., Boyd, D. & Schaechter, M. (1980). Isolation of a replication origin complex from *Escherichia coli*. *Proceedings of the National Academy of Sciences, USA*, **77**, 262–6.

Niki, H., Jaffe, A., Imamura, R., Ogura, T. & Hiraga, S. (1991). The new gene *mukB* codes for a 177 kD protein with coiled-coil domains involved in chromosome partitioning of *E. coli*. *EMBO Journal*, **10**, 183–93.

Nordstrom, K. & Austin, S. J. (1989). Mechanisms that contribute to the stable segregation of plasmids. *Annual Reviews of Genetics*, **23**, 37–69.

Nordstrom, K., Bernander, R. & Dasgupta, S. (1990). Analysis of the bacterial cell cycle using strains in which chromosome replication is controlled by plasmid R1. *Research in Microbiology*, **141**, 1–17.

Nordstrom, K., Wagner, E. G. H., Persson, C., Blomberg, P. & Ohman, M. (1988). Translational control by antisense RNA in control of plasmid replication. *Gene*, **72**, 237–40.

Norris, V., Alliotte, T., Jaffe, A. & D'Ari, R. (1986). DNA replication termination in *Escherichia coli parB* (a *dnaG* allele), *parA* and *gyrB* mutants affected in DNA distribution. *Journal of Bacteriology*, **168**, 494–504.

Norris, V. (1990). DNA replication in *Escherichia coli* is initiated by membrane detachment of *oriC*: A model. *Journal of Molecular Biology*, **215**, 67–71.

Novick, R. P. (1989). Staphylococcal plasmids and their replication. *Annual Reviews in Microbiology*, **43**, 537–65.

Novick, R. P., Iordanescu, S., Projan, S. J., Kornblum, J. & Edelman, I. (1989). pT181 plasmid replication is regulated by a countertranscript-driven transcriptional attenuator. *Cell*, **59**, 395–404.

Ogasawara, N., Moriya, S., Von Meyenburg, K., Hansen, F. G. & Yoshikawa, H. (1985). Conservation of genes and their organization in the chromosomal replication origin region of *Bacillus subtilis* and *Escherichia coli*. *EMBO Journal*, **4**, 3345–50.

Ogasawara, N., Seiki, M. & Yoshikawa, H. (1981). Initiation of DNA replication in *Bacillus subtilis*. V. Role of DNA gyrase and superhelical structure in initiation. *Molecular and General Genetics*, **181**, 332–7.

Ogawa, T., Wada, M., Kano, Y., Imamoto, F. & Okazaki, T. (1989). DNA replication in *Escherichia coli* mutants that lack protein HU. *Journal of Bacteriology*, **171**, 5672–9.

Ogden, G. B., Pratt, M. J. & Schaechter, M. (1988). The replicative origin of the *Escherichia coli* chromosome binds to cell membranes only when hemimethylated. *Cell*, **54**, 127–35.

Ogura, T. & Hiraga, S. (1983). Partition mechanism of F plasmid: two plasmid gene-encoded products and a cis-acting region are involved in partition. *Cell*, **32**, 351–60.

Ohmori, H., Murakami, Y. & Nagata, T. (1987). Nucleotide sequences required for a ColE1-type plasmid to replicate in *Escherichia coli* cells with or without RNaseH. *Journal of Molecular Biology*, **198**, 223–34.

Orr, E., Fairweather, N. F., Holland, I. B. & Pritchard, R.H. (1979). Isolation and characterization of a strain carrying a conditional lethal mutation in the *cou* gene of *Escherichia coli* K12. *Molecular and General Genetics*, **177**, 103–12.

Pal, S. K. & Chattoraj, D. K. (1988). P1 plasmid replication: initiator sequestration is inadequate to explain control by initiator binding sites. *Journal of Bacteriology*, **170**, 3554–60.

Persson, C., Wagner, E. G. H. & Nordstrom, K. (1988). Control of replication of plasmid R1: kinetics of *in vitro* interaction between the antisense RNA, CopA, and its target, CopT. *EMBO Journal*, **7**, 3279–88.

Persson, C., Wagner, E. G. H. & Nordstrom, K. (1990). Control of replication of plasmid R1: formation of an initial transient complex is rate-limiting for antisense RNA–target RNA pairing. *EMBO Journal*, **9**, 3777–85.

Polaczek, P. & Wright, A. (1990). Regulation of expression of the *dnaA* gene in *Escherichia coli*: role of the two promotors and the DnaA box. *New Biologist*, **2**, 574–582.

Polisky, B. (1988). ColE1 replication control circuitry: Sense from antisense, *Cell*, **55**, 929–32.

Polisky, B., Zhang, X.-Y. & Fitzwater, T. (1990). Mutations affecting primer RNA interaction with the replication repressor RNA I in plasmid ColE1: potential RNA folding pathway mutants. *EMBO Journal*, **9**, 295–304.

Pritchard, R. H., Barth, P. T. & Collins, J. (1969). Control of DNA synthesis in bacteria. *Symposium of the Society for General Microbiology*, **19**, 263–97.

Quinn, P. J. (1981). The fluidity of cell membranes and its regulation. *Progress in Biophysics and Molecular Biology*, **38**, 1–104.

Riise, E. & Molin, S. (1986). Purification and characterization of the CopB replication control protein, and precise mapping of its target site in the R1 plasmid. *Plasmid*, **15**, 163–71.

Russell, D. W. & Zinder, D. N. (1987). Hemimethylation prevents DNA replication in *E. coli. Cell*, **50**, 1071–9.

Schaefer, C. & Messer, W. (1989). Directionality of DnaA protein/DNA interaction. Active orientation of the DnaA protein/*dnaA* box complex in transcription termination. *EMBO Journal*, **8**, 1609–13.

Sekimizu, K., Bramhill, D. & Kornberg, A. (1987). ATP activates DnaA protein in initiating replication of plasmids bearing the origin of the *E.coli* chromosome. *Cell*, **50**, 259–65.

Sekimizu, K. & Kornberg, A. (1988). Cardiolipin activation of dnaA protein, the initiator protein of replication in *Escherichia coli. Journal of Biological Chemistry*, **263**, 7131–5.

Sekimizu, K., Yung, B. Y. & Kornberg, A. (1988). The dnaA protein of *Escherichia coli*: abundance, improved purification, and membrane binding. *Journal of Biological Chemistry*, **263**, 7136–40.

Shafferman, A., Flashner, Y., Hertman, I. & Menachem, L. (1987). Identification and characterization of the functional α origin of DNA replication of the R6K plasmid and its relatedness to the R6K β and gamma origins. *Molecular and General Genetics*, **208**, 263–70.

Shafferman, S. & Helinski, D. R. (1983). Structural properties of the β origin of replication of plasmid R6K. *Journal of Biological Chemistry*, **258**, 4083–90.

Shon, M., Germino, J. & Bastia, D. (1982). The nucleotide sequence of the replication origin β of the plasmid R6K. *Journal of Biological Chemistry*, **257**, 13823–7.

Siccardi, A. G., Shapiro, B. M., Hirota, Y. & Jacob, F. (1971). On the process of cellular division in *Escherichia coli*. IV. Altered protein composition and turnover of the membranes of thermosensitive mutants defective in chromosomal replication. *Journal of Molecular Biology*, **56**, 475–90.

Skarstadt, K., Boye, E. & Steen, H. B. (1986). Timing of initiation of chromosome replication in *E. coli. EMBO Journal*, **5**, 1711–17.

Skarstadt, K., Baker, T. A. & Kornberg, A. (1990). Strand separation required for initiation of replication at the chromosomal origin of *E. coli* is facilitated by a distant RNA–DNA hybrid. *EMBO Journal*, **9**, 2341–8.

Smith, D. W., Garland, A. M., Herman, G., Enns, R. E., Baker, T. A. & Zyskind, J. W. (1985). Importance of state of methylation of *oriC* GATC sites in initiation of DNA replication in *Escherichia coli*. *EMBO Journal*, **4**, 1319–26.

Smith, C. A. & Thomas, C. M. (1989). Relationships and evolution of IncP plasmids. In *Promiscuous Plasmids of Gram-negative Bacteria*, ed. C. M. Thomas, pp. 57–77, Academic Press, New York.

Sompayrac, L. & Maaloe, O. (1973). Autorepressor model for control of DNA replication. *Nature New Biology*, **241**, 133–5.

Stalker, D. M., Kolter, R. & Helinski, D. R. (1979). Nucleotide sequence of the region of an origin of replication of the antibiotic resistance plasmid R6K. *Proceedings of the National Academy of Sciences, USA*, **76**, 1150–4.

Stalker, D. M., Thomas, C. M. & Helinski, D. R. (1981). Nucleotide sequence of the region of the origin of replication of the broad host range plasmid RK2. *Molecular and General Genetics*, **181**, 8–12.

Steck, T. R. & Drlica, K. (1984). Bacterial chromosome segregation: evidence for DNA gyrase involvement in decatenation. *Cell*, **36**, 1081–8.

Stenzel, T. T., Patel, P. & Bastia, D. (1987). The integration host factor of *Escherichia coli* binds to bent DNA at the origin of replication of the plasmid pSC101. *Cell*, **49**, 709–17.

Sueoka, N., Hoshino, T., McKenzie, T. (1988). The *dnaB* operon of *Bacillus subtilis*: Anchorage-anchor model for the chromosome replication–initiation, partition and membrane attachment of *oriC* area. In *Genetics and Biotechnology of Bacilli*, eds A. T. Ganesan, J. A. Hoch, vol. 2, pp. 269–74. Academic Press, New York.

Swack, J. A., Pal, S. K., Mason, R. J., Abeles, A. L. & Chattoraj, D. K. (1987). P1 Plasmid replication: Measurement of initiator protein concentration *in vivo*. *Journal of Bacteriology*, **169**, 3737–42.

Tang, X. B., Womble, D. D. & Rownd, R. H. (1989). DnaA protein is not essential for replication of IncFII plasmid NR1. *Journal of Bacteriology*, **171**, 5290–5.

Thomas, C. D., Balson, D. & Shaw, W. V. (1988). Identification of the tyrosine residue involved in bond formation between replication origin and initiator protein of plasmid pC221. *Biochemical Society Transactions*, **16**, 758–9.

Thomas, C. M. (1987). Plasmid replication. In *Plasmids – A Practical Approach*, ed. K. Hardy, pp. 7–36, IRL Press.

Thomas, C. M. & Helinski, D. R. (1989). Vegetative replication and stable inheritance of IncP plasmids. In *Promiscuous Plasmids of Gram-negative Bacteria*, ed. C. M. Thomas, pp. 1–25, Academic Press, New York.

Thomas, C. M. & Hussain, A. A. K. (1984). The *korB* gene of broad host range plasmid RK2 is major copy number control element which may act together with *trfB* by limiting *trfA* expression. *EMBO Journal*, **3**, 1513–19.

Thomas, C. M., Meyer, R. & Helinski, D. R. (1980). Regions of broad-host range plasmid RK2 which are essential for replication and maintenance. *Journal of Bacteriology*, **141**, 213–22.

Thomas, C. M. & Smith, C. A. (1987). Incompatibility group-P plasmids: Genetics, evolution and use in genetic manipulation. *Annual Reviews in Microbiology*, **41**, 77–101.

Thomas, C. M., Theophilus, B. D. M., Johnston, L., Jagura-Burdzy, G., Schilf, W., Lurz, R. & Lanka, E. (1990). Identification of a seventh operon on plasmid RK2 regulated by the *korA* gene product. *Gene*, **89**, 29–35.

Tilly, K. & Yarmolinsky, M. (1989). Participation of *Escherichia coli* heat shock proteins DnaJ, DnaK and GrpE in P1 plasmid replication. *Journal of Bacteriology*, **171**, 6025–9.

Tokino, T., Murotsu, T. & Matsubara, K. (1986). Purification and properties of the mini-F plasmid-encoded E protein needed for autonomous replication control of the plasmid. *Proceedings of the National Academy of Sciences, USA*, **83**, 4109–13.

Tomizawa, J.-I. (1990). Control of ColE1 plasmid replication. Intermediates in the binding of RNAI and RNAII. *Journal of Molecular Biology*, **212**, 683–94.

Tomizawa, J., Itoh, T., Selzer, G. & Som, T. (1981). Inhibition of ColE1 RNA primer formation by a plasmid-specified small RNA. *Proceedings of the National Academy of Sciences, USA*, **78**, 1421–5.

Trawick, J. D. & Kline, B. C. (1985). A two-stage molecular model for control of mini-F replication. *Plasmid*, **13**, 59–69.

Vocke, C. & Bastia, D. (1985). The replication initiator protein of plasmid pSC101 is a transcriptional repressor of its own cistron. *Proceedings of the National Academy of Sciences, USA*, **82**, 2252–6.

Watanabe, E., Inamoto, S., Lee, M.-H., Kim, S. U., Ogura, T., Mori, H., Hiraga, S., Yamasaki, M. & Nagai, K. (1989). Purification and characterization of the *sopB* gene product which is responsible for stable maintenance of mini-F plasmid. *Molecular and General Genetics*, **218**, 431–6.

Wickner, S. (1990). Three *E.coli* heat shock proteins are required for P1 plasmid replication: formation of an active complex between *E.coli* DnaJ protein and the P1 initiator protein. *Proceedings of the National Academy of Sciences, USA*, **87**, 2690–4.

Wolf-Watz, H. & Masters, M. (1979). DNA and outer membrane: strains diploid for the *oriC* region show elevated levels of a DNA binding protein and evidence for specific binding of the *oriC* region to the outer membrane. *Journal of Bacteriology*, **140**, 50–8.

Womble, D. D. & Rownd, R. H. (1988). Genetic and physical map of plasmid NR1: comparison with other IncFII antibiotic resistance plasmids. *Microbiological Reviews*, **52**, 433–51.

Wong, E. M. & Polisky, B. (1985). Alternative conformations of the ColE1 replication primer modulate its interaction with RNA I. *Cell*, **42**, 959–66.

Yamaguchi, K. & Masamune, Y. (1985). Autogenous regulation of synthesis of the replication protein in plasmid pSC101. *Molecular and General Genetics*, **200**, 362–7.

Yasuda, S. & Hirota, Y. (1977). Cloning and mapping of the replication origin of *Escherichia coli*. *Proceedings of the National Academy of Sciences, USA*, **74**, 5458–62.

Yee, T. W. & Smith, D. W. (1990). *Pseudomonas* chromosomal replication origins: a bacterial class distinct from *Escherichia coli*-type origins. *Proceedings of the National Academy of Sciences, USA*, **87**, 1278–82.

Yoshikawa, H., Ogasawara, N. & Seiki, M. (1980). Initiation of DNA replication in *Bacillus subtilis*. IV. Effect of an intercalating dye, ethidium bromide, on the initiation. *Molecular and General Genetics*, **179**, 265–72.

Yoshimoto, M., Kambe-Honjoh, H., Nagai, K. & Tamura, G. (1986). Early replicative intermediates of *Escherichia coli* chromosome isolated from a membrane complex. *EMBO Journal*, **5**, 787–91.

Yung, B. Y.-M. & Kornberg, A. (1988). Membrane attachment activates DnaA protein of chromosome replication in *Escherichia coli*. *Proceedings of the National Academy of Sciences, USA*, **85**, 7202–5.

Yung, B. Y.-M. & Kornberg, A. (1989). The DnaA initiator protein binds separate domains in the replication origin of *Escherichia coli*. *Journal of Biological Chemistry*, **264**, 6146–50.

Zyskind, J. W., Clearly, J. M., Brusilow, W. S. A., Harding, N. E. & Smith, D. W. (1983). Chromosomal replication origin from the marine bacterium *Vibrio harveyi* functions in *Escherichia coli*: *oriC* consensus sequence. *Proceedings of the National Academy of Sciences, USA*, **80**, 1164–8.

STORAGE POLYMERS IN PROKARYOTES

EDWIN A. DAWES

Department of Applied Biology, University of Hull, Hull HU6 7RX, UK

INTRODUCTION

Many prokaryotes are able to accumulate polymeric materials which can serve as reserves of carbon, phosphate and/or energy, and which, under appropriate stress, may be degraded to provide the cell with carbon for resynthesis of essential, degraded compounds, and with energy for maintenance of viability (Dawes & Senior, 1973). The three major classes of storage material are polyglucans (glycogen, granulose), polyesters (poly-3-hydroxybutyrate (PHB) and related polyhydroxyalkanoates (PHA)), and polyphosphates. Additionally, some cyanobacteria can accumulate cyanophycin, which is believed to function as a nitrogen reserve (Allen, 1984), while the *Thiorhodaceae* and certain other apochlorotic sulphur bacteria store sulphur granules or globules.

Some bacteria accumulate more than one type of storage material and, in these cases, environmental conditions and the regulatory mechanisms involved determine the proportion of the different polymers synthesized.

If a storage function is to be assigned to a cell component, three criteria, first enumerated by Wilkinson (1959), should be satisfied. They are (1) the compound should accumulate intracellularly when the exogenous energy supply is in excess of that needed by the cell for growth and maintenance; (2) the compound should be utilized when the exogenous energy supply is no longer sufficient to sustain growth and the maintenance energy needed for preservation of viability; and (3) the compound should be degraded to furnish energy in a form utilizable by the cell and by this process confer a biological advantage for survival over a corresponding cell that does not possess the reserve.

Storage materials accumulate, usually in cytoplasmic granules, in response to a nutrient limitation in the presence of excess of the relevant substrate, e.g. at the end of the growth phase in batch culture, or under an appropriate nutrient limitation in continuous culture. However, some bacteria are known to synthesize reserves during unrestricted growth. The quantity of reserve accumulated varies widely depending upon the organism and the prevailing environmental conditions, but may be very high. Thus *Alcaligenes eutrophus* can accumulate PHB up to 90% of its biomass. The dependence, both qualitatively and quantitatively, on growth phase and environment for accumulation of storage materials, does emphasize though the caution which should be exercised concerning generalized statements about their 'absence' from any given bacterium. As storage compounds are polymers of high molecular weight their synthesis has a minimal effect on the intracellular osmotic pressure.

Research on storage compounds during the past decade has made much headway. There has been a major thrust in the genetics of glycogen biosynthesis, the structural genes of *Escherichia coli* have been characterized and the factors regulating their expression investigated (Preiss & Romeo, 1989). Intense industrial interest has been aroused by PHB and related polymers on account of their properties as biodegradable, biocompatible thermoplastics, and their potential commercial applications have stimulated considerable research effort world-wide (Dawes, 1990; Anderson & Dawes, 1990; Doi, 1990). Current developments embrace genetic engineering and the future possibility of introducing the PHB biosynthetic genes into plants (Pool, 1989). Another intriguing advance has been the discovery of PHB-polyphosphate complexes in bacterial cytoplasmic membranes where their role is not that of storage but apparently of membrane channels (Reusch & Sadoff, 1988); this function seemingly extends to eukaryotes (Reusch, 1989). Knowledge of the mechanism of polyphosphate synthesis has been greatly aided by the development and application of more sensitive analytical techniques (Wood & Clark, 1988), while the application of polyphosphate-accumulating bacteria to problems of wastewater treatment has also been a focus of much recent attention.

In this present survey, the three major classes of storage polymer are treated individually in succeeding sections with respect to their occurrence, biosynthesis, degradation, regulation and physiological function.

There have been several general reviews concerning bacterial reserve materials (Dawes & Senior, 1973; Dawes, 1976, 1985, 1989; Preiss, 1989), and specialist surveys of glycogen (Preiss, 1984; Preiss & Walsh, 1981; Preiss & Romeo, 1989), polyphosphate (Harold, 1966; Kulaev, 1979; Kulaev & Vagabov, 1983; Wood & Clark, 1988) and polyhydroxyalkanoates (Steinbüchel, 1989; Anderson & Dawes, 1990; Doi, 1990; Steinbüchel & Schlegel, 1991). A spectrum of current research on polyhydroxyalkanoates and polyphosphates was also reported at a NATO Advanced Workshop in 1990 devoted to Novel Biodegradable Microbial Polymers (Dawes, 1990). Prokaryotic inclusion bodies were reviewed by Shively (1974), and those of cyanobacteria more recently by Allen (1984).

GLYCOGEN

Occurrence in bacteria

The presence of glycogen has been reported in over 40 bacterial species (for a recent listing see Preiss (1989)). Glycogen accumulates in the cytoplasm

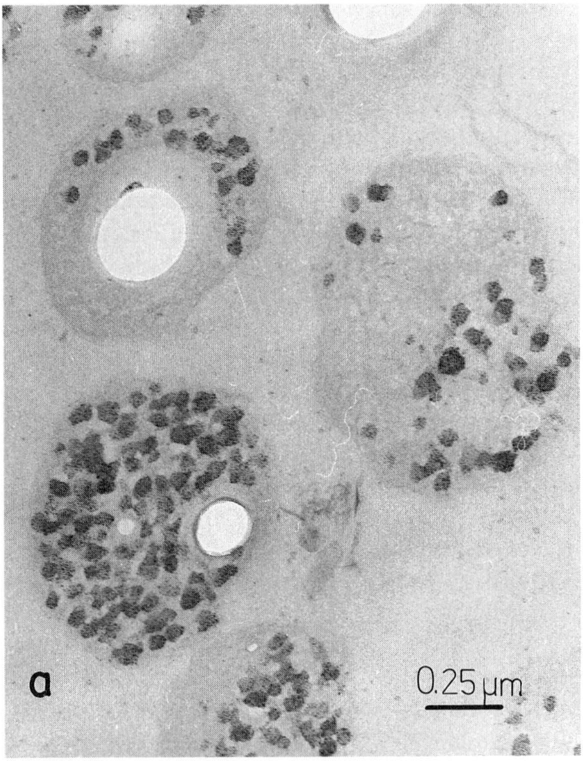

Fig. 1. Glycogen granules in *Methanolobus tindarius* grown under nitrogen limitation (2.5 mg $NH_4Cl\ l^{-1}$). Electron micrograph of a thin section stained for glycogen. Reproduced, with permission, from König, Nusser & Stetter (1985).

as granules, the nature of which seems to have attracted relatively little attention. Studies by König, Nusser & Stetter (1985) with *Methanolobus* and *Methanococcus* have shown by staining and electron microscopy that the granules are apparently distributed within the cytoplasm (Fig. 1) and not predominantly located near the cytoplasmic membrane; however, examination of lysed cells of *Methanococcus vannieli* disclosed glycogen mainly attached to the envelope.

Bacterial granules are 20 to 100 nm in diameter (Shively, 1974) but in cyanobacteria may occur as crystals, spheres or rods, usually seen between the thylakoid membranes (Wolk, 1973). Granules isolated from *Oscillatoria rubescens* may be as long as 300 nm and are composed of 7 nm discs with a central pore (Jost, 1965). There does not appear to be a granule-bounding membrane in those bacteria studied but Preiss and his colleagues have reported that glycogen synthase may be found in soluble or particulate form, the latter usually being bound to glycogen (surveyed by Preiss & Walsh,

1981). The granulose synthetase of *Clostridium pasteurianum* is also intimately associated with the granules (Robson, Robson & Morris, 1972) and the glycogen phosphorylase of *Streptococcus mitis* likewise binds to glycogen granules (Pulkownik & Walker, 1976).

Glycogen is accumulated usually when growth is limited by the supply of utilizable nitrogen in the presence of excess exogenous carbon but exceptions are known. For example, glycogen is synthesized during the exponential growth of *Rhodopseudomonas capsulata* (Eidels & Preiss, 1970) and *Sulfolobus solfataricus* (König *et al.*, 1982), and during carbon-limited continuous culture of *Streptococcus sanguis* (Keevil, Marsh & Ellwood, 1984). Glycogen accumulation is light-dependent in *Anabaena* sp.: it does not occur in the dark, or in the presence of light and an inhibitor of photophosphorylation (Sarma & Kanta, 1979).

There is an inverse relationship between growth rate and the quantity of glycogen accumulated by nitrogen-limited, glucose-grown cultures of *Escherichia coli* B (Holme, 1957); this is a general type of behaviour that has since been observed with other bacteria. Other, earlier fundamental investigations on the effects of environmental factors on glycogen storage in a variety of bacteria have been reviewed in detail by Dawes & Senior (1973).

Chemical structure

The polyglucans that function as storage compounds are composed of α-1,4 linked α-D-glucose units with some α-1,6 branches. Following the synthesis of a linear α-1,4 chain of glucosyl units, a branching enzyme transfers glucosyl units from the non-reducing end of the chain to the 6-position of some units to produce α-1,6-glucosyl linkages. The chain-lengths and degree of branching vary considerably. About 10% of the linkages in bacterial glycogen are α-1,6 but the precise value depends upon the bacterium and possibly the stage of synthesis, e.g. in *Mycobacterium smegmatis* the degree of branching decreases and the sedimentation coefficient increases as the glycogen granules increase in size (Antoine & Tepper, 1969*a*). The molecular weights of glycogen from different bacteria also vary; *Mycobacterium phlei* possesses a glycogen of molecular weight $1–2 \times 10^8$ (Antoine & Tepper, 1969*a*) compared with 8.2×10^7 for *Escherichia coli* (Holme, Laurent & Palmstierna, 1957).

The chain length of most bacterial glycogens is about 10 to 13 glycosyl units but Arthrobacter (Zevenhuizen, 1966), mycobacteria (Antoine & Tepper, 1969*a*,*b*) and some thermoacidic archaebacteria (König *et al.*, 1982) form glycogen of shorter chain lengths, in the range of 7 to 9 units. Less-branched glycogens resembling amylopectin and termed granulose, have been found in the Clostridia (Strasdine, 1968; Mackey & Morris, 1971), with chain lengths of some 21 to 41 glucosyl units in *Clostridium pasteurianum* (Darvill *et al.*, 1977). Similarly, the α-glucan from *Selenomonas ruminantium*,

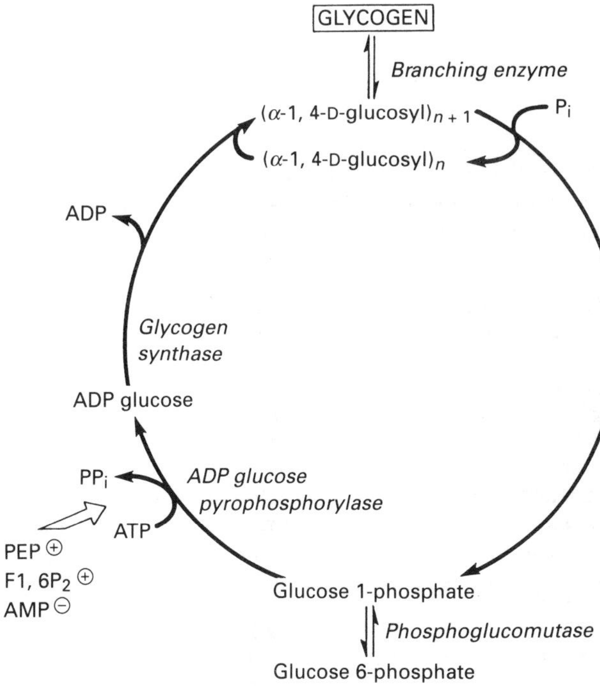

Fig. 2. Prokaryotic biosynthesis and degradation of glycogen and its regulation. Control of ADPglucose pyrophosphorylase activity is effected positively by intermediates of glycolysis, e.g. phosphoenolpyruvate (PEP) and fructose 1,6-bisphosphate ($F1,6P_2$), and negatively by AMP.

with a chain length of 23 units, is of the amylopectin type (Kamio *et al.*, 1981).

Enzymology of glycogen biosynthesis

The bacterial biosynthetic pathway

Bacterial glycogen synthesis occurs principally via a sugar nucleotide pathway involving the polymerization of glucosyl units derived from glucose 1-phosphate (Fig. 2). The actual glucosyl donor for polysaccharide synthesis in prokaryotes is ADPglucose (in contrast to eukaryotic micro-organisms which employ UDPglucose) which is formed from glucose 1-phosphate and ATP by the action of ADPglucose pyrophosphorylase. The glucosyl unit is then donated to an α-glucan primer to elongate the α-1,4 linear chain, catalysed by glycogen synthase, and finally, a branching enzyme introduces α-1,6 branches into the structure. The details of these reactions will now be considered.

ADPglucose pyrophosphorylase

This enzyme, initially reported in bacteria by Shen & Preiss (1964) and since identified in some 46 bacterial species (Preiss & Romeo, 1989), catalyses the reaction

$$\alpha\text{-glucose 1-phosphate} + \text{ATP} \rightleftharpoons \text{ADPglucose} + \text{PP}_i$$

It is the first enzyme unique to the glycogen biosynthetic sequence and plays a key regulatory role in the process. In most bacteria it is activated by intermediates of glycolysis and inhibited by AMP, ADP and/or inorganic phosphate (P_i). A report that pyrophosphate (PP_i), a product of the reaction, is an inhibitor of the enzyme at the physiological concentrations which obtain during growth and may thus coordinate glycogen synthesis with growth (Preiss & Greenberg, 1983; Preiss, 1984), does not appear to have been sustained. As ATP is a substrate, the adenylate energy charge of the cell may be regarded as a regulator of glycogen synthesis (Shen & Atkinson, 1970) while glycolytic intermediates may be viewed as indicators of intracellular carbon excess. In most cases the glycolytic intermediate activator increases the apparent affinity of the enzyme for its substrates, glucose 1-phosphate and ATP. The inhibition caused by AMP, ADP and/or P_i can be reversed or prevented by increasing the concentration of the activator, thus ensuring optimal glycogen formation only under conditions of excess carbon and high energy charge.

Preiss (1984) has divided the activators of ADPglucose pyrophosphorylase into seven groups based on differences in their specificity of activation by glycolytic intermediates, and a correlation has been suggested between activator specificity and the carbon assimilatory pathways operative in a given bacterium. Thus when glycolysis is the major pathway, e.g. in *E. coli*, the enzyme is usually activated by fructose 1,6-bisphosphate whereas in the cyanobacteria, which assimilate carbon via the ribulose bisphosphate carboxylase pathway, 3-phosphoglycerate, the primary product of CO_2 fixation in oxygenic photosynthesis, is the activator (Preiss, 1988). Considerable research has been carried out on the ADPglucose pyrophosphorylase of *E. coli*, including studies of the activator and inhibitor sites by chemical modification and mutagenesis, and a comparison of the amino-terminal sequence with that of other bacterial ADPglucose pyrophosphorylases (for review see Preiss, 1984; Preiss & Romeo, 1989). The general conclusion reached is that fructose 1,6-bisphosphate is the physiological activator of this enzyme.

While glycogen synthesis in the majority of enterobacteria displays a similar regulatory pattern to that of *E. coli*, there are two notable exceptions. The ADPglucose pyrophosphorylase of *Enterobacter hafniae* and *Serratia* spp, in common with that of *Clostridium pasteurianum*, appear not to be activated by any of the recognized modulators; they are, however, extremely sensitive to inhibition by AMP.

In oral streptococci it has been suggested that the site of regulation of glycogen synthesis might be at the stage of formation of glucose 1-phosphate rather than of ADPglucose (Keevil *et al.*, 1984). These bacteria take up sugars via both phosphoenolpyruvate-dependent phosphotransferase systems and proton symport. The intracellular glucose accumulated by the latter process is phosphorylated to glucose 6-phosphate by glucokinase or to glucose 1-phosphate using as phosphoryl donors carbamoyl phosphate (derived from arginine metabolism) and/or acetyl phosphate (arising from glucose metabolism under conditions when lactate formation is decreased, as when growing under carbon limitation). *Streptococcus sanguis*, at specific growth rates in excess of 0.1 h^{-1} with glucose limitation, expresses combined PEP phosphotransferase and glucokinase activities inadequate for the phosphorylation of all the available glucose, and consequently the other reactions assume prominence, leading to glucose 1-phosphate formation and increased glycogen synthesis.

Glycogen synthase

This enzyme catalyses the transfer of a glucosyl unit from ADPglucose to an α-1,4-glucan primer, thus

$$\text{ADPglucose} + (\alpha\text{-1,4-D-glucosyl})_n \rightarrow (\alpha\text{-1,4-D-glucosyl})_{n+1} + \text{ADP}$$

Glycogen synthase is specific for ADPglucose in the eubacteria and is the preferred substrate in some archaebacteria (König *et al.*, 1982). It displays no regulatory properties and does not exist in active and inactive forms. It thus differs in substrate and characteristics from the mammalian enzyme. The enzyme is inhibited by ADP and is susceptible to *p*-hydroxymercuribenzoate inhibition. Magnesium ions stimulate activity, probably by binding to ADP and thereby decreasing the inhibitory effect of the nucleotide. The *E. coli* B glycogen synthase has been purified to homogeneity; it has a monomeric molecular mass of about 49 kD and exists as dimers, trimers and tetramers. The reaction is reversible and displays a K_{eq} of approx. 46 over the pH range of 5.27 to 6.82 (Fox *et al.*, 1976). As previously noted, glycogen synthase may be found in both soluble and particulate (glycogen-associated) forms in the bacterial cell. Comparisons of the properties of the enzyme from various sources have been made by Preiss & Walsh (1981).

Branching enzyme

This enzyme transfers glucosyl units from the non-reducing end of an α-1, 4-D-glucan to the 6-position of certain glucosyl units within the chain, thereby creating α-1,6 branch points. As previously noted, the extent of branching is a characteristic of the bacterial species and hence some specificity of action is involved. The *E. coli* B enzyme has been purified to homogeneity and its monomeric (active form) molecular weight determined to be about 84 kD (Boyer & Preiss, 1977). Its structural gene has been cloned (Okita, Rodriguez

& Preiss, 1982) and the complete nucleotide sequence and deduced amino-acid sequence determined (Baecker, Greenberg & Preiss, 1986). The gene comprised 2181 base pairs specifying a protein of 727 amino acids with a deduced molecular mass of 84.23 kD, thus agreeing well with the determined value. A close relationship was adduced by immunological techniques between the branching enzymes of enterobacteria (Holmes, Boyer & Preiss, 1982).

Structural genes for glycogen synthesis

Significant advances have been made in our knowledge of the structural genes involved in glycogen synthesis in *Escherichia coli*, principally by the researches of Preiss and his group (reviewed by Preiss & Romeo, 1989). The genes for ADPglucose pyrophosphorylase (*glg C*), glycogen synthase (*glg A*) and branching enzyme (*glg B*) are located together in the order *glg B–glg C–glg A* (Latil-Damotte & Lares, 1977). Recombinant DNA techniques were subsequently applied with a mutant lacking the *glg* genes and the order of the genes was confirmed. However, two additional open reading frames, designated *glg X* and *glg Y*, were found in the gene cluster, in the order *glg B–glg X glg C–glg A–glg Y* (transcription from left to right), and evidence adduced for *glg X* coding for enzymes of the type which hydrolyse α-1,4-glucans or catalyse α-1,4-glucan transferase reactions (α-amylase, pullulanase etc); the product of *glg* Y is glycogen phosphorylase (Romeo, Kumar & Preiss, 1988; Yu *et al.*, 1988), designated by the latter workers as *glg* P. Thus enzymes for glycogen synthesis and degradation appear to be encoded on the same operon.

Control of glycogen gene expression

In addition to allosteric regulation of glycogen synthesis, control of enzyme formation also occurs. During rapid growth of *E. coli* B the glycogen biosynthetic enzymes are repressed and glycogen accumulation is minimal. Derepression occurs when batch culture cells enter the stationary phase on nitrogen exhaustion or when the growth rate of nitrogen-limited continuous cultures is decreased; the effect is more pronounced in rich media than in minimal salts media. Derepression of the three glycogen biosynthetic enzymes in *Salmonella typhimurium* LT-2 is co-ordinate (Steiner & Preiss, 1977).

There have been various postulated inhibitors of ADPglucose synthesis, including 5-aminoimidazole-4-carboxamide, an intermediate of purine biosynthesis (Leckie *et al.*, 1981), but there is now good evidence that cAMP and guanosine tetraphosphate (ppGpp) are involved in the regulation of glycogen gene expression. Following the original observation of Dietzler *et al.* (1977) that cAMP and cAMP-receptor protein (CRP) are required for optimal synthesis of glycogen, Urbanowski *et al.* (1983) obtained evidence for a direct effect of cAMP on the expression of the *glg* C gene. Subsequently Romeo & Preiss (1989) found that cAMP and CRP together stimulated the synthesis of ADPglucose pyrophosphorylase up to 25-fold and glycogen synthase up to 10-fold in an *in vitro* system but were without significant effect on branching enzyme. cAMP does not mediate its effect by direct allosterism or by affecting concentrations of modulators of ADPglucose pyrophosphorylase.

Bacterial mutants that are altered in the *rel A* gene (which encodes the enzyme needed for synthesis of the guanosine nucleotides pppGpp and ppGpp during the stringent response) are unable to synthesize glycogen as efficiently as the isogenic rel A^+ strain, and direct enhancement of the synthesis of ADPglucose pyrophosphorylase and glycogen synthase by ppGpp has now been demonstrated (Romeo & Preiss, 1989) although there was little effect on the branching enzyme. The activation effected by ppGpp was increased in the presence of cAMP and CRP.

Preiss & Romeo (1989) have proposed that the sites on the genome required for response to cAMP, CRP and ppGpp lie in the 0.5 kb non-coding DNA region that separates *glg X* and *glg C*.

Glycogen degradation

Degradation of intracellular glycogen occurs when the exogenous carbon source is seriously depleted or exhausted. While a number of enzymes that are capable of hydrolysing or phosphorylysing α-1,4 glucosidic bonds and of hydrolysing α-1,6 linkages have been identified in bacteria, the precise details of glycogen degradation and its control are still incomplete. It is a general observation that the rate of glycogen utilization is significantly lower than its rate of synthesis.

Escherichia coli K-12 possesses a constitutive glycogen phosphorylase that contains pyridoxal 5-phosphate and has a molecular weight of about 250 kD (Chen & Segel, 1968 *a*,b). It catalyses the phosphorolysis of α-1,4 bonds of glycogen yielding glucose 1-phosphate and eventually a product that requires the activity of a debranching enzyme for its further metabolism. Glucose 1-phosphate enters the major metabolic pathways after conversion to glucose 6-phosphate by phosphoglucomutase action. ADPglucose competitively inhibits the phosphorylase with respect to glucose 1-phosphate, displaying a K_i of 0.2 mM. Unlike the mammalian enzyme, there is no evidence for control of bacterial phosphorylase activity by chemical modification.

Debranching (α-1,6) enzymes of two types have been found in bacteria, namely pullulanase which debranches exogenous branched maltrodextrins but has little effect on glycogen and is dubiously present in *E. coli*, and one which debranches glycogen and amylopectin but not pullulan. The latter enzyme probably functions in glycogen degradation in *E. coli*. The proposed scheme of Palmer, Wober & Whelan (1973) for this organism encompasses a dual attack on glycogen by glycogen phosphorylase and a debranching isoamylase which yields linear maltodextrins, from which glucose is produced by amylomaltase action and glucose 1-phosphate by a maltodextrin phosphorylase.

However, *E. coli* K-12 strains with deletions of the genes coding amylomaltase and maltodextrin phosphorylase degrade glycogen at the same rate as the parent strain, casting doubt on the involvement of these enzymes in debranching (Creuzat-Sigal & Frixon, 1977). A purified debranching enzyme from this organism was inactive toward glycogen but hydrolysed the α-1,6 bonds of glycogen phosphorylase limit dextrin, and of β-amylase limit dextrin prepared from amylopectin. It was proposed that in *E. coli* K-12 glycogen phosphorylase converts glycogen to glucose 1-phosphate plus phosphorylase limit dextrin. The latter is then subjected to the debranching enzyme action yielding maltotetraose plus limit maltodextrins which are subjected to further action by glycogen phosphorylase to give glucose 1-phosphate. The maltotetraose is assumed to be converted to glucose but the details are not known.

Other enzymes involved in polysaccharide metabolism and their occurrence in bacteria have been considered by Preiss (1989). Currently, the observation (Chen & Segel, 1968*a*) that ADPglucose, the substrate for glycogen synthase, competitively inhibits glycogen phosphorylase appears to offer the only rational means of control of glycogen degradation, by preventing phosphorylase action when active glycogen synthesis is occurring. However, it has been suggested that the rate at which bacteria degrade their glycogen might be correlated with the structure of their individual polyglucans (Zevenhuizen & Ebbink, 1974). Organisms such as *Arthrobacter*, *Mycobacterium*, with highly branched glycogen (i.e. with short chain lengths of 7 to 9 glucosyl units), degrade it more slowly than those such as *Escherichia coli* and *Enterobacter aerogenes* with glycogen that has longer chain lengths (12–15 units). It was proposed that glycogen degradation is limited by the low activity of the debranching enzyme.

Physiological roles of glycogen

The possession of an intracellular store of glycogen enables some bacteria to survive better under starvation conditions than corresponding organisms not so endowed, but a crucial factor is believed to be the rate at which the reserve material is utilized (reviewed by Dawes, 1976, 1985, 1989). Those bacteria that are able to match closely the rate of glycogen degradation to their maintenance energy requirement survive for the longest periods. Glycogen often exerts a sparing action on the degradation of cellular protein and RNA under these conditions by providing energy and carbon. However this behaviour is not universal and there are examples of polyglucan-containing bacteria that die at a faster rate than their polyglucan-less counterparts, e.g. *Sarcina lutea*.

Polysaccharide also plays a role in sporulation of clostridial species. The α-1,4 glucan, granulose, may accumulate up to 60% of their biomass prior to sporulation and is then rapidly utilized concomitantly with spore formation. This suggests that granulose serves as a source of carbon and energy for the process of sporulation in these organisms (Strasdine, 1968; Mackey & Morris, 1971) and this is probably true of glycogen in certain other spore-forming bacteria.

In summary, although the available evidence is not entirely unequivocal, there is now good reason to consider that glycogen and related polysaccharides serve in many bacteria as reserves of carbon and energy.

POLYHYDROXYALKANOATES

Occurrence and structure

The polyester poly-3-hydroxybutyrate (PHB), discovered by Lemoigne (1926) in cytoplasmic granules of *Bacillus megaterium*, was found subsequently to have a widespread distribution in Gram-positive and Gram-negative aerobic bacteria, and in photosynthetic anaerobic species, lithotrophs, organotrophs and archaebacteria (Dawes & Senior, 1973; Dawes, 1989; Preiss, 1989). The first indication that this polymer may contain proportions of 3-hydroxy acids other than 3-hydroxybutyrate (3HB) was given by Wallen and Rohwedder (1974) who reported the presence of significant amounts of 3-hydroxyvalerate (3HV; 3-hydroxypentanoate) with minor amounts of C_6 and possibly C_7 3-hydroxyacids in extracts of sewage sludge. These observations were extended (Findlay & White, 1983; Odham *et al.*, 1986) and a significant development was the finding that *Pseudomonas oleovorans*, when grown on 50% (v/v) n-octane, accumulated granules that resembled PHB inclusions but consisted principally of a polyester of 3-hydroxyoctanoate (de Smet *et al.*, 1983). Since that time burgeoning research, fuelled by the commercially attractive properties of these polymers which behave as biodegradable thermoplastics, has led to the discovery of a wide range of polyesters of various compositions and the recognition that PHB is but one example, albeit the most abundant, of a general class of compound termed a polyhydroxyalkanoate (PHA; for review see Anderson & Dawes, 1990; Doi, 1990), with general formula

$$\left[-O-\underset{\substack{| \\ R}}{CH}-CH_2-\underset{\substack{\| \\ O}}{C}- \right]_n$$

These polyesters are accumulated in cytoplasmic granules which are usually spherical and vary in size according to organism: diameters are typically in the range 0.2 to 0.7 μm. Electron microscopy and freeze-etching of various organisms indicated the presence of a non-unit bounding membrane some 2.5 to 4.5 nm thick, and analysis disclosed that 2% (w/w) of the native PHB granule was protein and about 0.5% (w/w) lipid, mainly phosphatidic acid (Griebel, Smith & Merrick, 1968). The PHB synthase (and in some cases PHB depolymerase) appears to be associated with the membrane and thus it is interesting that the hydropathy profile of the PHB synthase of *Alcaligenes eutrophus*, sequenced by Peoples and Sinskey (1989*b*), does not conform to a typical membrane-spanning protein; these authors suggest that PHB synthesis may not require a complex membrane-bound polymerization system.

Studies with *A. eutrophus* have shown that the number of granules per cell is apparently fixed at the earliest stages of polymer accumulation (Ballard, Holmes & Senior, 1987). The number and size of granules were determined in two different-scale experiments with nitrogen-limited organisms using freeze-fracture and transmission electron microscopy. The average number of granules remained constant at 12.7 ± 1.0 and 8.6 ± 0.6, respectively, and increased uniformly in average diameter from 0.24 to 0.50 μm to accommodate the PHB synthesized. Polymer production in the organism ceased at a PHB content of about 80% (w/w) although PHB synthase activity remained high and substrate was available. It was concluded, therefore, that physical constraints operate and the cell is unable to accommodate more polymer within the fixed existing amount of cell wall material.

X-ray diffraction and conformational studies by Marchessault and his group showed that PHB is a compact right-handed helix with a two-fold screw axis and fibre repeat of 59.6 nm (Cornibert & Marchessault, 1972). It is optically active with the chiral centre of the monomer unit always in the *R* absolute [D-(−)] configuration. However, the earlier belief that granules *in vivo* contain crystalline polymer has been revised recently. Barham *et al.* (1989), using a range of physical techniques, found that the granules in *A. eutrophus* were completely amorphous and could be induced to crystallize only on heating or on removal of all water, a view supported by ^{13}C NMR spectroscopic studies on *in vivo* PHB granules of *Methylobacterium* sp. strain AM1 (Barnard & Sanders, 1989). [Mas, Pedrós-Alió & Guerrero (1985) concluded from volume and density measurements that PHB granules from *A. eutrophus* contain some 40% water.] The mobility of the polymer *in vivo* might therefore be due to the plasticizing effect of this water and/or lipid components (Kawaguchi & Doi, 1990), or the polymer may exist in the granules in a different form, e.g. enolic (Barham *et al.*, 1989). These observations offer some clues to the problem of how the enzymes of polymerization and depolymerization might operate in the highly hydrophobic environment of the polymer granule. Further, the lability of the granule which leads to loss of susceptibility to depolymerization (denaturation) can be correlated with loss of mobility of the polymer and its solidification.

Heteropolymer formation

It was discovered that when a glucose-utilizing mutant of *Alcaligenes eutrophus* is presented with glucose and propionic acid, a random copolymer containing both 3HB and 3HV monomer units is produced, a process patented by ICI plc (see Holmes, 1985). The 3HV content of the PHA is determined by the ratio of propionic acid to glucose in the medium during the polymer accumulation stage and the production of 70% (w/w) polymer containing 33 mol.% 3HV was reported; with valeric (pentanoic) acid as sole substrate, Doi *et al.* (1987) recorded formation of a copolymer containing 90 mol.% 3HV from *A. eutrophus*. (3HB-3HV) copolymers (which are more flexible than PHB) are marketed by ICI plc as 'Biopol' and are currently used for the manufacture of biodegradable bottles for packaging toiletries and motor oil.

Subsequently the incorporation into PHA of 4-hydroxybutyrate or 5-hydroxyvalerate monomer units, together with 3HB, was demonstrated by Doi and his colleagues [for review see Doi (1990)]; the longer backbone unit influenced the physical properties of the polymer which displayed greater elasticity and higher biodegradability than (3HB–3HV) copolymers.

Pathways and enzymology of PHA biosynthesis

Survey of pathways

A cyclic pathway for PHB synthesis and degradation, and its control (Fig. 3), were established for *Azotobacter beijerinckii* and *A. eutrophus* in 1973 (Senior & Dawes, 1973; Oeding & Schlegel, 1973). The biosynthetic route for PHB synthesis from glucose, and substrates which yield acetyl-CoA, in these and other organisms involves the sequential action of three enzymes: 3-ketothiolase, NADPH-dependent acetoacetyl-CoA reductase and PHB synthase (polymerase) which catalyse, respectively,

$$2\ \text{acetyl-CoA} \rightleftharpoons \text{acetoacetyl-CoA} + \text{CoA}$$
$$\text{acetoacetyl-CoA} + \text{NADPH} + \text{H}^+ \rightleftharpoons \text{3-hydroxybutyryl-CoA} + \text{NADP}^+$$
$$\text{3-hydroxybutyryl-CoA} + \text{P(3HB)}_n \rightarrow \text{P(3HB)}_{n+1} + \text{CoA}$$

A variant of this pathway occurs in *Rhodospirillum rubrum* involving two additional enzymes, stereospecific enoyl-CoA hydratases, which convert the *S*-3-hydroxyacyl-CoA (formed in this organism by the action of an NADH-dependent acetoacetyl-CoA-reductase) to the *R*-enantiomer required for PHA synthase activity (Moskowitz & Merrick, 1969).

Two other pathways for PHA synthesis have been discovered in the pseudomonads. Organisms of the rRNA homology group I synthesize polyesters consisting of medium-chain-length 3-hydroxy-alkanoates derived from fatty acyl-CoA intermediates of the β-oxidation of alkanes, alkanoic acids and alkanols (de Smet *et al.*, 1983; Brandl *et al.*, 1988; Huisman *et al.*, 1989; Haywood, Anderson & Dawes, 1989*a*). With the exception of *P. oleovorans*, almost all of these organisms possess another pathway which enables the synthesis of medium-chain-length polyesters from acetyl-CoA, e.g. *Pseudomonas aeruginosa* accumulates a polymer composed principally of 3-hydroxy-

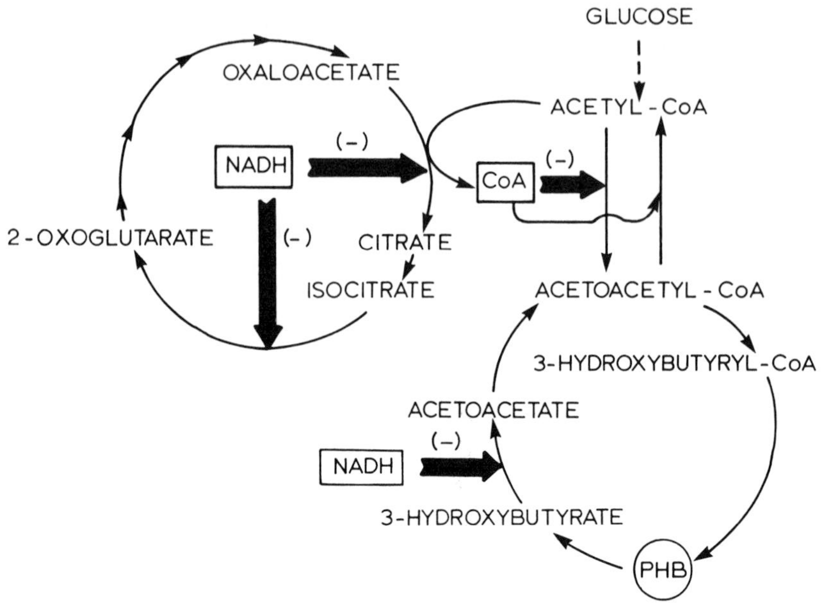

Fig. 3. Inter-relationship of the tricarboxylic acid and PHB metabolic cycles, and their control, in *A. beijerinckii* and *A. eutrophus*.

decanoate (PHD) when grown on a carbohydrate such as gluconate (Haywood *et al.*, 1990; Timm & Steinbüchel, 1990).

The photosynthetic bacterium *Rhodospirillum rubrum* exhibits another pattern of PHA accumulation, producing polyesters containing C_4 to C_6 3-hydroxy acids from a range (C_2 to C_{10}) of alkanoic acids (Fig. 4). Terpolyesters containing these hydroxy acids were formed from C_6, C_7 and C_{10} substrates. The shorter-chain substrates yielded polyesters containing C_4 and/or C_5 3-hydroxy acids but less than 1 per cent of the C_6 monomer unit (Brandl *et al.*, 1989).

Enzymology of PHA synthesis

3-Ketothiolase is the controlling enzyme for PHB synthesis with CoA the key modulator (Oeding & Schlegel, 1973; Senior & Dawes, 1973). There are two isoenzymes, with specificities predominantly for C_4 and C_5 3-keto-acyl-CoAs and for C_4 to C_{10} substrates, respectively. The former is regarded as a biosynthetic enzyme and the latter as a degradative enzyme mainly involved in fatty acid metabolism although both can, in fact, function in PHB synthesis in *A. eutrophus*; in this organism the biosynthetic enzyme is a tetramer of identical (M_r 44 kD) subunits (Haywood *et al.*, 1988*a*). The biosynthetic enzymes of *Zoogloea ramigera* and *A. eutrophus* have been puri-

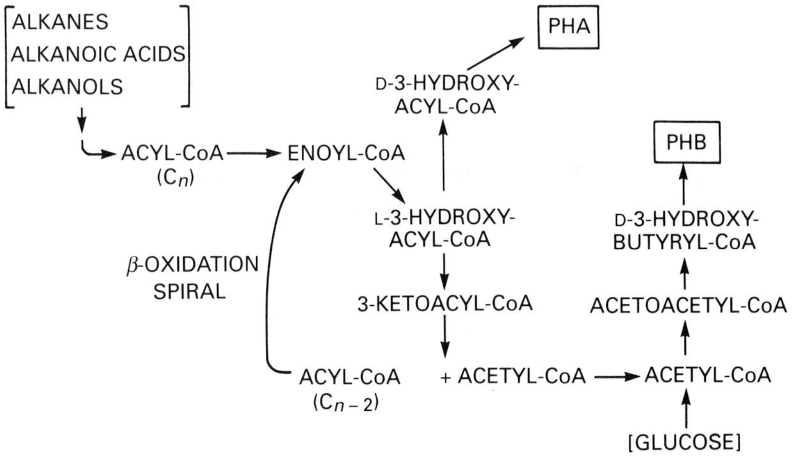

Fig. 4. Proposed pathways of biosynthesis of polyhydroxyalkanoates (PHA) and polyhydroxybutyrate (PHB) from alkanes, alkanoic acids and alkanols. Epimerization of the L(+), i.e. S, 3-hydroxyacyl-CoA, derived from the β-oxidation spiral, to the D(−), i.e. R, enantiomer is an essential prelude to polymerization. Pseudomonads of the rRNA I homology group synthesize only PHA from these substrates whereas *Rhodospirillum rubrum* accumulates both PHA and PHB from n-alkanoic acids.

fied and subjected to mechanistic studies by Sinskey and his colleagues (Davis *et al.*, 1987; Masumane *et al.*, 1989).

Isoenzymes of acetoacetyl-CoA reductase also exist in *Z. ramigera* and *A. eutrophus*, possessing different substrate and coenzyme specificities. Only the NADPH-specific enzyme (a tetramer of identical M_r 23 kDa subunits) reacting with *R*-C_4 to C_6 substrates operates in PHB synthesis in *A. eutrophus* (Haywood *et al.*, 1988*a*,*b*); the NAD enzyme is specific for *S*-substrates and thus cannot function in direct PHB formation.

PHB synthase is associated with the granules of polymer-synthesizing bacteria and has proved difficult to solubilize in native form. However, soluble forms of the enzyme have been found in both *Z. ramigera* (Fukui *et al.*, 1976) and *A. eutrophus* (Haywood *et al.*, 1989*a*). The location of the synthase depended on prevailing growth conditions: organisms not synthesizing PHB, or growing in carbon-limited continuous culture, contained mainly the soluble form. Upon switching to nitrogen limitation PHB accumulation began, accompanied by a rapid disappearance of soluble synthase and appearance of granule-associated enzyme (Haywood *et al.*, 1989*a*). Both forms of the enzyme have been partially purified but proved increasingly unstable as purification progressed.

The PHB synthase of *A. eutrophus* is specific for *R* enantiomers and active only with C_4 and C_5 3-hydroxyacyl-CoAs, although the discovery that this organism can incorporate 4-hydroxybutyrate and 5-hydroxyvalerate into polymer (reviewed by Doi, 1990) suggests that the synthase is also active with these substrates, unless an additional synthase is present. These observations correlate with the fact that the polyester compositions of *A. eutrophus* are confined to C_4 and C_5 hydroxyacids.

The mechanism of action of PHB synthase remains a current challenge. Active thiol groups are present and Griebel & Merrick (1971) proposed a two-stage polymerization reaction involving an acyl-S-enzyme intermediate, while Ballard *et al.* (1987) have proposed a model involving two thiol groups and a four-membered transition state. It is presumed that the chain transfer function performed by the synthase must in some way control the molecular weight of the polymer produced, which is characteristic of a given organism. Comparative studies with PHA synthases from organisms that synthesize heteropolymers from medium-chain length 3-hydroxyacyl-CoAs should prove instructive and investigations of control of molecular weight are in progress.

Pseudomonas species of the rRNA I homology group provide some interesting examples of polyester accumulation for although they cannot synthesize PHB they do produce PHAs containing C_6 to C_{12} 3-hydroxyacyl units from straight-chain alkanes, alkanoic acids or alkanols of the corresponding chain length (see Fig. 4; Lageveen *et al.*, 1988; Gross *et al.*, 1989; Haywood, Anderson & Dawes, 1989b). It has been suggested that this ability may be of taxonomic value for the fluorescent pseudomonads (Huisman *et al.*, 1989). A general observation is that minor constituents of the PHA are 3-hydroxy acids differing in chain length from the substrate by two carbon atoms.

A more surprising discovery was that certain pseudomonads produce PHAs from chemically-unrelated substrates, e.g. the synthesis of polymer containing principally 3-hydroxydecanoate (3HD) from gluconate, glucose, fructose, glycerol, lactate or acetate as the sole carbon source (Haywood *et al.*, 1990). This ability was shown to be characteristic of many strains of *Pseudomonas aeruginosa* and certain other *Pseudomonas* species (Timm & Steinbüchel, 1990). The biosynthetic pathways involved are under investigation and it seems likely that the precursors of the polymers are derived from reactions involved in fatty acid synthesis. Examples of other PHA-accumulating bacteria are surveyed by Anderson and Dawes (1990).

Regulation of PHA synthesis

In bacteria which synthesize PHB from glucose via the acetyl-CoA route, 3-ketothiolase is the key regulatory biosynthetic enzyme, modulated by intracellular CoA concentration. Diversion of acetyl-CoA from entry to the tricarboxylic acid cycle is effected by modulation of citrate synthase activity by NADH; the acetyl-CoA concentration increases while that of CoA decreases, and polymer synthesis via 3-ketothiolase proceeds (Senior & Dawes 1973; Jackson & Dawes, 1976). The NADH:NAD ratio of the cell may be influenced by oxygen limitation or by cessation of protein synthesis caused by nitrogen or other nutrient limitation. But a mutant strain of *Azotobacter vinelandii* deficient in respiratory NADH oxidase, and thus unable to reoxidize NADH via the electron transfer chain, synthesizes PHB during unrestricted growth (Page & Knosp, 1989).

The regulation of PHB biosynthesis from substrates that are not metabolized via acetyl-CoA is of interest because 3-ketothiolase is not involved. Thus the addition of $(NH_4)_2SO_4$ to nitrogen-limited, PHB-accumulating *A. eutrophus* suspensions inhibits polymer synthesis from glucose, but not from butyric acid which is converted to 3-hydroxybutyryl-CoA and incorporated directly (Doi *et al.*, 1988). The accumulation of PHA by *P. oleovorans* from n-alkanes and n-alkanoic acids, under apparently unrestricted growth conditions, suggests that effective control of polymer synthesis during exponential growth may not occur in this bacterium (Lageveen *et al.*, 1988; Gross *et al.*, 1989). However, further research is needed to elucidate the regulation of PHA synthesis from non-carbohydrate substrates.

The simultaneous synthesis and degradation of PHAs have been demonstrated with nitrogen-free suspensions of *A. eutrophus* (Doi *et al.*, 1990). When bacteria containing 55% (w/w) PHB (derived from butyric acid as substrate) were incubated with valeric acid, the gross composition of the accumulated polyester changed as a function of time with the 3HV content increasing from 0 to 49 mol% over 96 h, while the total PHA content, after an initial decrease, showed little change at 52%. The number-average molecular weight (M_n) of the polyester decreased during the experiment from 5.78×10^5 to 2.69×10^5 and the polydispersity [ratio of the weight-average (M_w) to the number-average molecular weight, M_w/M_n] increased from 2.0 to 4.2.

In the converse experiment, bacteria containing 50% (w/w) PHA (3HB–3HV copolymer, derived from valeric acid substrate) were incubated with butyric acid for 48 h, during which time the 3HB content of the polymer rose from 44 to 81 mol%, the M_n fell from 7.16×10^5 to 5.41×10^5, the M_w/M_n ratio remained unchanged at 1.7 to 1.8. The total polymer content, after a transient increase to 66%, remained at 50%.

In contrast, when *A. eutrophus* was grown under steady-state nitrogen limiting conditions in continuous culture, no evidence for turnover of PHB in glucose-grown cells was obtained (Haywood *et al.*, 1989*a*).

Genetics of PHB and PHA biosynthesis

The structural genes for the PHB biosynthetic pathway of *A. eutrophus* have now been cloned and expressed in *Escherichia coli* by three independent groups (Slater, Voige & Dennis, 1988; Schubert, Steinbüchel & Schlegel, 1988; Peoples & Sinskey, 1989*a*,*b*). The genes are clustered, are organized in one operon, and have been sequenced and shown to be transcribed in the order *phbC* (synthase), *phbA* (thiolase), *phbB* (reductase) (Peoples & Sinskey, 1989*b*; Steinbüchel *et al.*, 1990; Janes, Hollar & Dennis, 1990). Molecular analysis of the PHB biosynthetic operon has permitted identification of the N-terminus of the synthase and has also revealed the promoter and translational start site of the *A. eutrophus phbC* gene (Schubert, Kruger & Steinbüchel, 1991). However, expression of the *phbC* gene alone in *E. coli* does not produce PHB or significant levels of PHB synthase activity (Peoples & Sinskey, 1989*b*) and the reason(s) for this await explanation.

The *A. eutrophus* PHB-biosynthetic genes have also been expressed in *P. oleovorans*; the resulting recombinant strain synthesized a blend of PHB and polyhydroxyoctanoate (Steinbüchel & Schubert, 1989; Steinbüchel *et al.*, 1990). These workers also obtained recombinant strains of *P. aeruginosa* and other pseudomonads able to produce blends of PHB and polyhydroxydecanoate from gluconate.

Investigations of the PHA synthase locus of *P. oleovorans* (Witholt, Huisman & Preusting, 1990; Peoples & Sinskey, 1990) secured evidence for the presence of two PHA synthase genes, with a PHA depolymerase gene located between them. When the synthase genes were introduced in high copy number the ensuing increase in level of PHA synthase permitted faster chain elongation but not an increased accumulation of polymer, indicating that other, still unknown, factors are involved in the control or limitation of total PHA production (Witholt *et al.*, 1990).

Enzymology of PHB and PHA degradation

Intracellular PHB degradation

PHB degradation usually commences when the available exogenous carbon source is very limited or exhausted, but little has been recorded in the literature on the enzymology and control of the process since the subject was reviewed by Dawes & Senior (1973). Two different types of PHB depolymerase had then been recognized, in *Rhodospirillum rubrum* and *B. megaterium* yielding *R*-3-hydroxybutyrate as the product of the reactions. Native granules from *R. rubrum* are self-hydrolysing whereas those from *B. megaterium* are quite stable, although a soluble extract from *R. rubrum* was active in the degradation of native granules from *B. megaterium*; purified polymer or denatured granules could not serve as substrates (Merrick, Delafield & Doudoroff, 1962; Merrick & Doudoroff, 1964). A soluble activator protein was isolated from *R. rubrum* extracts which, in the presence of depolymerase, activated PHB hydrolysis, yielding (*R*)-3-hydroxybutyrate as the main product together with some dimeric ester (Merrick & Yu, 1966).

The soluble PHB depolymerase of *A. eutrophus* yielded *R*-3HB as the sole product of hydrolysis (Hippe & Schlegel, 1967) whereas the soluble enzyme of *B. megaterium* gave a mixture of dimer and monomer (Gavard *et al.*, 1967). A detailed study of the *B. megaterium* system disclosed that depolymerization required a heat-labile factor associated with the granules together with three soluble components, namely a heat-stable protein activator, PHB depolymerase, and a hydrolase. It was concluded that although the depolymerase is not granule bound, there is a granule-associated protein that inhibits depolymerase activity (Griebel, Smith & Merrick, 1968). Although it is assumed that interplay between inhibition and activation controls PHB degradation, the detailed mechanism awaits elucidation and, to date, there have been no reports of the depolymerases acting on heteropolymers containing medium-chain-length 3-hydroxyacyl units.

Extracellular PHA degradation

In contrast to the paucity of recent research on the intracellular degradation of PHA, there has been a flurry of interest in extracellular degradation of these polymers, spurred by their commercial exploitation as environmentally-friendly plastics. This work lies outwith the role of PHAs as bacterial storage compounds, however, and the reader is referred to recent reviews for information (Anderson & Dawes, 1990; Doi, 1990).

Physiological functions of PHB and other PHAs

Carbon and energy reserves

The role of PHB as a carbon and energy reserve has been reviewed in detail elsewhere (Dawes & Senior, 1973; Dawes, 1985; 1989; Preiss, 1989) and here a summary must suffice. The accumulation of large quantities of high molecular weight, reduced carbon compounds in some 8 to 12 granules per cell ensures minimal osmotic disturbance. The possession of PHB frequently, but not universally, retards the degradation of cell components such as protein and RNA during nutrient starvation, although there is not a common pattern of behaviour and, depending upon the bacterium, there may be sequential or simultaneous utilization of these macromolecules. PHB enhances the survival of some, but not all, of the bacteria investigated and serves as a carbon and energy source for sporulation in some *Bacillus* species, although its accumulation is not mandatory for the sporulation process. Likewise, PHB can serve as a carbon and energy source for the encystment of *Azotobacters.* It has also been proposed that PHB furnishes an oxidizable substrate to afford respiratory protection to the nitrogenase of *Azotobacters* when suitable exogenous substrate(s) is not immediately available for oxidation (Senior & Dawes, 1971). A similar role for PHB has been advanced for *Rhizobium* sp. strain ORS 571 by Stam *et al.* (1986).

Although there are not yet any reported studies of the survival characteristics of bacteria possessing PHAs other than PHB, it has been noted that PHA is degraded when the external carbon source is exhausted (Knee *et al.*, 1990) and it may reasonably be inferred that they, too, can be utilized as carbon and energy reserves. It seems likely that, for the majority of bacteria growing in their natural environments, the proportion of PHAs relative to PHB is fairly low and their role would therefore be minor. However, there are examples, such as *P. oleovorans* growing in an environment rich in n-octane and accumulating polyhydroxyalkanoate as the principal PHA, when it may be presumed that this material could be utilized by the organism as a source of carbon and energy. The crucial experiments to test the presumption have not yet been reported.

PHB in symbiotic nitrogen fixation

PHB has been implicated as an energy source in the symbiotic nitrogen fixation process that occurs between *Rhizobium* and *Bradyrhizobium* and leguminous plants (Karr *et al.*, 1984; McDermott *et al.*, 1989). The energy required for nitrogen fixation by bacteroids is largely met by the metabolism of photosynthetic compounds transported to the root nodules. The bacteroids accumulate up to 50% of their biomass as PHB despite the fact that both nitrogen fixation and PHB synthesis compete for the available reducing equivalents. It has been suggested that, as with *Azotobacter beijerinckii* (Senior *et al.*, 1972), PHB accumulation serves a redox regulatory role (McDermott *et al.*, 1989).

Application of PHAs as environmental markers

PHB and PHA metabolism have been used as a measure of unbalanced growth of estuarine detrital microbiota (Nickels, King & White, 1979), based on the sensitivity of polymer content and its metabolism to the environment in laboratory monocultures (Herron, King & White, 1978). By comparing the ratio of phospholipid to PHA synthesis, a means of estimating disturbance to the sediment is available, i.e. whether balanced or unbalanced growth ensues (Findlay & White, 1983, 1984, 1987).

PHB in prokaryotic and eukaryotic membranes

A comparatively recent development in the biochemistry of the two storage polymers, PHB and polyphosphate, has been the discovery of their presence in association, in the plasma membranes of some Gram-positive and Gram-negative bacteria (Reusch & Sadoff, 1983) and subsequently in eukaryotic membrane fragments (Reusch, 1989). As these complexes span the membrane, it has been suggested that they may play a role in the regulation of intracellular calcium concentrations and in calcium signalling (Reusch, 1989).

Reusch & Sadoff (1983) first recorded the presence of PHB in the cytoplasmic membranes of *Azotobacter vinelandii*, *Bacillus subtilis* and *Haemophilus influenzae*, three bacteria that undergo natural transformation but differ structurally and metabolically. Although *A. vinelandii* synthesizes substantial amounts of PHB in cytoplasmic granules, *B. subtilis* accumulates significantly less of the polymer, and *H. influenzae* none. However, within each species the concentrations of PHB in their membranes correlated with their transformability (Reusch *et al.*, 1987). Evidence that the membrane PHB exists as a labile, organized gel structure was adduced from fluorescence analysis of thermotropic lipid phase transitions in *A. vinelandii* and *B. subtilis*, and it was hypothesized that PHB synthesis is a prerequisite for transport of exogenous DNA through the membrane, i.e. the acquisition of genetic competence.

The calcium-dependent development of competence in *E. coli* was associated with *de novo* synthesis and incorporation of PHB into the plasma membrane, accompanied by modification of the bilayer structure, as indicated by an irreversible sharp new lipid phase transition at about 56 °C (Reusch, Hiske & Sadoff, 1986). It proved possible to extract a PHB complex from plasma membranes of genetically-competent *E. coli* and analysis of chloroform extracts revealed the presence of PHB, inorganic polyphosphate and calcium ions in the approximate molar ratios of 1:1:0.5. The PHB chain lengths were estimated as 120 to 200 and those of polyphosphate as 130 to 170 monomer units. The complex had sufficient structural integrity to permit its incorporation into liposomes (Reusch & Sadoff, 1988).

A model structure was proposed for the complex (Reusch & Sadoff, 1988; Reusch, 1989) which would function as a transmembrane channel (Fig. 5). An outer lipophilic coat, comprising a PHB helical cylinder with 14 monomer units per turn, encloses an inner polar helix of polyphosphate with 7 monomer units per turn. Only these particular configurations furnish a suitable geometry for accommodating the Ca^{2+} in the complex. The Ca^{2+} ions (3.5 per turn) link the two polymers, each ion being ligated to four phosphoryl oxygens and four PHB ester carbonyl oxygens. This putative structure has a diameter of 240 nm and a helical rise of 40 nm, with an average length of 450 nm. In addition to its value as a calcium store, Reusch & Sadoff (1988) suggest the transmembrane complex functions in the transport of calcium, phosphate and DNA. For example, export of Ca^{2+} and phosphate might be effected by elongation of the polyphosphate chain at the cytoplasmic interface by the action of polyphosphate kinase, in concert with polyphosphatase activity at the periplasmic face. Conversely, Ca^{2+} import might stem from changes in membrane potential or an increase in the concentration gradients.

Interestingly, Reusch (1989) has now secured evidence that similar complexes exist in eukaryotic membranes, demolishing the former belief that PHB is characteristic of prokaryotes.

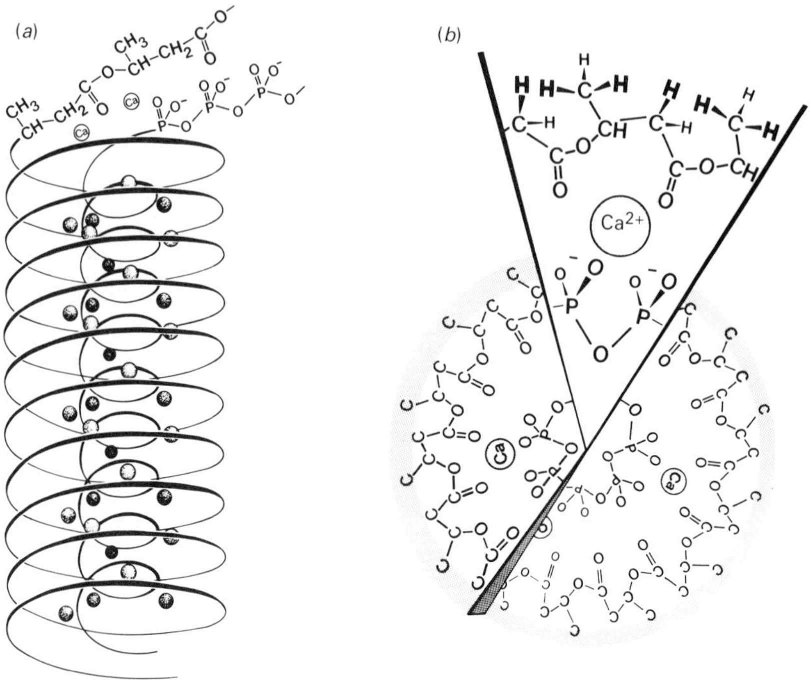

Fig. 5. (*a*) Proposed structure of a channel spanning the membrane, depicting the relationship between PHB, Ca^{2+}, and polyphosphate. PHB forms an outer helical channel around a core helix of polyphosphate with Ca^{2+} bridging the two polymers. (*b*) Interior view down the axis of the membrane channel structure. The PHB outer helix has 14 monomer units per turn. The methyl and methylene groups form a lipophilic shell (stippling represents the hydrogens), and the carbonyl ester oxygens form a polar lined cavity. The polyphosphate helix has seven monomer units per turn, with the phosphoryl oxygens facing outwards and the chain oxygens alternately facing in and out. There are 3.5 Ca^{2+} ions per turn; each is coordinated to four phosphoryl oxygens and four ester carbonyl oxygens (two of which are from the turn below and are not shown). From Reusch (1989), reproduced with permission.

POLYPHOSPHATES

Occurrence and chemical structure

Although polyphosphates are not universal constituents of prokaryotic cells their presence in metachromatically-staining granules of various microorganisms has been recognised for many years (Harold, 1966; Kulaev, 1979). They present an intriguing aspect of microbial energetics. Lipmann (1965) suggested that the earliest organisms on this planet used polyphosphate or pyrophosphate as their prime energy intermediary, the role of ATP as the universal energy mediator in contemporary organisms having arisen during the course of evolution. Polyphosphates might thus be regarded as a metabolic fossil which, over the centuries, has lost its primordial role in polymer synthesis and assumed new functions (Harold, 1966). It is now known, for

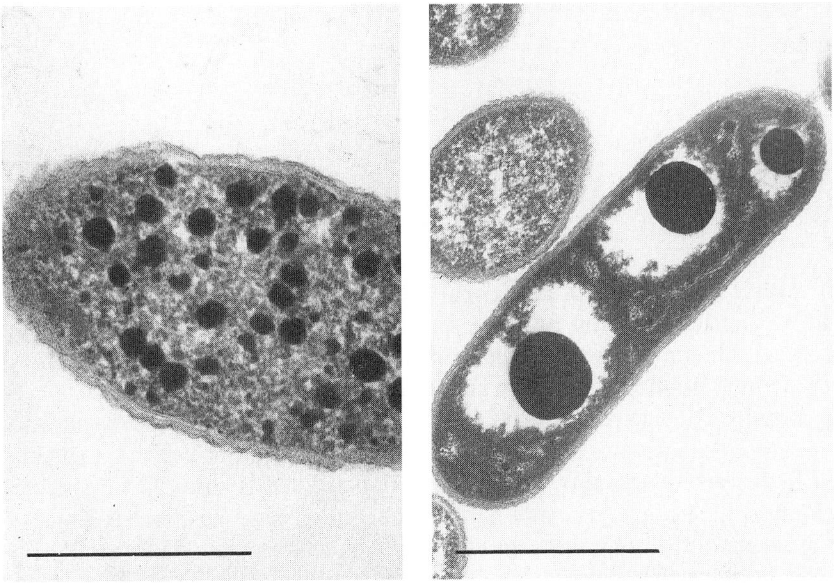

Fig. 6. Polyphosphate granules in *Acinetobacter* spp. isolated from an activated sludge pilot plant. Electron micrographs of (*a*) young cells (beginning of the exponential growth phase and (*b*) older cells (onset of the stationary phase). The bar represents 0.5 μm in (*a*) and 1.0 μm in (*b*). Reproduced, with permission, from Streichan, Golecki & Schön (1990).

example, that in some bacteria polyphosphates can occupy two cellular locations, namely in the cytoplasm as granules and in the periplasmic space associated with the inner membrane (Ostrovsky *et al.*, 1980; Halvorson *et al.*, 1987).

Polyphosphates are usually accumulated when bacteria are subjected to a nutrient deficiency in the presence of excess phosphate but this behaviour is not universal (Fig. 6). Thus *Propionibacterium shermanii* accumulates the polymer during exponential growth and 100-fold more with lactate as the carbon source than with glucose (Clark, Beegen & Wood, 1986). The concept of a competitive relationship between polyphosphate synthesis and nucleic acid metabolism has been reviewed by Harold (1966) and Dawes & Senior (1973).

The physiological role(s) of polyphosphate in the microbial cell is not, however, entirely unequivocal. Despite much recent research, its several putative functions as an energy-reserve compound, phosphorus reserve, regulator of metabolism, component of membrane channels, cation-binding compound, and possibly serving as an alarmone, require careful assessment for individual organisms, although there is no question that polyphosphate can, in some bacteria, fulfil the function of ATP.

The chemical structure of bacterial inorganic polyphosphates is that of a linear anhydride of orthophosphate (below) varying in chain length from three

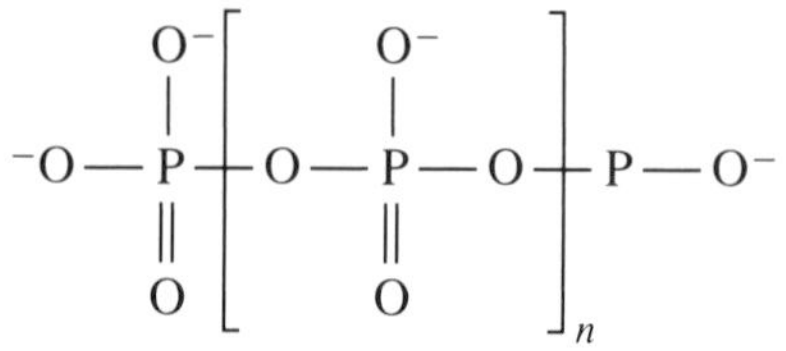

to over 10^3 units and usually consisting of mixtures of different molecular sizes. The introduction of more sensitive analytical techniques, including polyacrylamide gel and enzymic sizing methods (reviewed by Wood & Clark, 1988), has permitted significant advances in our knowledge of polyphosphate metabolism during recent years.

Polyphosphates are high-energy compounds and the standard free energy of hydrolysis of the phosphoanhydride linkage yields some 38 kJ per mole of phosphate released at pH 5. Their energy-storage function depends on the ability of the bond-cleavage reaction to effect phosphorylation and thereby conserve energy associated with the reaction

$$(\mathrm{P})_n + \mathrm{H_2O} \rightarrow (\mathrm{P})_{n-1} + \mathrm{P_i} \qquad \Delta G^{o'} \simeq -38 \text{ kJ mole}^{-1}$$

where (P) represents the monomer unit.

Around 60 diverse prokaryotic species are known to contain polyphosphate granules (Kulaev, 1979) which generally do not appear to possess a bounding membrane and may in some cases be associated with polysaccharide, polyhydroxybutyrate or RNA. Electron microscopy of cyanobacteria disclosed that the granules are mainly located in the region of DNA fibrils and associated ribosomes (Jensen & Sicko, 1974), and near the subcellular structures involved with photosynthesis (Stewart & Codd, 1975). X-ray energy-dispersion microanalysis has been combined with electron microscopy (Coleman *et al.*, 1972) to study the chemical nature of the granules in the cyanobacterium *Plectonema boryanum* and demonstrate the presence of phosphate (although not its degree of polymerization) together with cations such as potassium, calcium and magnesium (Baxter & Jensen, 1980*a*,*b*). When the medium contained excess concentrations of a particular cation, e.g. Mg^{2+}, Mn^{2+}, Ba^{2+} or Zn^{2+}, they accumulated in the polyphosphate granules in high quantity. [^{31}P]-NMR has been used to localize soluble polyphosphate in bacterial cells (Ferguson, Gadian & Kell, 1979) and Ostrovsky *et al.* (1980) discovered their presence in the periplasm of *Mycobacterium smegmatis*, thus reflecting similarities with some eukaryotic organisms including yeasts and fungi (reviewed by Kulaev & Vagabov, 1983). Interestingly, the granules observed in the sulphate-reducing bacterium *Desulfovibrio gigas* have been shown to contain magnesium tripolyphosphate (Jones & Chambers, 1975).

Polyphosphate granules in *Acinetobacter lwoffi* are intensely birefringent under a polarizing microscope, suggesting they contain polymers with a parallel orientation (Halvorson, 1990).

Polyphosphate biosynthesis

Polyphosphate kinase

The principal direct mechanism of polyphosphate biosynthesis in bacteria involves the transfer of the terminal phosphate group of ATP to polyphosphate, catalysed by a Mg^{2+}-dependent ATP-polyphosphate phosphotransferase (polyphosphate kinase)

$$(P)_n + ATP \rightarrow (P)_{n+1} + ADP$$

The enzyme, which is specific for ATP, was first discovered in a prokaryote (*Escherichia coli*) by Kornberg, Kornberg & Simms (1956) and has since been found in diverse aerobic, anaerobic and facultative bacteria (Dawes & Senior, 1973; Kulaev & Vagabov, 1983). Mutants of *Enterobacter aerogenes* which lack polyphosphate kinase are unable to synthesize the polymer (Harold & Harold, 1963; Harold, 1964). A glycogen-bound polyphosphate kinase has been reported in the archaebacterium *Sulfolobus acidocaldarius*; it displayed enzymatic activity only as a native complex with glycogen (Skórko, Osipiuk & Stetter, 1989).

Robinson & Wood (1986) investigated polyphosphate kinase from *Propionibacterium shermanii* and found the enzyme was monomeric with molecular mass of 83 kDa. By appropriate [^{32}P]-labelling experiments they demonstrated that the reactions in both directions, i.e. polyphosphate elongation at the expense of ATP and ADP phosphorylation by polyphosphate, are processive reactions. That is, in polymer synthesis the substrate polyphosphate binds to the enzyme and chain elongation occurs without repetitive dissociation from the enzyme, so that the polyphosphate species accumulating in the cell are all of high molecular weight. Similarly, in the reverse reaction, the phosphorylation of ADP occurs without the repetitive release of polyphosphate. Such processes are probably beneficial because the absence of low molecular weight intermediates will obviate any disturbance of osmotic balance and because the continuous binding of polyphosphate ensures that the overall rate of reaction will be a function of only one substrate (ATP in the case of polyphosphate synthesis or ADP for the reverse reaction). Short-chain polyphosphate serves as a primer for the kinase but experimentally it has been concluded that a mechanism of initiation not involving added primer also exists although its nature is uncertain (Robinson, Clark & Wood, 1987).

1,3-Bisphosphoglycerate:polyphosphate phosphotransferase

The evidence for the occurrence and the physiological significance of a second putative polyphosphate-synthesizing system, first reported in *Neurospora crassa* (Kulaev, Szymona & Bobyk, 1968) and then later in *E. coli*, *Micrococcus lysodeikticus* and *P. shermanii* (Kulaev *et al.*, 1971) has been questioned by Wood & Clark (1988). This system involves an enzyme, 1,3-bisphosphoglycerate : polyphosphate phosphotransferase, which catalyses the elongation of the polyphosphate chain at the expense of the potential energy of the high energy phosphate group of 1,3-bisphosphoglycerate:

$$\text{1,3-bisphosphoglycerate} + (\text{P})_n \rightleftharpoons \text{3-phosphoglycerate} + (\text{P})_{n+1}$$

The specific activities of the enzyme were calculated and found to be extremely low (Wood & Clark, 1988) and its presence in *P. shermanii* could not be confirmed by Wood & Goss (1985).

Polyphosphate utilization

General considerations

The degradation of polyphosphate in bacteria can be catalysed by several enzymes and consequently the assessment of their relative contribution to the overall rate of polymer breakdown under physiological conditions may present problems. Although the polyphosphate kinase reaction is reversible, probably being controlled by the ATP:ADP ratio of the cell on account of the high affinity of the enzyme for ADP (Kornberg, 1957), available evidence suggests that two other classes of enzyme play significant roles in bacteria. These comprise enzymes for the transfer of a terminal phosphate from polyphosphate to glucose, fructose or AMP, employing the group transfer potential, and exopolyphosphatases which hydrolyse inorganic phosphate from the ends of the polymer chains. The discovery of, and earlier work on, these enzymes has been reviewed elsewhere (Dawes & Senior, 1973; Kulaev & Vagabov, 1983; Dawes, 1989; Preiss, 1989) and here more recent studies will be considered.

Polyphosphate glucokinase

Polyphosphate glucokinase, which is Mg^{2+}-dependent, catalyses the reaction

$$(\text{P})_n + \text{glucose} \rightarrow (\text{P})_{n-1} + \text{glucose 6-phosphate}$$

and is specific for glucose and glucosamine. It has recently been purified to near homogeneity from *P. shermanii* and its characteristics determined (Clark, 1990). The enzyme was found to be active as a monomer unit with a molecular weight of about 31 kD. An interesting feature is that, even after high purification, the enzyme retains some activity with ATP as the phosphorylating agent, e.g. V_{max} values for $(\text{P})_n$ and ATP were 400 and 55 μmol $\text{min}^{-1}\text{mg}^{-1}$, respectively.

The mechanism of action has been investigated and, correcting their earlier report (1986), Pepin & Wood (1987) demonstrated that it is a non-processive or quasi-processive reaction. Measurement of the K_m disclosed a sharp increase at a polymer chain length of approximately 100. Thus a 2000-fold difference in K_m exists for polyphosphate of chain-length 32 (4.3 μM) compared with that for an average chain length of 724 (0.002 μM). The K_m increased gradually with decrease in the size of the polymer and then, at a chain length of about 100 residues, the K_m increased spectacularly. A consequence is that, in the presence of a mixture of polyphosphates of different chain lengths, the enzyme reacts preferentially with the long-chain polyphosphates, which are shortened and accumulate at chain lengths of about 100. The short chains accumulate because their low affinity (high K_m) precludes them from competing for the active site in the presence of the high-affinity long chains. This phenomenon has been exploited for determination of the average size of long-chain polyphosphates (Pepin, Wood & Robinson, 1986).

The distribution of polyphosphate glucokinase is of considerable taxonomic interest. Kulaev (1971) noted that it was confined to a limited group of organisms, including *Actinomyces*, *Proactinomyces*, *Mycobacterium*, *Corynebacterium*, *Propionibacterium*, *Micrococcus*, *Tetracoccus*, *Mycococcus* and related genera, and has discussed these findings in the context of the evolution of phosphorus metabolism.

Polyphosphate: AMP phosphotransferase

The enzyme which catalyses the reaction

$$(P)_n + AMP \rightleftharpoons (P)_{n-1} + ADP$$

was purified from *Corynebacterium xerosis* (Dirheimer & Ebel, 1965), shown to be free of polyphosphate kinase and adenylate kinase, and to be specific for polyphosphate and AMP. Subsequently the enzyme was found in *Acinetobacter* sp. (van Groenestijn, Deinema & Zehnder, 1987) and is now believed to function in ATP generation from polyphosphate in these organisms via the action of adenylate kinase (2ADP ⇌ ATP + AMP). Thus van Groenestijn & Deinema (1987) envisage the energy-requiring reactions of the cell being coupled to polyphosphate degradation in the following way:

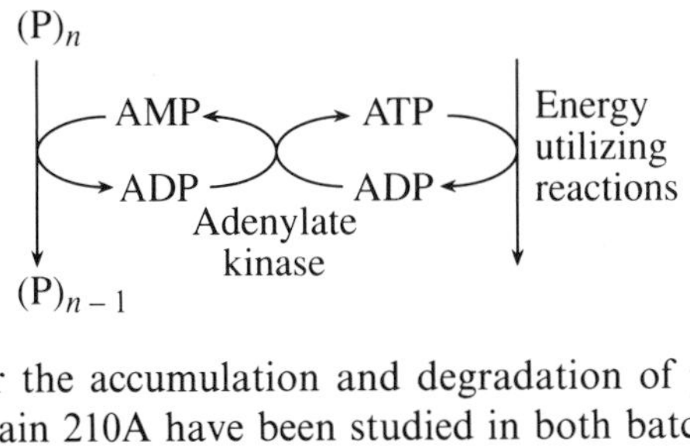

Conditions for the accumulation and degradation of polyphosphate by *Acinetobacter* strain 210A have been studied in both batch and continuous culture (van Groenestijn *et al.*, 1989); the highest phosphorus content was found in organisms grown under sulphur limitation at a growth rate of

$0.04\,h^{-1}$. Magnesium and potassium were taken up concomitantly with orthophosphate, and potassium was essential for significant phosphate uptake.

Phosphate removal from waste water by Acinetobacter *spp.*

The importance of *Acinetobacter* spp. which are usually the dominant micro-organisms in sewage treatment plants employing the active sludge process, has been recognized for some time (Fuhs & Chen; 1975; Nicholls & Osborn, 1979). Some have the ability to accumulate polyphosphate up to about 24% of their biomass under appropriate conditions, namely subjection to alternating periods of anaerobiosis and aerobiosis such as occur in the anaerobic and aerobic zones of waste water treatment plants. In the former zone, the sludge releases inorganic phosphate and in the latter the *Acinetobacters* accumulate polyphosphate. This behaviour correlates with their status as obligate aerobes; under anaerobic conditions they maintain their ATP levels at the expense of stored polyphosphate whereas aerobically they secure energy from extracellular substrates, take up phosphate, and store it as a polyphosphate. A good correlation exists between the measured activities of polyphosphate: AMP phosphotransferase and adenylate kinase in activated sludge and the effectiveness of the phosphorus removal process (Zehnder & van Groenestijn, 1990).

These properties of *Acinetobacter* spp. are now constructively exploited in waste water treatment for removal of the troublesome concentrations of phosphate in run-off waters, resulting from use of fertilizers and detergents (tripolyphosphates), which cause eutrophication of lakes. One such recent technique, the Renpho process, which reflects closely the findings made with *Acinetobacter* strain 210A, is described by Zehnder & van Groenestijn (1990). However, these authors also emphasize that our knowledge of the central role of polyphosphate in *Acinetobacter* spp. and the regulation of its formation and degradation in biological phosphate removal are not yet entirely elucidated.

Polyphosphatases

There are two types of polyphosphatase in micro-organisms that effect degradation of polyphosphates. In prokaryotes, only exopolyphosphatases, catalysing the release of inorganic phosphate from the ends of the polymer chains,

$$(P)_n + H_2O \rightarrow (P)_{n-1} + P_i$$

are important; the endo enzymes, which cleave within the polymer chain (depolymerases or polyphosphorylases), are found almost exclusively in eukaryotes. The high energy function associated with the anhydride linkage is dissipated in this hydrolytic cleavage. Nonetheless the reaction apparently has a physiologically important role in *Enterobacter aerogenes* because a mutant that accumulated polyphosphate in the normal way but was unable to degrade it, lacked the enzyme (Harold & Harold, 1965).

An exopolyphosphatase purified some 100-fold from *Corynebacterium xerosis* was shown to be active only with long-chain polyphosphates, to be activated by EDTA and inhibited by Mg^{2+} and all bivalent metal ions tested (Muhammed, Rodgers & Hughes, 1959). Similarly, the exopolyphosphatase of *E. coli* does not require bivalent metal ions (Dassa *et al.*, 1982), in contrast to other polyphosphatases.

Periplasmic polyphosphate

It has been shown that two pools of polyphosphate exist in *Acinetobacter lwoffi*, one in the cytoplasm and the other in the periplasm (Suresh *et al.*, 1985). The latter is metabolically more active and represents about 1 to 2% of the total cell polyphosphate. It is preferentially degraded with release of inorganic phosphate when anaerobiosis or inhibitors of energy transduction are imposed, or if the cells are treated with Cd^{2+}, which is actively transported by a proton symport system (Suresh *et al.*, 1986). The implication is that periplasmic polyphosphate can serve as an alternative energy source for nutrient uptake when ATP is limiting or not available (Suresh *et al.*, 1985).

There is now good evidence from work with alkaline phosphatase regulatory mutants of *E. coli* that polyphosphatase is controlled by the same regulatory genes as alkaline phosphatase and hence is part of the *pho* regulon which specifies the various proteins involved in phosphate metabolism (Rao & Torriani, 1988). Although evidence for the existence of the *pho* regulon in *A. lwoffi* is not yet complete, regulation of alkaline phosphomonoesterase, alkaline phosphodiesterase, a phosphate-binding protein, polyphosphate kinase, exopolyphosphatase and polyphosphate concentration are all effected by inorganic phosphate (Halvorson, 1990).

Physiological roles of polyphosphate

As previously discussed, in some bacteria, polyphosphate can function as an energy source in reactions in which it effectively replaces ATP as a phosphorylating agent. A more widespread and probably major role is that of a polymeric phosphate reserve which permits the maintenance of the intracellular phosphate concentration as well as serving to minimize osmotic fluctuations that would otherwise occur; this function is consistent with the coordination of polyphosphate regulation with those of phosphate transport and alkaline phosphatase. The importance of inorganic phosphate concentration for the control of various enzyme activities necessitates its regulation within acceptable limits and polyphosphate can assist in this homoeostasis. Further, because the phosphate content of many natural environments is low due to the insolubility of calcium phosphate, the existence of an intracellular phosphorus reserve and the derepression of the enzymes concerned with its metabolism during phosphate starvation, seem eminently reasonable (Harold, 1966). The proposed role of polyphosphate as a chelator of metals, arising from its established association with metal ions, is currently a focus of attention to determine whether the polymer might function in the control of intracellular metal ion concentrations and/or as a detoxication agent. The location of some helical polyphosphate in bacterial membranes, where it appears to be linked to a polyhydroxybutyrate helix by calcium ions, and its possible involvement in transmembrane channels, is discussed on p. 100.

However, the role of polyphosphate in bacterial survival under starvation conditions remains enigmatic and there is still no conclusive evidence that possession of the polymer confers an advantage. A fairly recent proposal that polyphosphates may serve as 'alarmones' derives from observations made with *E. coli* and *Salmonella typhimurium*. These enteric bacteria synthesize heat-shock proteins in response to an appropriate temperature shift, under which conditions adenylylated nucleotides (AppppA and related compounds) accumulate in the cell (Lee, Bochner & Ames, 1983). It has been suggested that these nucleotides may be 'alarmones', i.e. regulatory molecules that signal the onset of oxidation stress or hyperthermia, and trigger the heat-shock response. Kjeldstad *et al.* (1989) investigated hyperthermia-induced changes in the polyphosphate content of *Propionibacterium acnes* by [^{31}P]-NMR, a technique which permits measurement of short-chain polyphosphates but not granule-bound long-chain material. The concentration of short-chain polyphosphates, presumed to be derived from long-chain polymers, increased in hyperthermia and it was suggested that an alarmone function might be involved. The chemical nature of the short-chain polyphosphates detected was not reported and this interesting role proposed for polyphosphate must await further scrutiny.

SULPHUR

Elemental sulphur (S^o) is accumulated intracellularly by *Chromatiaceae* (purple sulphur bacteria) in the presence of sulphide or other reduced forms of sulphur during either phototrophic or chemotrophic growth. The green sulphur bacteria (*Chlorobiaceae*) deposit sulphur extracellularly (Trüper, 1989). Environmental conditions determine the quantity of sulphur accumulated: excess of electron donor usually results in sulphur accumulation whereas under starvation conditions S^o is oxidized to SO_4^{2-}. Glycogen is generally accumulated, too, by these bacteria under sulphur-depositing conditions, but phosphate limitation stimulates only glycogen synthesis (van Gemerden, Visscher & Mas, 1990).

Earlier work (reviewed by Shively (1974)) had reported that sulphur globules vary in diameter from about 100 nm to 1.0 μm. They possess a monolayer membrane consisting of globular subunits some 2.5 nm in diameter and composed entirely of protein with a molecular weight of 13.5 kD; it was suggested that the membrane provides binding sites for the enzymes involved in sulphur metabolism (Schmidt, Nicholson & Kamen, 1971). Steudel (1989) has proposed a vesicle model for S^o globules and, on the basis of experimental data, argues that S^o most likely consists of long-chain polythionates (sulphane bis-sulphonates, $^{-}O_3S{-}S_n{-}SO_3^{-}$) or 'hydrophilic sulphur'. Previously, Guerrero, Mas & Pedrós-Alió (1984) had determined the buoyant density of S^o produced by *Chromatium* as 1.22 gcm^{-3} and described the globules as 'hydrated sulphur'.

Van Gemerden *et al.* (1990), employing chemostat studies, found that sulphur deposition in the purple and green sulphur bacteria is associated with high growth rates and light-saturating conditions. Sulphur accumulation during light and phosphate limitation was comparable whereas glycogen deposition was enhanced by the latter constraint. It was concluded that, under conditions of energy excess, glycogen is the preferred storage compound because it can be used both as a fixed carbon source during light limitation and as an electron donor during depletion of reduced sulphur compounds.

CYANOPHYCIN GRANULE PEPTIDE

Bacteria do not generally possess identifiable reserves of nitrogen but cyanobacteria accumulate a branched polypeptide, termed cyanophycin granule peptide (CGP), which can be drawn upon under conditions of nitrogen limitation. However, Gupta & Carr (1981) envisage a wider role, serving as a reservoir between nitrogen fixation and the export of glutamine from heterocysts into vegetative cells, thus separating the constant requirement for fixed nitrogen for biosynthesis from the possibly non-constant rates of nitrogen fixation in the heterocysts.

CGP contains arginine and aspartic acid residues in the ratio 1:1, displaying a molecular mass in the range 25 to 125 kD (Simon, 1971). CGP from *Anabaena cylindrica* consists of a polyaspartic acid backbone to which arginyl residues are attached via their α-amino groups to each carboxyl group of the polyaspartate backbone and it is proposed that aspartate occupies both amino and carboxyl termini (Simon & Weathers, 1976). The auxiliary photosynthetic pigment phycocyanin has also been implicated as a nitrogen reserve (Simon, 1973; Foulds & Carr, 1977). The polymer is synthesized by a chloramphenicol-insensitive, ribosome and a t-RNA-independent enzyme system (Simon, 1976).

REFERENCES

Allen, M. M. (1984). Cyanobacterial cell inclusions. *Annual Review of Microbiology*, **38**, 1–25.

Anderson, A. J. & Dawes, E. A. (1990). Occurrence, metabolism, metabolic role and industrial uses of bacterial polyhydroxyalkanoates. *Microbiological Reviews*, **54**, 450–72.

Antoine, A. D. & Tepper, B. S. (1969*a*). Characterization of glycogens from *Mycobacteria. Archives of Biochemistry and Biophysics*, **134**, 207–13.

Antoine, A. D. & Tepper, B. S. (1969*b*). Environmental control of glycogen and lipid content of *Mycobacterium phlei. Journal of General Microbiology*, **55**, 217–26.

Baecker, P. A., Greenberg, E. & Preiss, J. (1986). Biosynthesis of bacterial glycogen: Primary structure of *Escherichia coli* α-1,4 glucan: α-1,4 glucan 6-glycosyl transferase as deduced from the nucleotide sequence of the *glg*B gene. *Journal of Biological Chemistry*, **261**, 8738–43.

Ballard, D. G. H., Holmes, P. A. & Senior, P. J. (1987). Formation of polymers of β-hydroxybutyric acid in bacterial cells and a comparison of the morphology of growth with the formation of polyethylene in the solid state. In *Recent Advances in Mechanistic and Synthetic Aspects of Polymerization*, ed. M. Fontanille & A. Guyot, vol. 215, pp. 293–314. Reidel (Kluwer) Publishing Co, Lancaster, UK.

Barham, P. J., Bennett, P., Fawcett, T., Hill, M. J., Stejny, J., & Webb, J. (1989). The structure of native PHB granules. Abstract (not paginated) Biological and Engineered Polymers Conference, Cambridge, UK.

Barnard, G. N. & Sanders, J. K. M. (1989). The poly-β-hydroxybutyrate granule *in vivo*. A new insight based on NMR spectroscopy of whole cells. *Journal of Biological Chemistry*, **264**, 3286–91.

Baxter, M. & Jensen, T. E. (1980*a*). A study of methods for *in situ* X-ray energy dispersive analysis of polyphosphate bodies in *Plectonema boryanum*. *Archives of Microbiology*, **126**, 213–15.

Baxter, M. & Jensen, T. E. (1980*b*). Uptake of magnesium, strontium, barium and manganese by *Plectonema boryanum* (*Cyanophyceae*) with special reference to polyphosphate bodies. *Protoplasma*, **104**, 81–9.

Boyer, C. & Preiss, J. (1977). Biosynthesis of bacterial glycogen: purification and properties of the *Escherichia coli* B α-1,4-glucan 6-glycosyl transferase. *Biochemistry*, **16**, 3693–9.

Brandl, H., Gross, R. A., Lenz, R. W. & Fuller, R. C. (1988). *Pseudomonas oleovorans* as a source of poly(β-hydroxyalkanoates) for potential applications as biodegradable polyesters. *Applied and Environmental Microbiology*, **54**, 1977–82.

Brandl, H., Knee, Jr. E. J., Fuller, R. C., Gross, R. A. & Lenz, R. W. (1989). Ability of the phototrophic bacterium *Rhodospirillum rubrum* to produce various poly(β-hydroxyalkanoates): potential sources for biodegradable polyesters. *International Journal of Biological Macromolecules*, **11**, 49–55.

Chen, G. S. & Segal, I. H. (1968*a*). *Escherichia coli* polyglucose phosphorylases. *Archives of Biochemistry and Biophysics*, **127**, 164–74.

Chen, G. S. & Segel, I. H. (1968*b*). Purification and properties of glycogen phosphorylase from *Escherichia coli. Archives of Biochemistry and Biophysics*, **127**, 175–86.

Clark, J. E. (1990). Purification of polyphosphate glucokinase from *Propionibacterium shermanii*. In *Novel Biodegradable Microbial Polymers*, ed. E. A. Dawes, pp. 213–21. Kluwer Academic Publishers, Dordrecht.

Clark, J. E., Beegen, H. & Wood, H. G. (1986). Isolation of intact chains of polyphosphate from *Propionibacterium shermanii* grown on glucose or lactate. *Journal of Biological Chemistry*, **168**, 1212–19.

Coleman, J. R., Nilsson, J. R., Warner, R. R. & Batt, P. (1972). Qualitative and quantitative electron probe analysis of cytoplasmic granules in *Tetrahymena pyriformis. Experimental Cell Research*, **74**, 207–19.

Cornibert, J. & Marchessault, R. H. (1972). Physical properties of poly-β-hydroxybutyrate. IV. Conformational analysis and crystalline structure. *Journal of Molecular Biology*, **71**, 735–56.

Creuzat-Sigal, N. & Frixon, C. (1977). Catabolism of glycogen in *Escherichia coli*. *FEMS Microbiology Letters*, **1**, 235–8.

Darvill, R. G., Hall, M. A., Fish, J. P. & Morris, J. G. (1977). The intracellular reserve polysaccharide of *Clostridium pasteurianum*. *Canadian Journal of Microbiology*, **23**, 947–53.

Dassa, E., Cahu, M., Desjoyaux-Cherel, B. & Boquet, P. (1982). The acid phosphatase with optimum pH of 2.5 of *Escherichia coli*. Physiological and biochemical study. *Journal of Biological Chemistry*, **257**, 6669–76.

Davis, J. T., Moore, R. N., Imperali, B., Pratt, A., Kobayashi, K., Masamune, S., Sinskey, A. J. & Walsh, C. T. (1987). Biosynthetic thiolase from *Zoogloea ramigera*. I. Preliminary characterization and analysis of proton transfer reaction. *Journal of Biological Chemistry*, **262**, 82–9.

Dawes, E. A. (1976). Endogenous metabolism and the survival of starved prokaryotes. In *The Survival of Vegetative Microbes*, ed. T. R. G. Gray & J. R. Postgate, pp. 19–53. Cambridge University Press, Cambridge.

Dawes, E. A. (1985). Starvation, survival and energy reserves. In *Bacteria in Their Natural Environments*, ed. M. Fletcher & G. D. Floodgate, pp. 43–79. Academic Press, London.

Dawes, E. A. (1989). Growth and survival of bacteria. In *Bacteria in Nature*, vol. 3, ed. J. S. Poindexter & E. R. Leadbetter, pp. 67–187. Plenum Press, New York.

Dawes, E. A. (1990) ed. *Novel Biodegradable Microbial Polymers*. Kluwer Academic Publishers, Dordrecht.

Dawes, E. A. & Senior, P. J. (1973). The role and regulation of energy reserve polymers in micro-organisms. *Advances in Microbial Physiology*, **10**, 135–266.

De Smet, M. J., Eggink, G., Witholt, B., Kingma, J. & Wynberg, H. (1983). Characterization of intracellular inclusions formed by *Pseudomonas oleovorans* during growth on octane. *Journal of Bacteriology*, **154**, 870–8.

Dietzler, D. N., Leckie, M. P., Sternheim, W. L., Taxman, T. L., Unger, J. M. & Porter, S. E. (1977). Evidence for the regulation of bacterial glycogen synthesis by cyclic AMP. *Biochemical and Biophysical Research Communications*, **77**, 1468–77.

Dirheimer, G. & Ebel, J. P. (1965). Caracterisation d'une polyphosphate-AMP-phosphotransferase dans *Corynebacterium xerosis*. *Compte rendu des séances de la Société de biologie (Paris)* **260**, 3787–90.

Doi, Y. (1990). *Microbial Polyesters*. VCH Publishers, Inc, New York.

Doi, Y., Segawa, A., Kawaguchi, Y. & Kunioka, M. (1990). Cyclic nature of poly(3-hydroxyalkanoate) metabolism in *Alcaligenes eutrophus*. *FEMS Microbiology Letters*, **67**, 165–70.

Doi, Y., Tamaki, A., Kunioka, M. & Soga, K. (1987). Biosynthesis of an unusual polyester (10 mol% 3-hydroxybutyrate and 90 mol% 3-hydroxyvalerate units) in *Alcaligenes eutrophus* from pentanoic acid. *Journal of the Chemical Society, Chemical Communications*, 1635–6.

Doi, Y., Tamaki, A., Kunioka, M. & Soga, K. (1988). Production of copolyesters of 3-hydroxybutyrate by *Alcaligenes eutrophus* from butyric and pentanoic acids. *Applied Microbiology and Biotechnology*, **28**, 330–4.

Eidels, L. & Preiss, J. (1970). Carbohydrate metabolism in *Rhodopseudomonas capulata*: enzyme titers, glucose metabolism and polyglucose polymer synthesis. *Archives of Biochemistry and Biophysics*, **140**, 75–89.

Ferguson, S. J., Gadian, D. G. & Kell, D. B. (1979). Evidence from ^{31}P nuclear magnetic resonance that polyphosphate synthesis is a slip reaction in *Paracoccus denitrificans*. *Biochemical Society Transactions*, **7**, 176–9.

Findlay, R. H. & White, D. C. (1983). Polymeric beta-hydroxyalkanoates from environmental samples and *Bacillus megaterium*. *Applied and Environmental Microbiology*, **45**, 71–8.

Findlay, R. H. & White, D. C. (1984). *In situ* determination of metabolic activity in aquatic environments. *Microbiological Science*, **1**, 90–5.

Findlay, R. H. & White, D. C. (1987). A simplified method for bacterial nutritional status based on the simultaneous determination of phospholipid and endogenous storage lipid poly beta-hydroxyalkanoate. *Journal of Microbiological Methods*, **6**, 113–20.

Foulds, I. J. & Carr, N. G. (1977). A proteolytic enzyme degrading phycocyanin in the cyanobacterium *Anabaena cylindrica*. *FEMS Microbiology Letters*, **2**, 117–9.

Fox, J., Kawaguchi, K., Greenberg, E. & Preiss, J. (1976). Biosynthesis of bacterial glycogen. Purification and properties of the *Escherichia coli* B ADPglucose: 1,4-α-D-glucan 4-α-glucosyltransferase. *Biochemistry*, **15**, 849–57.

Fuhs, G. W. & Chen, M. (1975). Microbiological basis of phosphate removal in the activated sludge process for the treatment of waste water. *Microbial Ecology*, **2**, 119–38.

Fukui, T., Yoshimoto, A., Matsumoto, M., Hosokawa, S., Saito, T., Nishikawa, H. & Tomita, K. (1976). Enzymatic synthesis of poly-β-hydroxybutyrate in *Zoogloea ramigera*. *Archives of Microbiology*, **110**, 149–56.

Gavard, R., Dahinger, A., Hauttecoeur, B. & Reynaud, C. (1967). Degradation of β-hydroxybutyric acid lipid by an enzyme extract of *Bacillus megaterium*. *Compte rendu, Academie des Sciences, Paris*, **265**, 1557–9.

Griebel, R. J. & Merrick, J. M. (1971). Metabolism of poly-β-hydroxybutyrate: effect of mild alkaline extraction on native poly-β-hydroxybutyrate granules. *Journal of Bacteriology*, **108**, 782–9.

Griebel, R., Smith, Z. & Merrick, J. M. (1968). Metabolism of poly-β-hydroxybutyrate. I. Purification, composition and properties of native poly-β-hydroxybutyrate granules from *Bacillus megaterium*. *Biochemistry*, **7**, 3676–81.

Gross, R. A., De Mello, C., Lenz, R. W., Brandl, H. & Fuller, R. C. (1989). Biosynthesis and characterization of poly(β-hydroxyalkanoates) produced by *Pseudomonas oleovorans*. *Macromolecules*, **22**, 1106–15.

Guerrero, R., Mas, J. & Pedrós-Alió, C. (1984). Buoyant density changes due to intracellular content of sulfur in *Chromatium warmingii* and *Chromatium vinosum*. *Archives of Microbiology*, **137**, 350–6.

Gupta, M. & Carr, N. G. (1981). Enzyme activities related to cyanophycin metabolism in heterocysts and vegetative cells of *Anabaena* spp. *Journal of General Microbiology*, **125**, 17–23.

Halvorson, H. (1990). Some possible roles of polyphosphate in microorganisms. In *Novel Biodegradable Microbial Polymers*, ed. E. A. Dawes, pp. 205–11. Kluwer Academic Publishers, Dordrecht.

Halvorson, H. O., Suresh, N., Roberts, M. F., Coccia, M. & Chikarmane, H. M. (1987). Metabolically active surface polyphosphate pool in *Acinetobacter lwoffi*. In *Phosphate Metabolism and Cellular Regulation in Microorganisms*, ed. Torriani-Gorini, A., Rothman, F. G., Silver, S., Wright, A. & Yagil, E., pp. 220–4. American Society for Microbiology Publishers, Washington.

Harold, F. M. (1964). Enzymic and genetic control of polyphosphate accumulation in *Aerobacter aerogenes*. *Journal of General Microbiology*, **35**, 81–90.

Harold, F. M. (1966). Inorganic polyphosphates in biology: structure, metabolism and functions. *Bacteriological Reviews*, **30**, 772–94.

Harold, R. L. & Harold, F. M. (1963). Mutants of *Aerobacter aerogenes* blocked in the accumulation of inorganic polyphosphate. *Journal of General Microbiology*, **31**, 241–6.

Harold, F. M. & Harold, R. L. (1965). Degradation of inorganic polyphosphates in mutants of *Aerobacter aerogenes*. *Journal of Bacteriology*, **89**, 1262–70.

Haywood, G. W., Anderson, A. J., Chu, L. & Dawes, E. A. (1988a). The role of NADH- and NADP-linked acetoacetyl-CoA reductases in the poly-3-hydroxyalkanoate-synthesizing organism *Alcaligenes eutrophus*. *FEMS Microbiology Letters*, **52**, 259–64.

Haywood, G. W., Anderson, A. J., Chu, L. & Dawes, E. A. (1988b). Accumulation of polyhydroxyalkanoates by bacteria and substrate specificities of the biosynthetic enzymes. *Biochemical Society Transactions*, **16**, 1046–7.

Haywood, G. W., Anderson, A. J. & Dawes, E. A. (1989a). The importance of PHB-synthase substrate specificity in polyhydroxyalkanoate synthesis by *Alcaligenes eutrophus*. *FEMS Microbiology Letters*, **57**, 1–6.

Haywood, G. W., Anderson, A. J. & Dawes, E. A. (1989b). A survey of the accumulation of novel polyhydroxyalkanoates by bacteria. *Biotechnology Letters*, **11**, 471–6.

Haywood, G. W., Anderson, A. J., Ewing, D. F. & Dawes, E. A. (1990). Accumulation of a polyhydroxyalkanoate containing primarily 3-hydroxydecanoate from simple carbohydrate substrates by *Pseudomonas* sp. strain NCIMB 40135. *Applied and Environmental Microbiology*, **56**, 3354–9.

Herron, J. S., King, J. D. & White, D. C. (1978). Recovery of poly-β-hydroxybutyrate from environmental sludge. *Applied and Environmental Microbiology*, **35**, 251–7.

Hippe, H. & Schlegel, H. G. (1967). Hydrolyse von PHBS durch intracellulare depolymerase von Hydrogenomonas H16. *Archiv für Mikrobiologie*, **56**, 278–99.

Holme, T. (1957). Continuous culture studies on glycogen synthesis in *Escherichia coli* B. *Acta Chemica Scandinavica*, **11**, 763–75.

Holme, T., Laurent, T. & Palmstierna, H. (1957). On the glycogen in *Escherichia coli* B; variations in molecular weight during growth. *Acta Chemica Scandinavica*, **11**, 757–62.

Holmes, E., Boyer, C. & Preiss, J. (1982). Immunological characterization of *Escherichia coli* B glycogen synthase from other bacteria and branching enzyme and comparison with enzymes from other bacteria. *Journal of Bacteriology*, **151**, 1444–53.

Holmes, P. A. (1985). Applications of PHB – a microbially produced biodegradable thermoplastic. *Physics in Technology*, **16**, 32–6.

Huisman, G. J., De Leeuw, O., Eggink, G. & Witholt, B. (1989). Synthesis of poly-3-hydroxyalkanoates is a common feature of fluorescent pseudomonads. *Applied and Environmental Microbiology*, **55**, 1949–54.

Jackson, F. A. & Dawes, E. A. (1976). Regulation of the tricarboxylic acid cycle and poly-*β*-hydroxybutyrate metabolism in *Azotobacter beijerinckii* grown under nitrogen or oxygen limitation. *Journal of General Microbiology*, **97**, 303–12.

Janes, B., Hollar, J. & Dennis, D. (1990). Molecular characterization of the poly-*β*-hydroxybutyrate biosynthetic pathway of *Alcaligenes eutrophus* H16. In *Novel Biodegradable Microbial Polymers*, ed. E. A. Dawes, pp. 175–90. Kluwer Academic Publishers, Dordrecht.

Jensen, T. E. & Sicko, L. M. (1974). Phosphate metabolism in blue-green algae. I. Fine structure of the polyphosphate 'overplus' phenomenon in *Plectonema boryanum*. *Canadian Journal of Microbiology*, **20**, 1235–9.

Jones, H. E. & Chambers, L. A. (1975). Localized intracellular polyphosphate formation by *Desulfovibrio gigas*. *Journal of General Microbiology*, **89**, 67–72.

Jost, M. (1965). Die Ultrastruktur von *Oscillatoria rubescens* D.C. *Archiv für Mikrobiologie*, **50**, 211–45.

Kamio, Y., Terawaki, Y., Nakajima, T. & Matsuda, K. (1981). Structure of glycogen produced by *Selenomonas ruminantium*. *Agricultural and Biological Chemistry*, **45**, 209–16.

Karr, D. B., Waters, J. K., Suzuki, F. & Emerich, D. W. (1984). Enzymes of the poly-β-hydroxybutyrate and citric acid cycles of *Rhizobium japonicum* bacteroids. *Plant Physiology*, **75**, 1158–62.

Kawaguchi, Y. & Doi, Y. (1990). Structure of native poly(3-hydroxybutyrate) granules characterized by X-ray diffraction. *FEMS Microbiology Letters*, **79**, 151–6.

Keevil, C. W., March, P. D. & Ellwood, D. C. (1984). Regulation of glucose metabolism in oral streptococci through independent pathways of glucose 6-phosphate and glucose 1-phosphate formation. *Journal of Bacteriology*, **157**, 560–7.

Kjelstad, B., Johnsson, A., Furuheim, K. M., Bergan, A. S. & Krane, J. (1989). Hyperthermia induced polyphosphate changes in *Propionibacterium acnes* as studied by [^{31}P] NMR. *Zeitschrift für Naturforschung*, **44**, 45–48.

Knee, Jr, E. J., Wolf, M., Lenz, R. W. & Fuller, R. C. (1990). Influence of growth conditions on production and composition of PHA by *Pseudomonas oleovorans*. In *Novel Biodegradable Microbial Polymers*, ed. E. A. Dawes, pp. 439–40. Kluwer Academic Publishers, Dordrecht.

König, H., Nusser, E. & Stetter, K. O. (1985). Glycogen in *Methanolobus* and *Methanococcus*. *FEMS Microbiology Letters*, **28**, 265–9.

König, H., Skorko, R., Zillig, W. & Reiter, W. D. (1982). Glycogen in Thermoacidophilic archaebacteria of the genera *Sulfolobus*, *Thermoproteus*, *Desulfurococcus* and *Thermococcus*. *Archives of Microbiology*, **132**, 297–303.

Kornberg, S. R. (1957). Adenosine triphosphate synthesis from polyphosphate by an enzyme from *Escherichia coli*. *Biochimica et Biophysica Acta*, **26**, 294–300.

Kornberg, A., Kornberg, S. R. & Simms, E. S. (1956). Metaphosphate synthesis by an enzyme from *Escherichia coli*. *Biochimica et Biophysica Acta*, **26**, 215–27.

Kulaev, I. S. (1971). Inorganic polyphosphates in evolution of phosphorus metabolism. In *Molecular Evolution*, ed. R. Buvet & C. Ponnamperuma, vol. 1, pp. 458–465. North Holland, Amsterdam.

Kulaev, I. S. (1979). *The Biochemistry of Inorganic Polyphosphates*; Wiley & Sons, Chichester & New York.

Kulaev, I. S., Bobyk, M. A., Nikolaev, N. N., Sergeev, N. S. & Uryson, S. O. (1971). The polyphosphate-synthesizing enzymes of some fungi and bacteria (in Russian). *Biokhimiya*, **36**, 943–9.

Kulaev, I. S., Szymona, O. & Bobyk, M. A. (1968). The biosynthesis of inorganic polyphosphates in *Neurospora crassa* (in Russian). *Biokhimiya*, **33**, 419–34.

Kulaev, I. S. & Vagabov, V. M. (1983). Polyphosphate metabolism in microorganisms. *Advances in Microbial Physiology*, **24**, 83–171.

Lageveen, R. G., Huisman, G. W., Preusting, H., Ketelaar, P., Eggink, G. & Witholt, B. (1988). Formation of polyesters by *Pseudomonas oleovorans*: effect of substrates on formation and composition of poly-(*R*)-3-hydroxyalkanoates and poly-(*R*)–hydroxyalkenoates. *Applied and Environmental Microbiology*, **54**, 2924–32.

Latil-Damotte, M. & Lares, C. (1977). Relative order of *glg* mutations affecting glycogen biosynthesis in *Escherichia coli* K12. *Molecular and General Genetics*, **150**, 325–9.

Leckie, M. P., Porter, S. E., Tieber, V. L. & Dietzler, D. N. (1981). Regulation of the basal and cyclic AMP-stimulated rates of glycogen synthesis in *Escherichia* coli by an intermediate of purine biosynthesis. *Biochemical and Biophysical Research Communications*, **99**, 1433–42.

Lee, P. C., Bochner, B. R. & Ames, B. N. (1983). AppppA, heat-shock stress, and cell oxidation. *Proceedings, National Academy of Sciences, Washington*, **80**, 7496–500.

Lemoigne, M. (1926). Products of dehydration and of polymerization of *β*-hydroxybutyric acid. *Bulletin de la Société de chimie biologique, (Paris)*, **8**, 770–82.

Lipmann, F. (1965). Projecting backward from the present stage of evolution of biosynthesis. In *The Origins of Prebiological Systems*, ed. S. W. Fox, pp. 259–280. Academic Press, New York.

McDermott, T. R., Griffith, S. M., Vance, C. P. & Graham, P. H. (1989). Carbon metabolism in *Bradyrhizobium japonicum* bacteroids. *FEMS Microbiology Reviews*, **63**, 327–40.

Mackey, B. M. & Morris, J. G. (1971). Ultrastructural changes during sporulation of *Clostridium pasteurianum*. *Journal of General Microbiology*, **66**, 1–13.

Mas, J., Pedrós-Alió, C. & Guerrero, R. (1985). Mathematical model for determining the effects of intracytoplasmic inclusions on volume and density of microorganisms. *Journal of Bacteriology*, **164**, 749–56.

Masumane, S., Walsh, C. T., Sinskey, A. J. & Peoples, O. P. (1989). Poly-(*R*)-3-hydroxybutyrate (PHB) biosynthesis: mechanistic studies on the biological Claisen condensation catalysed by *β*-ketoacyl thiolase. *Pure and Applied Chemistry*, **61**, 303–12.

Merrick, J. M., Delafield, F. P. & Doudoroff, M., (1962). Hydrolysis of poly-*β*-hydroxybutyrate by intracellular and extracellular enzymes. Federation Proceedings of the American Societies for Experimental Biology (Washington), **21**, 228.

Merrick, J. M. & Doudoroff, M. (1964). Depolymerization of poly-*β*-hydroxybutyrate by an intracellular enzyme system. *Journal of Bacteriology*, **88**, 60–71.

Merrick, J. M. & Yu, C. I. (1966). Purification and properties of a D(−)-*β*-hydroxybutyric dimer hydrolase from *Rhodospirillum rubrum*. *Biochemistry*, **5**, 3563–8.

Moskowitz, G. J. & Merrick, J. M. (1969). Metabolism of poly-*β*-hydroxybutyrate. II. Enzymatic synthesis of D(−)-*β*-hydroxybutyryl coenzyme A by an enoyl hydrase from *Rhodospirillum rubrum*. *Biochemistry*, **8**, 2748–55.

Muhammed, A., Rodgers, A. & Hughes, D. E. (1959). Purification and properties of a polymetaphosphatase from *Corynebacterium xerosis*. *Journal of General Microbiology*, **20**, 482–95.

Nicholls, A. & Osborne, D. W. (1979). Bacterial stress: prerequisite for biological removal of phosphorus. *Journal of Water Pollution Control Federation*, **51**, 557–69.

Nickels, J. S., King, J. D. & White, D. C. (1979). Poly-*β*-hydroxybutyrate accumulation as a measure of unbalanced growth of the estuarine detrital microbiota. *Applied and Environmental Microbiology*, **37**, 459–65.

Odham, G., Tunlid, A., Westerdahl, G. & Marden, P. (1986). Combined determination of poly-*β*-hydroxyalkanoic and cellular fatty acids in starved marine bacteria and sewage sludge by gas chromatography with flame ionization or mass spectrometry detection. *Applied and Environmental Microbiology*, **52**, 905–10.

Oeding, V. & Schlegel, H, G. (1973). *β*-Ketothiolase from *Hydrogenomonas eutropha* H16 and its significance in the regulation of poly-*β*-hydroxybutyrate metabolism. *Biochemical Journal*, **134**, 239–48.

Okita, T. W., Rodriguez, R. & Preiss, J. (1982). Isolation of *Escherichia coli* structural genes coding for the glycogen biosynthetic enzymes. *Methods in Enzymology*, **83**, 549–53.

Ostrovsky, D. N., Sepetov, N. F., Reshetnyak, V. I. & Sibel Dina, L. A. (1980).

Study of the localization of polyphosphates in cells of micro-organisms by high-resolution phosphorus-31 NMR at 145.78 MHz. *Biokhimiya*, **45**, 517–25 (in Russian).

Page, W. J. & Knosp, O. (1989). Hyperproduction of poly-β-hydroxybutyrate during exponential growth of *Azotobacter vinelandii* UWD. *Applied and Environmental Microbiology*, **55**, 1334–9.

Palmer, T. N., Wober, G. & Whelan, W. J. (1973). The pathway of exogenous and endogenous carbohydrate utilization in *Escherichia coli*: a dual function for the enzymes of the maltose operon. *European Journal of Biochemistry*, **39**, 601–12.

Peoples, O. P. & Sinskey, A. J. (1989a). Poly-β-hydroxybutyrate synthesis in *Alcaligenes eutrophus* H16. Characterization of the genes encoding β-ketothiolase and acetoacetyl-CoA reductase. *Journal of Biological Chemistry*, **264**, 15293–7.

Peoples, O. P. & Sinskey, A. J. (1989b). Poly-β-hydroxybutyrate (PHB) synthesis in *Alcaligenes eutrophus* H16. Identification and characterization of the PHB polymerase gene (phbC). *Journal of Biological Chemistry*, **264**, 15298–303.

Peoples, O. P. & Sinskey, A. J. (1990). Polyhydroxybutyrate (PHB): a model system for biopolymer engineering: II. In *Novel Biodegradable Microbial Polymers*, ed. E. A. Dawes, pp. 191–202. Kluwer Academic Publishers, Dordrecht.

Pepin, C. A. & Wood, H. G. (1986). The mechanism of utilization of polyphosphate by polyphosphate glucokinase from *Propionibacterium shermanii*. *Journal of Biological Chemistry*, **261**, 4476–80.

Pepin, C. A. & Wood, H. G. (1987). The mechanism of utilization of polyphosphate by polyphosphate glucokinase from *Propionibacterium shermanii*. *Journal of Biological Chemistry*, **262**, 5223–6.

Pepin, C. A., Wood, H. G. & Robinson, N. A. (1986). Determination of the size of polyphosphates with polyphosphate glucokinase. *Biochemistry International*, **12**, 111–23.

Pool, R. (1989). In search of the plastic potato. *Science*, **245**, 1187–9.

Preiss, J. (1984). Bacterial glycogen synthesis and its regulation. *Annual Review of Microbiology*, **38**, 419–58.

Preiss, J. (1988). Biosynthesis of starch and its regulation. In *The Biochemistry of Plants, Carbohydrates*, vol. XIV, ed. J. Preiss, pp. 181–254. Academic Press, New York.

Preiss, J. (1989). Chemistry and metabolism of intracellular reserves. In *Bacteria in Nature*, vol. 3, ed. J. S. Poindexter & E. R. Leadbetter, pp. 189–258. Plenum Press, New York.

Preiss, J. & Greenberg, E. (1983). Pyrophosphate may be involved in regulation of bacterial glycogen synthesis. *Biochemical and Biophysical Research Communications*, **115**, 820–6.

Preiss, J. & Romeo, T. (1989). Physiology, biochemistry and genetics of bacterial glycogen synthesis. *Advances in Microbial Physiology*, **30**, 183–238.

Preiss, J. & Walsh, D. A. (1981). The comparative biochemistry of glycogen and starch. In *Biology of Carbohydrates*, vol. 1, ed. V. Ginsburg & P. Robbins, pp. 199–314. Wiley, New York.

Pulkownik, A. & Walker, G. J. (1976). Metabolism of reserve polysaccharide of *Streptococcus mitior* (*mitis*): is there a second α-1,4-glucan phosphorylase? *Journal of Bacteriology*, **127**, 281–90.

Rao, N. N. & Torriani, A. (1988). Utilization by *Escherichia coli* of a high molecular weight linear polyphosphate: roles of phosphatases and pore proteins. *Journal of Bacteriology*, **170**, 5216–23.

Reusch, R. N. (1989). Poly-β-hydroxybutyrate/calcium polyphosphate complexes in

eukaryotic membranes. *Proceedings of the Society for Experimental Biology and Medicine*, **191**, 377–81.
Reusch, R. N., Hiske, T. W. & Sadoff, H. L. (1986). Poly-β-hydroxybutyrate membrane structure and its relationship to genetic transformability in *Escherichia coli*. *Journal of Bacteriology*, **168**, 553–62.
Reusch, R. N., Hiske, T. W. & Sadoff, H. L., Harris, R. & Beveridge, T. (1987). Cellular incorporation of poly-β-hydroxybutyrate into plasma membranes of *Escherichia coli* and *Azotobacter vinelandii* alters native membrane structure. *Canadian Journal of Microbiology*, **33**, 435–44.
Reusch, R. N. & Sadoff, H. L. (1983). D(−)-Poly-β-hydroxybutyrate in membranes of genetically competent bacteria. *Journal of Bacteriology*, **156**, 778–88.
Reusch, R. N. & Sadoff, H. L. (1988). Putative structure and functions of a poly-β-hydroxybutyrate/calcium polyphosphate channel in bacterial plasma membranes. *Proceedings of the National Academy of Sciences, USA*, **85**, 4176–80.
Robinson, N. A., Clark, J. E. & Wood, H. G. (1987). Polyphosphate kinase from *Propionibacterium shermanii*. Demonstration that polyphosphates are primers, and determination of the size of the synthesized polyphosphate. *Journal of Biological Chemistry*, **262**, 5216–22.
Robinson, N. A. & Wood, H. G. (1986). Polyphosphate kinase from *Propionibacterium shermanii*. *Journal of Biological Chemistry*, **261**, 4481–5.
Robson, R. L., Robson, R. M. & Morris, J. G. (1972). Regulation of granulose synthesis in *Clostridium pasteurianum*. *Biochemical Journal*, **130**, 4P–5P.
Romeo, T., Kumar, A. & Preiss, J. (1988). Analysis of the *Escherichia coli* glycogen gene cluster suggests that catabolic enzymes are encoded amongst the biosynthetic genes. *Gene*, **70**, 363–76.
Romeo, T. & Preiss, J. (1989). Genetic regulation of glycogen biosynthesis in *Escherichia coli*: in vitro effects of cyclic AMP and guanosine 5′-diphosphate 3′-diphosphate and analysis of in vivo transcripts. *Journal of Bacteriology*, **171**, 2773–82.
Sarma, T. A. & Kanta, S. (1979). Biochemical studies on sporulation in blue-green algae. I. Glycogen accumulation. *Zeutschrift für Allgemeine Mikrobiologie*, **19**, 571–5.
Schmidt, G. L., Nicolson, G. L. & Kamen, M. D. (1971). Composition of the sulfur particle of *Chromatium vinosum* strain D. *Journal of Bacteriology*, **105**, 1137–41.
Schubert, P., Kruger, N. & Steinbüchel, A. (1991). Molecular analysis of the *Alcaligenes eutrophus* poly(3-hydroxybutyrate) biosynthetic operon: identification of the N terminus of poly(3-hydroxybutyrate) synthase and identification of the promoter. *Journal of Bacteriology*, **173**, 168–75.
Schubert, P., Steinbüchel, A. & Schlegel, H. G. (1988). Cloning of the *Alcaligenes eutrophus* genes for synthesis of poly-β-hydroxybutyric acid (PHB) and synthesis of PHB in *Escherichia coli*. *Journal of Bacteriology*, **170**, 5837–47.
Senior, P. J., Beech, G. A., Ritchie, G. A. F. & Dawes, E. A. (1972). The role of oxygen limitation in the formation of poly-β-hydroxybutyrate during batch and continuous culture of *Azotobacter beijerinckii*. *Biochemical Journal*, **128**, 1193–201.
Senior, P. J. & Dawes, E. A. (1971). Poly-β-hydroxybutyrate biosynthesis and the regulation of glucose metabolism in *Azotobacter beijerinckii*. *Biochemical Journal*, **125**, 55–66.
Senior, P. J. & Dawes, E. A. (1973). The regulation of poly-β-hydroxybutyrate synthesis in *Azotobacter beijerinckii*. *Biochemical Journal*, **134**, 225–38.
Shen, L. C. & Atkinson, D. E. (1970). Regulation of adenosine diphosphate glucose synthase from *Escherichia coli*. *Journal of Biological Chemistry*, **245**, 3996–4000.

Shen, L. & Preiss, J. (1964). The activation and inhibition of bacterial adenosine-diphosphoglucose pyrophosphorylase. *Biochemical and Biophysical Research Communications*, **17**, 424–9.

Shively, J. M. (1974). Inclusion bodies of prokaryotes. *Annual Review of Microbiology*, **28**, 167–87.

Simon, R. D. (1971). Cyanophycin granules from the blue-green alga *Anabaena cylindrica*: a reserve material consisting of copolymers of aspartic acid and arginine. *Proceedings of the National Academy of Sciences, USA*, **68**, 265–7.

Simon, R. D. (1973). Measurement of the cyanophycin granule peptide contained in the blue-green alga *Anabaena cylindrica*. *Journal of Bacteriology*, **114**, 1213–16.

Simon, R. D. (1976). The biosynthesis of multi-L-arginyl poly(L-aspartic acid) in the filamentous cyanobacterium *Anabaena cylindrica*. *Biochimica et Biophysica Acta*, **422**, 407–18.

Simon, R. D. & Weathers, P. (1976). Determination of the structure of the novel polypeptide containing aspartic acid and arginine which is found in cyanobacteria. *Biochimica et Biophysica Acta*, **420**, 165–76.

Skórko, R., Osipiuk, J. & Stetter, K. O. (1989). Glycogen-bound polyphosphate kinase from the archaebacterium *Sulfolobus acidocaldarius*. *Journal of Bacteriology*, **171**, 5162–4.

Slater, S. C., Voige, W. H. & Dennis, D. E. (1988). Cloning and expression in *Escherichia coli* of the *Alcaligenes eutrophus* H16 poly-β-hydroxybutyrate biosynthetic pathway. *Journal of Bacteriology*, **170**, 4431–6.

Stam, H., Van Verseveld, H. W., De Vries, W. & Stouthamer, A. H. (1986). Utilization of poly-β-hydroxybutyrate in free-living cultures of *Rhizobium* ORS571. *FEMS Microbiology Letters*, **35**, 215–20.

Steinbüchel, A. (1989). Poly(hydroxyalkanoates) – storage materials of bacteria: biosynthesis and genetics. *Forum Mikrobiologie*, **12**, 190–8.

Steinbüchel, A. & Schlegel, H. G. (1991). Physiology and molecular genetics of poly(β-hydroxyalkanoic acid) synthesis in *Alcaligenes eutrophus*. *Molecular Microbiology* **5**, 535–42.

Steinbüchel, A. & Schubert, P. (1989). Expression of the *Alcaligenes eutrophus* poly(β-hydroxybutyric acid) – synthetic pathway in *Pseudomonas* sp. *Archives of Microbiology*, **153**, 101–4.

Steinbüchel, A., Schubert, P., Timm, A. & Pries, A. (1990). Genetic and molecular analysis of the *Alcaligenes eutrophus* polyhydroxybutyrate-biosynthetic genes and accumulation of PHA in recombinant bacteria. In *Novel Biodegradable Microbial Polymers*, ed. E. A. Dawes, pp. 143–59. Kluwer Academic Publishers, Dordrecht.

Steiner, K. E. & Preiss, J. (1977). Biosynthesis of bacterial glycogen: genetic and allosteric regulation of glycogen synthesis in *Salmonella typhimurium* LT-2. *Journal of Bacteriology*, **129**, 246–53.

Steudel, R. (1989). On the nature of 'elemental sulfur' (S^o) produced by sulfur-oxidising bacteria – a model for S^o globules. In *Autotrophic Bacteria*, ed. H. G. Schlegel & B. Bowien, pp. 289–303. Science Tech Publishers, Madison, Springer, Berlin.

Stewart, W. D. P. & Codd, G. A. (1975). Polyhedral bodies: carboxysomes of nitrogen fixing blue-green algae. *British Phycological Journal*, **10**, 273–8.

Strasdine, G. A. (1968). Amylopectin accumulation in *Clostridium botulinum* type E. *Canadian Journal of Microbiology*, **14**, 1059–62.

Streichan, M., Golecki, J. R. & Schön, G. (1990). Polyphosphate-accumulating bacteria from sewage plants with different processes for biological phosphorus removal. *FEMS Microbiology Ecology*, **73**, 113–24.

Suresh, N., Roberts, M. F., Coccia, M., Chikarmane, H. M. & Halvorson, H. O. (1986). Cadmium-induced loss of surface polyphosphate in *Acinetobacter lwoffi*. *FEMS Microbiology Letters*, **36**, 91–4.

Suresh, N., Warburg, R., Timmerman, M., Wells, J., Coccia, M., Roberts, M. F. & Halvorson, H. O. (1985). New strategies for the isolation of microorganisms responsible for phosphate accumulation. *Water Science Technology*, **17**, 99–111.

Timm, A. & Steinbüchel, A. (1990). Formation of polyesters consisting of medium chain-length 3-hydroxyalkanoic acids from gluconate by *Pseudomonas aeruginosa* and other fluorescent pseudomonads. *Applied and Environmental Microbiology*, **56**, 3360–7.

Trüper, H. G. (1989). Physiology and biochemistry of phototrophic bacteria. In *Autotrophic Bacteria*, ed. H. G. Schlegel & B. Bowien, pp. 267–81. Science Tech Publishers, Madison, Springer, Berlin.

Urbanowski, J., Leung, P., Weissbach, H. & Preiss, J. (1983). The *in vitro* expression of the gene for *Escherichia coli* ADP glucose pyrophosphorylase is stimulated by cyclic AMP and cyclic-AMP receptor protein. *Journal of Biological Chemistry*, **258**, 2782–4.

Van Gemerden, H., Visscher, P. T. & Mas, J. (1990). Environmental control of sulfur deposition in anoxygenic purple and green sulfur bacteria. In *Novel Biodegradable Microbial Polymers*, ed. E. A. Dawes, pp. 247–62. Kluwer Academic Publishers, Dordrecht.

Van Groenestijn, J. W. & Deinema, M. H. (1987). The utilization of polyphosphate as an energy reserve in *Acinetobacter* sp. and activated sludge. In *Biological Phosphate Removal from Waste Waters*, ed. R. Ramadori. Proceedings of IAWPRC Specialized Conference, Rome, pp. 1–6. Pergamon Press, Oxford.

Van Groenestijn, J. W., Deinema, M. H. & Zehnder, A. J. B. (1987). ATP production from polyphosphate in *Acinetobacter* strain 210A. *Archives of Microbiology*, **148**, 14–19.

Van Groenestijn, J. W., Zuidema, M., Worp, J. J. M., van der, Deinema, M. H. & Zehnder, A. J. B. (1989). Influence of environmental parameters on polyphosphate accumulation in *Acinetobacter* spp. *Antonie van Leeuwenhoek*, **55**, 67–82.

Wallen, L. L. & Rohwedder, W. K. (1974). Poly-β-hydroxyalkanoate from activated sludge. *Environmental Science and Technology*, **8**, 576–9.

Witholt, B., Huisman, G. W. & Preusting, H. (1990). Bacterial poly(3-hydroxyalkanoates). In *Novel Biodegradable Microbial Polymers*, ed. E. A. Dawes, pp. 161–73. Kluwer Academic Publishers, Dordrecht.

Wilkinson, J. F. (1959). The problem of energy-storage compounds in bacteria. *Experimental Cell Research, Supplement*, **7**, 111–30.

Wolk, C. P. (1973). Physiology and cytological chemistry of blue-green algae. *Bacteriological Reviews*, **37**, 32–101.

Wood, H. G. & Clark, J. E. (1988). Biological aspects of inorganic polyphosphates. *Annual Review of Biochemistry*, **57**, 235–60.

Wood, H. G. & Goss, N. H. (1985). Phosphorylation enzymes of the propionic acid bacteria and the roles of ATP, inorganic pyrophosphate, and polyphosphates. *Proceedings of the National Academy of Sciences, USA*, **82**, 312–15.

Yu, F., Jen, Y., Takeuchi, E., Inouye, M., Nakayama, H., Tagaya, M. & Fukui, T. (1988). α-Glucan phosphorylase from *Escherichia coli*. Cloning of the gene and purification and characterization of the protein. *Journal of Biological Chemistry*, **263**, 13706–11.

Zehnder, A. J. B. & Groenestijn, J. W. van (1990). Accumulation of polyphosphate by *Acinetobacter* sp: physiology, ecology, and application. In *Novel Biodegradable Microbial Polymers*, ed. E. A. Dawes, pp. 235–43. Kluwer Academic Publishers, Dordrecht.

Zevenhuizen, L. P. T. M. (1966). Formation and function of the glycogen-like polysaccharide of *Arthrobacter*. *Antonie van Leeuwenhoek Journal of Microbiology and Serology*, **32**, 356–72.

Zevenhuizen, L. P. T. M. & Ebbink, A. G. (1974). Interrelations between glycogen, poly-β-hydroxybutyric acid and lipids during accumulation and subsequent utilization in a *Pseudomonas*. *Antonie van Leeuwenhoek Journal of Microbiology and Serology*, **40**, 103–20.

GENETICS OF BACTERIAL CELL DIVISION

ERFEI BI and JOE LUTKENHAUS

Department of Microbiology, Molecular Genetics and Immunology, University of Kansas Medical Center, Kansas City, KS 66103, USA

INTRODUCTION

Cell division is an essential process in any cellular organism. However, the mechanisms by which this process is temporally and spatially regulated are poorly understood. The specific problems that need to be solved in the study of bacterial cell division are as follows: 1. How many components are specifically involved in the division process? 2. What is the activity of each component? 3. How are the activities of these components coordinated to ensure the ordered events occurring in the division process? 4. How is the frequency of division regulated so that the cell divides only once per cell cycle? 5. How is the division site formed and selected? 6. How is the division process regulated to adapt to specific physiological conditions? In this chapter information about cell division in the Gram-negative bacterium, *Escherichia coli* and the Gram-positive bacterium, *Bacillus subtilis* is reviewed.

A BRIEF HISTORY OF DIVISION GENES OF *E. COLI*

In the narrowest sense, division genes are defined as genes that are essential and specific for the separation process during the normal cell cycle. Inactivation of such a gene should only inhibit the cell division process without affecting cell growth (increase in cell mass and elongation), DNA replication or cell physiology, resulting in cell filamentation with regularly spaced nucleoids. This definition may prove to be too narrow since it is possible that a missing cell division gene product could lead to cell lysis specifically at the division site (for a review of genes affecting morphology see Donachie, Begg & Sullivan, 1984). Since cell division is a vital process to all cellular organisms, genes directly involved in the division process are presumed to be essential. Therefore, identification of division genes has relied on isolation of conditional-lethal mutants. One approach was simply to screen a collection of temperature-sensitive mutants for a filamentation phenotype at the non-permissive temperature. Fourteen genes were identified by this approach. They were designated *fts* genes (*fts* for the *f*ilamentation *t*emperature *s*ensitive phenotype). These include *ftsA-G* (Hirota, Ryter & Jacob, 1968; Ricard

& Hirota, 1973), *ftsH* (Santos & Almeida, 1975), *ftsI* or *sep* (Allen *et al.*, 1974; Fletcher *et al.*, 1978; Irwin *et al.*, 1979; Spratt, 1975, 1977), *ftsM* (Drapeau, Chausseau & Gariepy, 1983), *ftsS*, *ftsT* (Dwek *et al.*, 1984), *fts36* and *ftsW* (Ishino *et al.*, 1989). A modified version of this approach was to isolate a collection of amber mutations in the presence of a temperature-sensitive tRNA suppressor and then screen for cell filamentation at the non-permissive temperature. One of the mutations isolated in this way mapped in the *ftsA* gene(Lutkenhaus & Donachie, 1979). Another approach to identify cell division genes was to enrich directly for filamentous cells from a mutagenized population after expression of the phenotype. The limitation of this approach is that it requires the mutant to remain viable during the period when the cell is exposed to the non-permissive temperature. Several *fts* mutations, including *ftsQ* (Begg, Hatfull & Donachie, 1980) and *ts-20* (Nagai & Tamura, 1972), were isolated in this way. However, mutations in the *ftsA* gene are most frequently isolated by this approach (Van de Putte, Van Dillewijn & Rorsch, 1964; Reeve, Groves & Clark, 1970; Wijsman, 1972).

Among a collection of previously classified *ftsA* mutants, the *ftsZ* gene was identified by a series of complementation tests with a series of transducing phages which carry different alleles of the *ftsA* gene or different fragments from the *ftsA* region (Lutkenhaus, Wolf-Watz & Donachie, 1980). Only one temperature sensitive allele of this gene is known, although it has recently been shown that the null phenotype is similar (Dai & Lutkenhaus, 1991).

In another approach, the *ftsH** mutant (this designation is used to distinguish it from *ftsH*, a designation used by Santos and Almeida, 1975) was isolated following mutagenesis and selection for tolerance to colicin E2 at low temperature. A temperature-sensitive growth and filamentation phenotype was screened among these colicin E2 tolerant mutants. This led to the identification of the *ftsH** locus (Holland & Darby, 1976).

The principle which is common to all of the approaches is the screening for the cell filamentation phenotype. However, cell filamentation is not a strict criterium to justify whether or not a gene is specifically involved in cell division. It is known that inhibition of DNA synthesis by treating cells with chemical and physical agents, or incubation of a temperature-sensitive mutant defective in DNA synthesis, such as *dnaA*(Ts), at the non-permissive temperature leads to cell filamentation (Helmstetter & Pierucci, 1968; Mulder & Woldringh, 1989). Also some mutants defective in protein secretion form long filaments at the non-permissive temperature (Oliver & Beckwith, 1981). Also, many mutants defective in the heat shock response display a filamentation phenotype (Tsuchido, VanBogelen & Neidhardt, 1986). Therefore, all *fts* mutants obtained through the above approaches should be carefully examined for the possible defects in other biological processes.

It was observed that nucleoid segregation was affected in the *ftsB*(Ts) and *ftsC*(Ts) mutants at the non-permissive temperature (Ricard & Hirota,

1973) and the *ftsB* gene was shown to be allelic to *nrdB* which encodes the B2 subunit of ribonucleoside-diphosphate reductase (Kren & Fuchs, 1987; Taschner, Verest & Woldringh, 1987). In addition, *ftsT*(Ts) is temperature sensitive for DNA replication (Ben-Neria & Ron, 1991). Therefore, these three genes are no longer considered division genes. The original *ftsH* mutant (Santos & Almeida, 1975) was shown to contain a decreased level of penicillin-binding protein 3 (PBP3) which can be corrected by overexpressing PBP3 in *trans* with the simultaneous suppression of the filamentation phenotype (Ferreira *et al.*, 1987). However, overexpression of PBP3 does not suppress the temperature-sensitive growth phenotype. In addition, when the *ftsH* mutation was transduced into some other strains, temperature-sensitive growth was observed but not the filamentation phenotype (Ogura *et al.*, 1990). It is likely that the original mutant contains two mutations: one affects PBP3 activity, the other affects cell growth. Most likely the latter defines the *ftsH* gene. Therefore, the *ftsH* gene might not have anything to do with cell division. However, it is still of interest to study the other mutation which affects PBP3 activity in the original mutant. The *ftsM* gene was recently shown to be allelic to *serU* gene, therefore it is no longer considered a cell division gene (Leclerc, Sirard & Drapeau, 1989). Since growth rates of the *fts36* and *ftsW* at the non-permissive temperature are somewhat affected (Ishino *et al.*, 1989), their classification as cell division genes is debatable. The *ftsE*(Ts) mutant filaments only in rich medium at the non-permissive temperature (Taschner *et al.*, 1988*a*), however, it does not rule out that *ftsE* is an essential cell division gene. The *ftsE* gene was located at 76 min of the *E. coli* genetic linkage map (Salmond & Plakidou, 1984). Sequence analysis of the *ftsE* region revealed a putative cell division operon, which consists of three genes *ftsY*, *ftsE* and *ftsX* (Gill, Hatfull & Salmond, 1986). Conditional-filamentation mutations were isolated in the *ftsE* and *ftsX* genes, but not yet in the *ftsY* gene. Both the FtsE and FtsY proteins have a nucleotide-binding domain based on amino acid sequence comparisons. Interestingly, the FtsE protein was shown to be homologous with a small group of proteins involved in transport including the putative cystic fibrosis gene product (Crickmore & Salmond, 1986; Gill & Salmond, 1990). The FtsY protein exhibits homology with the SRα protein of eucaryotes, which is part of the protein secretion machinery (Romisch *et al.*, 1989; Bernstein *et al.*, 1989). The FtsY, FtsE and FtsX proteins have been shown to be cytoplasmic membrane proteins (Gill & Salmond, 1987). All the above information is consistent with the hypothesis put forward by Salmond's group that the FtsE protein, perhaps in conjunction with FtsY and FtsX, functions in a hypothetical septalsome to couple ATP hydrolysis to the division process. Also they may be involved in the transport of some of the cell division gene products.

In summary, within the framework of the definition, *ftsA*, *D*, *F*, *G*, *H**, *I*, *Q*, *S*, *ts-20*, *X* and *Z* are considered cell division genes, while the *ftsB*, *ftsC*, *ftsH*, *ftsT* and *ftsM* do not belong to this category. Whether or not *ftsE*, *fts36* and *ftsW* are cell division genes is debatable. Classification of *ftsY* remains undetermined. Also, there has been no further study on *ftsD*, *F*, *G*, *H**, *S*, and *ts-20* mutants after the initial report on their isolation.

The frequency of obtaining conditional-lethal mutations in the known division genes varies with mutations in *ftsA* being predominant. Alternative approaches, fundamentally different from the previous ones would have to be taken, to isolate additional division genes.

CELL DIVISION INHIBITION SYSTEMS

The study of cell division inhibitors and isolation of suppressors of these inhibitors have proven useful in emphasizing the essential components of the division process as well as revealing aspects of the regulation of the division process. Here, a few better studied division inhibitors are discussed.

The SOS response inducible cell division inhibitor SulA (SfiA)

The *E. coli* B strain was found to be unusually sensitive to ultraviolet (UV) radiation. When the gene responsible for this phenotype was transduced into a wild-type *E. coli* K-12 strain, the transductant became UV sensitive and mucoid (Donch, Chung & Greenberg, 1969). A UV sensitive mutant of *E. coli* K-12 was isolated, which displayed a mucoid phenotype and grew as long filaments after exposure to UV. The gene responsible for these phenotypes was designated *lon* (*lon*g form; Howard-Flanders, Simson & Theriot, 1964). Because of the striking similarity in phenotypes and genetic location of the mutation in the *E. coli* B strain and the *E. coli* K-12 *lon*, *deg* (Bukhari & Zipser, 1973) and *capR* mutants (Markovitz, 1964) they were all thought to contain a mutation in the same gene (Donch *et al.*, 1969; Chung & Goldberg, 1981). This gene was designated *lon* and encodes a 94 kD protein which is an ATP-dependent serine protease (Charette, Henderson & Markovitz, 1981; Zehnbauer *et al.*, 1981).

An elegant hypothesis put forward by Witkin (1967) to explain both filament formation and prophage induction upon UV irradiation predicted all the regulatory components of today's known SOS response (for review see Walker, 1984). In the hope of defining the genes encoding the division inhibitor and its target, a UV resistant mutant of the *E. coli* B strain was isolated (Witkin, 1947). The UV resistant *E. coli* B/r strain no longer formed filaments upon UV irradiation. The mutation responsible for the UV resistant phenotype was mapped near the *fabA* marker (22 min on the *E. coli* genetic linkage map; Johnson & Greenberg, 1975). This type of suppressor was designated *sul* (*su*ppressor of *l*on; Donch *et al.*, 1969). Subsequently, more suppressors of *lon* were isolated and most were found to map at two genetic loci, *sulA* and *sulB* (Gayda, Yamamoto & Markovitz, 1976; Johnson, 1977; Gottesman, Halpern & Trisler, 1981). The *sul* mutation in *E. coli* B/r is most likely a *sulA* mutation. Another set of suppressor mutations, designated *sfiA* and *sfiB* (*s*uppressor of *fi*lamentation), were isolated by selecting temperature resistant survivors of a *tif lon* double mutant (George, Castellazzi & Buttin, 1975; Huisman, D'Ari & George, 1980). These mutations are now known to be allelic to *sulA* and *sulB*. The *sulA* (*sfiA*) locus was shown to encode the SOS response-inducible cell division inhibitor SulA (SfiA; Huisman & D'Ari, 1981; Huisman, D'Ari & Gottesman, 1984). The division inhibition by SulA is amplified in a *lon* genetic background where even a mild induction of SulA causes cell filamentation and eventually cell death (George *et al.*, 1975; Gayda *et al.*, 1976). The explanation is that SulA is stabilized in a *lon* mutant, suggesting that SulA is a presumed *in vivo* substrate of the La protease (*lon* gene product; Mizusawa & Gottesman, 1983), however this enzyme-substrate relationship has not been demonstrated *in vitro*. Two independently isolated *sulB* mutations, *sulB9* and *sulB25*, were mapped in the essential cell division gene *ftsZ* (Lutkenhaus, 1983) as was the *sfiB114* mutation (Jones & Holland, 1984). These results led to the suggestion that SulA may inhibit cell division by directly interacting with FtsZ and inhibiting its essential activity. Surprisingly, the two independently isolated *sulB* mutations resulted in an identical change in amino acid sequence. The same was true for another two independently isolated *sfiB* mutations, *sfiB103* and *sfiB114* (Bi & Lutkenhaus, 1990*b*), suggesting that the diversity of *sulB* mutations might be quite limited. However, the isolation of more *sulB*-like mutations in a strain with two copies of the *ftsZ* gene demonstrated that *sulB* mutations were dispersed in the *ftsZ* gene (Bi & Lutkenhaus, 1990*b*). For the sake of standardization, all the *sulB* or *sfiB* alleles were redesignated *ftsZ*(Rsa) to refer to resistance to SulA.

Biochemical evidence suggestive of an interaction between FtsZ and SulA came from examination of the half-life of SulA. The half-life of SulA in the presence of *ftsZ*(Rsa) alleles such as *ftsZ114*(Rsa) (Jones & Holland, 1985) and *ftsZ9*(Rsa) (Bi & Lutkenhaus, 1990*b*) is shorter than in the presence of the wild-type *ftsZ*. These data are consistent with the interpretation by Jones and Holland (1985) that FtsZ interacts with SulA, protecting SulA from degradation by La protease while the FtsZ(Rsa) protein cannot interact with SulA so that SulA is exposed to degradation. In addition to mutational alterations of *ftsZ*, increasing the level of FtsZ also suppresses the division inhibition by SulA (Lutkenhaus, Sanjanwala & Lowe, 1986). All the genetic and biochemical data agree with the simplest model that SulA inhibits cell division by directly interacting with FtsZ, therefore coordinating DNA replication with cell division during the SOS response (Lutkenhaus, 1983). The discovery of a *sulA* homologue among several closely related bacteria suggests that this type of coordination mechanism is probably conserved among these bacterial species (Freudl *et al.*, 1987).

MinCD cell division inhibition system

The *minB* locus, which functions during the normal cell cycle, is thought to govern the selection of the division site by inhibiting polar sites (Teather *et al.*, 1974; de Boer *et al.*, 1989). The *minB* locus consists of three genes, *minC*, *minD* and *minE* (Fig. 1). By expressing different combinations of these genes and examining the effects on the division process, de Boer *et al.* (1989) proposed that MinC and MinD function together to inhibit division at all potential division sites while MinE confers topological specificity to the MinCD inhibitor such that the poles are blocked for division while the medial site is accessible. The molecular mechanism by which these three proteins achieve the topological specificity is unknown. However, it is known that the ratio of MinE to MinC + MinD is critical to achieve the specificity. Excess MinC + MinD results in inhibition at all sites whereas excess MinE results in no inhibition. Interestingly, the division arrest caused by overproduction of MinCD can be overcome by either increasing the level of FtsZ (de Boer, Crossley & Rothfield, 1990; Bi & Lutkenhaus, 1990*c*), an essential cell division protein, or by some of the *ftsZ*(Rsa) alleles, which were isolated by selecting for resistance to SulA (Bi & Lutkenhaus, 1990*b*). These results suggest that MinCD may antagonize FtsZ to inhibit its essential division activity.

The fact that the division arrest caused by both SulA and MinCD can be suppressed by increasing the level of FtsZ, as well as *ftsZ*(Rsa) alleles, suggests that FtsZ might be the common target for these two inhibitors (Fig. 2). However, these two inhibitors do not depend upon each other for their division inhibition function (de Boer, Crossley & Rothfield, 1990; Bi & Lutkenhaus, 1990*c*).

SfiC cell division inhibition system

Another division inhibition mechanism, termed the SfiC pathway, acts independent of the SulA system and the inhibition is irreversible (D'Ari & Huisman, 1983; Maguin, Lutkenhaus & D'Ari, 1986). The *sfiC* gene is part of an excisable element *e14*, which is only present in certain *E. coli* strains (Maguin et al., 1986). The SfiC phenotype is controlled by the *recA* but not the *lexA* gene. The division inhibition by SfiC is amplified in a *lon* genetic background but to a lesser degree than by SulA. The division arrest by SfiC can be suppressed by a *sulB* mutation, suggesting that FtsZ might be the target of SfiC (D'Ari & Huisman, 1983).

DicB inhibition system

Several genes in the terminus region of the chromosome are involved in the control of cell division. The genetic organization and regulation of these genes are quite similar to the immunity loci of a lambdoid bacteriophage,

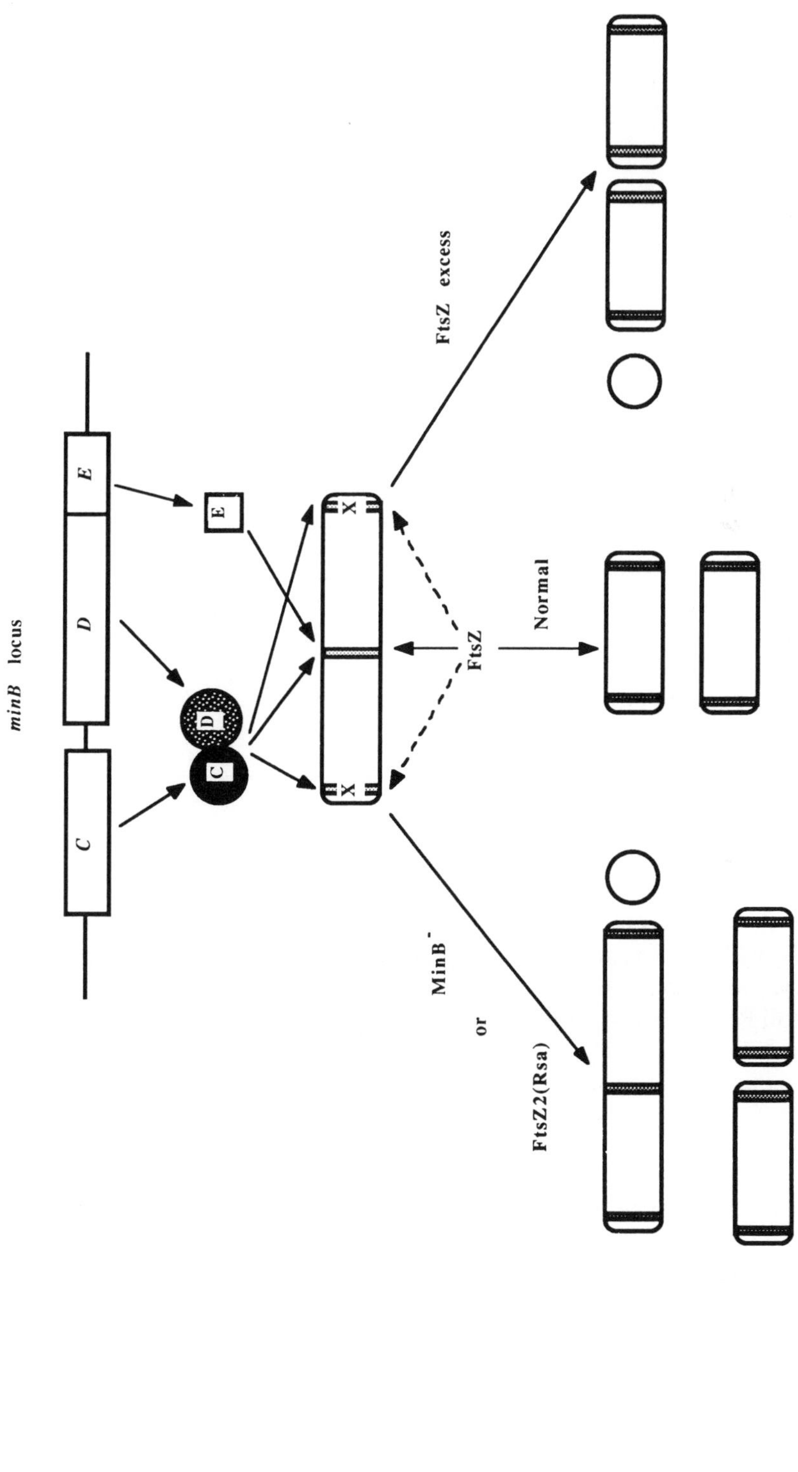

Fig. 1. Selection of the division site and the quantal behavior of FtsZ in the division process.

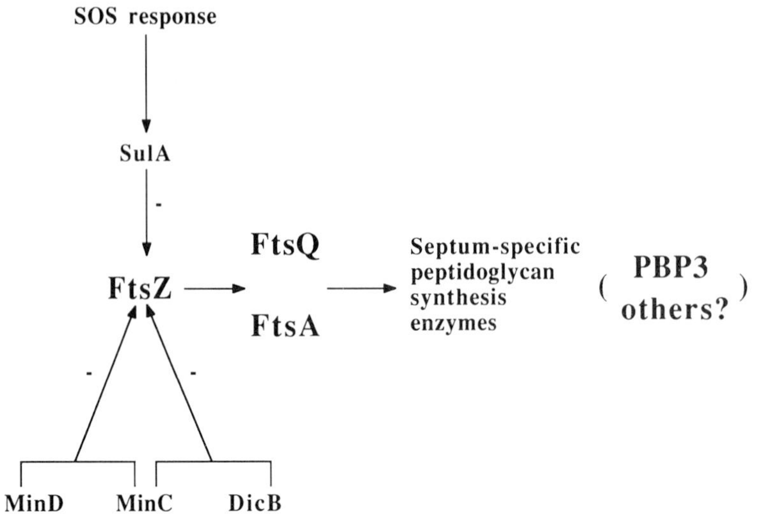

Fig. 2. The activity of FtsZ in initiation of septation is regulated by division inhibitors.

however, the physiological condition under which these genes are expressed *in vivo* is unknown. The functions of these genes, designated *dic* (*di*vision *c*ontrol), have been inferred from genetic study (Bejar & Bouche, 1985; Bejar, Bouche & Bouche, 1988). The *dicA* gene encodes a 15.5 kD protein, which functions as a transcriptional repressor for the *dicC* and *dicB* operons. There are operator sites for DicA protein in the promoter region of the *dicB* and *dicC* operons. Immediately upstream of the *dicA* is the *dicC* gene, which encodes an 8.5 kD protein and is transcribed in an opposite direction to the *dicA* gene. Increased expression of the *dicC* gene can complement the *dicA* mutant (Bejar, Cam & Bouche, 1986). Downstream of the *dicA* gene is the *dicB* operon, which is transcribed in the same direction as the *dicA* gene. The *dicB* operon consist of two *trans*-acting cell division inhibitors, DicB and DicF. The *dicB* gene encodes a 7 kD protein, whereas *dicF*, upstream of the *dicB* gene in the nontranslated region of the *dicB* operon, encodes a small RNA species (Cam *et al.*, 1988; Bouche & Bouche, 1989; Faubladier, Cam & Bouche, 1990). To determine how these inhibitors blocked cell division, mutations which confer resistance to the DicB inhibitor were isolated. Many of them resulted in the production of minicells to varying degrees and were mapped within or very close to the *minB* locus (Labie, Bouche & Bouche, 1989). Sequence analysis of some of the *minB* mutations showed that all the DicB resistant *minB* mutations were located in the *minC* gene and all DicB sensitive *minB* mutations were in the *minD* gene (Labie, Bouche & Bouche, 1990). This suggests that DicB requires MinC for its division inhibition function. The requirement of MinC for the DicB-mediated division inhibition was shown more directly by de Boer, Crossley & Rothfield (1990).

Both the MinCD and MinC-DicB division inhibition can be suppressed by an increased level of FtsZ, suggesting that they might target FtsZ for their division inhibition function (de Boer, Crossley & Rothfield, 1990). A difference between them is that MinCD responds to MinE while MinC-DicB does not (de Boer, Crossley & Rothfield, 1990).

These division inhibitors are not essential for cell growth and viability. Physiologically speaking, the inducible inhibitors probably play a role in fine-tuning the division process under specific conditions, while the MinCD inhibitor is required for proper site selection. Extensive study on these inhibitors will certainly enhance the understanding of the regulation of the division process.

ORGANIZATION AND PROPERTIES OF THE DIVISION GENES IN THE 2 MIN REGION OF THE CHROMOSOME

It is a striking feature that a large number of genes involved in cell wall synthesis and the cell division process are clustered at 2 min on the *E. coli* genetic linkage map (Fig. 3). A fragment of about 18 kb containing genes from *ftsI* to *envA* has been sequenced (Yi *et al.*, 1985; Yi & Lutkenhaus, 1985; Robinson *et al.*, 1986; Beall & Lutkenhaus, 1987; Ishino *et al.*, 1989; Jung, Ishino & Matsuhashi, 1989; Gomez, Merchante & Ayala, 1990; Ikeda *et al.*, 1991). Most of these genes overlap and all of them are transcribed in the same direction. Only one transcriptional terminator was located in the region just downstream of the *envA* gene which would appear to define the downstream boundary of this gene cluster (Beall & Lutkenhaus, 1987). The upstream boundary of this cluster has not yet been defined. All promoters upstream of a gene in this cluster should contribute to its expression. A complex cluster of promoters preceding a gene might incorporate more flexibility in terms of regulation of that gene. Within this cluster the *ftsI*, *ftsQ*, *ftsA*, *ftsZ*, and *envA* are the most extensively studied cell division genes.

The *ftsI* gene encodes the cell division-specific penicillin-binding protein 3 (PBP3) with a molecular weight of 64 kD (Nakamura *et al.*, 1983). It has two enzymic activities: transglycosylase and transpeptidase. Cells treated with cephalexin or furazlocillin, which preferentially target PBP3, form long filaments with regularly distributed constriction sites; a phenotype similar to the *ftsI*(Ts) mutant at the non-permissive temperature (Spratt, 1975). The transpeptidase has been shown to be critical for cell division (Broome-Smith, Hedge & Spratt, 1985). The number of PBP3 molecules has been estimated to be around 50 per cell. The cellular location of PBP3 has been studied with β-lactamase fusions and with Tn*phoA*. The results demonstrate that PBP3 is an integral membrane protein with the N-terminus in the cytoplasm, a hydrophobic domain spanning the membrane and the large C-terminus containing the enzymatic activities in the periplasmic region (Bowler & Spratt, 1989; Carson *et al.*, 1991). PBP3 is processed at the C-

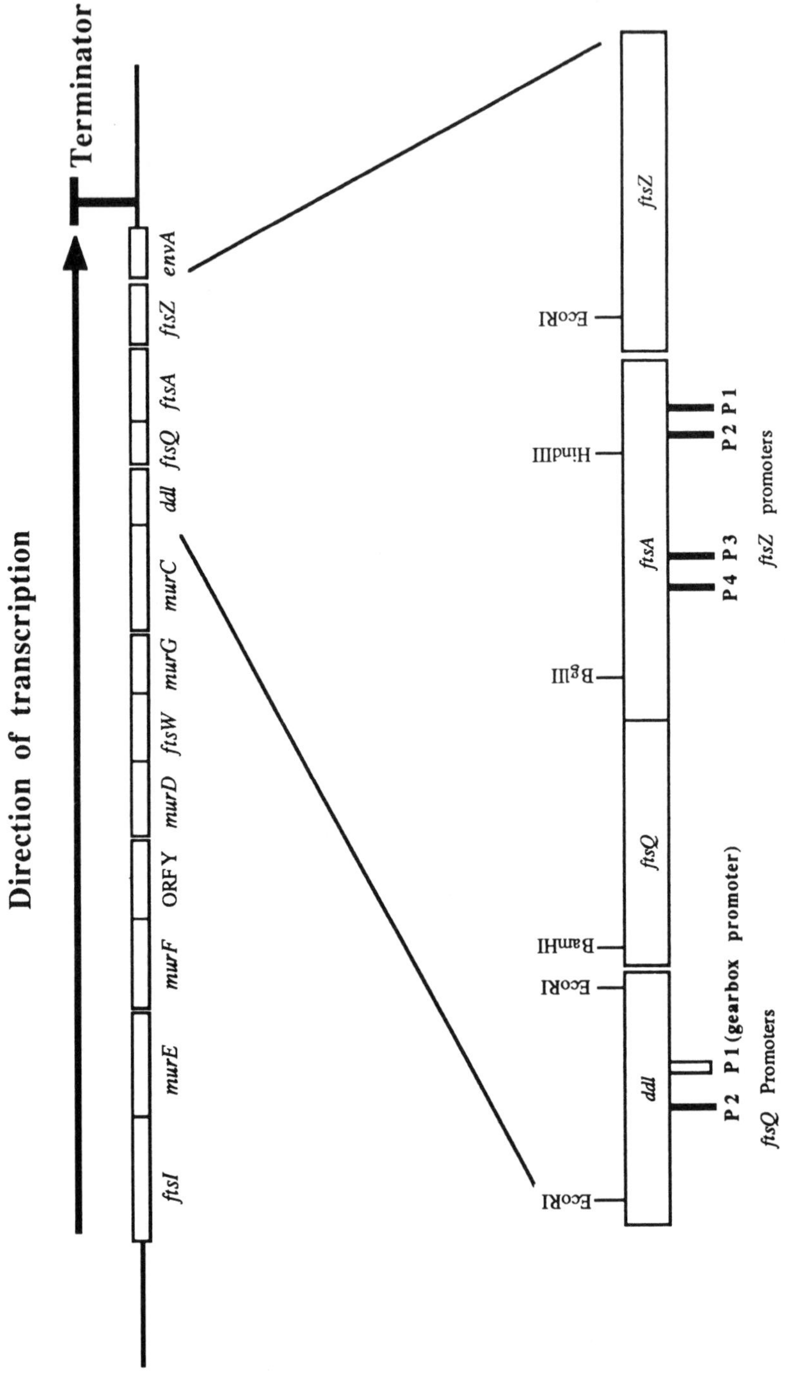

Fig. 3. Organization of the cell division genes within the large 2 min gene cluster

terminus for its maturation (Hara *et al.*, 1989; Nagasawa *et al.*, 1989). Also cells grown at 37 °C or at 42 °C but not at 30 °C contain a modified form of PBP3 (Prats *et al.*, 1989). The significance of these posttranslational modifications is not understood. Possibly, they are involved in the regulation of PBP3 activity.

The *ftsQ* gene product is a 31 kD integral membrane protein with a topology like that of PBP3 (Yi *et al.*, 1985; Storts *et al.*, 1989; Carson *et al.*, 1991). It is estimated to be present at about 25 molecules per cell and is required throughout septation (Carson *et al.*, 1991). *ftsQ*(Ts) mutants form long filaments at the non-permissive temperature without affecting other biological processes such as DNA synthesis and segregation (Begg, Hatfull & Donachie, 1980). Elucidation of FtsQ membrane location with respect to the septal site is certainly important in understanding its function.

FtsA is a 45 kD protein (Lutkenhaus & Donachie, 1979; Yi *et al.*, 1985) and is cytoplasmic membrane-associated (Pla, Dopazo & Vicente, 1990). Several physiological studies indicate that FtsA is part of the septum structure and is required throughout septation (Tormo, Matinez-Salas & Vicente, 1980; Tormo & Vicente, 1984). At the non-permissive temperature, all the *ftsA*(Ts) mutants grow as long filaments with constriction sites regularly distributed along the cell length. Other aspects of cell physiology are not affected (Lutkenhaus & Donachie, 1979; Tormo & Vicente, 1984). The biochemical function of FtsA remains unknown.

The essential cell division gene *ftsZ* is thought to be involved in the initiation of septum formation. Either incubation of a *ftsZ*(Ts) mutant at the non-permissive temperature or decreasing the FtsZ level in wild type cell results in the formation of filaments lacking constrictions (Ricard & Hirota, 1973; Lutkenhaus *et al.*, 1980; Dai & Lutkenhaus, 1991). The product of the *ftsZ* gene has been purified (Ward Jr. & Lutkenhaus, unpublished observations). It is a cytoplasmic protein with a molecular weight of 40 kD (Bi & Lutkenhaus, unpublished observations). FtsZ is relatively abundant in the cell compared with other Fts proteins. What is most intriguing is that increasing the FtsZ level in a wild-type strain in the range of two to sevenfold leads to minicell formation, a phenotype indicative of hyperdivision activity, suggesting that FtsZ might be the rate-limiting factor for the division process (Ward Jr. & Lutkenhaus, 1985).

The *envA* gene was identified on the basis of a single non-temperature-sensitive mutation, *envA1*. Multiple phenotypes are associated with this mutation including the increased permeability to hydrophobic and hydrophilic antibiotics and a defect in cell separation resulting in chain formation (Normark, Boman & Matsson, 1969; Normark, 1970). The increased permeability to antibiotics is unlikely to be specifically associated with the septal sites (Wijsman & Koopman, 1976). A six-fold decrease in *N*-acetylmuramyl-L-alanine amidase activity was observed in the mutant (Wolf-Watz & Normark, 1976). This enzyme is thought to be in the periplasmic region or associated with the outer membrane so *envA* cannot be the gene coding for this enzyme since the predicted amino acid sequence does not show any signal-like sequence. EnvA is a 34 kD protein and essential for cell viability (Beall & Lutkenhaus, 1987).

REGULATION OF THE DIVISION GENES AT 2 MIN REGION

Cell division is a strictly controlled process so that the cell divides only once per cell cycle irrespective of growth-rate. The organization of the cell wall synthesis and cell division genes in the 2 min region might be evolutionarily selected for coordinate regulation of their expression. The promoters for the *ftsQAZ-envA* cluster have been studied by S1 mapping and the results are summarized in Fig. 3 (Aldea *et al.*, 1990; Corton & Lutkenhaus, unpublished observations). The transcriptional organization in this region clearly does not define a classical operon. Multiple promoters with different strengths lie in this region and possible regulation of these promoters is only beginning to be explored.

Cell size varies with growth rate such that a cell present in a fast-growing culture is larger than that in a slow-growing culture. However, only one division event occurs in both cases per mass doubling. Assuming that one division event consumes the same amount of division potential independent of growth-rate, the concentration of division potential in a fast-growing cell should be lower than that in a slow-growing cell. Transcripts of the *ftsQAZ* genes and the FtsZ protein concentration have been measured in a study by Aldea *et al.* (1990). A promoter in front of the *ftsQ* gene (P1) is classified as a member of the so-called gearbox promoters. This promoter is regulated inversely by growth-rate so that a constant amount of Fts proteins per cell is produced at all growth rates. Transcriptional regulation of this cluster of genes has also been studied following the shift of a culture between two media supporting different growth rates (Dewar *et al.*, 1989; Robin, Joseleau-Petit & D'Ari, 1990). A transducing phage, which carries an operon fusion of some of the *ftsQAZ* promoters to the *lacZ* gene, was lysogenized in a wild-type strain. The expression of these *fts* genes was shown to decrease per cell mass upon the shift from a medium supporting a slow growth rate to a medium supporting a fast growth rate, although the expression of *fts* genes per cell increased slightly. The transducing phage used in this study does not contain the P1 promoter yet expresses a similar regulatory pattern for the *fts* genes. It is likely that both the P1 promoter and the promoter(s) carried on the transducing phage contribute to the overall inverse growth-rate regulation of the *fts* genes.

A gene named *sdiA*, (*s*uppressor of *d*ivision *i*nhibition), has been isolated as a result of screening a plasmid library of chromosomal DNA for suppression of the cell division inhibitor MinCD (Rothfield, Wang & de Boer, 1990). Recently it has been shown that SdiA can activate the P2 promoter

(Wang, de Boer & Rothfield, personal communication; Fig. 3) thereby increasing the concentration of Fts proteins. An increase in the level of FtsZ is known to suppress the inhibitory effect of MinCD on cell division (de Boer, Crossley & Rothfield, 1990; Bi & Lutkenhaus, 1990*c*). No apparent division defect was observed in a *sdiA* null mutant. Therefore, it is likely that SdiA exerts its effect on the division process through the essential cell division gene *ftsZ*, however, the physiological significance of regulation of the *fts* genes by SdiA is not clear.

Several observations suggest that expression of the *fts* genes is fine-tuned under certain circumstances, however the significance of it is not understood. Inactivation of any one of the proteins DnaA, DnaB and DnaC, which are essential for the initiation of chromosome replication, results in an increased expression of the *fts* genes (Masters *et al.*, 1989). A null mutant of the heat shock gene *htrM*, which is essential for cell growth at temperatures above 42 °C, produces minicells at permissive temperatures (Raina *et al.*, 1990). The level of FtsZ was increased twofold in this mutant, which is consistent with the low frequency of minicell formation that was observed (Bi, unpublished observations). Inactivation of another heat shock gene *dnaK* results in pleiotropic phenotypes such as division inhibition, slow cell growth, poor DNA segregation and poor cell viability (Bukau & Walker, 1989*a,b*). These defects could be the consequence of overexpressing heat shock proteins in the *dnaK* mutant. The division inhibition can be overcome by increasing the level of FtsZ while the defect in cell viability remains the same. How the *dnaK* locus affects the division process is not known. The simplest possibility is that DnaK affects the functional level of FtsZ by either assisting the conversion of the FtsZ protein from an inactive to an active state or affecting its stability.

INTERACTION OF Fts PROTEINS AND PBPs

When a cell divides, the peptidoglycan layer invaginates together with the cytoplasmic membrane. Electronmicroscopy of thin-sections of dividing cells shows the physical juxtaposition of these two layers at the division site (Burdett & Murray, 1974; Bi, unpublished observations) and a specific designation, the septal attachment site (SAS), was suggested by MacAlister *et al.* (1987). When the septation process begins, the main synthetic activity for the murein sacculus is shifted from the longitudinal mode to the septal mode (Woldringh *et al.*, 1987). It is conceivable that the cytoplasmic membrane and the murein sacculus layer have a mechanism to communicate with each other at the septation site to coordinate all these changes. It is known that the *ftsQAZ* genes are specifically involved in the septation process and PBPs are involved in murein sacculus synthesis. The effect of functional inactivation of the *fts* genes on the activity of PBPs or vice versa needs to be determined in order to explore the functional relationship between the Fts proteins and the PBPs. One of the *ftsA*(Ts) alleles was found to confer some resistance to cell lysis caused by benzylpenicillin at the non-permissive temperature. Interestingly, at the non-permissive temperature PBP3 was not labelled by [^{125}I] ampicillin. However, immunoblotting showed that PBP3 was present, suggesting that PBP3 is not accessible to ampicillin in this mutant at the non-permissive temperature (Tormo *et al.*, 1986). Since FtsA and PBP3 are known to function specifically during division, this result indicates that these two proteins might interact at the division site.

Morphological changes specifically associated with division sites were observed in a strain containing only the *ftsZ2*(Rsa) allele, suggesting that FtsZ can somehow communicate with the septal specific peptidoglycan synthesis machinery (Bi & Lutkenhaus, in preparation). It has been reported that an *ftsI*(Ts) mutant grown at the non-permissive temperature for two mass doublings forms aberrant constrictions during its recovery at the permissive temperature (Taschner *et al.*, 1988*a*). In addition, an amino acid substitution in *ftsI* gene product (PBP3) caused pointed polar caps. Slightly pointed polar caps were also observed when PBP3 was partially inhibited by subinhibitory concentration of antibiotics, which preferentially target PBP3. It was suggested that the shape of the polar caps was correlated with PBP3 activity (Taschner *et al.*, 1988*b*). In light of these facts, it is conceivable that FtsZ2(Rsa) might affect PBP3 activity directly or through other intermediate proteins, such as FtsA, resulting in the abnormal morphology at the division sites.

COORDINATION OF DNA REPLICATION AND CELL DIVISION

The *Escherichia coli* cell cycle consists of two major periodic events: DNA replication and cell division. Immediately following DNA replication and segregation cell division ensues, which results in the symmetrical division of a mother cell at its longitudinal midpoint. For a wild-type growing population, division occurs only once per cell cycle and it can occur only after DNA replication and segregation are completed and the cell attains a critical cell length (Donachie & Begg, 1989). Therefore, the division process must be regulated temporally and spatially. By replacing the chromosomal replication origin with a controllable plasmid replication origin, Bernander and Nordstrom (1990) showed that increasing DNA content did not affect the cell size distribution, leading to the conclusion that chromosome replication does not trigger cell division. It seems that DNA replication and cell division are regulated independently within the cell cycle although both must somehow be coupled to the growth rate.

Since one round of DNA replication is usually followed by one cell division event and inhibition of DNA replication results in division arrest, it was suggested that these two processes are likely to be coupled (Helmstetter & Pierucci, 1968). Inhibition of DNA replication often induces the SOS response leading to division inhibition through SulA. However, even in the absence of the SOS response division is inhibited although not completely (Burton & Holland, 1983; Jaffe, D'Ari & Norris, 1986). This incomplete inhibition is seen with many *dna* (Ts) mutants at the non-permissive temperature (Mulder & Woldringh, 1989). The mechanism of this inhibition is not known but it helps to coordinate DNA replication with cell division.

FORMATION AND SELECTION OF THE DIVISION SITE

Specific structures, designated the periseptal annuli (PSA), have been identified and appear to flank the future division site before the onset of the invagination process (MacAlister, MacDonald & Rothfield, 1983). The development of PSA is proposed to occur as follows: a pair of new PSAs emerge by duplicating the midcell PSA of a newborn cell and are laterally displaced and arrested at ¼ and ¾ cell lengths. These PSAs are located at the midpoint of the future daughter cells and mark where the next round of division is going to take place (Anba *et al.*, 1984; Cook, MacAlister & Rothfield, 1986; Cook *et al.*, 1987; de Boer, Cook & Rothfield, 1990). The mechanisms for lateral displacement and arrest are unknown. Following septation, each daughter cell retains a PSA as a polar annulus (PA) from the mother cell. The fact that minicells can form at the cell ends under some conditions suggests that a remnant structure from a previous division, presumably the PA, can be used for division (Teather, Collins & Donachie, 1974; de Boer, Cook & Rothfield, 1990).

The selection of the site for division is thought to be governed by the *minB* locus (Teather *et al.*, 1974; de Boer, Crossley & Rothfield, 1989). Deletion of the *minB* locus results in a minicell phenotype indicating that this locus only affects site selection and does not encode part of the septation machinery. The normal physiological inhibition of polar sites by the *minB* locus is overcome by increasing the level of FtsZ as evidenced by the induction of a minicell phenotype by an increase in the FtsZ level (Ward & Lutkenhaus, 1985). In addition, the mutations *ftsZ2*(Rsa) and *ftsZ3*(Rsa), selected for resistance to the SOS division inhibitor, SulA, can overcome the *minB* polar inhibition (Bi & Lutkenhaus, 1990*c*). These results suggest that the *minB* locus functions to channel the division activity of FtsZ to the medial site by preventing its usage at polar sites (Fig. 1).

An alternative theory was recently postulated to explain positioning of the division site. It is called the nucleoid occlusion model (Mulder & Woldringh, 1989, 1990; Woldringh *et al.*, 1990). It states that division sites are signalled and positioned *in situ* as a result of both termination of DNA replication and nucleoid segregation. The model predicts a strict coupling between DNA replication and cell division and excludes the independent development of envelope sites for division. In molecular terms, two hypothetical factors are proposed in this hypothesis. One is a short-range, negative acting factor which surrounds the replicating chromosome. Upon the termination of DNA replication, a second positive factor is produced in the cell centre and is able to initiate a constriction anywhere in the bacterial envelope where the nucleoids have segregated and the influence of the negative factor has weakened sufficiently. It is assumed that the positive factor is degraded before it reaches the poles. However, to explain the minicell phenotype caused by overproduction of MinE, the positive factor is assumed to be stabilized by MinE allowing it to reach the cell poles before segregation of the nucleoids (Mulder & Woldringh, 1990). In our view, this model is not consistent with the following experimental results: 1. The *gyr*(Ts) mutant at the non-permissive temperature produces minicells and anucleate rods, yet it is known that nucleoids cannot segregate under such a condition. How are the division sites signalled and positioned and why can the hypothetical positive factor reach the division site away from nucleoids? 2. The recently isolated *mukB* mutant does not segregate its DNA properly and yet produces normal sized cells (Hiraga *et al.*, 1989). 3. The fact that anucleate rods can form in the *dnaA*(Ts), *dnaX*(Ts) and *gyr*(Ts) mutants at the non-permissive temperature simply means that division sites can develop independent of DNA replication and nucleoid segregation.

FREQUENCY OF CELL DIVISION

Frequency of cell division is defined as the number of division events per cell cycle. Normally, a cell divides only once per cell cycle. This limitation of the frequency of division is perhaps best seen in the *minB* mutant (Adler *et al.*, 1967). This mutant grows as a population of cells of different lengths, from anucleate minicells to multinucleated filaments. Quantitative analysis of this phenotype led Teather *et al.* (1974) to propose that a limited division potential behaved as a quantum that has equal access to the three potential sites in the *minB* mutant: the medial one and the two polar sites. Division at the polar sites is at the expense of the normal division, therefore resulting in a population of cells with an increased average cell length. This model also implies that the cell must have a system to control the frequency of division.

Inactivation of the *ftsZ* gene product or its elimination results in the formation of filaments lacking constrictions. This led to the suggestion that the FtsZ protein is involved in the earliest stage of the division process. Molecular cloning and expression of the genes in the *ftsZ* region revealed a very interesting property of FtsZ. When the level of FtsZ was increased two to sevenfold in a wild-type strain minicells were produced at the cell ends along with a medial division, suggesting that the FtsZ protein might regulate the frequency of division (Ward Jr. & Lutkenhaus, 1985). Frequency of a macromolecular synthetic pathway is often controlled at the first unique step in the pathway. By analogy the division process might be regulated by modulating a component activity acting early in the process. Is this the case for the FtsZ protein and the division process?

The *minB* deletion mutant was used to address this question. This deletion mutant expresses an identical phenotype to that seen in the original minicell mutant: a population of cells from minicells to filaments (Adler *et al.*, 1967; de Boer *et al.*, 1989). The minicell formation observed with *minB* mutants is very different from minicell formation as a consequence of increasing the amount of FtsZ protein in the cell in one important aspect. In the latter case, the medial and polar divisions can both occur within a cell cycle, although asynchronously. Also, cell division is observed in cells that are a shorter length than wild type dividing cells. As a result, the median cell length is shortened (Ward Jr. & Lutkenhaus, 1985). If the FtsZ protein were the proposed limiting division factor, increasing the amount of FtsZ protein in the *minB* mutant should prevent the phenotype of increased average cell length. This is exactly what was observed (Bi & Lutkenhaus, 1990*a*).

The medial division and the polar divisions do not occur simultaneously, even in the presence of sufficient FtsZ, indicating that there is no mechanism to govern the synchrony of the availability or utilization of potential division sites as there is for initiation or replication at all *oriC* sites present in a cell (Helmstetter & Leonard, 1987). At a given growth rate, the FtsZ protein is synthesized continuously (Robin, Joseleau-Petit & D'Ari, 1990). Whenever the FtsZ protein level or activity is sufficient to drive one division event, one of the available potential division sites is used. The availability of the polar sites is most likely controlled by the duration of the septation process (i.e. minicell formation may have to be completed before the pole can be reused) while access to the medial site is presumably subject to the control of cell growth and the nucleoid segregation process. The sequential (asynchronous) usage of the division sites suggests that FtsZ behaves as a quantum.

A MODEL OF CELL DIVISION

A hypothetical model of cell division (Fig. 4), based upon mutant phenotypes, the cellular location of different Fts proteins and their relative quantity in the cell, is formulated as follows: when a cell is ready to divide, FtsZ protein might form a ring-like structure at the potential division site. This structure interacts with the cytoplasmic domain of FtsQ and FtsI to activate two septal specific peptidoglycan synthetic systems: one system, controlled by FtsQ, acts early and is followed by the second system consisting of FtsI (PBP3). The interaction between the FtsZ structure and FtsI could be mediated by FtsA. The FtsZ ring-like structure contracts to initiate the invagination process. Activated PBP3 and the other septal specific peptidoglycan synthetic enzymes synthesize the septal peptidoglycan by using the invaginated cytoplasmic membrane as the template.

This model assigns the cytoplasmic membrane an active role in the division

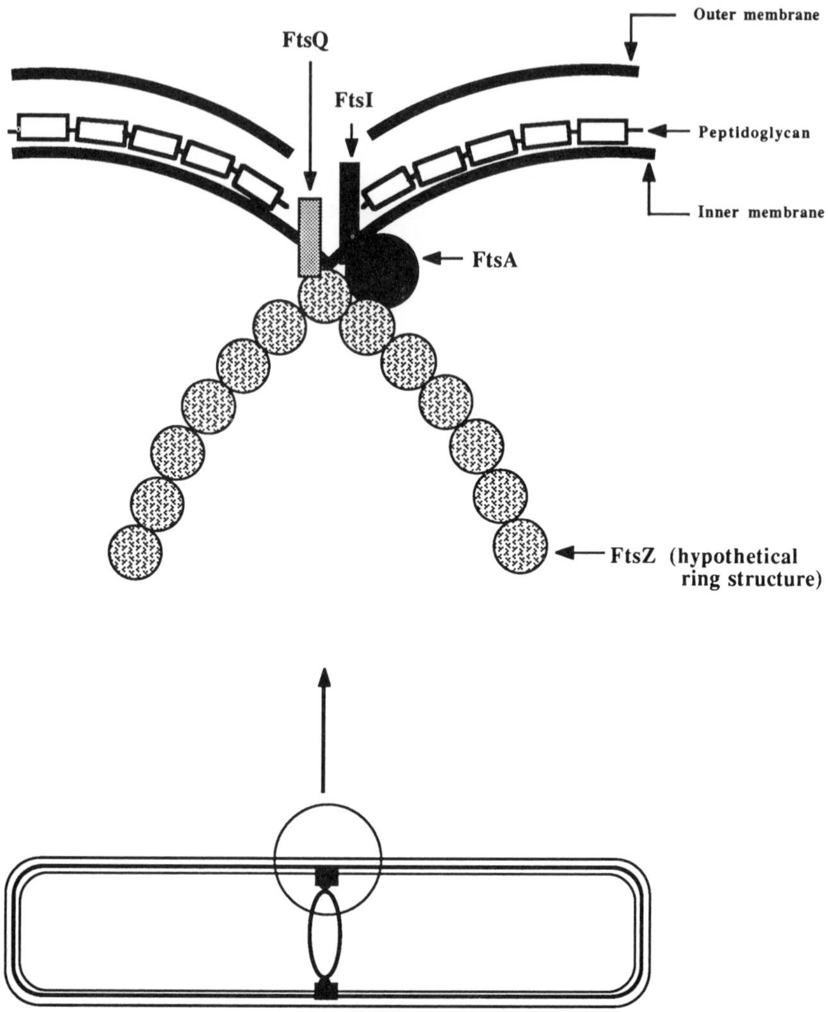

Fig. 4. A model for cell division. The diagram indicates the location and interactions of proteins as described in the text.

process. Since the cytoplasmic membrane is a common structure to all cells, while the peptidoglycan and outer membrane layers are confined to certain bacterial species, it is possible that the mechanism for cellular division process is conserved in all types of cells.

The following facts are consistent with the model:

1. FtsZ is a cytoplasmic protein, FtsA is a cytoplasmic membrane associated protein and FtsQ and FtsI (PBP3) are both integral membrane proteins.

2. FtsZ is relatively abundant in the cell, FtsA is much less than FtsZ. FtsQ and FtsI are less than FtsA.
3. Inactivation of FtsZ by either decreasing its concentration in the cell, or a temperature-sensitive allele at the non-permissive temperature, results in smooth filamentation. *ftsA*(Ts) and *ftsI*(Ts) mutants at the non-permissive temperature all form constricted filaments. In *ftsZ* mutants the hypothetical ring would either not form (null mutants) or be aberrant (*ftsZ2* mutant) whereas it should form in *ftsA* and *ftsI* mutants that form filaments with constrictions.
4. FtsZ is the rate-limiting factor for the septation process.
5. PBP3 in the presence of the *ftsA2*(Ts) allele at the non-permissive temperature loses its ability to bind ampicillin.
6. FtsZ2(Rsa) affects the septal specific peptidoglycan synthesis.
7. There might be at least two septal specific peptidoglycan synthetic systems: one acting before PBP3, the other is PBP3 (Wientjes & Nanninga, 1989).

GENETIC ANALYSIS OF *B. SUBTILIS* CELL DIVISION

Bacillus subtilis is an interesting organism for the study of cell division. First, it is a gram positive organism and its cell wall structure is quite different from gram negative organisms. Comparative genetic analysis of *B. subtilis* and *E. coli* should reveal the common and presumed essential components of the division machinery. Secondly, during sporulation cells of *B. subtilis* form a unique asymmetric septum (see chapter by A. Moir for details). This septum forms at a site displaced to one cell pole and therefore is a model for control of site selection. In addition, this septum is morphologically distinct suggesting that the septal development is uniquely modified.

The genetic analysis of cell division in *B. subtilis* was initiated with the isolation of a number of temperature sensitive, conditional lethal mutations. As with *E. coli* these were isolated by screening the morphology of temperature sensitive mutants at the non-permissive temperature (Nukushina & Ikeda, 1969; Miyakawa & Komano, 1981; Breakefield & Landman, 1973) or by enrichment for filamentous cells following a period of growth at the non-permissive temperature (van Alstyne & Simon, 1971). Analysis of these mutations has led to the identification of eight genetic loci (Piggot & Hoch, 1985). In addition to these conditional lethal mutations, mutations leading to minicell formation have been isolated and mapped to three different loci (Reeve *et al.*, 1973; van Alstyne & Simon, 1971) with mutations at the *divIVB* locus most closely resembling *minB* mutations in *E. coli*. Thus, the phenomenon of minicell formation is widespread among bacteria.

In another approach to identify cell division genes Corton, Ward Jr. & Lutkenhaus (1987) used antibodies against the *E. coli* FtsZ protein to screen a diverse group of bacteria for the presence of an FtsZ protein. A positive result for all bacteria tested led to the cloning of the *B. subtilis ftsZ* homologue

(Beall, Lowe & Lutkenhaus, 1988). The two *ftsZ* genes are highly conserved with the gene products having an overall 50% amino acid identity. Located just upstream of the *ftsZ* gene is a gene that is highly homologous to the *E. coli ftsA* gene. The *ftsZ* gene was mapped to 135°, a region containing several cell division loci. The *ts1* mutation, a well characterized cell division mutation, was located in the *ftsZ* gene indicating that the gene function was conserved in this organism (Beall & Lutkenhaus, 1987; Harry & Wake, 1989). Despite the degree of conservation between these two genes they can not substitute for each other. Introduction of the *B. subtilis* homologue into *E. coli* leads to filamentation and cell death. This failure of the genes to functionally replace each other between these two organisms may not be that unusual. Other genes, which encode proteins that form multisubunit complexes, can not be functionally interchanged between these two organisms.

FtsZ *in* B. subtilis *is required for vegetative and sporulation (asymmetric) septation*

The level of FtsZ has to be maintained within a narrow range for the septation machinery to function normally in *B. subtilis*, as was previously shown for *E. coli*. This was demonstrated by utilizing the inducible *spac* promoter (Yansura & Henner, 1984). Positioning this promoter just upstream of the *ftsZ* gene on the chromosome results in an inducer dependent phenotype (Beall & Lutkenhaus, 1991). The morphology in the absence of inducer is extreme filamentation followed by cell lysis. Overproduction of FtsZ results in filamentation, as was shown for *E. coli*, however, minicell formation was not observed.

The availability of a conditional lethal mutation and the inducer-dependent strain also allowed an examination of the role of *ftsZ* in sporulation. The results show that *ftsZ* is required for sporulation and the examination of *spo–lacZ* fusions revealed that *ftsZ* is required at a specific step in the developmental programme (Beall & Lutkenhaus, 1991). Expression of stage II genes *spoIIA*, *spoIIG* and *spoIIE* (mutations in which block development after the formation of the asymmetric septum; refer to chapter by P. Stragier for details) occurred normally whereas expression of *spoIID* was not turned on when the FtsZ level was reduced. Examination of cells induced to sporulate under conditions of reduced FtsZ revealed a complete absence of asymmetric septation with the DNA in an apparent condensed form as though the cells were blocked at Stage I when the DNA condenses into an axial filament. These observations were intriguing in light of the model proposed by Stragier, Bonamy and Karmazyn-Campelli (1988) for activation of *spoIID* expression, which is known to depend on a sporulation specific sigma factor, σ^E (Rong, Rosenkrantz & Sonenshein, 1986; Stragier, this volume). This sigma factor is synthesized as a precursor along with a protease that is responsible for its cleavage (Labell, Trempy & Haldenwang, 1987; Stragier *et al.*, 1988). The Stragier model proposes that the protease is not active until it is inserted into the newly formed asymmetric septum. Thus, formation of the asymmetric septum acts as a morphological control over the developmental programme. This model is entirely consistent with the results described above. The results also indicate that at least some of the vegetative division machinery is utilized for formation of the asymmetric septum.

Selection of the division site during sporulation

Sporulation presents a challenging problem to the cell since it must shift the division machinery from the midpoint of the cell to the polar region. This may involve the unmasking of polar sites and perhaps the masking of the medial site. This is reminiscent of the *min* system in *E. coli.* One possibility is that a homologue of the *E. coli min* system exists in *B. subtilis* and that this plays a role in site selection during sporulation. The existence of a *min* system is suggested by the *min* mutants that have been isolated in this organism.

A model for control of site selection during sporulation has been proposed (Beall & Lutkenhaus, 1991). In this model it was assumed that *B. subtilis* has a *min* system homologous to that of *E. coli.* It is proposed that an early event in sporulation is the inactivation of the *min* system or perhaps an inversion of its specificity such that the medial site is blocked and the polar sites are rendered accessible to the division machinery. Unmasking the polar sites allows the division machinery to operate at either of the polar sites. One important difference between minicell formation and asymmetric septation is that only the latter involves DNA segregation to the polar compartment. However, the induction of sporulation involves axial filament formation and possibly novel segregation ability that does not exist in vegetative cells. Once septation is initiated it is modified by the *spoII* genes into a unique asymmetric septum. Such a model suggests that both polar sites would be accessible. This does appear to be the case since certain *spoII* mutants have a terminal disporic phenotype in which an abnormal (or not completely processed) asymmetric septum is formed at each cell pole. In addition, it is consistent with the random placement of the asymmetric septum at either cell pole during normal sporulation. Asymmetric septation may require additional *ftsZ* expression from a sporulation promoter (σ^{H} dependent) located upstream of the vegetative promoter in the *ftsAZ* operon (Beall & Lutkenhaus, 1991; Stragier & Losick, 1990). This expression may be required to boost the level of FtsZ in cells entering stationary phase to increase the division potential. Recent examination of electron micrographs indicate that medial sites may be blocked during sporulation as opposed to another possibility, that the site was not yet constructed (Lutkenhaus, unpublished observations).

FUTURE WORK

The proposed model for bacterial cell division is just a sketch drawn according to limited genetic studies. Aspects of the model need to be confirmed. In addition, genes specifically involved in the division process need to be extensively searched for with novel approaches as it is unlikely that all division genes have yet been identified. On the other hand, additional genetic and biochemical characterization of protein interactions that have been suggested to date needs to be further addressed. Ultimately, an *in vitro* system for peptidoglycan synthesis needs to be established in which the interactions of PBP3 and Fts proteins can be explored. The role of the *min* system in site selection appears well established, however the mechanism by which the gene products achieve topological specificity remains to be elucidated. Whether or not this site selection system operates during sporulation also will have to be examined. FtsZ appears to play a key role in initiating division and suggested interactions with various inhibitors needs to be confirmed. Also, a thorough examination of the regulation of the expression of the *ftsZ* gene and the resultant effect of altering the expression on the frequency of cell division would confirm the key regulatory role indicated for *ftsZ*. Furthermore, both the *min* system and FtsZ must interact with the division sites and perhaps this can be explored to identify components of the site.

ACKNOWLEDGEMENTS

The authors thank the members of the laboratory for stimulating discussions. The work in the laboratory of J.L. is supported by grants from the National Institutes of Health and the Wesley Foundation.

REFERENCES

Adler, H. I., Fisher, W. D., Cohen, A. & Hardigree, A. A. (1967). Miniature *Escherichia coli* cells deficient in DNA. *Proceedings of the National Academy of Sciences, USA*, **57**, 321–6.

Aldea, M., Garrido, T., Pla, J. & Vicente, M. (1990). Division genes in *Escherichia coli* are expressed coordinately to cell septum requirements by gearbox promoters. *EMBO Journal*, **9**, 3787–94.

Allen, J. S., Filip, C. C., Gustafson, R. A., Allen, R. G. & Walker, J. R. (1974). Regulation of bacterial cell division: genetic and phenotype analysis of temperature-sensitive, multinucleate, filament-forming mutants of *Escherichia coli*. *Journal of Bacteriology*, **117**, 978–86.

Anba, J., Bernadac, A., Pages, J. & Lazdunski, C. (1984). The periseptal annulus in *Escherichia coli*. *Biology of the Cell*, **50**, 273–8.

Beall, B., Lowe, M. & Lutkenhaus, J. F. (1988). Cloning and characterization of *Bacillus subtilis* homologs of *Escherichia coli* cell division genes *ftsZ* and *ftsA*. *Journal of Bacteriology*, **170**, 4855–64.

Beall, B. & Lutkenhaus, J. F. (1987). Sequence analysis, transcriptional organization, and insertional mutagenesis of the *envA* gene of *Escherichia coli*. *Journal of Bacteriology*, **169**, 5408–15.

Beall, B. & Lutkenhaus, J. (1991). *ftsZ* in *Bacillus subtilis* is required for vegetative septation and for asymmetric septation during sporulation. *Genes and Development*, **5**, 447–55.

Begg, K. J., Hatfull, G. F. & Donachie, W. D. (1980). Identification of new genes in a cell envelope-cell division gene cluster of *Escherichia coli*: cell division gene *ftsQ*. *Journal of Bacteriology*, **144**, 435–7.

Bejar, S. & Bouche, J.-P. (1985). A new disposable genetic locus of the terminus region involved in control of cell division in *Escherichia coli*. *Molecular and General Genetics*, **201**, 146–50.

Bejar, S., Bouche, F. & Bouche, J.-P. (1986). Cell division inhibition gene *dicB* is regulated by a locus similar to lambdoid bacteriophage immunity loci. *Molecular and General Genetics*, **212**, 11–19.

Bejar, S., Cam, K. & Bouche, J.-P. (1986). Control of cell division in *Escherichia coli*: DNA sequence of *dicA* and of a second gene complementing mutation *dicA1*, *dicC*. *Nucleic Acids Research*, **14**, 6821–83.

Ben-Neria, R. & Ron, E. (1991). A DNA replication gene maps near *terC* in *Escherichia coli* K-12. *Molecular and General Genetics*, **226**, 315–17.

Bernander, R. & Nordstrom, K. (1990). Chromosome replication does not trigger cell division in *Escherichia coli*. *Cell*, **60**, 365–74.

Bernstein, H. D., Poritz, M. A., Strub, K., Hoben, P. J., Brenner, S. & Walter, P. (1989). Model for signal sequence recognition from amino-acid sequence of 54K subunit of signal recognition particle. *Nature*, London, **340**, 482–6.

Bi, E. & Lutkenhaus, J. F. (1990*a*). FtsZ regulates frequency of cell division in *Escherichia coli*. *Journal of Bacteriology*, **172**, 2765–8.

Bi, E. & Lutkenhaus, J. F. (1990*b*). Analysis of *ftsZ* mutations that confer resistance to the cell division inhibitor SulA (SfiA). *Journal of Bacteriology*, **172**, 5602–9.

Bi, E. & Lutkenhaus, J. F. (1990*c*). Interaction between the *min* locus and *ftsZ*. *Journal of Bacteriology*, **172**, 5610–16.

Bouche, F. & Bouche, J.-P. (1989). Genetic evidence that DicF, a second division inhibitor, encoded by the *Escherichia coli dicB* operon, is probably RNA. *Molecular Microbiology*, **3**, 991–4.

Bowler, L. D. & Spratt, B. G. (1989). Membrane topology of penicillin-binding protein 3 of *Escherichia coli*. *Molecular Microbiology*, **3**, 1277–86.

Breakefield, X. O. & Landman, O. E. (1973). Temperature-sensitive divisionless mutant of *Bacillus subtilis* defective in the initiation of septation. *Journal of Bacteriology*, **113**, 985–98.

Broome-Smith, J. K., Hedge, P. J. & Spratt, B. G. (1985). Production of thiol-penicillin-binding protein 3 of *Escherichia coli* using a two primer method of site-directed mutagenesis. *EMBO Journal*, **4**, 231–5.

Bukau, B. & Walker, G. C. (1989*a*). Cellular defects caused by deletion of the *Escherichia coli dnaK* gene indicate roles for heat shock proteins in normal metabolism. *Journal of Bacteriology*, **171**, 2337–46.

Bukau, B. & Walker, G. C. (1989*b*). Δ*dnaK*52 mutants of *Escherichia coli* have defects in chromosome segregation and plasmid maintenance at normal growth temperatures. *Journal of Bacteriology*, **171**, 6030–8.

Bukhari, A. I. & Zipser, D. (1973). Mutants of *Escherichia coli* with a defect in the degradation of nonsense fragments. *Nature (London) New Biology*, **243**, 238–41.

Burdett, I. D. J. & Murray, R. G. E. (1974). Septum formation in *Escherichia coli*: Characterization of septal structure and the effects of antibiotics on cell division. *Journal of Bacteriology*, **119**, 303–24.

Burton, P. & Holland, I. B. (1983). Two pathways of division inhibition in UV-irradiated *E. coli*. *Molecular and General Genetics*, **189**, 128–32.

Callister, H., McGinnis, M. & Wake, R. G. (1983). Timing and other features of the action of the *ts1* division initiation gene product of *Bacillus subtilis*. *Journal of Bacteriology*, **154**, 537–46.

Cam, K., Bejar, S., Gil, D. & Bouche, J.-P. (1988). Identification and sequence of gene *dicB*: translation of the division inhibitor from an in-phase internal start. *Nucleic Acids Research*, **16**, 6327–38.

Carson, M., Barondess, J. & Beckwith, J. (1991). The FtsQ protein of *Escherichia coli*: membrane topology, abundance, and cell division phenotypes due to overproduction and insertion mutations. *Journal of Bacteriology*, **173**, 2187–95.

Charette, M. F., Henderson, G. W. & Markovitz, A. (1981). ATP hydrolysis-dependent protease activity of the *lon* (*capR*) protein of *Escherichia coli* K12. *Proceedings of the National Academy of Sciences, USA*, **78**, 4728–32.

Chung, C. H. & Goldberg, A. L. (1981). The product of the *lon* (*capR*) gene in *Escherichia coli* is the ATP-dependent protease, protease La. *Proceedings of the National Academy of Sciences, USA*, **78**, 4931–5.

Cook, W. R., MacAlister, T. J. & Rothfield, L. I. (1986). Compartmentalization of the periplasmic space at division sites in Gram-negative bacteria. *Journal of Bacteriology*, **168**, 1430–8.

Cook, W. R., Kepes, F., Joseleau-Petit, D., MacAlister, T. J. & Rothfield, L. I. (1987). A proposed mechanism for the generation and location of new division sites during the division cycle of *Escherichia coli*. *Proceedings of the National Academy of Sciences, USA*, **84**, 7144–8.

Corton, J. C., Ward Jr., J. E. & Lutkenhaus, J. F. (1987). Analysis of cell division gene *ftsZ* (*sulB*) from Gram-negative and Gram-positive bacteria. *Journal of Bacteriology*, **169**, 1–7.

Crickmore, N. & Salmond, G. P. C. (1986). The *Escherichia coli* heat shock regulatory gene is immediately downstream of a cell division operon: the *fam* mutation is allelic with *rpoH*. *Molecular and General Genetics*, **205**, 535–9.

Dai, K. & Lutkenhaus, J. (1991). *ftsZ* is an essential cell division gene in *Escherichia coli*. *Journal of Bacteriology*, **173**, in press.

D'Ari, R. & Huisman, O. (1983). Novel mechanism of cell division inhibition associated with the SOS response in *Escherichia coli*. *Journal of Bacteriology*, **156**, 243–50.

de Boer, P. A. J., Cook, W. R. & Rothfield, L. I. (1990). Bacterial cell division. *Annual Review of Genetics*, **24**, 249–74.

de Boer, P. A. J., Crossley, R. & Rothfield, L. I. (1989). A division inhibitor and a topological specificity factor coded for by the minicell locus determine proper placement of the division septum in *Escherichia coli*. *Cell*, **56**, 641–9.

de Boer, P. A. J., Crossley, R. & Rothfield, L. I. (1990). Central role for the *Escherichia coli minC* gene product in two different cell division inhibition systems. *Proceedings of the National Academy of Sciences, USA*, **87**, 1129–33.

Dewar, S. J., Kagan-Zur, V., Begg, K. J. & Donachie, W. D. (1989). Transcriptional regulation of cell division genes in *Escherichia coli*. *Molecular Microbiology*, **3**, 1371–7.

Donachie, W. D. & Begg, K. J. (1989). Cell length, nucleoid segregation, and cell division of rod-shaped and spherical cells of *Escherichia coli*. *Journal of Bacteriology*, **171**, 4633–9.

Donachie, W. D., Begg, K. J. & Sullivan, N. F. (1984). Morphogenes of *Escherichia coli*, pp. 27–62. In *Microbial Development*, eds, R. Losick and L. Shapiro. Cold Spring Harbor Laboratory, Cold Spring Harbor, NY.

Donch, J., Chung, Y. S. & Greenberg, J. (1969). Locus for radiation resistance in *Escherichia coli* B/r. *Genetics*, **61**, 363–70.

Drapeau, G. R., Chausseau, J. P. & Gariepy, F. (1983). Unusual properties of a new division mutant of *Escherichia coli. Canadian Journal of Microbiology*, **29**, 694–9.

Dwek, R., Or-Gad, S., Rozenhak, S. & Ron, E. Z. (1984). New cell division mutations in *Escherichia coli* map near the terminus of chromosome replication. *Molecular and General Genetics*, **193**, 379–81.

Faubladier, M., Cam, K. & Bouche, J.-P. (1990). *Escherichia coli* cell division inhibitor DicF-RNA of the *dicB* operon. Evidence for its generation *in vivo* by transcription termination and by RNase III and RNase E-dependent processing. *Journal of Molecular Biology*, **212**, 461–71.

Ferreira, L. C. S., Keck, W., Betzner, A. & Schwartz, U. (1987). *In vivo* cell division gene product interactions in *Escherichia coli* K12. *Journal of Bacteriology*, **169**, 5776–81.

Fletcher, G., Irwin, C. A., Henson, J. M., Fillingim, C., Malone, M. M. & Walker, J. R. (1978). Identification of the *Escherichia coli* cell division gene *sep* and organization of the cell division-cell envelope genes in the *sep-mur-ftsA-envA* cluster as determined with specialized transducing lambda bacteriophages. *Journal of Bacteriology*, **133**, 91–100.

Freudl, R., Braun, G., Honore, N. & Cole, S. T. (1987). Evolution of the enterobacterial *sulA* gene: a component of the SOS system encoding an inhibitor of cell division. *Gene*, **52**, 31–40.

Gayda, R. C., Yamamoto, L. T. & Markovitz, A. (1976). Second-site mutations in *capR* (*lon*) strain of *Escherichia coli* K12 that prevent radiation sensitivity and allow bacteriophage lambda to lysogenize. *Journal of Bacteriology*, **127**, 1208–16.

George, J., Castellazzi, M. & Buttin, G. (1975). Prophage induction and cell division in *Escherichia coli* III. Mutations *sfiA* and *sfiB* restore division in *tif* and *lon* strains and permit the expression of mutator properties of *tif. Molecular and General Genetics*, **140**, 309–22.

Gill, D. R., Hatfull, G. F. & Salmond, G. P. C. (1986). A new cell division operon in *Escherichia coli. Molecular and General Genetics*, **205**, 134–45.

Gill, D. R. & Salmond, G. P. C. (1987). The *Escherichia coli* cell division proteins FtsY, FtsE and FtsX are inner membrane-associated. *Molecular and General Genetics*, **210**, 504–8.

Gill, D. r. & Salmond, G. P. C. (1990). The identification of the *Escherichia coli* FtsY gene product: an unusual protein. *Molecular Microbiology*, **4**, 575–83.

Gomez, M. J., Merchante, R. & Ayala, J. A. (1990). Is *dinA* an allele of *orfB*? EMBO Workshop. The Bacterial Cell Cycle: structural and molecular aspects, 90.

Gottesman, S., Halpern, E. & Trisler, P. (1981). Role of *sulA* and *sulB* in filamentation by *lon* mutants of *Escherichia coli* K12. *Journal of Bacteriology*, **148**, 265–73.

Hara, H., Nishumura, Y., Kato, J.-I., Suzuki, H., Nagasawa, H., Suzuki, A. & Hirota, Y. (1989). Genetic analysis of processing involving C-terminal cleavage in penicillin-binding protein 3 of *Escherichia coli. Journal of Bacteriology*, **171**, 5882–9.

Harry, E. J. & Wake, R. G. (1989). Cloning and expression of a *Bacillus subtilis* division initiation gene for which a homolog has not been found in another organism. *Journal of Bacteriology*, **171**, 6835–9.

Helmstetter, C. E. & Leonard, A. C. (1987). Coordinate initiation of chromosome and minichromosome replication in *Escherichia coli. Journal of Bacteriology*, **169**, 3489–94.

Helmstetter, C. E. & Pierucci, O. (1968). Cell division during inhibition of deoxyribonucleic acid synthesis in *Escherichia coli. Journal of Bacteriology*, **95**, 1627–33.

Hiraga, S., Niki, H., Ogura, T., Ichinose, C., Mori, H., Ezaki, B. & Jaffe, A. (1989). Chromosome partitioning in *Escherichia coli*: novel mutants producing anucleate cells. *Journal of Bacteriology*, **171**, 1496–505.

Hirota, Y., Ryter, A. & Jacob, F. (1968). Thermosensitive mutants of *Escherichia coli* affected in the process of DNA synthesis and cellular division. *Cold Spring Harbor Symposium on Quantitative Biology*, **33**, 677–93.

Holland, I. B. & Darby, V. (1976). Genetical and physiological studies on a thermosensitive mutant of *Escherichia coli* defective in cell division. *Journal of General Microbiology*, **92**, 156–66.

Howard-Flanders, P., Simson, E. & Theriot, L. (1964). A locus that controls filament formation and sensitivity to radiation in *Escherichia coli* K12. *Genetics*, **49**, 237–46.

Huisman, O. & D'Ari, R. (1981). An inducible DNA replication–cell division coupling mechanism of *Escherichia coli. Nature*, London, **290**, 797–9.

Huisman, O., D'Ari, R. & George, J. (1980). Further characterization of *sfiA* and *sfiB* mutations in *Escherichia coli. Journal of Bacteriology*, **144**, 185–91.

Huisman, O., D'Ari, R. & Gottesman, S. (1984). Cell division control in *Escherichia coli*: specific induction of the SOS function SfiA protein is sufficient to block septation. *Proceedings of the National Academy of Sciences, USA*, **81**, 4490–4.

Ikeda, M., Wachi, M., Jung, H. K., Ishino, F. & Matsuhashi, M. (1991). The *Escherichia coli mraY* gene encoding UDP-*N*-acetylmuramoyl-pentapeptide: undecaprenyl-phosphatephospho-*N*-acetylmuramoyl-pentapeptidetransferase. *Journal of Bacteriology*, **173**, 1021–6.

Irwin, C. A., Fletcher, G., Gills, C. L. & Walker, J. R. (1979). Expression of the *Escherichia coli* cell division gene *sep* cloned in a λ Charon phage. *Science*, **206**, 220–2.

Ishino, F., Jung, H. K., Ikeda, M., Doi, M., Wachi, M. & Matsuhashi, M. (1989). New mutations *fts-36*, *lts-33*, and *ftsW* clustered in the *mra* region of the *Escherichia coli* chromosome induce thermosensitive cell growth and division. *Journal of Bacteriology*, **171**, 5523–30.

Jaffe, A., D'Ari, R. & Norris, V. (1986). SOS-independent coupling between DNA replication and cell division in *Escherichia coli. Journal of Bacteriology*, **165**, 66–71.

Johnson, B. F. (1977). Fine structure mapping and properties of mutations suppressing the *lon* mutation in *Escherichia coli* K12 and B strains. *Genetic Research*, **30**, 273–286.

Johnson, B. F. & Greenberg, J. (1975). Mapping of *sul*, the suppressor of *lon* in *Escherichia coli. Journal of Bacteriology*, **122**, 570–4.

Jones, C. A. & Holland, I. B. (1984). Inactivation of essential genes *ftsA*, *ftsZ*, suppresses mutations at *sfiB*, a locus mediating division inhibition during the SOS response in *Escherichia coli. EMBO Journal*, **3**, 1181–6.

Jones, C. A. & Holland, I. B. (1985). Role of SfiB (FtsZ) protein in division inhibition during the SOS response in *Escherichia coli.* FtsZ stabilizes the inhibitor SfiA in maxicells. *Proceedings of the National Academy of Sciences, USA*, **82**, 6045–9.

Jung, H. K., Ishino, F. & Matsuhashi, M. (1989). Inhibition of growth of *ftsQ*, *ftsA*, and *ftsZ* mutant cells of *Escherichia coli* by amplification of a chromosomal region encompassing closely aligned cell division and cell growth genes. *Journal of Bacteriology*, **171**, 6379–82.

Kren, B. & Fuchs, J. A. (1987). Characterization of the *ftsB* gene as an allele of the *nrdB* gene in *Escherichia coli. Journal of Bacteriology*, **169**, 14–18.

LaBell, T. L., Trempy, J. E. & Haldenwany, W. G. (1987). Sporulation-specific factor[29] of *Bacillus subtilis* is synthesized from a precursor protein, P31. *Proceedings of the National Academy of Sciences, USA*, **84**, 1784–8.

Labie, C., Bouche, F. & Bouche, J.-P. (1989). Isolation and mapping of *Escherichia coli* mutations conferring resistance to division inhibitor DicB. *Journal of Bacteriology*, **171**, 4315–19.

Labie, C., Bouche, F. & Bouche, J.-P. (1990). Minicell-forming mutants of *Escherichia coli*: suppression of both DicB and MinD-dependent division inhibition by inactivation of the *minC* gene product. *Journal of Bacteriology*, **172**, 5852–5.

Leclerc, G., Sirard, C. & Drapeau, G. (1989). The *Escherichia coli* cell division mutation *ftsM1* is in *serU*. *Journal of Bacteriology*, **171**, 2090–5.

Lutkenhaus, J. F. (1983). Coupling of DNA replication and cell division: *sulB* is an allele of *ftsZ*. *Journal of Bacteriology*, **154**, 1339–46.

Lutkenhaus, J. F. & Donachie, W. D. (1979). Identification of the *ftsA* gene product. *Journal of Bacteriology*, **137**, 1088–94.

Lutkenhaus, J. F., Wolf-Watz, H. & Donachie, W. D. (1980). Organization of genes in the *ftsA–envA* region of the *Escherichia coli* genetic map and identification of a new *fts* locus (*ftsZ*). *Journal of Bacteriology*, **142**, 615–20.

Lutkenhaus, J. F., Sanjanwala, B. & Lowe, M. (1986). Overproduction of FtsZ suppresses sensitivity of *lon* mutants to division inhibition. *Journal of Bacteriology*, **166**, 756–62.

MacAlister, T. J., MacDonald, B. & Rothfield, L. I. (1983). The periseptal annulus: an organelle associated with cell division in Gram-negative bacteria. *Proceedings of the National Academy of Sciences, USA*, **80**, 1372–6.

MacAlister, T. J., Cook, W. R., Weigard, R. & Rothfield, L. I. (1987). Membrane–murein attachment at the leading edge of the division septum: second membrane–murein structure associated with morphogenesis of the Gram-negative bacterial division septum. *Journal of Bacteriology*, **169**, 3945–51.

Maguin, E., Lutkenhaus, J. F. & D'Ari, R. (1986). Reversibility of SOS-associated division inhibition in *Escherichia coli*. *Journal of Bacteriology*, **166**, 733–8.

Maguin, E., Brody, H., Hill, C. W. & D'Ari, R. (1986). SOS-associated division inhibition gene *sfiC* is part of excisable element *e14* in *Escherichia coli*. *Journal of Bacteriology*, **168**, 464–6.

Markovitz, A. (1964). Regulatory mechanisms for synthesis of capsular polysaccharides in mucoid mutants of *Escherichia coli* K12. *Proceedings of the National Academy of Sciences, USA*, **51**, 239–46.

Masters, M., Paterson, T., Popplewell, A. G., Owen-Hughes, T., Pringle, J. H. & Begg, K. J. (1989). The effect of DnaA protein levels and the rate of initiation at *oriC* on transcription originating in the *ftsQ* and *ftsA* genes: *in vivo* experiments. *Molecular and General Genetics*, **216**, 475–83.

Miyakawa, Y. & Komano, T. (1981). Study on the cell cycle of *Bacillus subtilis* using temperature-sensitive mutants 1. Isolation and genetic analysis of the mutants defective in septum formation. *Molecular and General Genetics*, **181**, 207–14.

Mizusawa, S. & Gottesman, S. (1983). Protein degradation in *Escherichia coli*: the *lon* gene controls stability of SulA protein. *Proceedings of the National Academy of Sciences, USA*, **80**, 358–62.

Mulder, E. & Woldringh, C. L. (1989). Actively replicating nucleoids influence positioning of division sites in *Escherichia coli* filament-forming cells lacking DNA. *Journal of Bacteriology*, **171**, 4303–14.

Mulder, E. & Woldringh, C. L. (1990). The *min* phenotype: DNA conformation and cell division. EMBO Workshop. The Bacterial Cell Cycle: structural and molecular aspects, 118.

Nagai, K. & Tamura, G. (1972). Mutant of *Escherichia coli* with thermosensitive protein in the process of cellular division. *Journal of Bacteriology*, **112**, 959–66.

Nagasawa, H., Sakagami, Y., Suzuki, A., Suzuki, H., Hara, H. & Hirota, Y. (1989). Determination of the cleavage site involved in C-terminal processing of penicillin-binding protein 3 of *Escherichia coli*. *Journal of Bacteriology*, **171**, 5890–3.

Nakamura, M., Naruyama, I. N., Soma, M., Kato, J.-I., Suzuki, H. & Hirota, Y. (1983). On the process of cellular division in *Escherichia coli*: nucleotide sequence of the gene for penicillin-binding protein 3. *Molecular and General Genetics*, **191**, 1–9.

Normark, S. (1970). Genetics of a chain-forming mutant of *Escherichia coli*: transduction and dominance of the *envA* gene mediating increased penetration to some antibacterial agents. *Genetic Research*, **16**, 63–70.

Normark, S., Boman, H. G. & Matsson, E. (1969). Mutant of *Escherichia coli* with anomalous cell division and ability to decrease episomally and chromosomally mediated resistance to ampicillin and several other antibiotics. *Journal of Bacteriology*, **97**, 1334–42.

Nukushina, J.-I. & Ikeda, Y. (1969). Genetic analysis of the developmental processes during germination and outgrowth of *Bacillus subtilis* spores with temperature-sensitive mutants. *Genetics*, **63**, 63–74.

Ogura, T., Tomoyasu, T., Yuki, T., Morimura, S., Begg, K. J., Donachie, W. D., Mori, H., Niki, H. & Hiraga, S. (1990). Structure and function of the *ftsH* gene in *Escherichia coli*. EMBO Workshop. The Bacterial Cell Cycle: structural and molecular aspects, 100.

Oliver, D. B. & Beckwith, J. (1981). *E. coli* mutant pleiotropically defective in the export of secreted proteins. *Cell*, **25**, 765–72.

Piggot, P. & Hoch, J. A. (1985). Revised genetic linkage map of *Bacillus subtilis*. *Microbiological Reviews*, **49**, 158–79.

Pla, J., Dopazo, A. & Vicente, M. (1990). The native form of FtsA, a septal protein of *Escherichia coli*, is located in the cytoplasmic membrane. *Journal of Bacteriology*, **173**2, 5097–102.

Prats, R., Gomez, M., Pla, J., Blasco, B. & Ayala, J. A. (1989). A new β-lactam-binding protein derived from penicillin-binding protein 3 of *Escherichia coli*. *Journal of Bacteriology*, **171**, 5194–8.

Raina, S., Karow, M., Fayet, O., Lipinska, B. & Georgopoulos, C. (1990). Complex phenotypes of null mutations in the *htr* genes, whose products are essential for *Escherichia coli* at elevated temperatures. EMBO Workshop. The Bacterial Cell Cycle: structural and molecular aspects, 92.

Reeve, J. N., Groves, D. J. & Clark, D. J. (1970). Regulation of cell division in *Escherichia coli*: characterization of temperature-sensitive division mutants. *Journal of Bacteriology*, **104**, 1052–64.

Reeve, J. N., Mendelson, N. M., Coyne, S. I., Hallock, L. L. & Cole, R. M. (1973). Minicells of *Bacillus subtilis*. *Journal of Bacteriology*, **114**, 860–73.

Ricard, M. & Hirota, Y. (1973). Process of cellular division in *Escherichia coli*: physiological study on thermosensitive mutants defective in cell division. *Journal of Bacteriology*, **116**, 314–22.

Robin, A., Joseleau-Petit, D. & D'Ari, R. (1990). Transcription of the *ftsZ* gene and cell division in *Escherichia coli*. *Journal of Bacteriology*, **172**, 1392–9.

Robinson, A. C., Kenan, D. L., Sweeney, J. & Donachie, W. D. (1986). Further evidence for overlapping transcriptional units in an *Escherichia coli* cell envelope-cell division gene cluster: DNA sequence and transcriptional organization of the *ddl ftsQ* region. *Journal of Bacteriology*, **167**, 809–17.

Romisch, K., Webb, J., Herz, J., Prehn, S., Frank, R., Vingron, M. & Dobberstein,

B. (1989). Homology of 54kD protein of signal recognition particle, docking protein and two *E. coli* proteins with putative GTP-binding domains. *Nature*, London, **340**, 478–82.

Rong, S., Rosenkrantz, M. S. & Sonenshein, A. L. (1986). Transcriptional control of the *Bacillus subtilis spoIID* gene. *Journal of Bacteriology*, **165**, 771–9.

Rothfield, L. I., Wang, X. & De Boer, P. A. J. (1990). The *sdiA* gene product resembles the FtsZ protein in its effects on cell division, including the ability to restore division activity to FtsZ$^-$ cells. EMBO Workshop. The Bacterial Cell Cycle: structural and molecular aspects, 95–6.

Santos, D. & Almeida, D. F. (1975). Isolation and characterization of a new temperature-sensitive cell division mutant of *Escherichia coli* K12. *Journal of Bacteriology*, **124**, 1502–7.

Salmond, G. P. C. & Plakidou, S. (1984). Genetic analysis of essential genes in the *ftsE* region of the *Escherichia coli* genetic map and identification of a new cell division gene, *ftsS*. *Molecular and General Genetics*, **197**, 304–8.

Spratt, B. G. (1975). Distinct penicillin binding proteins involved in the division, elongation and shape of *Escherichia coli* K12. *Proceedings of the National Academy of Sciences, USA*, **72**, 2999–3003.

Spratt, B. G. (1977). Temperature-sensitive cell division mutants of *Escherichia coli* with thermolabile penicillin-binding protein. *Journal of Bacteriology*, **131**, 293–305.

Storts, D. R., Aprico, O. M., Schoemaker, L. M. & Markovitz, A. (1989). Overproduction and identification of the *ftsQ* gene product, an essential cell division protein in *Escherichia coli* K12. *Journal of Bacteriology*, **171**, 4290–7.

Stragier, P., Bonamy, C. & Karmazyn-Campelli, C. (1988). Processing of a sigma factor in *Bacillus subtilis*: How morphological structure could control gene expression. *Cell*, **52**, 697–704.

Stragier, P. & Losick, R. (1990). Cascades of sigma factors revisited. *Molecular Microbiology*, **4**, 1801–6.

Taschner, P. E. M., Verest, J. G. & Woldringh, C. L. (1987). Genetic and morphological characterization of *ftsB* and *nrdB* mutants of *Escherichia coli*. *Journal of Bacteriology*, **169**, 19–25.

Taschner, P. E. M., Huls, P. G., Pas, E. & Woldringh, C. L. (1988*a*). Division behavior and shape changes in isogenic *ftsZ*, *ftsQ*, *ftsA*, and *ftsE* cell division mutants of *Escherichia coli* during temperature shift experiments. *Journal of Bacteriology*, **170**, 1533–40.

Taschner, P. E. M., Ypenburg, N., Spratt, B. G. & Woldringh, C. L. (1988*b*). An amino acid substitution in penicillin-binding protein 3 creates pointed polar caps in *Escherichia coli*. *Journal of Bacteriology*, **170**, 4828–37.

Teather, R. M., Collins, J. F. & Donachie, W. D. (1974). Quantal behavior of a diffusible factor which initiates septum formation at potential division sites in *Escherichia coli*. *Journal of Bacteriology*, **118**, 407–13.

Tormo, A., Ayala, J. A., De Pedro, M. A., Aldea, M. & Vicente, M. (1986). Interaction of FtsA and PBP3 proteins in the *Escherichia coli* septum. *Journal of Bacteriology*, **166**, 985–92.

Tormo, A., Matinez-Salas, E. & Vicente, M. (1980). Involvement of the *ftsA* gene product in later stages of the *Escherichia coli* cell cycle. *Journal of Bacteriology*, **141**, 806–13.

Tormo, A. & Vicente, M. (1984). The *ftsA* gene product participates in formation of the *Escherichia coli* septum structure. *Journal of Bacteriology*, **157**, 779–84.

Tsuchido, T., VanBogelen, R. A. & Neidhardt, F. C. (1986). Heat shock response in *Escherichia coli* influences cell division. *Proceedings of the National Academy of Sciences, USA*, **83**, 6959–63.

Van Alstyne, D. & Simon, M. I. (1971). Division mutants of *Bacillus subtilis*: Isolation and PBS1 transduction of division-specific markers. *Journal of Bacteriology*, **108**, 1366–79.

Van de Putte, P., Van Dillewijn, J. & Rorsch, A. (1964). The selection of mutants of *Escherichia coli* with impaired cell division at elevated temperature. *Mutation Research*, **1**, 121–8.

Walker, G. C. (1984). Mutagenesis and inducible responses to deoxyribonucleic acid damage in *Escherichia coli*. *Microbiological Reviews*, **48**, 60–93.

Ward Jr., J. E. & Lutkenhaus, J. F. (1985). Overproduction of FtsZ induces minicell formation in *E. coli*. *Cell*, **42**, 941–9.

Wientjes, F. B. & Nanninga, N. (1989). Rate and topography of peptidoglycan synthesis during cell division in *Escherichia coli*: concept of a leading edge. *Journal of Bacteriology*, **171**, 3412–19.

Wijsman, H. J. W. (1972). A genetic map of several mutations affecting the mucopeptide layer of *Escherichia coli*. *Genetic Research*, **20**, 65–74.

Wijsman, H. J. W. & Koopman, C. R. M. (1976). The relation of the genes *envA* and *ftsA* in *Escherichia coli*. *Molecular and General Genetics*, **147**, 99–102.

Witkin, E. M. (1947). Genetics of resistance to radiation in *Escherichia coli*. *Genetics*, **32**, 221–48.

Witkin, E. M. (1967). The radiation sensitivity of *Escherichia coli* B: a hypothesis relating filament formation and prophage induction. *Proceedings of the National Academy of Sciences, USA*, **57**, 1275–9.

Woldringh, C. L., Huls, P., Pas, E., Brakenhoff, G. J. & Nanninga, N. (1987). Topography of peptidoglycan synthesis during elongation and polar cap formation in a cell division mutant of *Escherichia coli* MC4100. *Journal of General Microbiology*, **133**, 575–86.

Woldringh, C. L., Mulder, E., Valkenberg, J. A. C., Wientjes, F. B., Zaritsky, A. & Nanninga, N. (1990). Role of the nucleoid in the toporegulation of division. *Research in Microbiology*, **141**, 39–49.

Wolf-Watz, H. & Normark, S. (1976). Evidence for a role of *N*-acetylmuramyl-L-alanine amidase in septum separation in *Escherichia coli*. *Journal of Bacteriology*, **128**, 580–6.

Yansura, D. G. & Henner, D. J. (1984). Use of the *Escherichia coli lac* repressor and operator to control gene expression in *Bacillus subtilis*. *Proceedings of the National Academy of Sciences, USA*, **81**, 439–43.

Yi, Q.-M., Rockenbach, S., Ward Jr., J. E. & Lutkenhaus, J. F. (1985). Structure and expression of the cell division genes *ftsQ*, *ftsA*, and *FtsZ*. *Journal of Molecular Biology*, **184**, 399–412.

Yi, Q.-M. & Lutkenhaus, J. F. (1985). The nucleotide sequence of the essential cell division gene *ftsZ* of *Escherichia coli*. *Gene*, **36**, 241–7.

Zehnbauer, B. A., Foley, E. C., Henderson, G. W. & Markovitz, A. (1981). Identification and purification of the lon^+ ($capR^+$) gene product, a DNA-binding protein. *Proceedings of the National Academy of Sciences, USA*, **78**, 2043–7.

COMPARISON OF THE PROKARYOTIC AND EUKARYOTIC CELL CYCLES

A. E. WHEALS

Microbiology Group, School of Biological Sciences, University of Bath, Bath, BA2 7AY, UK

INTRODUCTION

The cell cycle of a growing cell has been defined as the period between the formation of the cell by the division of its mother cell and the time when it itself divides to form two daughters (Mitchison, 1971). Such a definition is perhaps too restricted because it excludes consideration of cells which are not growing (i.e. increasing in biomass), those that do not divide, those that divide into more than two cells, and those that are undergoing terminal differentiation. A more recent definition is much broader defining the cell cycle as the period during which events required for successful cell reproduction are completed (Nurse, 1990). More evocatively, the cell cycle has been called the developmental biology of the cell (Mitchison, 1971).

The title of this chapter hides the fact that detailed cell cycle knowledge is only available for a very limited number of organisms. In relation to prokaryotes mention will only be made of *Escherichia coli*, *Bacillus subtilis*, *Enterococcus hirae* and *Caulobacter crescentus*, with not an archaebacterium in sight. With respect to microbial eukaryotes the spectrum of detailed knowledge is wider but only work with *Saccharomyces cerevisiae*, *Schizosaccharomyces pombe*, *Physarum polycephalum* and *Aspergillus nidulans* will be referred to. If prokaryotes and eukaryotes had a common origin, with the acquisition of aerobic metabolism in eukaryotes occurring by symbiotic association with ancestral eubacteria, then we must presume that solving the problem of the cell cycle predated this split. There is every prospect therefore that, on the principle of conservation of evolutionary function, we might find both structural and functional similarities between these two groups (Woese, Kandler & Wheelis, 1990; Mayr, 1990). On the other hand, we might also find major differences reflecting the fundamental divergence in their cell biology. Also the organisms we shall be looking at are free-living and we have a totally inadequate understanding of parasitic protozoa and symbiotic organisms whose rate of replication may be coupled to that of the host.

What are the problems that have to be solved in non-differentiating, growing and dividing cells? First, all cellular material has to be duplicated. With regard to material that is present in large amounts (such as enzymes, ribosomes, membrane components, etc.) exact duplication would be hard to achieve, and unnecessary provided gross deficiencies and excesses in one cycle are compensated for subsequently. With regard to material that is present in unique, or very low numbers (such as DNA molecules, nuclei, centrosomes, etc.) this has to be duplicated exactly, which means there must be a mechanism to duplicate these materials once, and only once, per cell cycle. These increases in amount and number need to be coordinated. Secondly, the cellular material has to be divided amongst the daughter cells. For essential genetic material this has to be a highly accurate process since errors could be lethal. For non-genetic material equipartition is not essential provided there is some mechanism which adjusts errors. Failure to do this, accompanied by additional errors at each generation, could lead to cells with a very atypical, and possibly deleterious, composition. Thirdly, since morphogenesis accompanies growth, it is essential that the correct structures are made in the right place, emphasising the spatial aspect of the cell cycle. Fourthly, cells should be able to respond appropriately to lack or withdrawal of nutrients by ceasing proliferation, and to other signals requiring a developmental response. Prolonged arrest may be accompanied by a reversible differentiation process or by terminal differentiation (e.g. sporulation). There is thus likely to be at least one decision point in the cell cycle where commitment to different cellular fates occurs.

After a brief historical account, there are three major sections comparing cell cycle biology. The third section covers the nature of dependency relationships, the fourth section describes our understanding of cell cycle transitions and the timing of events, and the fifth section highlights our inadequate knowledge of morphogenesis.

HISTORICAL PERSPECTIVE

Modern cell cycle biology stems from the work of Howard and Pelc (1953) who first demonstrated that DNA replication occurred in mammalian cells only during a discrete (S) phase and there was a gap (G2) between S phase and mitosis (M) and between M phase and S phase (G1), a nomenclature which is still useful today because it is descriptive rather than deductive and presumptive. The pioneering work of Helmstetter and Cooper (Helmstetter & Cooper, 1968; Cooper & Helmstetter, 1968) on *E. coli* revealed the more complicated pattern of DNA synthesis in bacteria due to a constant replication time (C) followed by a constant time to division (D) which together could exceed cell cycle time. The present era has just come of age since it stems from Lee Hartwell's seminal work starting in 1970 (Hartwell, 1974; Hartwell *et al.*, 1974) on isolating cell division cycle (*cdc*) mutants of the yeast *S. cerevisiae*. Similar approaches were made in *S. pombe* by Nurse (1975), and others, and in *E. coli* by Hirota (Hirota, Ryter & Jacob, 1968), Donachie (Lutkenhaus & Donachie, 1979) and others. Modern molecular genetics has enabled the cloning and sequencing of these genes to

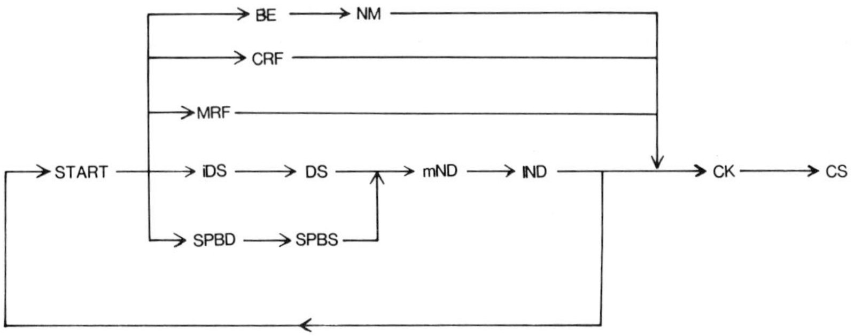

Fig. 1. Dependency relationships in the *Saccharomyces cerevisiae* cell cycle. The diagram indicates the order of cell cycle events and their dependency. Events on the right are dependent on events to the left if they are directly connected by arrows but mND, for example, is dependent on both DS and SPBS. Events on parallel lines are independent of each other even though they may be temporally simultaneous. All events are dependent on *start*. *Start* is not dependent on CK. BE, bud emergence; NM, nuclear migration; CRF, chitin ring formation; MRF, microfilament ring formation; iDS, initiation of DNA synthesis; DS, DNA synthesis; mND, medial nuclear division; lND, late nuclear division (mND and lND constitute mitosis); CK, cytokinesis, CS, cell separation; SPBD, spindle pole body duplication; SPBS, spindle pole body separation.

take place. *In vitro* mutagenesis, gene disruptions, gene replacements and over-expression systems have allowed, particularly in yeasts, a functional analysis of these genes to be made. This led to a major breakthrough in 1988 when it was possible to integrate much genetic, biochemical and physiological data from a wide range of eukaryotes into a coherent model of the regulation of mitosis (Lewin, 1990). In one aspect the work has progressed further in *E. coli* since it is now possible to study the initiation of DNA synthesis *in vitro* with defined components. Our understanding of cell cycle morphogenesis in both prokaryotes and eukaryotes is less well understood perhaps because much of it will require a global description of the process before we can understand the contribution of the parts (Harold, 1990).

DEPENDENCY RELATIONSHIPS

Order-of-function mapping

One of the key aims of cell cycle research has been to understand how the correct ordering of cell cycle events is achieved. The first major achievement after isolating *cdc* mutants in *S. cerevisiae* was to make such an objective attainable. Jarvik and Botstein (1973) invented an experimental procedure where, by sequentially applying two reversible blocks to the completion of a biological developmental process and then repeating the experiment in the reverse order, they were able to ascertain dependency relationships of discrete events. Hartwell and his colleagues (Hartwell, 1974) applied similar analyses to cell cycle events of *S. cerevisiae* and the results are shown in Fig. 1 (after Wheals, 1987). There are four important conclusions from the diagram: First, there is a clear beginning described by a single pathway including those events associated with the crucial decision step called *start*. *Start* is defined experimentally as a point of arrest which is interdependent with the arrest point of mating pheromones and traverse of which commits a cell to the mitotic cycle. Secondly, many of the genetic functions are interdependent, that is, all are involved in the same process, each is necessary for its completion but none is sufficient on its own to allow transit through that particular gate. The *start* genes comprise such a set. Thirdly, the pathways branch into several independent routes, events on one branch proceeding independently of those on another so a block on one pathway does not prevent progression of the other(s). The prime pathway is that of DNA synthesis and mitosis describing a closed circular route. Ancillary pathways are mostly to do with the events of budding. Fourthly, the pathways can converge at some points, most notably at mitosis which requires prior completion of both DNA synthesis and nuclear migration suggesting that the chromosomes have to be both replicated and in the correct location. Fifthly, it is possible to reinitiate the cycle without prior completion of cytokinesis and cell separation. This conclusion is in agreement with results from other organisms that do not have the dispensable events of cell division, such as the plasmodial stage of the Myxomycete *Physarum polycephalum*. It emphasises that in eukaryotes, unlike prokaryotes, cell division is not part of the cell cycle.

Having discovered the existence of dependency relationships the question arises as to the reason for this dependency. One of the early concepts of how to explain ordered cell cycle progression was that of a genetic program. That is, events A and B would occur in that order because gene A would execute A functions and then gene B would execute B functions and B would follow A because one of A functions would be to 'activate' gene B and/or its products. Two kinds of activation are possible. One is by transcriptional and/or translational control of the relevant genes leading to the *de novo* appearance of the gene product, and the other is by activation of pre-existing but inactive protein. There is increasing evidence that both methods of control are used in the cell but the former method is perhaps used more for production of proteins required in large amounts during particular phases of the cell cycle (pp. 156–61), and the latter method of control is more used for regulation of cell cycle progression (pp. 161–2). A third kind of dependency is possible, namely substrate limitation. In this case function B would not be executed until its substrate had appeared and this substrate is produced by function A (pp. 162–3).

Cell cycle periodicities and transcriptional control

The development of 2D-gel electrophoresis (O'Farrell, 1975) enabled large numbers of proteins to be separated and detected and, in suitable cases,

Table 1. S. cerevisiae *genes showing cell cycle periodic transcription*

Gene	Function	Reference
CDC6	DNA Replication	1
CDC8	Thymidylate kinase	2
CDC9	DNA ligase	3
CDC21	Thymidylate synthetase	4
RAD6	DNA repair/radiation sensitivity	5
POL1(CDC17)	DNA polymerase I	6
POL2	DNA polymerase II	7
POL3(CDC2)	DNA polymerase III	8
POL30	Proliferating Cell Nuclear Antigen	8
DPB2	DNA polymerase II subunit B	7
DPC2	DNA polymerase II subunit C	7
PRI1	DNA primase I	9
PRI2	DNA primase II	10
RNR1	Ribonucleotide reductase regulatory subunit	11
DBF2	A protein kinase	12
HO	HO endonuclease	13
SWI5	Mating type switching	14
TOP2	Topoisomerase II	15
CLN1	G1 cyclin	16
CLN2	G1 cyclin	16
HTA1 & *HTA2*	Histone 2A	17
HTB1 & *HTB2*	Histone 2B	17

References: 1. Zhou & Yong (1990); 2. White *et al.* (1987); 3. White *et al.* (1986); 4. Storms *et al.* (1984); 5. Kupiec & Simchen (1986); 6. Johnston *et al.* (1987); 7. L. H. Johnston & A. Sugino, personal communication; 8. Bauer & Burgers (1990); 9. Johnston *et al.* (1990*a*); 10. Foiani *et al.* (1989); 11. Elledge & Davis (1990); 12. Johnston *et al.* (1990*b*); 13. Nasmyth, (1983); 14. Nasmyth *et al.* (1987); 15. J. Nitiss quoted in Ref. 1; 16. Wittenberg *et al.* (1990); 17. Hereford *et al.* (1981).

their rate of synthesis to be quantified. Using appropriate synchronizing methodologies it was quite quickly shown for both *E. coli* and *S. cerevisiae* that most proteins showed no detectable variation or periodicities in amount or rate of synthesis through the cell cycle (Elliott & McLaughlin, 1978, 1979; Lutkenhaus *et al.*, 1979). The general limitation of this method was that it would not detect minor proteins (which could include important regulatory molecules) or very basic, very acidic or very insoluble proteins which are not recovered in the extraction system. However, there was a significant minority of proteins in *Saccharomyces cerevisiae* that did show clear periodicities although their function could not be assigned (Lorincz *et al.*, 1982). Thus, although bulk protein synthesis occurred at a constant differential rate, there was a minority of proteins whose periodic appearance might be important.

Modern gene cloning techniques have started to reveal the identity of these periodic proteins. The number of yeast genes showing periodic transcription is quite large (Table 1). All *S. cerevisiae* genes which have been isolated and which are involved in DNA synthesis or in synthesis of histones show periodic expression and cell cycle control. The other genes showing periodic transcription include some involved in DNA repair, two involved in traverse of *start* and one involved in mating type switching. At least four of the genes (encoding thymidylate kinase, thymidylate synthetase, DNA ligase and DNA polymerase I) are co-ordinately regulated (White *et al.*, 1987; Johnston *et al.*, 1987). It had previously been shown that all four histone genes were similarly regulated but White *et al.* (1987) showed that the H2A gene (and therefore by implication the other histones) was expressed at a distinctly later time in the cell cycle implying a different mechanism of cell cycle control (Johnston, 1990). The genes directly involved in DNA replication all share a common upstream element, an ACGCGT hexamer. Possession of this sequence can confer periodic expression on a non-cell cycle heterologous gene and this expression becomes co-incident with the other periodic genes containing this element. This element is not found in the histone genes or in *DBF2*, a cell cycle periodically expressed protein kinase that seems to have functions both in the initiation of DNA synthesis and in late nuclear division (Johnston *et al.*, 1990*b*). The element binds a factor called DSCl (DNA Synthesis Control) whose activity mirrors the expression of the genes to which it binds (Lowndes *et al.*, 1991).

Most *CDC* genes are not cell cycle regulated. The teleological explanation for periodic synthesis of DNA synthesis related genes could be a simple logistical one namely that the demands of DNA replication are so great during a small fraction of the cell cycle that extra substrate supplies are needed and this is best accommodated by a burst of synthesis of the relevant enzymes. Such a strategy would not necessarily be useful for bacteria that have overlapping rounds of DNA replication requiring a continuous (but not constant) supply of precursors. However, the *dnaC* gene of *Caulobacter crescentus* which has a role in DNA chain elongation, is periodically transcribed (Ohta *et al.*, 1990). In this case the periodic transcription is revealed because the cells divide asymmetrically into rapidly recycling stalk cells and swarmer cells which have a prolonged presynthetic phase before the onset of DNA synthesis (see Fig. 2a and Newton, 1989). Similar results might be seen in slow growing *E. coli* cells which have a similar long 'G1' period.

A periodic transcriptional event can also be found in *S. cerevisiae* cells which also show asymmetric division into mothers and daughters (Fig. 2b). In homothallic strains of *S. cerevisiae* the mating type can switch between MAT**a** and MATalpha due to transposition of 'silent' mating type genes into the *MAT* locus where they are then expressed. The initial transposition event is the creation of a site specific double stranded cut in the DNA by the HO (homothallic) endonuclease. This endonuclease is periodically expressed in the G1 phase of the cell cycle but only after execution of the *start* step as defined by the *CDC28* gene. This cell cycle specific regulation is at the level of transcription. Analysis of the complex HO endonuclease promoter reveals several upstream regulatory sites, one of which (URS2) confers on the gene cell cycle specific transcription. Two other products,

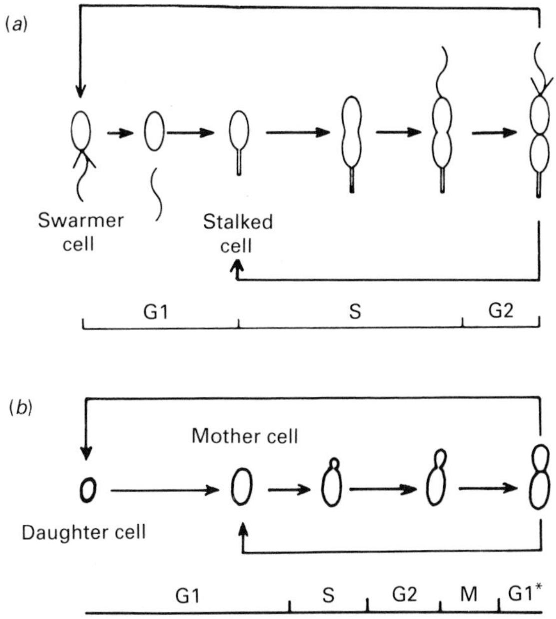

Fig. 2. Asymmetric cell cycles. a) *Caulobacter crescentus* cell cycle showing different fates of swarmer and stalked cells in relation to cell cycle stages (After Newton, 1989). b) *Saccharomyces cerevisiae* cell cycle showing fates of different sized mother and daughter cells. G1* is a G1 phase before cell separation. (After Wheals, 1987.)

of the *SWI4* and *SWI6* genes (Breeden & Nasmyth, 1987), are necessary for expression from this promoter but if this promoter is deleted then the HO endonuclease can still be expressed in a cell cycle manner but now during the whole of G1 rather than in post-*start* G1 (Nasmyth *et al.*, 1990). The means whereby mother and daughter cells behave differently is still unknown.

In bacteria, examples of cell cycle regulated transcriptional control are rare but two different kinds of promotion have been discovered. The first involves the *ftsZ* gene whose expression has been thought to be rate limiting for septation (Ward & Lutkenhaus, 1985). One study suggests that it is only transcribed at the time of septation (Dewar *et al.*, 1989). Another study suggests it is expressed throughout the cell cycle but shows a bilinear rate of transcription with a doubling in rate at a specific cell age close to that at which DNA replication is initiated and the rate of phospholipid synthesis doubles (Robin, Joseleau-Petit & d'Ari, 1990). A third study suggests that the promoter is a member of an unusual class of promoters, known as gearbox promoters, which regulate the transcription of the *ftsZ*, *ftsQ*, *ftsA*, and *bolA* genes all of which are involved in morphogenesis and septation (Aldea *et al.*, 1990). These promoters contain specific sequences at the −10 position upstream from the mRNA start point which are essential for the characteris-

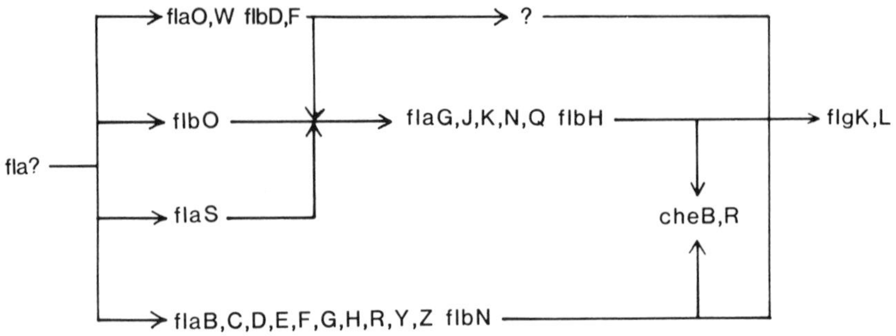

Fig. 3. Dependency relationship of flagellum biosynthesis in *Caulobacter crescentus*. The dependency relationships are as described in Fig. 1. The four letter codes define individual genes for flagellar biosynthesis. (After Newton 1989.)

tic of these promoters namely, activity that is inversely related to growth rate. The effect of this kind of control is that the gene products of these genes are synthesised at constant amounts per cell independently of cell size, that is, in a quantal fashion. This mode of expression would be expected for proteins that either play a regulatory role in cell division or form a stoichiometric component of the septum, a structure which is produced once per cell cycle. Since over-expression of *ftsZ* causes minicell production and over-expression of *bolA* causes a *ftsZ*-independent spherical morphology it is speculated that these four genes regulate the number of division sites. There is an analogous precedent in *S. cerevisiae* where altering the normal 1:1 stoichiometry of histone proteins causes up to a 50-fold increase in the frequency of mitotic loss of two different chromosomes (Meeks-Wagner & Hartwell, 1986).

The second example again involves *Caulobacter crescentus* (Newton, 1989). Flagellar formation occurs about 3/10ths of a cycle before division and the components of the flagellum are synthesized at that time. In part, this is due to transcriptional control. The relevant genes are organised in a number of separate gene clusters which are transcribed in a periodic manner. Furthermore, the exact timing of expression seems to correspond well with the order of assembly of the components of the flagellum. This is partly achieved by a hierarchy of controls, or a dependent sequence involving the four transcriptional units of the hook gene cluster, plus two flagellin genes (*flgK* and *flgL*). It is proposed that there are *trans* acting factors working in a cascade whose dependency relationships are shown in Fig. 3. Similar regulatory hierarchies for assembly have been discovered for the *fla* genes in *E. coli* and *Salmonella* but in these cases the regulation is also coupled to cell cycle progression. The nature of the signal is unknown. In *E. coli* the master control is a cyclic AMP/CRP complex (working in a non-cell cycle way). In *Caulobacter* there are suggestions that it is coupled to DNA synthesis. An analysis of the promoters of these genes reveals no canonical −10, −35 sequence but rather a consensus −12, −24 sequence also shared with a limited number of genes from other bacteria. These sequences are binding sites for sigma 54 and the gene activity is also controlled by an enhancer site at −100 which is thought to bind the transcription activator NRI. Although the sequence at −100 in the *fla* and *flg* genes is different it is also thought to act as an enhancer. As a working model it is proposed that the −100 site in transcription unit II requires a transcriptional activator encoded for by one of the genes of transcription unit III which is higher up in the hierarchy of control (Newton, 1989).

For any protein that does show periodicities in its amount through the cell cycle, a key question is whether the periodic pattern is necessary for cell cycle progression or whether it is sufficient that a certain amount be present at a particular cell cycle stage. Removal of the protein (by mutation) does not answer this question. The key experiment is to remove the cell cycle control such that expression is constitutive and levels are maintained high throughout the cycle. If this has no effect on cell cycle progression then it is possible to deduce that periodic expression is not regulatory for the cell cycle. One protein clearly does show this regulatory property. When the *Schizosaccharomyces pombe* gene *cdc25*$^{+}$ is overexpressed then cells prematurely enter mitosis (Russell & Nurse, 1986).

Protein activation

In addition to periodic appearance, periodic protein activation is another major aspect of cell cycle biology. This has been most clearly demonstrated in yeasts where a large number of cell division cycle genes, particularly those with a regulatory function, have been shown to encode protein kinases and/or be substrates for protein kinases. There are about 25 protein kinases so far described in *S. cerevisiae*. Several of these are associated with the cell cycle. All eukaryotic protein kinases have a similar structure (Hanks *et al.*, 1988) containing 11 definable domains, some of which can be assigned to specific reactions. The major differences between the kinases are in regions VI and VIII which determine whether they are protein–serine/threonine kinases or protein–tyrosine kinases, and in the N and C termini which presumably define their substrate specificity. Protein–serine/threonine kinases are abundant in lower eukaryotes and include the cell division cycle genes *CDC28*^{Sc}, *CDC7*^{Sc} and *cdc2*^{Sp} (see Wheals, 1987). Tyrosine kinases have been harder to find, although they are known to be present, because of their enzymic activity and by the existence of phosphotyrosine residues. Four have been found (encoded by the genes *wee1*$^{+}$ in *Schizosaccharomyces pombe* (Featherstone & Russell, 1991), *YPK1* in *S. cerevisiae* (Dailey *et al.*, 1990), *DPYK1* and *DPYK2* in *Dictyostelium discoideum* (Tan & Spudich, 1990)) and all have atypical structures perhaps representing a distinct subfamily of lower eukaryotic protein–tyrosine kinases. Most unusual is p107^{wee1}

which is a unique protein–tyrosine/serine kinase although the tyrosine phosphorylation activity has not been demonstrated *in vivo*.

A significant homology has been found between two *E. coli* genes involved in septation (*ftsA* and *fic*) and two yeast genes ($CDC28^{Sc}$ and $cdc2^{+Sp}$) involved crucially in both the initiation of DNA synthesis and the initiation of mitosis (Robinson, Collins & Donachie, 1987; Kawamukai *et al.*, 1989). Although convergent evolution cannot be eliminated, an alternative possibility is that the 60 amino acid region arose in a common ancestor and has been conserved. The yeast genes are both serine/threonine protein kinases. The region of homology with the *E. coli* genes is outside the ATP binding and phosphorylation sites but it could be part of a domain specifying either the interaction with an activating protein kinase or the site of the substrate specificity. The major role of protein kinases in cell cycle transitions is described below (pp. 165–73).

Checkpoints

The third mechanism of dependency mentioned on p. 156 is that of substrate limitation. There is at least one example where, although it had strong theoretical appeal, there is evidence against it. In both *E. coli* and yeasts it is possible to uncouple the dependency of division on DNA synthesis by mutations in particular genes. This suggests that it is not substrate limitation which determines the dependency relationship of division on DNA synthesis but that there are particular genes whose function is to maintain it. These have been described as checkpoints (Weinert & Hartwell, 1988) and several examples are known.

When wild-type *S. cerevisiae* G1 cells and most temperature sensitive *rad* mutants suffer radiation damage to their DNA they irreversibly arrest in G2 as single cells with a large bud. The exception is cells containing *rad9* which continue to divide and form microcolonies. Furthermore, many double *cdc rad9* mutants fail to show cell cycle arrest. Weinert and Hartwell (1988) suggested that the *RAD9* gene controls mitosis in response to DNA replication, serving as a checkpoint that assures complete DNA replication of DNA before mitosis is permitted. Mutation of the *RAD9* gene relieves cells of this dependency. However, there may be at least another checkpoint involving supply of DNA precursors. Ribonucleotide reductase (encoded by *RNR* genes) catalyses the reduction of ribonucleotides into their corresponding deoxyribonucleotides. Defects in the *RNR1* or *RNR2* genes (or addition of hydroxyurea, a specific inhibitor of ribonucleotide reductase) cases *RAD9* independent arrest and Elledge and Davis (1989*a*, *b*) suggest that any cells arrested with a cell cycle phenotype due to a block in DNA replication should be *rad9* independent, as is the case with a *cdc8* (thymidylate kinase) *rad9* strain.

In *E. coli* it is part of the design of the cell cycle that septation can be initiated while DNA replication is continuing, and new DNA replication initiation can start before old initiations have terminated. Nevertheless, there are checkpoints to prevent cell division if DNA replication is arrested, and three independent mechanisms intervene to block cell division: the SfiA (or SulA) and SfiC division inhibitors, induced as part of the SOS response (Huisman & d'Ari, 1981; d'Ari & Huisman, 1983), and a poorly understood SOS independent mechanism (Jaffe, d'Ari & Norris, 1986). One way in which this coupling between DNA synthesis and division is achieved is via induction of the *sfiA* gene (because of inactivation of the LexA repressor) whose gene product either inactivates or represses the *ftsZ* gene product which is essential for septum formation.

In fission yeast an elegant study of the role of the *cdc25* gene has revealed that it too has checkpoint capabilities (Enoch & Nurse, 1990). Mutations that circumvent the *cdc25*$^{+}$ pathway result in lack of dependency of mitosis on DNA synthesis, that is, cells arrested in S phase still attempt an abortive, and lethal, mitosis. However, circumvention of the *cdc25*$^{+}$ pathway has no effect on cells that are *not* committed to the mitotic cycle (and have thus not traversed the G1/S *start* step). The *cdc25*$^{+}$ gene product is thought to interact directly with the mitotic control gene *cdc2*$^{+}$.

There is also a negative regulator of mitosis in *Aspergillus nidulans* encoded by the *bimE* gene (Engle *et al.*, 1990). Mutations in this gene will not prevent entry into mitosis when S and G2 phase events are incomplete. However, unlike *RAD9* mutations in *S. cerevisiae*, *bimE* mutations also cause the cells to arrest in mitosis. The key difference between prokaryotes and eukaryotes is that there are regulatory circuits that ensure that there is an alternation of *initiation* of S and M phases in eukaryotes.

CELL CYCLE TRANSITIONS

G0-G1 transitions

When budding yeast cells are deprived of nutrients they arrest after mitosis and cell division at a stage which, in actively growing cells, would be called G1. However, there are a number of lines of evidence that strongly suggest that these cells are in a resting state and 'out of cycle' showing a number of analogies with a similar condition in mammalian cells termed G0 (Fig. 4). The major features of this state are (i) a general cessation of protein and other macromolecular synthesis, (ii) a specific induction of a number of proteins including heat shock (hsp) proteins (Boucherie, 1985), (iii) increased thermotolerance (Walton, Carter & Pringle, 1979), (iv) physiological changes in the cell wall making it more resistant to enzymic degradation (Deutch & Parry, 1974) and (v) accumulation of reserve carbohydrate (Lillie & Pringle, 1980). It is the response to stress that induces the properties of the G0 state rather than entry into the state itself (Drebot *et al.*, 1990).

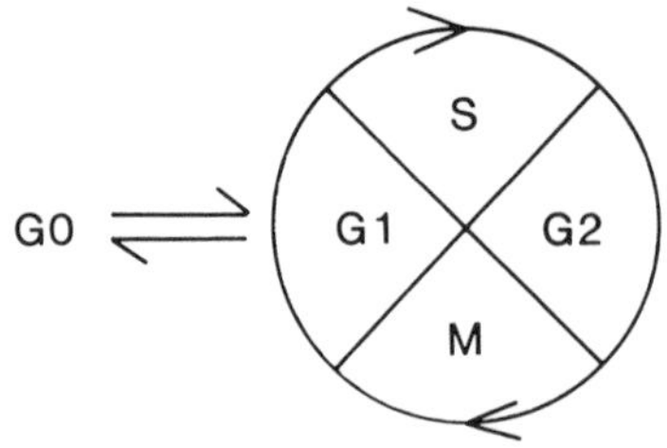

Fig. 4. Eukaryotic cell cycle. The diagram, which is read clockwise, illustrates the dependency relationships of the cell cycle and the reversible transition into the quiescent G0 phase from G1.

Something similar to this process in prokaryotes is the formation of the endospore in Gram positive bacteria (Losick *et al.*, 1987) or the myxospore in myxobacteria (Kaiser, 1987). However, the analogy may not be appropriate because these structures are the terminal differentiation stage of a vegetative phase created at the expense of a sibling cell at the last division or at the expense of colony lysis (see Dworkin, 1985). Something more akin to G0 in bacteria is the resting or somni-cell. The clearest evidence for this is in the budding bacteria which produce on division two different cell types, one of which can survive starvation conditions (Newton, 1989). Similar proposals have been made for those cells which do not have a clearly differentiated resting cell stage, such as *E. coli* (Matin *et al.*, 1989). Both pro- and eukaryotes enter the resting stage after division and before DNA replication although the chromosome in the unreplicated state would be at its most prone to DNA damaging agents and most difficult to repair in the absence of intact copies of the gene. In prokaryotes commitment to DNA synthesis is probably also commitment to division and arrest after DNA synthesis, but before division, may neither be prudent nor possible. In eukaryotes there is a second commitment point at the G2/M transition and it is curious that arrest does not happen here when there are two copies of a gene. Perhaps the retention of a pre-replicative decision point is an ontogenetic remnant of phylogenetic relatedness.

In *Saccharomyces cerevisiae* the evidence suggests that the effects of starvation are mediated through the cAMP pathway (Wheals, 1987; Drebot *et al.*, 1990) leading to the expression of G0 characteristics. One of the discoveries during the analysis of this system in *S. cerevisiae* was the existence of multiple (alternative?) forms of genes (see Gibbs & Marshall, 1989 and Table 2). It is by no means restricted to *CDC* genes and is not shown by the majority of *CDC* genes. It has been argued that it is a device to limit developmental error and maximise evolutionary flexibility (see Nasmyth, 1990). Since there are only data for *S. cerevisiae*, over interpretation should be eschewed. One of the more trivial causes of redundancy could be an over-active transposition system which is probably the basis of the processed pseudogene nature of

Table 2. *Genetic functions in* Saccharomyces cerevisiae *present in multiple copies.*

Genes	Protein	Reference
RAS1 & *RAS2*	G-proteins	[1]
PDE1 & *PDE2*	Phosphodiesterases	[1]
CLN1 & *CLN2* & *CLN3* (= *WHI1* = *DAF1*)	Cyclins	[2]
TPK1 & *TPK2* & *TPK3*	Protein kinases	[1]

References: 1. Gibbs and Marshall (1989); 2. Richardson *et al.* (1989).

the budding yeast genome.

It has become apparent that in other organisms cAMP has different roles. In *E. coli* the role of cAMP is to mediate catabolite repression without having a direct effect on growth (Beckwith, 1987). In *S. pombe* the role of cAMP is primarily to regulate sexual development although cells with no intracellular cAMP do have a reduced growth rate (Maeda, Mochizuki & Yamamoto, 1990). In the case of *S. cerevisiae* its activity is regulated by the G-proteins encoded by *RAS* genes (Gibbs & Marshall, 1989). The *ras* homologue in *S. pombe* does not control cAMP levels (Fukui *et al.*, 1986). It is perhaps an example of a biological sensing and transduction system which has been recruited for different purposes because of evolutionary pressure.

The question of return to the G1 state from the G0 state has been less well studied but at least in *S. cerevisiae* it is known to be genetically controlled (Drebot, Johnston & Singer, 1987) and mutants defective in re-entering the cell cycle can be isolated (Johnston & Singer, 1990). Experiments with chemostats, where it is possible to hold cells under conditions of very low nutrient availability, have shown that rather than all cells proliferating at low rates, a fraction of the cells ceases proliferating leaving the remainder of the cells to proliferate at a faster rate than is indicated by the dilution rate. However, the non-proliferating population has a finite probability of re-entering the cycle (Britton & Wheals, 1987). It is likely that this effect also applies to batch cultures (Plesset *et al.*, 1987). A related effect has been seen in *E. coli* (Koch & Coffman, 1970).

G1-S Transition and the initiation of DNA synthesis

Lee Hartwell's pioneering analysis of the *CDC* genes of budding yeast led to a remarkable insight into the initiation of the cell cycle. He showed that there was a unique point in G1, known as *start*, which represented a point of commitment to the mitotic cycle (Hartwell *et al.*, 1974). At *start* the cell assesses a number of informational inputs and, if favourable, the *start* process will be executed and the cell will progress through to S phase. The key

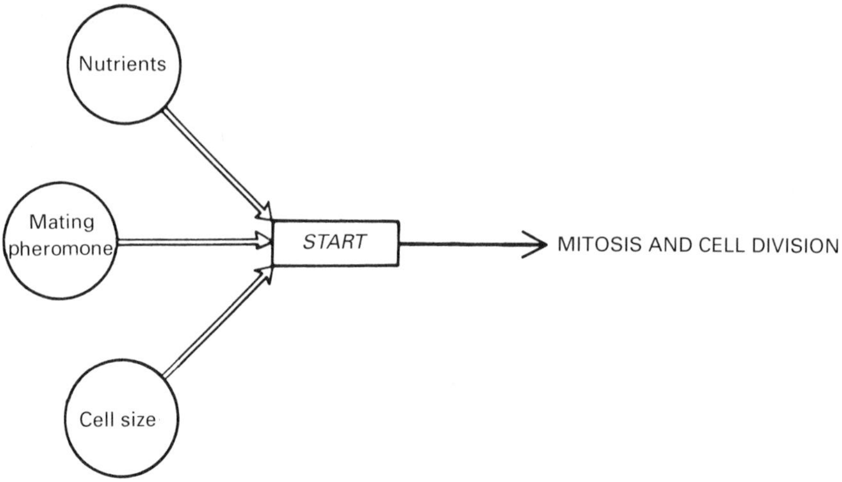

Fig. 5. Informational inputs into *start*. The diagram indicates the one internal and two external signals which have been identified as determining the rate at which cells traverse the physiological step known as *start* leading to mitosis and cell division.

inputs seem to be nutrient status, assessed by the cAMP pathway described above; presence or absence of mating pheromones, and cell size (Fig. 5).

Haploid *S. cerevisiae* cells secrete one of two mating pheromones which arrest cells of opposite mating type at *start*. Teleologically this is viewed as a synchronizing step which ensures that when two cells undergo cell fusion followed rapidly by nuclear fusion, that the two nuclei have the chromosomes in the same replicated state. Failure to achieve this could lead to various kinds of problems in replication and/or mitosis. This arrest is achieved by binding of the pheromone to a receptor in haploid cells of opposite mating type. The pheromone ligand binding to the receptor causes a G-protein heterotrimer to dissociate into a G_{alpha}-like monomer and a $G_{beta/gamma}$-like dimer which, released from repression, activates the response pathway (Blinder, Bouvier & Jenness, 1989; Whiteway *et al.*, 1989).

Although the physiological existence of a size control step is well established (see Wheals, 1987), its genetic basis has proved more elusive to ascertain. One gene that does affect the size at traverse of *start* is *WHI1*(*CLN3*). Hyperactive (truncated) mutations of this gene create small-sized cells whereas disruptions of the gene lead to enlarged cells. The *WHI1* gene (whose transcript level remains constant) is a homologue of metazoan cyclin genes which were originally discovered (and named) as cell cycle periodic proteins. Two other homologues have been isolated (*CLN1* and *CLN2*; Hadwiger *et al.*, 1989) which show cell cycle periodicities. Since the *CLN2* gene product is very unstable it has been suggested that it might be the accumulation of Cln2 which triggers *start* in relation to cell size (Nasmyth, 1990). Since there are also many interactions between the pheromone response system and the *CLN* genes (see Nasmyth, 1990) it could be that these are two aspects of one regulatory system.

The key regulatory gene for the traverse of *start* seems to be the *CDC28* gene which functions both at *start* (Hartwell *et al.*, 1974; Nasmyth, 1990) and at the G2/M boundary controlling initiation of mitosis (Reed & Wittenberg, 1990). Cdc28 is protein–serine/threonine kinase which can exist in complexes with other proteins and whose activity is necessary for entry into S phase (Wittenberg & Reed, 1988). Dephosphorylation of the protein, which reduces its protein kinase activity on model substrates *in vitro*, occurs during stationary phase and on entry into other non-proliferative states (Mendenhall, Jones & Reed, 1987). There is suggestive evidence that the Cln2 cyclin interacts directly with the Cdc28 protein kinase (Wittenberg, Sugimoto & Reed, 1990) and this evidence has been inspired by prior work on the *cdc2*$^{+}$ gene of fission yeast which is a structural and functional homologue of the *CDC28* gene and which similarly acts both at the G1/S and G2/M transitions (Nurse & Bissett, 1981; pp. 169–73).

The *start* process occurs well before the actual initiation of DNA synthesis and it initiates a set of events which proceeds into S-phase. *S. cerevisiae* cells contain in their chromosomes approximately 400 origins of replication. Initiation of DNA synthesis is not a single event as in *E. coli* but a multiple event which occurs sequentially as revealed by the timing of replication of individual replicons (see Newlon, 1988). Thus, although the *start* event is the prime initiator of the onset of events leading to the initiation of DNA synthesis, there are a series of events functionally downstream of *start* before the onset of DNA synthesis. This has been most clearly demonstrated by order of function mapping which places events mediated by *CDC28* upstream of *CDC4*, *DBF4* and *CDC7* gene functions. The Cdc7 protein kinase may be part of the replication complex (Jazwinski, 1988). One of the consequences of traversing *start* may be to produce an amplification of the initial signal by creating or initiating many replication complexes in a kind of entraining auto(?)catalytic response. This would be in agreement with the observation that in fusions between G1 and S phase cells DNA synthesis is prematurely induced in the G1 nuclei suggesting the presence of positive initiators of DNA synthesis.

The situation in *E. coli* is quite different where there are no clearly identifiable events before the actual onset of DNA synthesis. It was proposed by Sompayrac and Maaløe (1973) that the rate of initiation of DNA replication was determined by the level of a hypothetical activator kept at constant concentration by autorepression. The *dnaA* gene product fits many of these requirements, in particular autoregulation, positive initiation, stability, no cell cycle fluctuations (Atlung, Løbner-Olesen & Skarstad, 1987) and altering the amount of DnaA alters the initiation mass of *E. coli* (Løbner-Olesen *et al.*, 1989). One intriguing result is that an excess of DnaA protein leads to over-initiation of origins but they do not lead to complete replication events. This suggests that there are checks on DNA replication to prevent precocious replications. In addition, there is remarkable synchrony between initiations of different origins. Two explanations for this are possible. First that there is a rapid autocatalytic firing of all replicons after the first has fired, or secondly that they are all physically part of the same structure (the primasome) and this causes simultaneous initiation of all origins. If there is such a structure it may well be associated with the membrane because there is some evidence in *E. coli* that the DnaA protein is activated by the membrane component cardiolipin (Sekimizu & Kornberg, 1988), that the replication origin sequences of the DNA are hemi-methylated, that they can selectively associate with the membrane (Ogden, Pratt & Schaechter, 1988) and unmethylated sequences are asynchronous (Bakker & Smith, 1989). Indeed, the activity of the *dam* methyltransferase has to be precisely controlled to ensure a high degree of synchrony (Boye & Løbner-Olesen, 1990).

One of the paradoxes of the *E. coli* cell cycle is that it takes at least 60 minutes to complete one cell cycle from the initiation of DNA synthesis to the completion of septation even though cells can grow and divide every 20 minutes. One possible reason for this could be that prokaryotes evolved under conditions of nutrient limitation when rapid cell cycles would not have occurred and genome size may have been smaller than at present. As nutrients became locally abundant more rapid cell cycles became necessary for r-selected prokaryotes. Rather than accelerate the whole process *E. coli* has evolved overlapping cell cycles and dichotomous DNA replication rather than extra origins of replication. The problem is perhaps a topological one. Having a single chromosome origin which separates on replication defines the prokaryotic equivalent of the centromere. Segregation is then achieved by separation of the origins. To have additional origins could speed up the process of replication but one origin would have to be structurally and functionally designated as the partitioning replicon. In eukaryotes initiation of replication and segregation are events which are spatially distinct and could thus become temporally distinct.

The 'sizer and timer' model proposed that there was a size monitoring device which, when a critical size is achieved, initiates DNA synthesis, to be followed a constant time later by cell division, a model devised in prokaryotes and subsequently applied to eukaryotes (Fig. 6; Fantes & Nurse (1981) and Cooper (1981) for further discussion on this point). It successfully accounted for homeostatic regulation of cell size by alteration of cycle time at constant growth rate and fulfilled the requirement for at least one event coordinating cell growth and cell proliferation. It also neatly explained the asymmetrical cell cycles in *S. cerevisiae*. The model is still useful but needs reappraisal in the light of new data. First, the critical size in eukaryotes is growth rate modulated rather than a constant (Wheals, 1987). Secondly, the constant time is not invariant but elongates at slower growth rates in

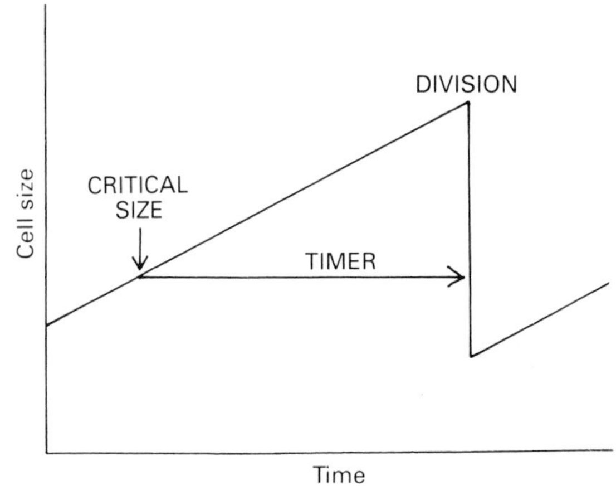

Fig. 6. Sizer + Timer model. The diagram indicates the growth of a cell with time which, when it reaches a critical size, initiates events which lead to division a fixed time later. The size of a cell at division is thus a function of growth rate, the critical size at initiation and the duration of the timer.

E. coli (Helmstetter, 1987) and *S. cerevisiae* (Wheals, 1981). Thirdly, there is not one size control point in yeasts but two (Veinot-Drebot, Johnston & Singer, 1991), although one of the regulatory points may be cryptic under many conditions. Two size control points can explain a variable 'timer' but less is known about the size control process itself.

G2-M transition

The central role of the Cdc2 protein kinase in regulating entry into mitosis of *Schizosaccharomyces pombe* has been established in a remarkable series of experiments from the laboratory of Paul Nurse (see Nurse, 1990). The *cdc2* gene has a central role in a regulatory network involving the activity of a large number of gene products. All were initially identified genetically and the biochemical function of some of them has been established. Entry into mitosis requires the activation of the *cdc2*$^+$ gene product. This can occur either by the positive action of *cdc25*$^+$ or by the relief from repression of *wee1*+, which itself can be inhibited by *nim1*$^+$, and its interaction with *mik1* (Fig. 7). Epistatic and allele specific interactions with other genes have also revealed *suc1*$^+$ and *cdc13*$^+$ to be important players in this drama.

The key to biochemical understanding was the purification of maturation (or M-phase) promoting factor (MPF), a protein complex found in developing amphibian oocytes, that, when injected into quiescent but otherwise competent oocytes, could cause them rapidly and irreversibly to enter mitosis. Independently two groups ascertained that one of the components of MPF was a functional homologue of the *cdc2*$^+$ gene product (Nurse, 1990).

The Cdc2 protein, known as p34, is a protein–serine/threonine kinase

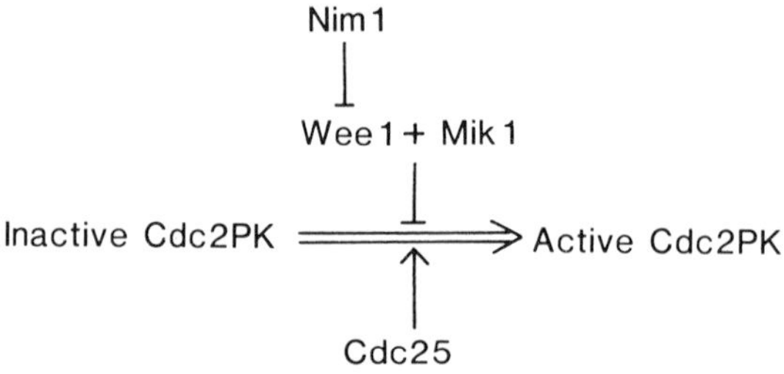

Fig. 7. Regulation of the mitotic activator in *Schizosaccharomyce pombe*. The protein kinase encoded by the *cdc2⁺* gene (Cdc2PK) exists in an inactive phosphoprotein form and an active form after specific dephosphorylation. The active form can initiate mitosis. The *cdc25* gene product is a positive activator of Cdc2 and the *wee1* gene product is an inhibitor which itself can be inhibited by the product of the *nim1* gene. Advancement into mitosis can be achieved by overexpression of Cdc25, inactivation of Wee1 or activation of Nim1.

which can itself be phosphorylated and the phosphorylation state determines the protein kinase activity. To enter mitosis the Cdc2 kinase has to be dephosphorylated on Tyr^{15}, although in amphibia an adjacent phosphorylated threonine has also to be dephosphorylated (Gould & Nurse, 1989). In addition, p34 interacts with several other proteins including members of the cyclin group, encoded in *S. pombe* by *cdc13*. MPF contains both $cdc2^+$ and $cdc13^+$ gene products. In budding yeast three cyclin genes were discovered associated not with the G2/M transition but with the G1/S transition (Richardson *et al.*, 1989) although G2/M cyclins have subsequently been discovered (Steve Reed, personal communication). The other interacting protein was defined by the $suc1^+$ gene of fission yeast, initially isolated as a suppressor of $cdc2^+$ but its exact function is unknown.

There are a number of proteins which act on $p34^{cdc2}$ regulating its activity. The $cdc25^+$ gene product (the phosphoprotein $p80^{cdc25}$) activates $p34^{cdc2}$ by dephosphorylation. Since the sequence of $cdc25^+$ suggests that the protein is neither a phosphatase or kinase it may act by inhibiting the former or activating the latter, either directly or indirectly. A second gene, $wee1^+$, acts as a repressor of $cdc2^+$ and $p107^{wee1}$ has protein–tyrosine/serine kinase activity at least *in vitro* and presumably acts by directly or indirectly maintaining the phosphorylated (inactive) state of $p34^{cdc2}$. The *mik1* gene has only been discovered recently (Lundgren *et al.*, 1991) because null alleles have no discernible phenotype but *mik1 wee1* double mutants are hypermitotically lethal, bypassing all normal cellular controls including the requirement for *start* and S-phase. The data suggest that *mik1* and *wee1* act cooperatively on *cdc2*. The $wee1^+$ gene in turn is inhibited by the $nim1^+$ gene which is a protein–serine/threonine kinase. There is thus a complex network of activators and repressors, each one of which can alter the phosphorylation state of $p34^{cdc2}$ and thus alter its activity and consequently the timing or rate of entry into mitosis. Since homologues of $cdc2^+$, $cdc25^+$ and $cdc13^+$

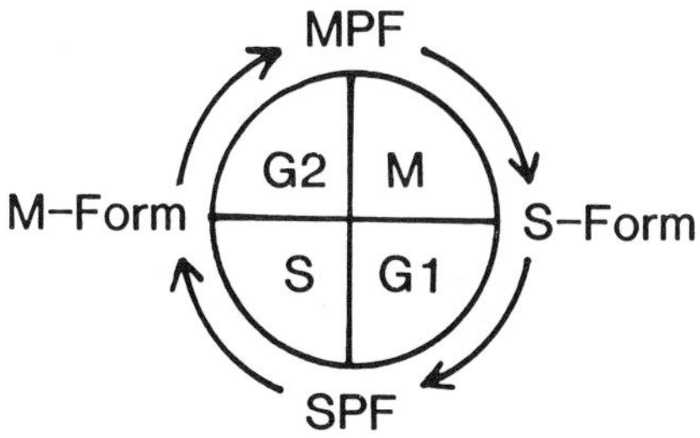

Fig. 8. The p34^{cdc2} cycle. The diagram indicates current views on the state of the p34^{cdc2} protein through the cell cycle. It can exist in four states. The active states are MPF (M-phase promoting factor) and SPF (*start* promoting factor). The inactive forms are the S-form (which is the default form after *de novo* synthesis) and the M-form. Activation of the S- and M-forms allows transition from G1 to S, and from G2 to M respectively. The difference in the states is due to post-translational modification either by association with other molecules (probably proteins) and by secondary modification, such as the phosphorylation state. The cell cycle can be defined in molecular terms by the state of the p34^{cdc2} protein. (After Broek *et al.*, 1991; Murray, 1991.)

have been found in a wide variety of eukaryotes and probably function in analogous ways (Featherstone & Russell, 1991) they are probably members of a universal mitotic control mechanism (Nurse, 1990).

A recent landmark paper from the laboratory of Paul Nurse strongly suggests that the cell cycle of eukaryotes can be simply described as a p34^{cdc2} cycle (Broek *et al.*, 1991; Fig. 8). They showed that the p34^{cdc2} protein undergoes post-translational modification and/or association which defines its state in the cell. It can exist in two forms called the S and M forms which define whether it is in G1 (S-form) or in G2 (M-form). These forms are proposed to be inactive and require additional modifications/interactions to become the active MPF form and, by analogy, the SPF (*start* promoting factor) form. The G2 form is not defined by the phosphorylation state on Tyr^{15} but by some other modification or association. Degradation of p34^{cdc2} (by heat-inactivation of temperature sensitive forms or by nitrogen starvation) reveals that the form which is created *de novo* is the S-form, a result consistent with the return to cycling of G0 cells (quiescent spores or starved vegetative cells) into G1 when the first event is *start* and S phase. In addition, passage through *start* resets the replication block allowing another round of DNA synthesis, thus segregation of the chromosomes is not a prerequisite for S phase.

The substrate of p34^{cdcd2} *in vitro* are many and various and include H1 histone kinase, cyclin and nuclear lamins. Active H1 kinase is known to be a cause of chromosome condensation and the *cdc2* gp could be the activator of chromosome condensation *in vivo*. Disaggregation of nuclear lamin structure is a prelude to dissolution of the nuclear membrane. However the principal targets *in vivo* remain to be established.

What determines the timing of mitosis? The fission yeast p80^{cdc25} has two of the essential attributes of 'division initiator proteins' that have been postu-

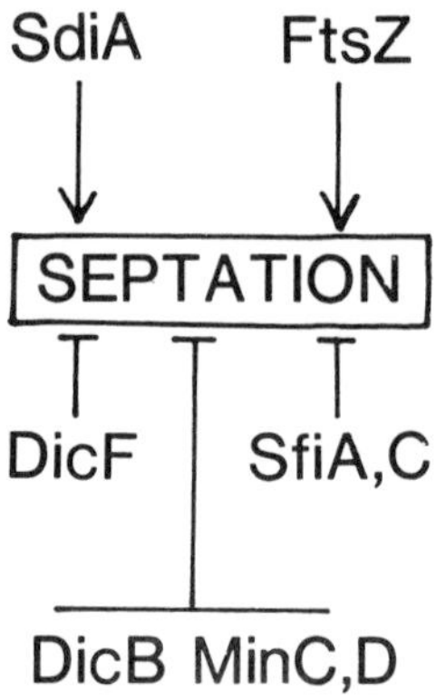

Fig. 9. Regulatory network controlling septation in *Escherichia coli*. The diagram illustrates that septation is controlled by a number of positive acting genes (with arrows) and negative acting genes (with bars) to ensure proper location and timing of the process. (Modified and simplified from de Boer, Cook & Rothfield, 1990).

lated to drive mitosis in somatic cells: its level contributes to the rate at which cells progress through G2 to initiate mitosis, and its cellular concentration increases as cells grow. The periodic accumulation of $p80^{cdc25}$ mitotic inducer, integrated with the activities of other rate-limiting mitotic inducers and inhibitors, must play an important part in determining the timing of mitotic initiation in fission yeast (Moreno *et al.*, 1990).

In yeasts, small-celled mutants have been selected in the hope that they represented mutations in regulatory genes allowing more rapid transit through the cell cycle. This strategy has been successful and both hyperactive genes or genes that have lost repressors of progress through the cycle have been isolated. In prokaryotes small (minicells) are caused not by premature initiation of division but by a failure to repress previous potential septation sites which exist at the poles. Septation at these places leads to the production of small cells although the actual number of septation events and their timing is unaltered – what changes is their distribution within and between cells. Although the details of division in bacteria are quite different from yeasts it is clear that the control of septation in *E. coli* is similarly under the control of both positive and negative regulators (Fig. 9) but the molecular details remain to be determined.

The failure to find directly comparable regulatory circuits in *E. coli* by genetic analysis has led to some novel thinking about the nature of such control. In eukaryotic cells intracellular calcium levels, through the medium of calmodulin, play an essential role in many molecular processes. Some of these are associated with cell cycle events such as spindle pole body duplication (in *S. cerevisiae*), microtubule polymerization, myosin phosphorylation and perhaps DNA replication (for review see Norris *et al.*, 1988). These observations have led to a hypothesis that in prokaryotes a single calcium flux, abruptly raising the intracellular concentration of free calcium, is responsible for triggering the major cell cycle events such as initiation of DNA synthesis, chromosome partition and cell division (Norris *et al.*, 1988). These ideas stem from observations on prokaryotes that include (i) the existence of several calcium transport systems, (ii) an intracellular concentration of free calcium identical to that of higher organisms which seems to fluctuate during the cell cycle, and (iii) the existence of proteins which are known (in eukaryotes) to be susceptible to calcium flux activation such as a myosin-like protein (Casaregola *et al.*, 1990, 1991) and a protein kinase C-like protein (Sandler *et al.*, 1989). The circumstantial evidence is quite extensive but direct proof is still lacking. Indeed, the case for the role of Ca^{2+} ions in eukaryotic cell cycle biology is perhaps stronger for aspects of morphogenesis, localized growth and development (see Harold, 1990) and this may also be true of prokaryotes.

MORPHOGENESIS

The interesting events of morphogenesis – how forms actually arise and how a dividing or developing cell models itself upon itself – take place on a higher plane than that of the genes. It is not sufficient to identify morphogenes and to work out their primary function. The pertinent level is the epigenetic one. (Harold, 1990).

Although the details of the mode of wall growth of yeasts and filamentous fungi are different there are several fundamental aspects which are very similar (Wessels *et al.*, 1990; Adams *et al.*, 1990; May & Mitchison, 1986). First there can be zonal growth sites which can be identified as apical extensions in filamentous fungi, as polar extensions in fission yeast or as deposition solely at the point of bud emergence in budding yeast. Secondly, there is the possibility of generalized overall diffuse growth as seen in the mother cell of budding yeast before bud emergence and in the hyphae of some basidiomycete fruiting bodies. Thirdly, there must be developmental switches between the two modes of growth and in yeasts these are clearly cell cycle regulated as for example during the emergence of a bud in budded yeast and from unipolar to bipolar extension occurring in fission yeast at 0.35 of a cycle. Fourthly, in yeasts, sections of wall are inherited from the mother cell which remain essentially unchanged while new material is deposited elsewhere such that each progeny cell has a mosaic cell wall made up of wall of different cell ages. Fifthly, deposition of wall material is associated with the underlying presence of filamentous proteins such as actin which shows cell cycle related changes in distribution and type (F- and G- actins) (Adams & Pringle, 1984; Marks, Hagan & Hyams, 1987). Finally, septa are laid down as non-stress bearing structures before being exposed to turgor pressure. In *S. cerevisiae* four essential genes (*CDC24*, *CDC42*, *CDC43* and *BEM1*) are required for restricting growth of the yeast cell surface to the budding site and another set (*BUD1-5*) are involved in bud site selection

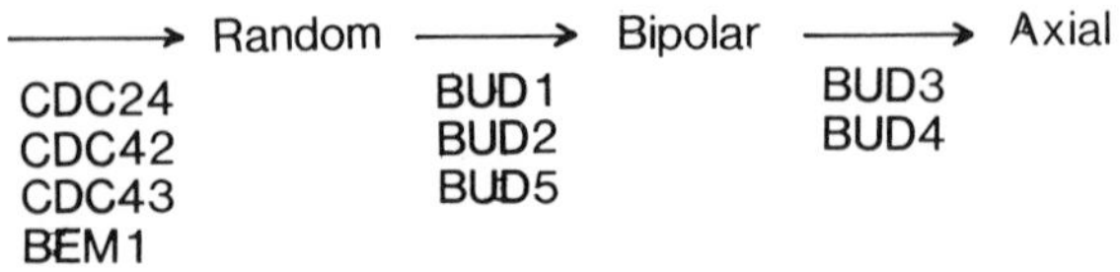

Fig. 10. Bud-site selection in *Saccharomyces cerevisiae*. It is proposed that there is a hierarchy of genes controlling the development and location of buds. In haploid cells the *CDC24*, *CDC42*, *CDC43* and *BEM1* genes control bud formation which occurs randomly. Functional *BUD1*, *BUD2* and *BUD5* genes are required for bipolar patterns and *BUD3* and *BUD4* genes can convert this pattern into the classical haploid axial array. (After Chant & Herskowitz, 1991 and Chant *et al.*, 1991.)

perhaps by guiding a complex necessary for bud formation (Adams *et al.*, 1990; Chant *et al.*, 1991; Chant & Herskowitz, 1991; Fig. 10).

How does this compare with prokaryotes? First, insertion of new cell wall material in *E. coli* occurs diffusely over the cell except for the polar caps. Secondly, zonal growth does occur during septation to create a stress-bearing cross wall (although the Gram positive *Bacillus subtilis* synthesizes a stress-free septum). By way of contrast, *Enterococcus hirae* only grows from a zone but when complete this zonal growth apparatus also divides the cell into two and thus may be envisaged as a continuous division process. Thirdly, the latter two points imply a switch from elongation to septation normally in response to the cell cycle regulated termination of DNA synthesis. Fourthly, the septum from a previous division becomes the polar cap and this will consist of old wall material. Fifthly, the location of the future septum is predetermined by the presence of a periseptal annulus acquired from its parent cell. This structure produces additional copies of itself and separation occurs providing new annuli. The mechanisms whereby these annuli move apart and are located at the right places is unknown (see de Boer, Cook & Rothfield, 1990). Thus, although the materials, mechanisms and forms are quite different, the strategy of solving cell division in rigid-walled unicellular free-living organisms is similar.

CONCLUSIONS

There are many similarities between prokaryotic and eukaryotic cell cycles. First, reversible transition out of the cell cycle can occur in response to adverse conditions and this occurs after cell division and before the onset of DNA synthesis. Secondly, the frequency of initiation of the cell cycle is coupled to growth. Thirdly, there are checkpoints that ensure that division does not occur unless the round of DNA synthesis to be partitioned by that division has been completed. Fourthly, the control of partitioning involves both positive and negative regulatory networks. Fifthly, wall biogenesis is normally tightly coupled to cell cycle progression and involves switches from generalized growth to localized growth associated with cell division. Lastly, the basic organization of cell cycle control is similar with

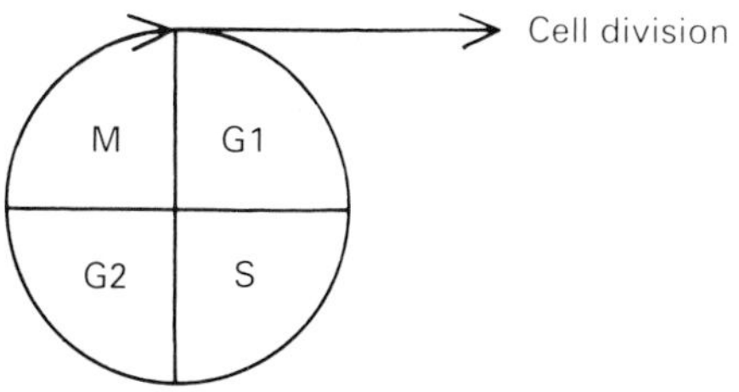

Fig. 11. Cell cycle dependency network in eukaryotes. The diagram illustrates the cyclical dependency of the four cell cycle phases, the dependency of cell division on nuclear division (mitosis) and the independency of all other events on cell division.

both cyclical and linear elements. The observation of syncytial formation in *Physarum polycephalum*, the dependency relationships deduced for *Saccharomyces cerevisiae* and the $p34^{cdc2}$ cycle of *Schizosaccharomyces pombe* all point to the view that the regulatory dependency network controlling the cell cycle in eukaryotes does not include cell division in the cyclical part of the circuit. By regulatory dependency network is meant the informational flow which determines whether the initiation of a later event is predicated on the completion of an earlier one and the minimum closed circuit of the events constitutes the cell cycle. Since the definition of a cell cycle described earlier referred to *cell* reproduction and the regulatory cycle here does not include cell division it is perhaps necessary to redefine some terms. To be most useful the term *cell cycle* should refer to those processes which give rise to new cells. It should be measured from some defined point in one generation to a point in the next generation where the cell has reached the biologically identical stage. Ordinarily this would be at birth and cell division but it could be measured from some other definable point. The closed circuit not including cell division defined by the regulatory dependency network and the reproductive process in *Physarum* plasmodia is better described as a *nuclear* or *mitotic* cycle. The dependency diagram thus has both a cyclical and a linear component (Fig. 11). The frequency and timing of G1/S and G2/M transitions is coupled to cell growth probably by the cyclical accumulation of G1 and G2 cyclins.

The regulatory dependency network in bacteria may only include a cyclical re-initiation of DNA synthesis because there is no evidence that termination, nucleoid partitioning or cell division exert any direct effect on a second initiation, as can be seen in wild type cells at fast growth rates when a second round of initiation can ensue before the completion of any of these processes initiated as a consequence of the first DNA initiation event. Except for a short delay, the frequency and timing of re-initiation is determined by biomass increase. Evidence from mutants agrees with this concept and extends it to such an extent that it has been suggested (Nordstrom *et al.*, 1991, and references therein) that a better description of the *E. coli* cell

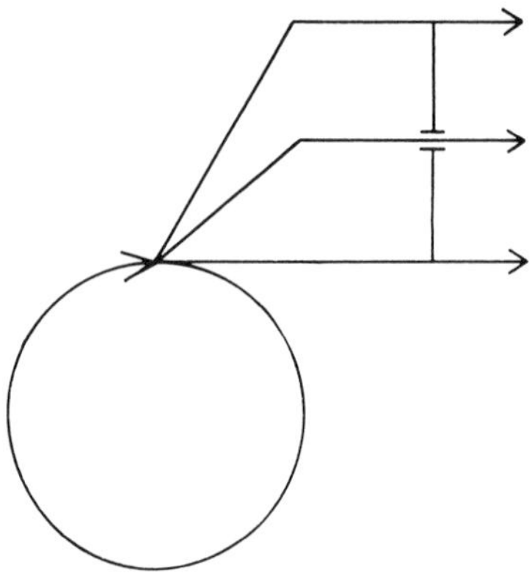

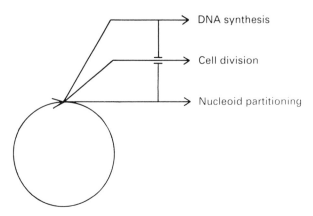

Fig. 12. Cell cycle dependency network in prokaryotes. The diagram illustrates the dependency of DNA synthesis, cell division and nucleoid partitioning on a cyclical initiation event which is independent of all other events. Failure to complete DNA synthesis or nucleoid partitioning can prevent cell division. A common event initiates all three events but the initiation of each is independent of the others.

cycle is of a cyclical process leading to a set of linear processes (Fig. 12). A single process may initiate DNA synthesis, nucleoid partitioning and cell division. Re-initiation is not dependent on the completion of any of these events. There are again checkpoints (as in yeasts) and cell division can be arrested by either failure of nucleoid segregation or the completion of DNA synthesis. The cyclical part of the dependency relationship could be called the *replicon cycle*. In *E. coli* re-initiation may only require remethylation of the hemi-methylated DNA created after origin replication, but this cannot be the explanation for unmethylated *B. subtilis* DNA regulation. Whether such complicated controls exist in other prokaryotes is unknown but it is unlikely that overlapping cell cycles occur in, for example, asymmetrically dividing budding bacteria where the fate of the two daughter cells is different.

There are many trivial differences between prokaryotes and eukaryotes involving the actual biochemistry of the components and the mechanisms of their interaction. There are only three important differences. First, eukaryotes undergo a true cyclical process made up of alternating states of the p34 protein kinase which define the events that a cell has completed and which it will do next and there are distinct regulatory points at the onset of both S and M phases. Secondly, periodic transcriptional control and protein activation are both used extensively in eukaryotes for both cell cycle progression and cell cycle regulation whereas in prokaryotes there is only good evidence for transcriptional control. Lastly, there is the difference stemming from the organization of the biology of these cells, namely that there is only a single origin of replication in prokaryotes. Initiation is probably triggered by the same process that also triggers nucleoid segregation and cell division making them very tightly coupled. In this sense the prokaryotic cell is rather like the embryonic eukaryotic cell in that there is only a single regulatory point which initiates all subsequent cell cycle events. In eukaryotes DNA synthesis, chromosome segregation and cell division are separately regulated enabling them to be both spatially and temporally distinct processes.

ACKNOWLEDGEMENTS

I thank Lee Johnston, Ira Herskowitz, Miguel Vicente and Barry Holland for sending me manuscripts before publication and Conrad Woldringh and Nanne Nanninga for helpful comments.

REFERENCES

Adams, A. & Pringle, J. R. (1984). Relationship of actin and tubulin distribution to bud growth in wild-type and a morphogenetic mutant of *Saccharomyces cerevisiae*. *Journal of Cell Biology*, **98**, 934–45.

Adams, A. E. M., Johnson, D. I., Longnecker, R. M., Sloat, B. F. & Pringle, J. R. (1990). *CDC42* and *CDC43*, two additional genes involved in budding and the establishment of cell polarity in the yeast *Saccharomyces cerevisiae*. *Journal of Cell Biology*, **111**, 131–42.

Aldea, M., Garrido, T., Pla, J. & Vicente, M. (1990). Division genes in *Escherichia coli* are expressed coordinately to cell septum requirements by gearbox promoters. *EMBO Journal*, **9**, 3787–94.

Atlung, T., Løbner-Olesen, A. & Skarstad, K. (1987). Overproduction of DnaA protein stimulates initiation of chromosome and minichromosome replication in *Escherichia coli*. *Molecular and General Genetics*, **206**, 51–9.

Bakker, A. & Smith, D. W. (1989). Methylation of GATC sites is required for precise timing between rounds of DNA replication in *Escherichia coli*. *Journal of Bacteriology*, **171**, 5738–42.

Bauer, G. A. & Burgers, P. M. J. (1990). Molecular cloning, structure and expression of the yeast proliferating cell nuclear antigen gene. *Nucleic Acids Research*, **18**, 261–5.

Beckwith, J. (1987). The lactose operon. In *Escherichia coli and Salmonella typhimurium*, vol. 2, ed. F. C. Neidhardt, J. L. Ingraham, B. Magasanik, K. B. Low, M. Schaechter & H. E. Umbarger, pp. 1444–52. Washington, American Society of Microbiology.

Blinder, D., Bouvier, S. & Jenness, D. D. (1989). Constitutive mutants in the yeast pheromone response: ordered function of the gene products. *Cell*, **56**, 479–86.

Boucherie, H. (1985). Protein synthesis during transition and stationary phases under glucose limitation in *Saccharomyces cerevisiae*. *Journal of Bacteriology*, **161**, 385–92.

Boye, E. & Løbner-Olesen, A. (1990). The role of *dam* methyltransferase in the control of DNA replication in *E. coli*. *Cell*, **62**, 981–9.

Breeden, L. & Nasmyth, K. (1987). Cell cycle control of the HO gene: *cis*- and *trans*-acting regulators. *Cell*, **48**, 389–97.

Britton, N. & Wheals, A. E. (1987). Mathematical models for a G0 phase in *Saccharomyces cerevisiae*. *Journal of Theoretical Biology*, **125**, 269–81.

Broek, D., Bartlett, R., Crawford, K. & Nurse, P. (1991). Involvement of $p34^{cdc2}$ in establishing the dependency of S phase on mitosis. *Nature*, London, **349**, 388–93.

Casaregola, S., Norris, V., Goldberg, M. & Holland, I. B. (1990). Identification of a 180 kD protein in *Escherichia coli* related to a yeast heavy-chain myosin. *Molecular Microbiology*, **4**, 505–11.

Casaregola, S., Chen, M., Bouquin, N., Norris, V., Jacq, A., Goldberg, M., Margarson, S., Tempete, M., McKenna, S., Sweetman, S., Bernard, S., McGurk, G., Seror, S. & Holland, I. B. (1991). Analysis of a myosin-like protein and the role of calcium in the *E. coli* cell cycle. *Research in Microbiology*, **142**, 201–7.

Chant, I., Corrado, K., Pringle, J. R. & Herskowitz, I. (1991). The yeast *BUD5* gene, which encodes a putative GDP-GTP exchange factor, is necessary for bud-site selection and interacts with bud-formation gene *BEM1*. *Cell*, **65**, 1213–24.

Chant, J. & Herskowitz, I. (1991). Genetic control of bud-site selection in yeast by a set of gene products that comprise a morphogenetic pathway. *Cell*, **65**, 1203–12.

Cooper, S. (1981). The central dogma of cell biology. *Cell Biology International Reports*, **5**, 539–51.

Cooper, S. & Helmstetter, C.E. (1968). Chromosome replication and the division cycle of *Escherichia coli* B/r. *Journal of Molecular Biology*, **31**, 519–40.

Dailey, D., Schieven, G. L., Lim, M. Y., Marquardt, H., Gilmore, T., Thorner, J. & Martin, G. S. (1990). Novel protein kinase (*YPK1* gene product) is a 40-kilodalton phosphorylated protein associated with protein–tyrosine kinase activity. *Molecular and Cellular Biology*, **10**, 6244–56.

d'Ari, R. & Huisman, O. (1983). Novel mechanism of cell division inhibition associated with the SOS response in *Escherichia coli*. *Journal of Bacteriology*, **156**, 243–50.

de Boer, P. A. J., Cook, W. R. & Rothfield, L. I. (1990). Bacterial cell division. *Annual Review of Genetics*, **24**, 249–74.

Deutch, C. E. & Parry, J. M. (1974). Sphaeroplast formation in yeast during the transition from exponential phase to stationary phase. *Journal of General Microbiology*, **80**, 259–66.

Dewar, S. J., Kagan-Zur, V., Begg, K. J. & Donachie, W. D. (1989). Transcriptional regulation of cell division genes in *Escherichia coli*. *Molecular Microbiology*, **3**, 1371–7.

Drebot, M. A., Johnston, G. C. & Singer, R. A. (1987). A yeast mutant conditionally defective only for reentry into the mitotic cell cycle from stationary phase. *Proceedings of the National Academy of Sciences, USA*, **84**, 7948–52.

Drebot, M. A., Barnes, C. A., Singer, R. A. & Johnston, G. C. (1990). Genetic assessment of stationary phase for cells of the yeast *Saccharomyces cerevisiae*. *Journal of Bacteriology*, **172**, 3584–9.

Dworkin, M. (1985). *Developmental Biology of the Bacteria*. Benjamin/Cummings, Menlo Park, California.

Elledge, S. J. & Davis, R. W. (1989*a*). DNA damage induction of ribonucleotide reductase. *Molecular and Cellular Biology*, **9**, 4932–40.

Elledge, S. J. & Davis, R. W. (1989*b*). Identification of the DNA damage-responsive element of *RNR2* and evidence that four distinct cellular factors bind it. *Molecular and Cellular Biology*, **9**, 5373–86.

Elledge, S. J. & Davis, R. W. (1990). Two genes differentially regulated in the cell cycle and by DNA-damaging agents encode alternative regulatory subunits of ribonucleotide reductase. *Genes and Development*, **4**, 740–51.

Elliott, S. G. & McLaughlin, C. S. (1978). The rate of macromolecular synthesis through the cell cycle of the yeast *Saccharomyces cerevisiae*. *Proceedings of the National Academy of Sciences, USA*, **75**, 4384–8.

Elliott, S. G. & McLaughlin, C. S. (1979). Synthesis and modification of proteins

during the cell cycle of the yeast *Saccharomyces cerevisiae*. *Journal of Bacteriology*, **137**, 1185–90.
Engle, D. B., Osmani, S. A., Osmani, A. H., Rosborough, S., Xiang, X. & Morris, N. R. (1990). A negative regulator of mitosis in *Aspergillus* is a putative membrane-spanning protein. *Journal of Biological Chemistry*, **265**, 16132–7.
Enoch, T. & Nurse, P. (1990). Mutation of fission yeast cell cycle control genes abolishes dependence of mitosis on DNA replication. *Cell*, **60**, 665–73.
Fantes, P. & Nurse, P. (1981). Division timing: controls, models and mechanisms. In *The Cell Cycle*, ed. P. C. L. John, pp. 11–33. Cambridge, Cambridge University Press.
Featherstone, C. & Russell, P. (1991). Fission yeast p107^{wee1} mitotic inhibitor is a tyrosine/serine kinase. *Nature*, London, **349**, 808–11.
Foiani, M., Santocanale, C., Plevani, P. & Lucchini, G., (1989). A single essential gene, *PRI2*, encodes the large subunit of DNA primase in *Saccharomyces cerevisiae*. *Molecular and Cellular Biology*, **9**, 3081–7.
Fukui, Y., Kozasa, T., Kaziro, Y., Takeda, T., & Yamamoto, M. (1986). Role of a *ras* homolog in the life cycle of *Schizosaccharomyces pombe*. *Cell*, **44**, 329–36.
Gibbs, J. B. & Marshall, M. S. (1989). The *ras* oncogene – an important regulatory element in lower eucaryotic organisms. *Microbiological Reviews*, **53**, 171–85.
Gordon, C. & Fantes, P. A. (1986). The cdc22 gene of *Schizosaccharomyces pombe* encodes a cell cycle-regulated transcript. *EMBO Journal*, **5**, 2981–5.
Gould, K. L. & Nurse, P. (1989). Tyrosine phosphorylation of the fission yeast *cdc2*$^+$ protein kinase regulates entry into mitosis. *Nature*, London, **342**, 39–45.
Hadwiger, J. A., Wittenberg, C., Richardson, H. E., de Barros Lopes, M. & Reed, S. I. (1989). A family of cyclin homologs that control the G_1 phase in yeast. *Proceedings of the National Academy of Sciences, USA*. **86**, 6255–9.
Hanks, S. K., Quinn, A. M. & Hunter, T. (1988). The protein-kinase family: conserved features and deduced phylogeny of the catalytic domains. *Science*, **241**, 42–52.
Harold, F. (1990). To shape a cell: an inquiry into the causes of morphogenesis in microorganisms. *Microbiological Reviews*, **54**, 381–431.
Hartwell, L. H. (1974). *Saccharomyces cerevisiae* cell cycle. *Bacteriological Reviews*, **38**, 164–98.
Hartwell, L. H., Culotti, J. R., Pringle, J. R., & Reid, B. J. (1974). Genetic control of the cell division cycle in yeast: a model. *Science*, **183**, 46–51.
Hereford, L. M., Osley, M. A., Ludwig, J. R. & McLaughlin, C. S. (1981). Cell cycle regulation of yeast histone mRNA. *Cell*, **24**, 367–75.
Helmstetter, C. E. (1987). Timing of synthetic activities in the cell cycle. In *Escherichia coli and Salmonella typhimurium*, vol. 2, ed. F. C. Neidhardt, J. L., Ingraham, B. Magasanik, K. B. Low, M. Schaechter & H. E. Umbarger, pp. 1594–605. Washington, American Society of Microbiology.
Helmstetter, C. E. & Cooper, S. (1968). DNA synthesis during the division cycle of rapidly growing *Escherichia coli* B/r. *Journal of Molecular Biology*, **31**, 507–18.
Hirota, Y., Ryter, A. & Jacob, F. (1968). Thermosensitive mutants of *E. coli* affected in the processes of DNA synthesis and cellular division. *Cold Spring Harbor Symposium for Quantitative Biology*, **33**, 677–93.
Howard, A. & Pelc, S. R. (1953). Synthesis of deoxyribonucleic acid in normal and irradiated cells and its relation to chromosome breakage. *Heredity, London (Supplement)*, **6**, 261–73.
Huisman, O. & d'Ari, R. (1981). An inducible DNA replication-cell division coupling mechanism in *E. coli*. *Nature*, London, **290**, 797–9.
Jaffe, A., d'Ari, R. & Norris, V. (1986). SOS-independent coupling between DNA replication and cell division in *Escherichia coli*. *Journal of Bacteriology*, **165**, 66–71.

Jarvik, J. & Botstein, D. (1973). A genetic method for determining the order of events in a biological pathway. *Proceedings of the National Academy of Sciences, USA*, **70**, 2046–50.
Jazwinski, S. M. (1988). *CDC7*-dependent protein kinase activity in replicative-complex preparations. *Proceedings of the National Academy of Sciences, USA*, **85**, 2101–5.
Johnston, G. C. & Singer, R. A. (1990). Regulation of proliferation by the budding yeast *Saccharomyces cerevisiae*. *Biochemistry and Cell Biology*, **68**, 427–35.
Johnston, L. H. (1990). Periodic events in the cell cycle. *Current Opinion in Cell Biology*, **2**, 274–9.
Johnston, L. H., White, J. H. M., Johnson, A. L., Lucchini, G. & Plevani, P. (1987). The yeast DNA polymerase I transcript is regulated both in the mitotic cycle and in meiosis and is also induced after DNA damage. *Nucleic Acids Research*, **15**, 5017–30.
Johnston, L. H., White, J. H. M., Johnson, A. L., Lucchini, G. & Plevani, P. (1990*a*). Expression of the yeast DNA primase gene, *PRI1*, is regulated within the mitotic cell cycle and in meiosis. *Molecular and General Genetics*, **221**, 44–8.
Johnston, L. H., Eberly, S. L., Chapman, J. W., Araki, H. & Sugino, A. (1990*b*). The product of the *Saccharomyces cerevisiae* cell cycle gene *DBF2* has homology with protein kinases and is periodically expressed in the cell cycle. *Molecular and Cellular Biology*, **10**, 1358–66.
Kaiser, D. (1987). Multicellular development in myxobacteria. In *Genetics of Bacterial Diversity*, ed. D. A. Hopwood & K. F. Chater, pp. 243–63. Academic Press, London.
Kawamukai, M., Matsuda, H., Fujii, W., Utsumi, R. & Komano, T. (1989). Nucleotide sequences of *fic* and *fic-1* genes involved in cell filamentation induced by cyclic AMP in *Escherichia coli*. *Journal of Bacteriology*, **171**, 4525–9.
Koch, A. L. & Coffman, R. (1970). Diffusion, permeation or enzyme limitation: a probe for the kinetics of enzyme induction. *Biotechnology and Bioengineering*, **12**, 651–77.
Kupiec, M. & Simchen, G. (1986). Regulation of the *RAD6* gene of *Saccharomyces cerevisiae* in the mitotic cycle and in meiosis. *Molecular and General Genetics*, **203**, 538–43.
Lewin, B. (1990). Driving the cell cycle: M phase kinase, its partners and substrates. *Cell*, **61**, 743–52.
Lillie, S. H. & Pringle, J. R. (1980). Reserve carbohydrate metabolism in *Saccharomyces cerevisiae*: response to nutrient limitation. *Journal of Bacteriology*, **143**, 1384–1344.
Løbner-Olesen, A., Skarstad, K., Hansen, F. G., von Meyenburg, K. & Boye, E. (1989). The DnaA protein determines the initiation mass of *Escherichia coli* K-12. *Cell*, **57**, 881–9.
Lorincz, A. T., Miller, M. J., Xuong, N-G. & Geiduschek, E. P. (1982). Identification of proteins whose synthesis is modulated during the cell cycle of *Saccharomyces cerevisiae*. *Molecular and Cellular Biology*, **2**, 1532–49.
Losick, R., Kroos, L., Errington, J. & Youngman, P. (1987). Pathways of developmentally regulated gene expression in *Bacillus subtilis*. In *Genetics of Bacterial Diversity*, ed. D. A. Hopwood & K. F. Chater, pp. 221–42. Academic Press, London.
Lowndes, N., Johnson, A. L., & Johnston, L. H. (1991). A cell cycle regulated *trans* factor DSC1 co-ordinates expression of DNA synthesis genes in budding yeast. *Nature*, London, **350**, 247–50.
Lundgren, K., Walworth, N., Booher, R., Dembski, M., Kirschner, M. & Beach,

D. (1991). mik1 and wee1 cooperate in the inhibitory tyrosine phosphorylation of cdc2. *Cell*, **64**, 1111–22.

Lutkenhaus, J. F. & Donachie, W. D. (1979). Identification of the *ftsA* gene product. *Journal of Bacteriology*, **137**, 1088–94.

Lutkenhaus, J. F., Moore, B. A., Masters, M., & Donachie, W. D. (1979). Individual proteins are synthesised continuously throughout the *Escherichia coli* cell cycle. *Journal of Bacteriology*, **138**, 352–60.

Maeda, T., Mochizuki, N. & Yamamoto, M. (1990). Adenylyl cyclase is dispensable for vegetative growth in the fission yeast *Schizosaccharomyces pombe*. *Proceedings of the National Academy of Sciences, USA*, **87**, 7814–18.

Marks, J., Hagan, I. & Hyams, J. (1987). Spatial association of F-actin with growth polarity and septation in the fission yeast *Schizosaccharomyces pombe*. In *Spatial Organization in Eukaryotic Microbes*, ed. R. K. Poole & A. P. J. Trinci, pp. 119–35. Oxford, IRL Press.

Matin, A., Auger, E. A., Blum, P. H. & Schultz, J. E. (1989). Genetic basis of starvation survival in nondifferentiating bacteria. *Annual Review of Microbiology*, **43**, 293–316.

May, J. W. & Mitchison, J. M. (1986). Length growth in fission yeast cells measured by two novel techniques. *Nature*, London, **322**, 752–4.

Mayr, E. (1990). A natural system of organisms. *Nature*, London, **348**, 491.

Meeks-Wagner, D. & Hartwell, L. H. (1986). Normal stoichiometry of histone dimer sets is necessary for high fidelity of mitotic chromosome transmission. *Cell*, **44**, 43–52.

Mendenhall, M. D., Jones, C. A. & Reed, S. I. (1987). Dual regulation of the yeast *CDC28*-p40 protein kinase complex: Cell cycle, pheromone and nutrient limitation effects. *Cell*, **50**, 927–35.

Mitchison, J. M. (1971). *The Biology of the Cell Cycle*. Cambridge University Press, Cambridge.

Moreno, S., Nurse, P. & Russell, P. (1990). Regulation of mitosis by cyclic accumulation of $p80^{cdc25}$ mitotic inducer in fission yeast. *Nature*, London, **344**, 549–52.

Murray, A. W. (1991). Remembrance of things past. *Nature*, London, **349**, 367–8.

Nasmyth, K. (1983). Molecular analysis of a cell lineage. *Nature*, London, **302**, 670–6.

Nasmyth, K. (1990). *FAR*-reaching discoveries about the regulation of START. *Cell*, **63**, 1117–20.

Nasmyth, K., Seddon, A. & Ammerer, G. (1987). Cell cycle regulation of *SWI5* is required for mother-cell-specific HO transcription in yeast. *Cell*, **49**, 549–58.

Nasmyth, K., Adolf, G., Lydall, D. & Seddon, A. (1990). The identification of a second cell cycle control on the HO promoter in yeast: cell cycle regulation of Swi5 nuclear entry. *Cell*, **62**, 631–47.

Newlon, C. (1988). Yeast chromosome replication and segregation. *Microbiological Reviews*, **52**, 568–601.

Newton, A. (1989). Differentiation in *Caulobacter*: Flagellum, development, motility and chemotaxis. In *Genetics of Bacterial Diversity*, pp. 199–220, ed. D. A. Hopwood & K. F. Chater. Academic Press, London.

Nordstrom, K., Bernander, R. & Dasgupta, S. (1991). The *Escherichia coli* cell cycle: one cycle or multiple independent processes that are co-ordinated? *Molecular Microbiology*, **5**, 769–74.

Norris, V., Seror, S. J., Casaregola, S. & Holland, I. B. (1988). A single calcium flux triggers chromosome replication, segregation and septation in bacteria: a model. *Journal of Theoretical Biology*, **143**, 341–50.

Nurse, P. (1975). Genetic control of cell size at cell division in yeast. *Nature*, London, **256**, 547–51.

Nurse, P. (1990). Universal control mechanism regulating onset of M-phase. *Nature*, London, **344**, 503–8.

Nurse, P. & Bissett, Y. (1981). A gene required in G1 for commitment to cell cycle and in G2 for commitment to mitosis in fission yeast. *Nature*, London, **292**, 558–60.

O'Farrell, P. H. (1975). High resolution two-dimensional electrophoresis of proteins. *Journal of Biological Chemistry*, **250**, 4007–21.

Ogden, G. B., Pratt, M. J. & Schaechter, M. (1988). The replicative origins of the *E. coli* chromosome bind to cell membranes only when hemimethylated. *Cell*, **54**, 127–35.

Ohta, N., Masurekar, M. & Newton, A. (1990). Cloning and cell-cycle dependent expression of DNA replication gene *dnaC* from *Caulobacter crescentus*. *Journal of Bacteriology*, **172**, 7027–34.

Plesset, J., Ludwig, J. R., Cox, B. S. & McLaughlin, C. S. (1987). Effect of cell cycle position on thermotolerance in *Saccharomyces cerevisiae*. *Journal of Bacteriology*, **169**, 779–84.

Reed, S. I. & Wittenberg, C. (1990). Mitotic role for the Cdc28 protein kinase of *Saccharomyces cerevisiae*. *Proceedings of the National Academy of Sciences, USA*, **87**, 5697–701.

Richardson, H. E., Wittenberg, C., Cross, F. & Reed, S. I. (1989). An essential G1 function for cyclin-like proteins in yeast. *Cell*, **59**, 1127–33.

Robin, A., Joseleau-Petit, D. & d'Ari, R. (1990). Transcription of the *ftsZ* gene and cell division in *Escherichia coli*. *Journal of Bacteriology*, **172**, 1392–9.

Robinson, A. C., Collins, J. F. & Donachie, W. D. (1987). Prokaryotic and eukaryotic cell-cycle proteins. *Nature*, London, **328**, 766.

Russell, P. & Nurse, P. (1986). *cdc25*$^{+}$ functions as an inducer in the mitotic control of fission yeast. *Cell*, **45**, 145–53.

Sandler, N., Dvir, A., Amos, S., Keynan, A. & Milner, Y. (1989). Cell cycle-dependent protein kinase C-like activity in *Bacillus subtilis*. Abstracts 7th International Conference, Tokyo. Cyclic nucleotides, calcium, protein phosphorylation.

Sekimizu, K. & Kornberg, A. (1988). Cardiolipin activation of dnaA protein, the initiator protein of replication in *Escherichia coli*. *Journal of Biological Chemistry*, **263**, 7131–5.

Sompayrac, L. & Maaløe, O. (1973). Autorepressor model for control of DNA replication. *Nature*, London, **241**, 133–5.

Storms, R. K., Ord, R. W., Greenwood, M. T., Miradammadi, B., Chu, F. K. & Belfort, M. (1984). Cell cycle-dependent expression of thymidylate synthase in *Saccharomyces cerevisiae*. *Molecular and Cellular Biology*, **4**, 2858–64.

Tan. J. L. & Spudich, J. A. (1990). Developmentally regulated protein-tyrosine kinase genes in *Dictyostelium discoideum*. *Molecular and Cellular Biology*, **10**, 3578–83.

Veinot-Drebot, L. M., Johnston, L. H. & Singer, R. A. (1991). A cyclin protein modulates mitosis in the budding yeast *Saccharomyces cerevisiae*. *Current Genetics*, **19**, 15–19.

Walton, E. F., Carter, B. L. A. & Pringle, J. R., (1979). An enrichment method for temperature-sensitive and auxotrophic mutants of yeast. *Molecular and General Genetics*, **171**, 111–14.

Ward, J. E. & Lutkenhaus, J. (1985). Overproduction of FtsZ induces minicell formation in *E. coli*. *Cell*, **42**, 941–9.

Weinert, T. A. & Hartwell, L. H. (1988). The *RAD9* gene controls the cell cycle response to DNA damage in *Saccharomyces cerevisiae*. *Science*, **241**, 317–22.

Wessels, J. G. H., Mol, P. C., Sietsma, J. H. & Vermuelen, C. A. (1990). Wall structure, wall growth, and fungal cell morphogenesis. In *Biochemistry of Cell*

Walls and Membranes in Fungi, ed. P. J. Kuhn, A. P. J. Trinci, M. J. Jung, M. W. Goosey & L. G. Copping, pp. 81–95. Berlin, Springer-Verlag.

Wheals, A. E. (1981). Timing of events in the *Saccharomyces cerevisiae* cell cycle. In *Fifth International Symposium on Yeasts*, ed. G. G. Stewart & I. Russell, pp. 549–54. London, Pergamon Press.

Wheals, A. E. (1987). Biology of the cell cycle in yeast. In *The Yeasts*, 2nd edn. vol. 1, ed. A. H. Rose & J. S. Harrison, pp. 283–390. Academic Press, London.

White, J. H. M., Green, S. R., Barker, D. G., Dumas, L. B. & Johnston, L. H. (1987). The *CDC8* transcript is cell cycle regulated in yeast and is expressed coordinately with *CDC9* and *CDC21* at a point preceding histone transcription. *Experimental Cell Research*, **171**, 223–31.

White, J. H. M., Barker, D. G., Nurse, P. M. & Johnston, L. H. (1986). Periodic transcription as a means of regulating gene expression during the cell cycle: contrasting modes of expression of DNA ligase genes in budding and fission yeast. *EMBO Journal*, **5**, 1705–9.

Whiteway, M., Hougan, L., Dignard, D., Thomas, D. Y., Bell, L., Saari, G. C., O'Hara, P. & MacKay, V. L. (1989). The *STE4* and *STE18* genes of yeast encode potential beta and gamma subunits of the mating factor receptor-coupled G protein. *Cell*, **56**, 467–77.

Wittenberg, C. & Reed, S. I. (1988). Control of the yeast cell cycle is associated with assembly/disassembly of the Cdc28 protein kinase complex. *Cell*, **54**, 1061–72.

Wittenberg, C., Sugimoto, K. & Reed, S. I. (1990). G1-specific cyclins of *S. cerevisiae*: Cell cycle periodicities, regulation by mating pheromone, and association with the $p34^{CDC28}$ protein kinase. *Cell*, **62**, 225–37.

Woese, C. R., Kandler, O. & Wheelis, M. L. (1990). Towards a natural system of organisms: proposal for the domains Archaea, Bacteria and Eucarya. *Proceedings of the National Academy of Sciences, USA*, **87**, 4576–9.

Zhou, C. & Jong, A. (1990). *CDC6* mRNA fluctuates periodically in the yeast cell cycle. *Journal of Biological Chemistry*, **265**, 19904–9.

ENVELOPE GROWTH IN *ESCHERICHIA COLI* – SPATIAL AND TEMPORAL ORGANIZATION

N. NANNINGA, F. B. WIENTJES, E. MULDER and C. L. WOLDRINGH

Section of Molecular Cytology, Department of Molecular Cell Biology, University of Amsterdam

INTRODUCTION

Envelope growth in *Escherichia coli* comprises the interplay of three layers: the cytoplasmic membrane, the peptidoglycan (PG) or murein layer and the outer membrane. Such a division in layers is certainly schematic. For instance, cytoplasmic membrane proteins with large periplasmic domains belong, at least operationally, to two layers. Furthermore, during growth, vectorial transport of proteins and peptidoglycan precursors occurs continuously. The PG layer is the essential shape-maintaining layer and it can be considered also as the assembly scaffold of the outer membrane components (Mizushima, 1985). It does not determine cell shape, because it is the end result of morphogenetic events (Henning, 1975). Morphogenesis takes place at different hierarchical levels of organization. On the molecular level there is the effect of growth conditions on gene expression. This interaction leads to the assembly of various macromolecular structures such as, for instance ribosomes, membranes, peptidoglycan (one single macromolecule) and so on. The whole of the macromolecular assemblies constitute a unicellular organism like *E. coli* (Ingraham, Maaløe & Neidhardt, 1983). In a simplified way, the end result of *E. coli* morphogenesis is a cylindrical tube with hemispherical caps. How does this shape come about? How is cylindrical growth (cell elongation) related to PG biochemistry? What happens when new hemispherical (polar) caps are being formed during cell division? This last question can be addressed at the envelope level, but also in relation to an internal cellular structure such as the nucleoid. In fact, division does not make sense if the duplicated genomes have not acquired separate domains. Does the envelope sense the physical presence (or absence) of the genome? In our view the various questions should be considered in the framework of the spatial and temporal organization of the envelope of which the PG layer is the essential component.

ORGANIZATION OF THE PEPTIDOGLYCAN LAYER

The PG layer or sacculus is an extremely dynamic structure. Bonds are continuously broken while precursors are being inserted. In addition, 'old

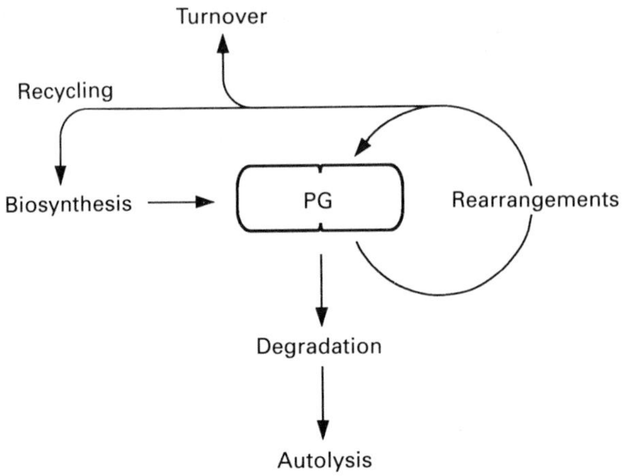

Fig. 1. Global overview of peptidoglycan (PG) metabolism in *E. coli*. (Courtesy of Dr J. van Heijenoort.)

material' is being discarded (turnover; Goodell & Schwarz, 1985; Greenway & Perkins, 1985) or re-used after degradation (recycling; Goodell, 1985; Fig. 1) (for review see Höltje & Glauner, 1990). The sacculus is a porous structure pervaded by numerous proteins. Firstly, there are the various penicillin-binding proteins (PBPs) which are the main agents for its extension. Proteins are there which are on their way to the outer membrane and, as more permanent residents, one has the numerous periplasmic proteins and the membrane-derived oligosaccharides (Kennedy, 1987). Together, all these substances might constitute a so-called periplasmic gel (Hobot *et al.*, 1984) of which PG is the dominant component. For a detailed treatment of the periplasm see chapter by Ferguson, this volume.

More recently it was proposed, on the basis of cytochemical staining, that *E. coli* PG is organized into a multilayered structure more similar to Gram-positive bacteria (Leduc *et al.*, 1989). This is in conflict with the classic notion that *E. coli* PG is monolayered (Preusser, 1969; Braun *et al.*, 1973). For this reason the problem has been re-investigated. Neutron small angle X-ray diffraction measurements (Labischinsky *et al.*, 1991) and biochemical determinations with cell size measurement (Wientjes, Woldringh & Nanninga, unpublished observations) independently showed that the thickness of the PG is between one and two layers. Possibly the cytochemical approach would need still more specificity with respect to PG staining.

Is the PG evenly distributed over the cell's total surface or do polar caps contain more PG than the lateral walls (Labischinsky *et al.*, 1991)? Intuitively, the latter would seem likely because one can imagine that the polar caps have to be somewhat sturdier in order to maintain the hemispherical shape. However, this does not seem to be the case. When isolated sacculi are flat-

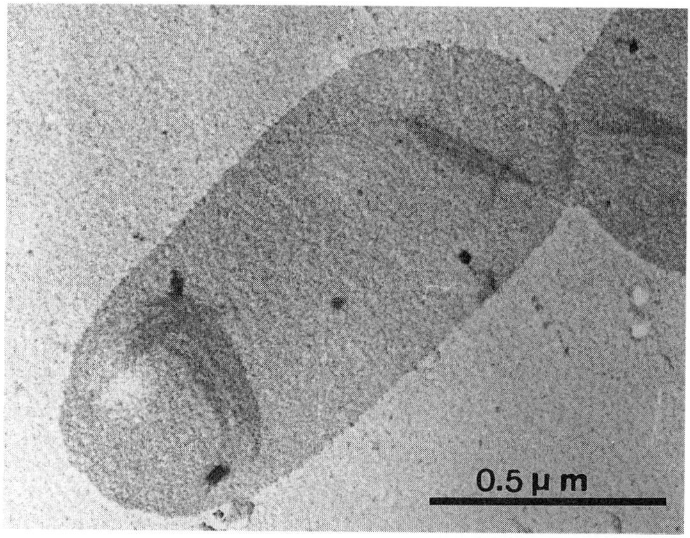

Fig. 2. Isolated sacculus. The specimen was stained with uranyl acetate and then lightly shadowed with Pt/Pd while rotating. Bar: 0.5 µm. (From Verwer *et al.*, 1978.)

tened during preparation for electron microscopy, typical half-moon like structures appear at the polar caps (Fig. 2). It can be inferred that in the half-moon regions there is twice as much PG as is in the surroundings; in other words, twice the increase in contrast (equivalent to mass) can easily be discerned. Because otherwise there is no clear-cut difference in contrast between the lateral part and the polar caps, it seems likely that there is no significant difference in thickness between lateral wall and poles.

This suggests that the central PG structure is a monolayer with ingoing and outgoing material on the cytoplasmic membrane side and outer membrane side, respectively (Fig. 3). On average this would lead to a thickness of about one and a half layers.

ARRANGEMENT OF GLYCAN CHAINS

Biophysical studies have indicated that glycan chains are not randomly organized in the PG layer; rather they are arranged in parallel (for review and references see Höltje & Schwarz,1985; Höltje & Glauner, 1990). These parallel arrays in the plane of the monolayer again run in a direction more or less perpendicular to the length axis of the cell (Fig. 4). This was shown in two ways. Breakage of peptide cross-bridges by endopeptidase followed by electron microscopy (Verwer *et al.*, 1978) and sonication of sacculi followed by chemical analysis (Verwer *et al.*, 1980) both revealed preferential weakening in a direction perpendicular to the cell's long axis. The anisotropic glycan strand organization could be the result of the fact that, in a rod-shaped

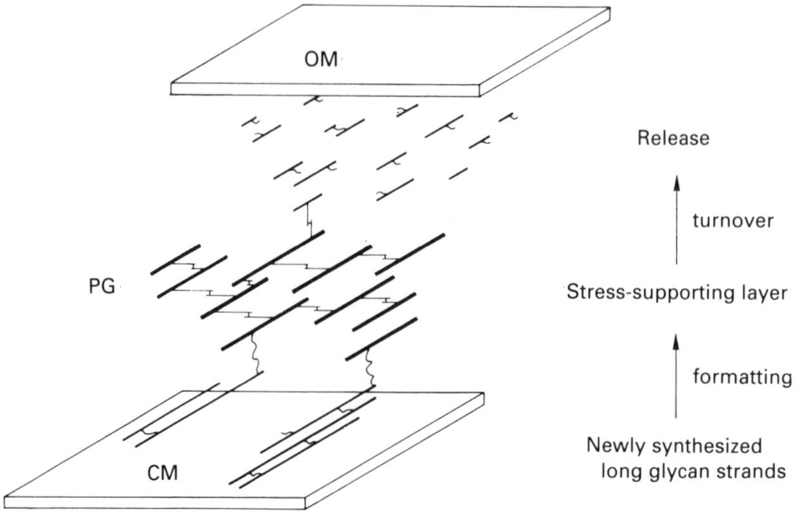

Fig. 3. The *E. coli* PG monolayer with material being inserted and discarded. (From A. Koch, 1988 as modified by J.-V. Höltje.)

cell, the circumferential surface stress is twice that in the axial direction (Chatterjee, Dasgupta & Chatterjee, 1988; Koch, 1988; Fig. 5).

CELL ENVELOPE EXTENSION

In the past, attention has often been focused on whether cells increase their length, surface or mass in an exponential, bilinear or other fashion. Due to the methodological limitations, no definite answer has been forthcoming (for review and earlier references see Nanninga & Woldringh, 1985). Recent experiments on envelope synthesis are compatible with an exponential increase in size (Cooper, 1990; Wientjes & Nanninga, 1989).

A more profitable field has been the study of insertion-topography of precursors into the outer membrane and of *meso*-diaminopimelic acid (Dap) into peptidoglycan. These investigations dealt with the question of whether insertion is diffuse or whether there is a zonal mode of insertion as required for the Jacob, Brenner & Cuzin (1963) DNA separation model (see also Thomas, this volume). Remarkably, the answers were not the same for the outer membrane and PG layer. Topographical immuno-labelling studies have shown insertion of lipopolysaccharides (Mühlradt, 1976), outer membrane proteins (Smit & Nikaido, 1978), LamB (Vos-Scheperkeuter *et al.*, 1984) and bound lipoprotein (Hiemstra *et al.*, 1987) to occur diffusely over the non-dividing cell. Whereas the same has been found by autoradiography of [^{3}H]-Dap incorporation (for review and references see Nanninga, 1991) a striking difference was observed with respect to the constriction process. Detailed autoradiographic analysis has shown that, upon division, a switch

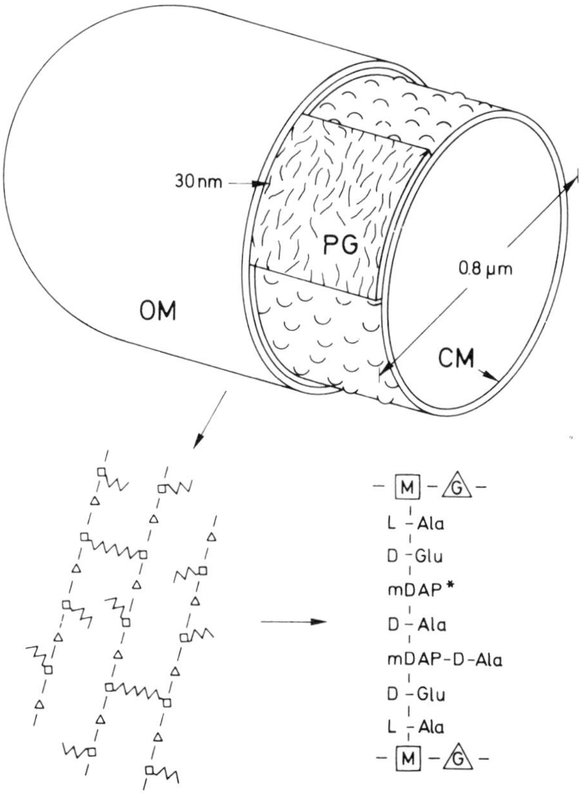

Fig. 4. Arrangement of glycan chains in the peptidoglycan (PG) layer and structure of peptidoglycan. CM, cytoplasmic membrane; OM, outer membrane; M, *N*-acetylmuramic acid; G, *N*-acetylglucosamine.

in Dap incorporation occurs from the lateral wall to the site of division at the expense of incorporation at the former (Woldringh *et al.*, 1987; Wientjes & Nanninga, 1989, Fig. 6(*a*). Such a switch does not take place with respect to LamB (Vos-Scheperkeuter *et al.*, 1984) and lipoprotein insertion (Fig. 6(*b*); Hiemstra *et al.*, 1987).

The labelling studies thus appear to indicate that the outer membrane has a passive role in cell division, i.e. it follows the more active lead of the PG during the constriction process. This is in line with earlier cytological data from electron microscopic thin sections where the invaginating PG layer was well ahead of the outer membrane (Burdett & Murray, 1974).

Concerning the cytoplasmic membrane, we are not aware of recent topographical studies (for earlier references see Nanninga & Woldringh, 1985). A basic experimental problem might be the cytoplasmic membrane's fluidity. The stagnation in knowledge concerning insertion topography in the cyto-

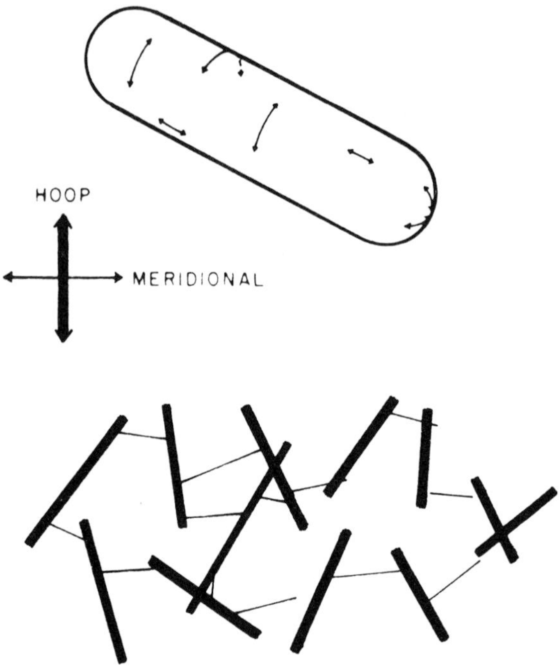

Fig. 5. Tensile stress in the surface of a rod-shaped organism and the effect on glycan chain orientation. In the hoop direction the stress is twice that in the meridional direction. Glycan chains and peptide-cross bridges have been drawn thick and thin, respectively. (From Koch, 1988.)

plasmic membrane is in strong contrast to the numerous studies that continue to appear on vectorial transport of proteins to the different envelope layers (for review see Pugsley, this volume).

CELL DIVISION AND DAP INCORPORATION

As mentioned above, cell division is accompanied by a local increase in Dap incorporation. Further comparison of autoradiographic data (Wientjes & Nanninga, 1989) at different stages of the division·process (slightly constricted, medium constricted and far constricted cells) indicated: (i) that the Dap-incorporation activity remained constant and (ii) that incorporation took place at the leading edge of constriction (Fig. 7). The latter observation brings *E. coli* in line with *Enterococcus faecium (S. faecalis, S. faecium)* where septal growth at the leading edge was deduced several years ago (Higgins & Shockmann, 1976). However, otherwise there are numerous differences between surface extension of the two organisms (for review see Koch, 1988).

A further differentiation within the division process can be made by use

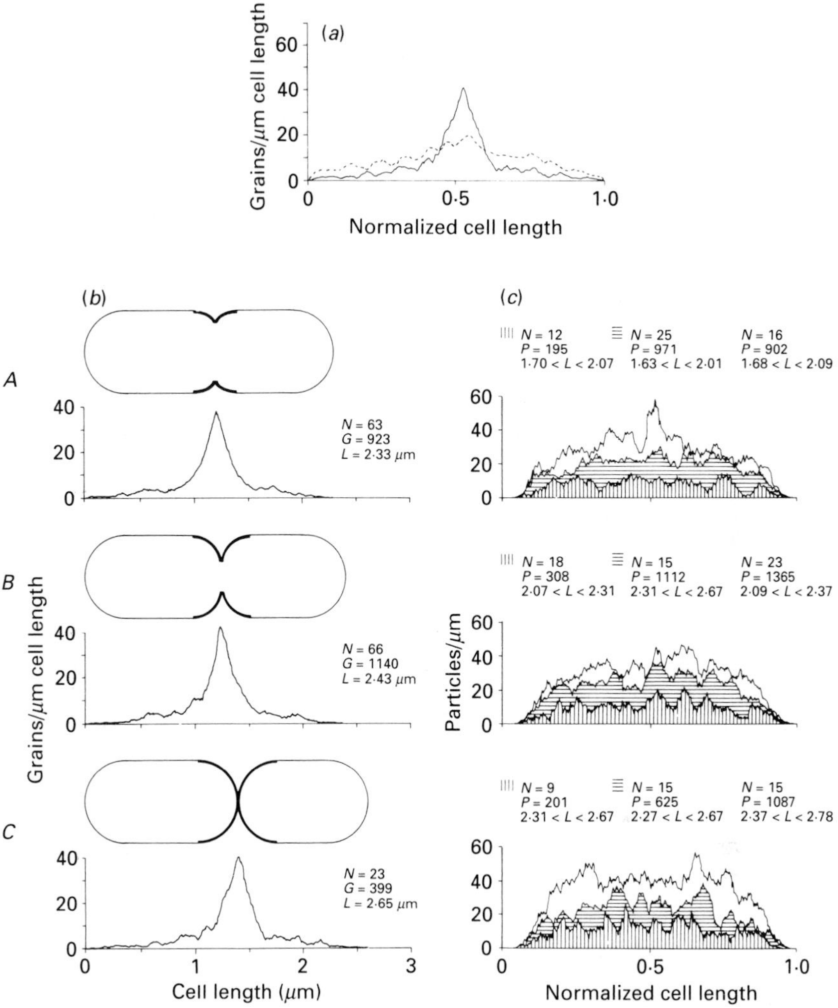

Fig. 6. Topography of Dap-incorporation (*a*), (*b*) and lipoprotein insertion (*c*). (*a*) Comparison of the topography of [^{3}H]-Dap incorporation in constricting (-) and non-constricting (. . .) cells of *E. coli* MC4100 *lysA*. The cells were pulse-labelled with [^{3}H]-Dap, autoradiographed, and after development, the number of silver grains per cell was determined. The grain distributions are plotted as grain densities versus cell length. (From Wientjes & Nanninga, 1989.) (*b*) Silver grain distribution over cells in progressing stages of constriction (cells with slight (A), medium (B), and deep (C) constrictions). Drawings of the cells of the different length classes are shown above the distribution, and the number of cells (N) and grains (G) as well as the average length of the cells (L) are given (From Wientjes & Nanninga, 1989.) (*c*) Distribution of protein A-gold particles (P) bound to the cell surface of constricting *E. coli* JA 221 (pKEN 115) after immunogold labelling with lipoprotein-specific antiserum. Each panel shows noninduced cells (lower curves, cells induced for 5 min with IPTG (middle curves), and cells induced for 10 min with IPTG (upper curves). (From Hiemstra *et al.*, 1987.)

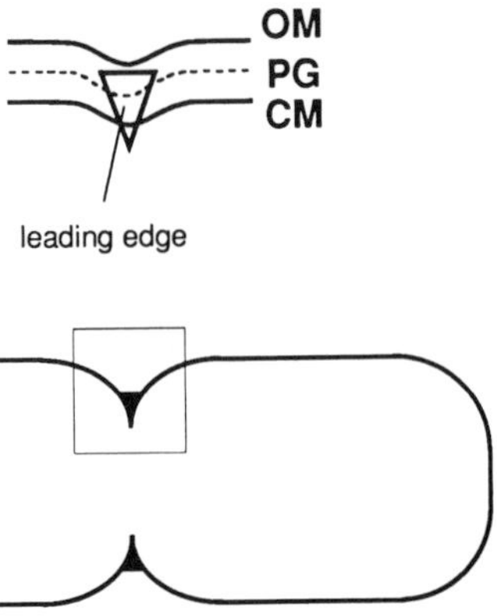

Fig. 7. Schematic representation of peptidoglycan assembly at the leading edge in nascent polar caps. This interpretation is based on data such as in Fig. 6 b. (For details see Wientjes & Nanninga, 1989; Nanninga, 1991.)

of antibiotics such as furazlocillin or cephalexin which specifically inhibit PBP3. Furazlocillin induces filamentation, whereby normal constrictions are transformed into so-called 'blunt' ones or 'bottle necks' (Fig. 8). Further growth of the filaments gives rise to a new constriction in the centre of the cellular compartments flanking the old impaired constriction. When such cells are pulse-labelled with [^{3}H]-Dap for autoradiography, it can be seen (Fig. 8) that the new ones incorporate Dap (in the presence of furazlocillin) whereas the old one appears inert (Woldringh *et al.*, 1988; Wientjes & Nanninga, 1989). The same result has been obtained when using a temperature-sensitive mutant (*pbpB*, or *ftsI*) of PBP3 (Wientjes & Nanninga, 1989). These observations indicate that peptidoglycan synthesis during initiation of division is independent of PBP3, but dependent upon PBP3 during its continuation (see below).

So far, our attention has been mainly focused on Dap-incorporation as such. Before discussing more chemical aspects of PG extension, we have to mention its main synthesizing enzymes: the penicillin-binding proteins.

PENICILLIN-BINDING-PROTEINS

PG biosynthesis is confined to three compartments: the cytoplasm, the cytoplasmic membrane and the periplasm (for review see Höltje & Schwarz, 1985). It is the last step that is carried out by PBPs. *E. coli* PBPs belong to two main groups, namely the high-molecular weight (HMW) PBPs and the low-molecular weight (LMW)-PBPs. The first comprises PBPs1A, 1B, 2 and 3 and they are anchored in the cytoplasmic membrane by their amino terminal parts (Fig. 9). All four are bifunctional enzymes carrying transglycosylase activity (glycan chain elongation) and transpeptidase activity (cross-linking of peptide side chains; for review and references see for instance: Spratt,

1977; Waxman & Strominger, 1983; Matsuhashi, Wachi & Ishino, 1990; Fig. 10). Although PBP1A and PBP1B appear to be exchangeable, the absence of both of them is lethal (Yousif, Broome-Smith & Spratt, 1985). Inhibition of PBP2 (gene product of *pbpA*) by mecillinam or growth at the non-permissive temperature of a *pbpA* temperature-sensitive mutant leads to rounded cells. Inhibition of PBP3 (gene product of *pbpB* or *ftsI*), as mentioned above, leads to filaments. These observations led to the following concept (Spratt, 1975; Matsuhashi *et al.*, 1990). PBPs 1A and 1B are needed for overall PG synthesis; PBP2 is involved in maintaining rod-shape and cell elongation and PBP3 effects division. In recent experiments, however, this notion has been modified somewhat (Wientjes & Nanninga 1991). PBPs 1A/1B are supposed to make primers for PG assembly only. Their activity is not cell-cycle dependent (for details Jacoby & Young, 1991; Wientjes & Nanninga, 1991). During cell elongation, PBP2 continues PG synthesis whereas PBP3 takes over during division. In this modified view, the role of PBP2 for overall PG synthesis has increased. This was deduced by measuring PG synthesis in the presence of mecillinam (PBP2-inhibition) and of furazlocillin or cephalexin (PBP3-inhibition) in synchronized cells (Fig. 11). Inhibition of PBP2 lowered PG synthesis considerably (Park & Burman, 1973; Wientjes & Nanninga, 1991), thus pointing to PBP2 as the major PG synthesizing enzyme. PBP2 was found to be needed throughout the cell cycle and, in line with earlier notions (Botta & Park, 1981), PBP3 appeared particularly sensitive during division. The roles of the various HMW-PBPs during cell elongation and cell constriction are schematically depicted in Fig. 12.

The LMW-PBPs have different orientations with respect to the cytoplasmic membrane. Whereas the HMW-PBPs have their membrane anchor at the aminoterminal part of molecule, the LMW-PBPs have it at the carboxyterminal part (Fig. 9). Enzymically they seem to be monofunctional. PBP4 has endopeptidase as well as carboxypeptidase I activity and PBP5 and PBP6 have carboxypeptidase I activity only (Fig. 10). Deletions of PBP4 (Iwaya & Strominger, 1977), of PBP5 (Spratt, 1980) and PBP6 (Broome-Smith & Spratt, 1982) are not lethal.

PEPTIDOGLYCAN CHEMISTRY

The recent application of high-pressure (performance) liquid chromatography (HPLC) has led to the insight that the PG composition of the sacculus is considerably more complex than had been anticipated (Glauner, 1982; Höltje & Schwarz, 1985; Pisabarro *et al.*, 1986; Glauner & Schwarz, 1987; Driehuis & Wouters, 1987; Glauner, 1988; De Jonge *et al.*, 1989). Numerous new compounds have been detected by HPLC. The various components that can be found upon degrading PG with *Chalaropsis* sp. muramidase are depicted in Fig. 13. The relative contents of the various muropeptides

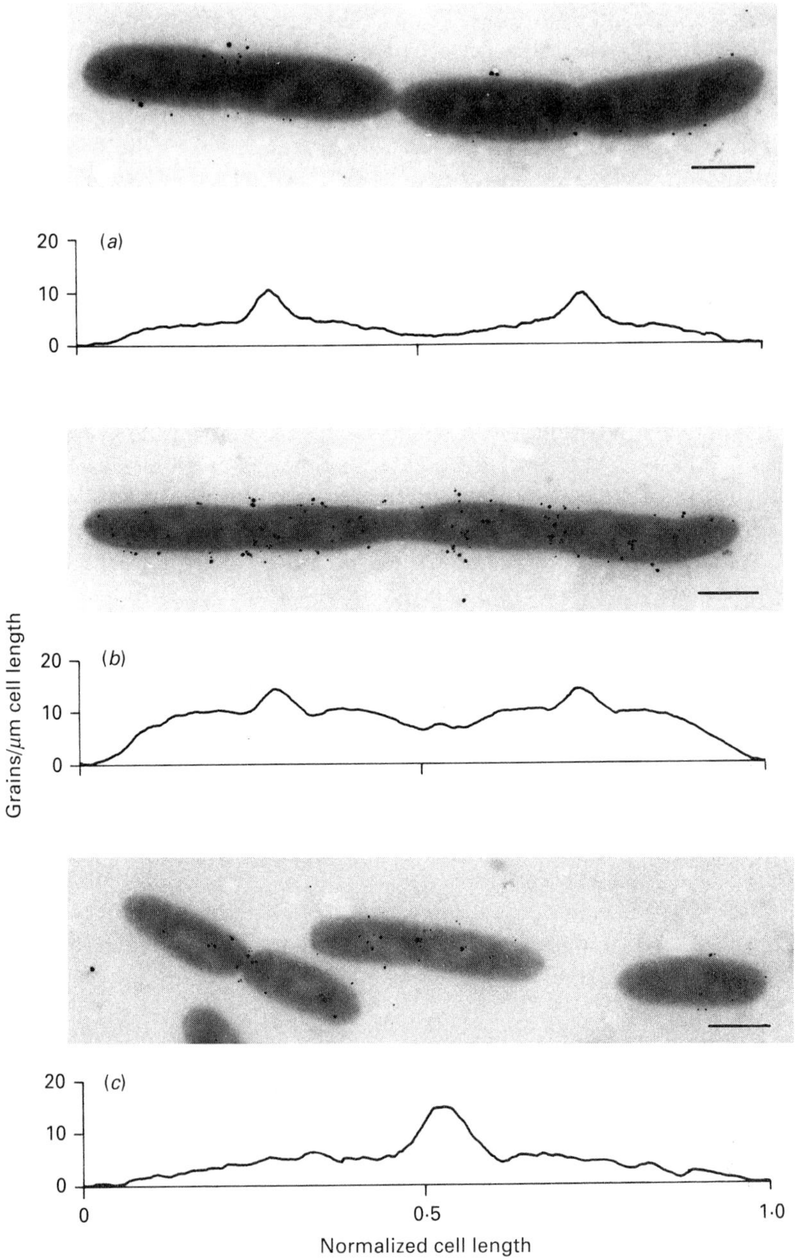
(a)
(b)
(c)
20
10
0
Grains/μm cell length
0
0·5
1·0
Normalized cell length

in exponentially growing cells are shown in Table 1. Dominant components are the *N*AcGlc-*N*AcMur-tetrapeptide (Tet) and its dimer (Tet–Tet). In addition, a new type of cross-link apart from the D-Ala–Dap bond has been found, i.e. the Dap–Dap cross-link (Glauner, 1982). The physiological significance of this component is not yet known.

CELL DIVISION AND PEPTIDOGLYCAN COMPOSITION

Many attempts have been made to study cell division by analysing the PG composition of morphological equivalents of polar caps and to compare this with cells enriched in lateral walls such as non-dividing filaments. Minicells or rounded cells produced by mecillinam have been considered as polar caps. Minicells appeared relatively deprived of PBP2, though not enriched in PBP3 (Buchanan, 1981). Alternatively, short cells with relatively large amounts of polar cap material have been compared with longer cells or with filaments. Generally these studies revealed no clear-cut clue with respect to a difference in chemical composition between rounded cells and rods (for review see Höltje & Glauner, 1990). Strikingly, the composition of the growth medium appeared to have greater influence on PG composition than cell shape (Driehuis & Wouters, 1987; Tuomanen & Cozens, 1987).

However, this does not exclude the possibility that differences exist between the lateral wall and polar cap in one and the same cell. Therefore, in a more straightforward approach, non-dividing and dividing cells have been analysed after synchronization using a centrifugal elutriation method (De Jonge *et al.*, 1989). Fractions containing few constricting cells have been compared with those containing a high percentage of constricting cells. The various components listed in Table 1 for exponentially growing cells were present in non-dividing as well as in dividing cells, indicating that no specific muropeptide is made for the division process. However, an increase in radioactivity was found in Tet–Tet which was accompanied by a decrease in the amount of Tet. This has been interpreted to mean that the extra radioactivity in Tet–Tet of dividing cells is due to donor–Dap as well as acceptor–Dap being labelled. This is reflected in the so-called

Fig. 8. Topography of [^{3}H]-Dap incorporation after inhibition of PBP3. The temperature-sensitive cell division mutant MC4100 *lysA pbpB* was grown at 28 °C (the permissive temperature) in the presence of 2 μg of furazlocillin per ml (*a*) at 37 °C (the restrictive temperature for this mutant) without furazlocillin (*b*) at 28 °C without furazlocillin (*c*). After two mass doubling times, the cells were pulse-labelled with [^{3}H]-Dap. (*a*) Grain distribution of furazlocillin-induced filaments with three constrictions ($N = 130$; $G = 4876$; $L = 10.68$ μm, where N is the number of cells, G is the number of grains, and L is the average length of the cells). (*b*) Grain distribution of temperature-induced filaments with three constrictions ($N = 164$; $G = 15{,}506$; $L = 12.16$ μm). (*c*) Grain distribution of normally dividing *pbpB* cells at the permissive temperature ($N = 92$; $G = 1533$; $L = 4.11$ μm). Electron micrographs are shown above the respective distributions. Bars, 1 μm. (From Wientjes & Nanninga, 1989.)

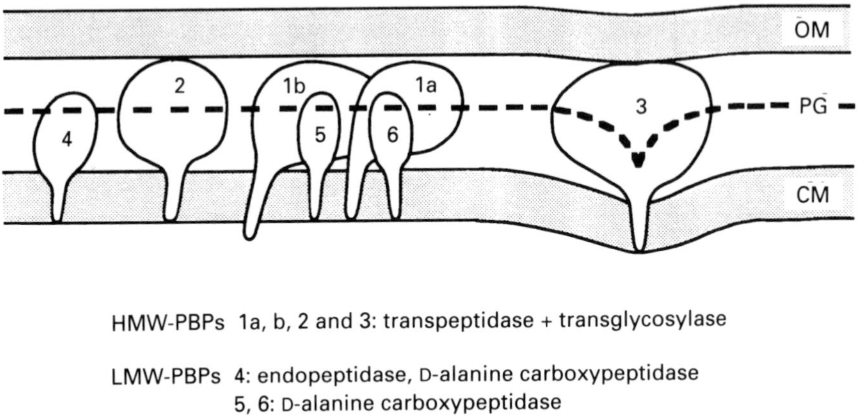

Fig. 9. Highly schematic representation of *E. coli* penicillin-binding proteins (PBPs) in the cell envelope. The high molecular-weight (HMW) PBPs: PBP1A/1B, PBP2 and PBP3 have a carboxyterminal anchor in the cytoplasmic membrane and the lower molecular weight (LMW) PBPs (PBP4, PBP5 and PBP6) an aminoterminal one. (For further details see text and references: Adachi, Ohta & Matsuzawa, 1987; Bowler & Spratt, 1989; Edelman *et al.*, 1987; Pratt, Jackson & Holland, 1986.)

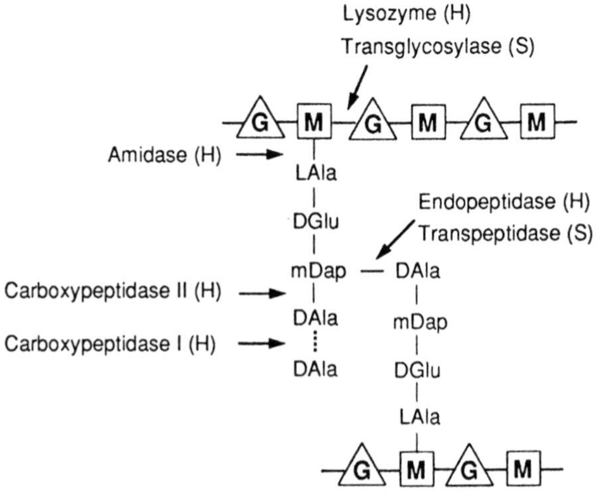

Fig. 10. Some enzymes involved in synthesis (S) and hydrolysis (H) of peptidoglycan. During transpeptidation (cross-linking) the terminal D-ala is being removed. (See also Fig. 14.)

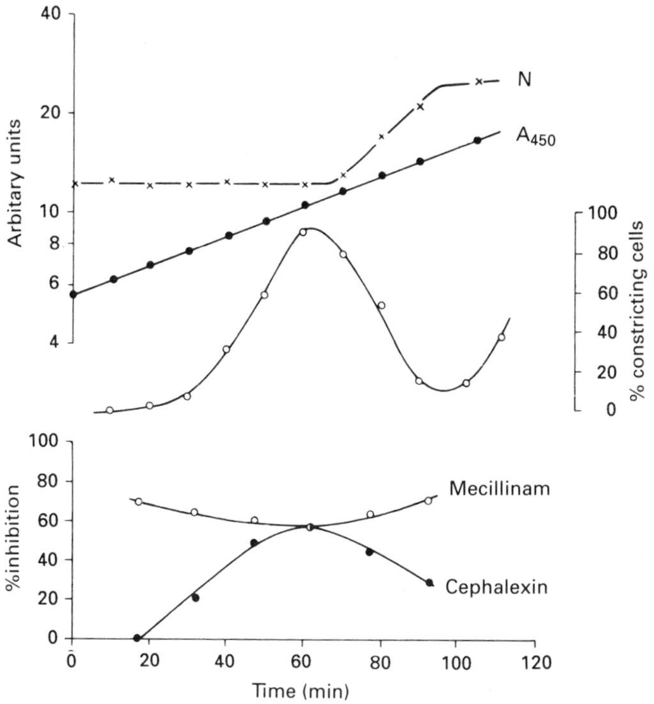

Fig. 11. The contributions of PBP2 and PBP3 to peptidoglycan synthesis in relation to the division cycle. The [^{3}H]-Dap incorporation was measured after inhibition of PBP2 (mecillinam) and PBP3 (cephalexin). The synchronous culture was obtained by selection of small cells by centrifugal elutriation. N, cell number; A_{450}, absorbance at 450 nm. (From Wientjes & Nanninga, 1991.)

acceptor–donor radioactivity ratio (ADRR; Burman & Park, 1984) by which the donor and acceptor muropeptide chains can be distinguished (Fig. 14). During the transpeptidation reaction, a donor pentapeptide is bound to an acceptor tetrapeptide. The ε-amino group of the donor peptide which remains free after formation of a cross-link can be dansylated. In this way the radioactive donor-peptide can be traced after it has formed a cross-link. The ADRR is zero, as long as the acceptor does not become radioactive. It was found that the ADRR correlated positively with the percentage of constricting cells (De Jonge *et al.*, 1989). The interpretation has been that the mode of glycan strand insertion is mainly single-stranded during cell elongation (a low ADRR; see also Cooper, Hsieh & Guenther, 1988) and mainly multi-stranded (or very fast single-stranded) at the leading edge in dividing cells (a high ADRR; De Jonge *et al.*, 1989; Fig. 15). Taking the ADRR into account, no change in cross-linking of nascent peptidoglycan was observed during the cell cycle. (For a recent discussion on ADRR and cross-linkage, see Cooper, 1990.) It thus appears that the construction of

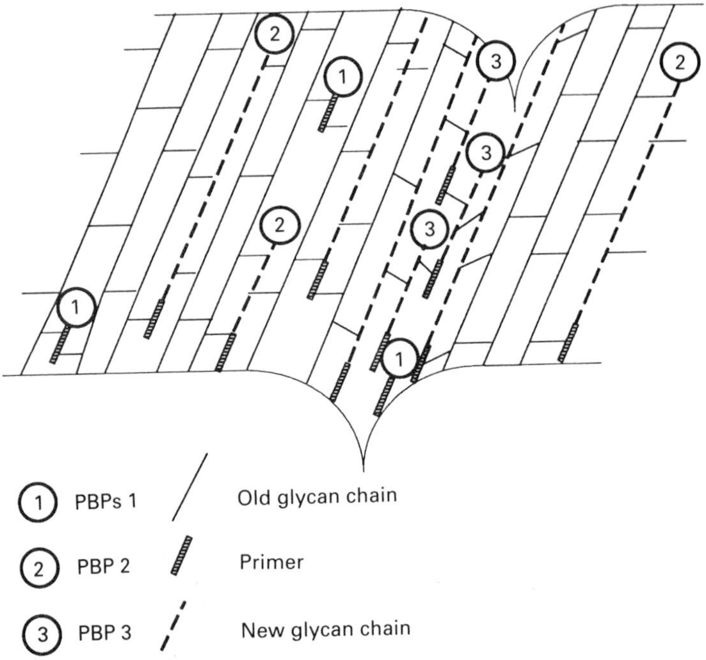

Fig. 12. Primer model of peptidoglycan synthesis. A two-dimensional sheet of peptidoglycan is schematically presented with a cylindrical part and a constriction. The glycan chains run perpendicular to the length axis of the cell. In between old glycan chains, new primers are inserted by PBPs 1. PBP2 and PBP3 need these primers for elongation of the glycan chains. PBP2 does so in the cylindrical part of the cell wall (elongation of the cell) and PBP3 at the constriction site. (From Wientjes & Nanninga, 1991.)

polar caps does not require a specific peptidoglycan composition. This leads to the question as to what extent specific enzymes are involved in cell division.

CELL DIVISION AND PEPTIDOGLYCAN ENZYMOLOGY

Apart from a change in the mode of glycan strand insertion from single-stranded (cell elongation) to multi-stranded (cell division) there is as yet no indication that specific muropeptides are produced during the division process. Of course, the possibility exists that enzymes are employed that are made and/or triggered only when division has to take place. As is well known, PBP3 is one of these enzymes (see above). Furthermore, it is present in a constant cellular concentration during the whole cell cycle (Wientjes *et al.*, 1983), suggesting that it is specifically activated during division. However, as argued above, PBP3 seems not to be involved in the initiation of the division process (Fig. 8). If so, this would require another PG synthesizing activity not related to the other HMW-PBPs. The latter conclusion has been drawn because the initiation process could not be inhibited by cefsulodin or moenomycin (PBP1A/B) and mecillinam (PBP2; Wientjes & Nanninga,

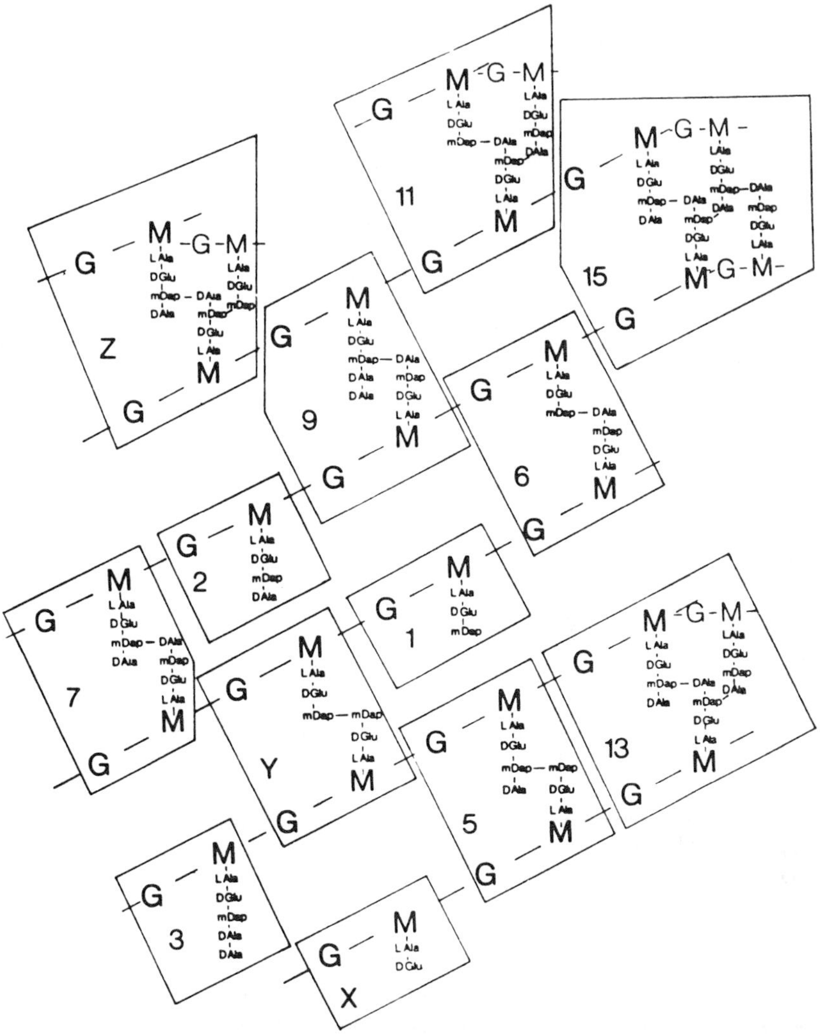

Fig. 13. Schematic drawing of the muropeptides present in the peptidoglycan. Compounds can have a lipoprotein molecule attached or can exist in the anhydro form. G, *N*-acetylglucosamine; Mg, *N*-acetylmuramic acid. The following abbreviations for muropeptides were used. Tri, Tet, and Pen, *N*-acetylglucosaminyl (NAcGlc)-*N*-acetylmuramyl (*N*AcMur)-tripeptide, *N*AcGlc-*N*AcMur-tetrapeptide, and *N*AcGlc-*N*AcMur-pentapeptide, respectively, where the tripeptide consists of L-Ala-D-Glu-*meso*-Dap, the tetrapeptide of LAla-D-Glu-*meso*-Dap-D-Ala, and the pentapeptide of L-Ala-D-Glu-*meso*-Dap-D-Ala; Lys, lysyl-arginine residue from a lipoprotein linked to *meso*-Dap of the amino side chain; Anh, NAcGlc- (1,6)-anhydro-NAcMur; DAP, cross bridge between the *meso*-Dap residues of the peptide side chains. All other oligomers are cross-linked between *meso*-Dap and D-Ala. Thus, Tet-Tri-Dap is the dimer of *N*AcGlc-*N*AcMur-tetrapeptide and *N*AcGlc-*N*AcMur-tripeptide cross-linked through a Dap–Dap cross bridge. Muropeptides: 3, Pen; 5, Tet-Tri-DAP; 6, Tet-Tri; 7, Tet–Tet; 9, Tet–Pen; 11, Tet–Tet–Tri; 13, Tet–Tet–Tet; 15, Tet–Tet–Tet–Tet; x, Di; Y, Tri–Tri–DAP; Z, Tet–Tet–Tri-DAP. (From De Jonge *et al.*, 1989.)

Table 1. *Muropeptide composition of an exponentially growing* E. coli *MC4100 culture.*

Muropeptide	Fraction of incorporated [^{3}]H-Dap
Tri	0.4(0.1)
Tet	54.4(0.9)
Pen	1.9(0.1)
Tri-Lys	3.1(0.1)
Tet-Tri-DAP	0.4(0.1)
Tet-Tri	2.4(0.1)
Tet-Tet	28.8(0.4)
Tet-Anh	1.0(0.3)
Tet-Pen	2.0(0.0)
Tri-Tri-DAP-Lys	0.3(0.0)
Tet-Tet-Tri	0.4(0.0)
Tet-Tri-Lys	0.8(0.2)
Tet-Tet-Tet	1.6(0.1)
Tet-Tri-Anh	0.4(0.1)
Tet-Tet-Tet-Tet	0.1(0.0)
Tet-Tet-Tri-Lys	0.3(0.2)
Tet-Tet-Anh	1.2(0.1)
Tet-Tet-Tri-Anh	0.2(0.1)
Tet-Tet-Tet-Anh	0.4(0.1)
	Degree of cross-linking
Dap-Dap	0.4(0.0)
	Disaccharide peptides (% of total) containing lipoprotein
	4.1(0.3)

The culture was pulse labelled for 4 min with [^{3}H]-Dap. The muropeptide composition was determined by HPLC. The values are presented as percentages of total [^{3}H]-Dap label and are averages of two experiments. Values between brackets are standard deviations (From De Jonge *et al.*, 1989). For nomenclature of the various components, see Fig. 13.

unpublished observations). This penicillin-independent peptidoglycan synthesizing system has been termed PIPS (Nanninga, 1991). Recent examples of penicillin-insensitive enzymes are a glycan polymerase (Hara & Suzuki, 1984). the *murH* gene product (Dai & Ishiguro, 1988) and the *mepA* product (DD-endopeptidase; Keck *et al.*, 1990). The functional significance of these enzymes, however, has still to be ascertained.

Since the initiation of division is inhibited in temperature-sensitive *ftsZ* mutants (for reviews see Donachie & Begg, 1990; Lutkenhaus, this volume) it is likely that FtsZ interacts with PIPS. FtsZ might exist in a cytoplasmic form and a membrane-bound form (Jones & Holland, 1985). Because it is logical that PIPS has a periplasmic location, it seems plausible that the interaction between cytoplasmic membrane-associated FtsZ is mediated by a transmembrane protein (X in Fig. 16). The interaction between X and FtsZ has to

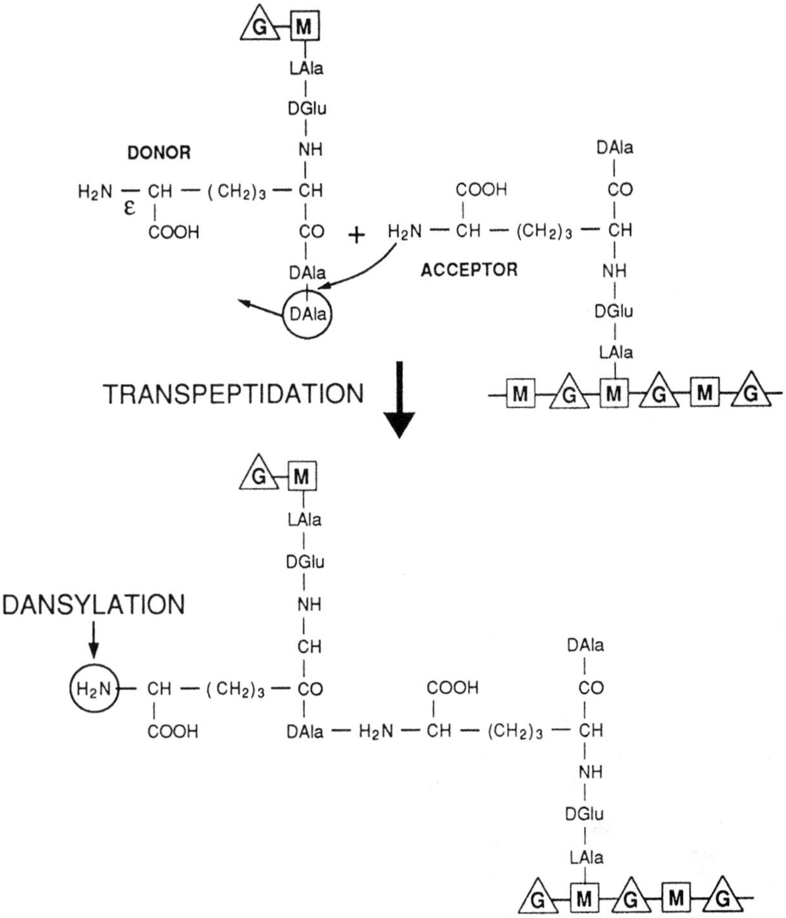

Fig. 14. Cross-link formation by the transpeptidase reaction of the disaccharide pentapeptide (donor) and a disaccharide tetrapeptide (acceptor) which forms part of existing murein. The terminal D-Ala of the pentapeptide is released during this process and the ε-amino group of donor Dap remains free. The free amino group can be labelled by dansylation. See also Fig. 10. (From De Jonge *et al.*, 1989.)

be cell-cycle dependent to serve as a topological switch in peptidoglycan synthesis (Nanninga, 1991). As pointed out above and in line with earlier genetic studies (Walker *et al.*, 1975; Begg & Donachie, 1985; Taschner *et al.*, 1988), PBP3 functions after the initiation of cell division (Fig. 8). Though PBP3 has the same bifunctional enzymic activities as PBP2, the basis for the division of labour is not at all clear. Possibly, the enzyme specificity with respect to substrate and site of action is determined, at least in part, through the interaction with other proteins (see also below). Whereas the

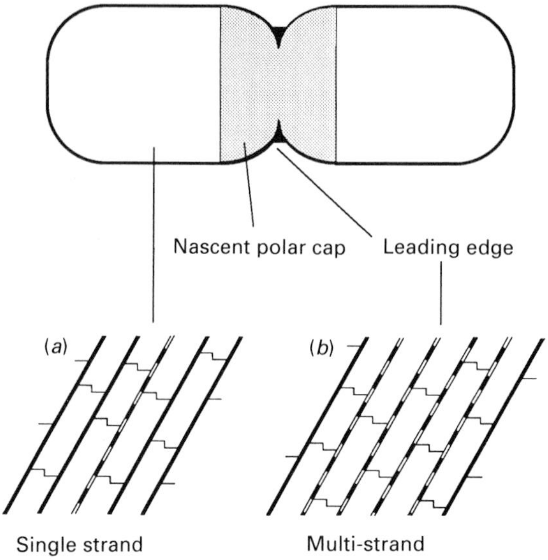

Fig. 15. Mode of glycan strand insertion during cell elongation and cell division.

PBP2 interacts with RodA (Ishino *et al.*, 1986), PBP3 interacts with FtsA (Tormo *et al.*, 1986) and possibly FtsW (Matsuhashi, Wachi & Ishino, 1990). *RodA* mutants, like *pbpA* mutants, are spherical. Rod A is a membrane protein (Stoker *et. al.*, 1983) and its hydropathy profile is strikingly similar to that of FtsW (Ikeda *et al.*, 1989). The suggestion that FtsA interacts with PBP3 is based on the fact that *ftsA* mutants when filamenting show blunt constrictions like those of *pbpB* and that binding of B-lactam antibiotics is affected by mutations in *ftsA* (Tormo *et al.*, 1986). Moreover, PBP3-specific furazlocillin does not inhibit [^{3}H]-Dap incorporation at the non-permissive temperature in a *ftsA* mutant (Nanninga *et al.*, 1990).

The final step in cell division, cell separation, is affected by *envA* (for review see Donachie & Begg, 1990). Unfortunately very little is known about its gene product.

CELL ELONGATION, CELL DIVISION AND CELL SHAPE

For the sake of simplicity the shape of *E. coli* can be conceived of as a cylinder with hemispherical caps at its two poles. The shape can then be described as a so-called aspect-ratio (L/2*R*; for review see Nanninga & Woldringh, 1985), where *L* represents the mean cell length and *R* the mean radius of the cylinder. The aspect ratio is determined by the relative contributions of the lateral wall and polar caps to the cells' surface. Thus the question can be asked, what is the biochemical basis of the above-mentioned contributions?

As mentioned in the previous section, some specificity exists with respect

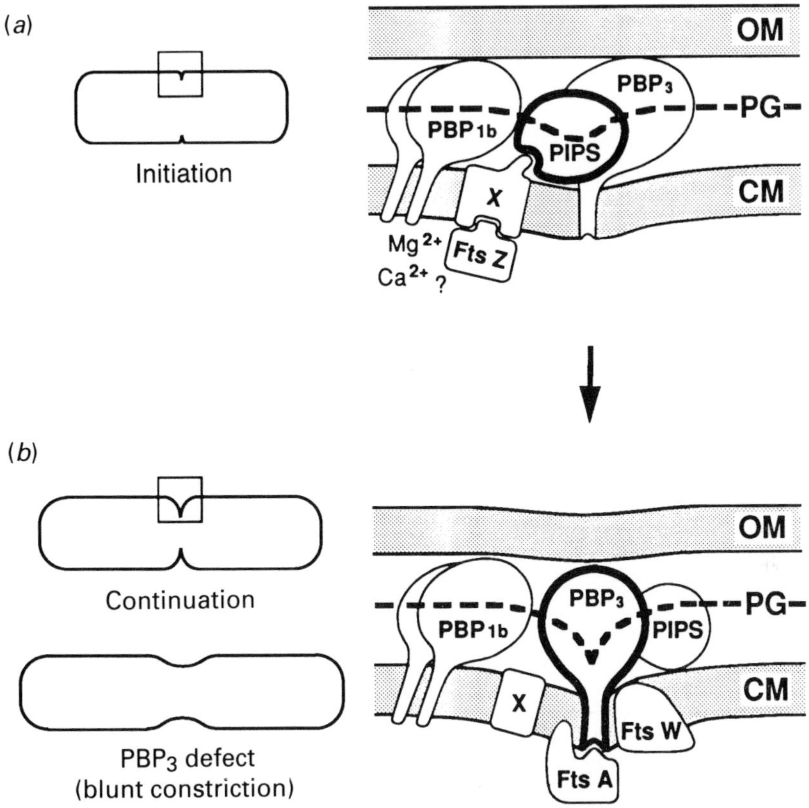

Fig. 16. Two steps in the division process: initiation (*a*) and continuation (*b*). During initiation, FtsZ acts through a hypothetical transmembrane protein X on penicillin-insensitive peptidoglycan synthesis (PIPS). Then (*b*) PBP3 takes over with its associated proteins FtsA (Tormo *et al.*, 1986) and FtsW (Matsuhashi *et al.*, 1990). Also shown is the morphological effect of PBP3 inhibition. (From Nanninga, 1991; slightly modified.)

to proteins involved in maintaining rod-shape (PBP2, Rod A) and in those needed for cell division (the putative PIPS, FtsZ, PBP3, FtsA and FtsW). The rod-shape system (cell elongation) can be specifically inhibited by mecillinam, the division system (at least in part) by cephalexin. A so-called PBP3/PBP2 inhibition ratio can then be obtained if for a particular culture [^{3}H] Dap-incorporation is measured for each antibiotic (Wientjes & Nanninga, 1991). If this approach is carried out for cultures in which the cells have different mean aspect ratios an inverse relation with the PBP3/PBP2-inhibition ratio is found (Fig. 17; see also Lleo, Canepari & Satta, 1990). Clearly then, the aspect ratio has a biochemical basis directly related to PG biosynthesis.

However, this does not yet answer the question about the origin and

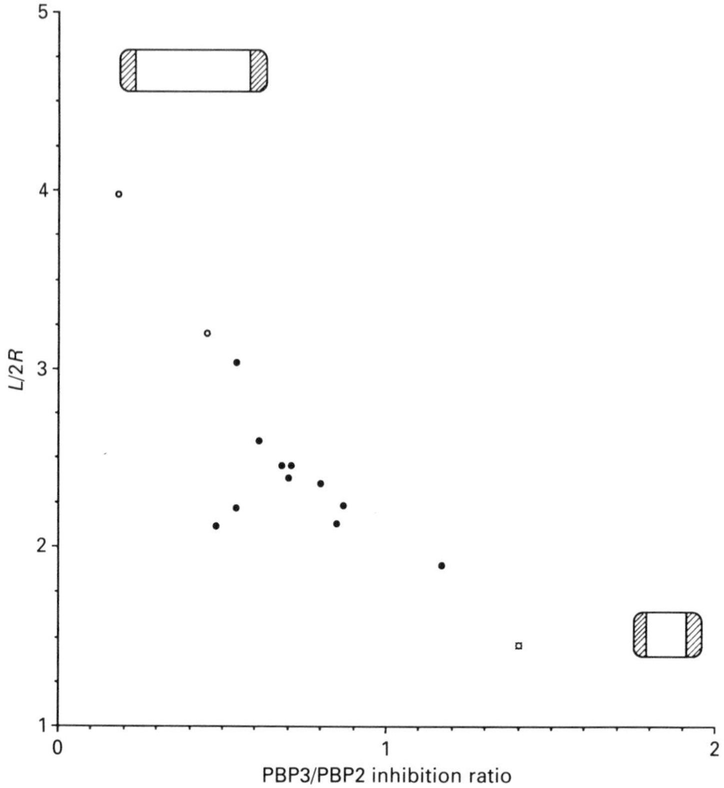

Fig. 17. Relation between cell shape and PBP3/PBP2 inhibition ratio. The average cell dimensions were measured and the length/width ratio ($L/2R$) was calculated. Each culture was probed for inhibition of [^{3}H]-Dap incorporation by cephalexin (10 μg/ml) and mecillinam (2 μg/ml). The ratio of these inhibition percentages were taken as the PBP3/PBP2 ratio. For further detail see Wientjes & Nanninga, 1991.

maintenance of rod shape as such. This question has been dealt with extensively by A. L. Koch (1985, 1988, 1990; for an earlier discussion see Henning, 1975). Briefly, geometrical bodies resembling bacterial shapes can be produced by soap bubbles. In this way, conditions have been outlined that would produce stable cylinders that spontaneously divide when the cell length equals $2\pi R$ (for review see Koch, 1985 and earlier references). In the case of *E. coli*, cylindrical growth (of constant width) would require physically stable polar caps and diffuse envelope extension in such a way that $P = T/R$ (P, internal hydrostatic pressure; T, surface tension; R, radius of the cell). An attractive feature of the model is also the modulating effect of internal hydrostatic pressure on the activities of PG synthetic and hydrolytic enzymes. However, there are still many outstanding questions (see Nanninga, 1988). (i) First, the emphasis placed on the idea of surface tension is perhaps too simple. The similarity between soap bubbles and bacterial shapes is very striking indeed, but the complexity of a bacterial cell is much greater than that of a simple soap film. Although the cytoplasmic membrane comes physically closest to a soap film, the argument has been based on stretchability of the PG layer, which does not possess membrane-like properties. This paradox might be solved by considering that the cytoplasmic membrane and the PG layer are closely linked and highly dynamic structures. Probably, even minute changes in the cytoplasmic membrane are immediately transferred to the function of various hydrolytic and synthetic enzymes in the PG layer. In this sense the PG layer and its underlying membrane form an integrated and dynamic organic structure. So, to a first approximation, it might not be unreasonable to treat the PG layer as the surface tension target. (ii) So far, it has not been possible to measure the surface tension of the bacterial cell envelope reliably, let alone that of the discontinuities implied in cell division. (iii) It is sufficiently known whether stretchability of the cell envelope represents a physiological parameter. We do not know whether stretchability resides in breaking and reestablishing cross-bridges between peptide chains and/or whether stretchability of the peptide side as such plays a role. (iv) There is no experimental evidence that stressed bonds are more easily split than unstressed ones. One could even argue in the opposite direction, in the sense that the spatial configuration of the stressed substrate is less well recognized by the relevant enzyme.

THE TRIGGER FOR DIVISION AT THE ENVELOPE LEVEL

In a foregoing section we have mentioned which enzyme system might specifically be involved in effecting division (Fig. 16). Also, mention has been made that apart from the mode of glycan stand insertion (Fig. 15) no indications exist for division-specific PG. What then might be the trigger that initiates division?

In the surface stress theory as developed for *E. coli* (Koch, 1990) a sufficient decrease in surface tension of the envelope in the cell centre would suffice to ensure division. An alternative explanation (Koch, 1990) could be the increased Dap-incorporation at the leading edge (Fig. 7; Wientjes & Nanninga, 1989). However, these possibilities do not have to be mutually exclusive. It is very feasible that an initial lowering of local surface tension stimulates PG synthesis at the site of constriction. Such a change in local surface tension could affect the spatial grouping of PG-specific enzymes and substrates in such a way that enzyme activities are modulated.

An interesting model has been proposed recently (Begg *et al.*, 1990) in which the alternation of PBP2 activities is based on substrate specificity and availability, the essence being that PBP3 (and PIPS?) would require acceptor tripeptide side chains to receive donor peptides (Botta & Park, 1981; Pisabarro *et al.*, 1986). The model combines several earlier results including the observation that overproduction of PBP5 (D-alanine carboxypeptidase I; Fig. 10) leads to cells with rounded morphology (Markiewicz *et al.*, 1982). An increase in the level of PBP5 could also reverse the cell division block in a *pbpB* (*ftsI23*) mutant at 42°C. Since PBP5 produces tetrapeptides, this would, in turn, increase the possibility to provide tripeptides by the action of D-alanine carboxypeptidase II (Beck & Park, 1976; Beck & Park, 1977). Tripeptides can also be produced by inhibition of the cytoplasmic step of PG synthesis by D-cycloserine (see Höltje & Schwarz, 1985 for details). Conditions of inhibition by D-cycloserine could be found which also reversed the *ftsI23* division block (Begg *et al.*, 1990). In line with these observations the activity (not the amount) of carboxypeptidase II has been shown to increase during cell division (Beck & Park, 1976). Conflicting results have been published with respect to carboxypeptidase I activity during the cell cycle. Beck & Park (1976) found no change in activity at division, whereas Mirelman *et al.* (1977) reported that the activity increased. The rounded cell shape (*pbpA* morphology) seen after overproduction of plasmid-encoded PBP5 could thus be explained by the reduced availability of acceptor muropeptides for the cell elongation system of PBP2/RodA. A decreased level of RodA could also increase the level of PBP5 (Begg *et al.*, 1990). It thus seems that the *rodA* operon (including the genes for PBP2 and PBP5) has an important role in the regulation of cell elongation. Unfortunately, no increase in the amount of tripeptide side chains could be found during division in synchronized cells (De Jonge *et al.*, 1989). This could mean that, under physiological conditions, only a slight change in tripeptide side chain 'concentration' would suffice. In addition, the carboxypeptidase II could be considerably lower *in vivo* because, in the assay system used, dividing cells were more prone to enzyme release than non-dividing ones (Beck & Park, 1977). Though the above discussion has been focused on the alternation between PBP2 and PBP3, it might be speculated that tripeptide side chain acceptors are also needed for PIPS (Fig. 14). Essential for division to occur would thus be the specific activation of carboxypeptidases for trimming peptide side chains to the correct acceptor size. It should be noted that the carboxypeptidase I (Mirelman *et al.*, 1977) and carboxypeptidase II (Beck & Park, 1977) are continuously present during the cell cycle. Activation would thus imply the removal of a barrier which inhibits the activities during cell elongation (Beck & Park, 1977; see also Hartmann, Bock-Hennig & Schwarz, 1974; Nanninga & Woldringh, 1985). What could be the barrier during cell elongation?

ENVELOPED GROWTH, NUCLEOID SEGREGATION AND CELL DIVISION

Thus far we have considered the envelope to be mainly represented by the dynamic peptidoglycan layer and we have argued how changes in the topography of its synthesis accompany cell division. How is this switch from lateral wall extension to localized synthesis of polar caps regulated? Does the growing envelope itself play a primary role in cell cycle regulation? Does it trigger events like the segregation of daughter chromosomes? Or is it rather the physiological and structural state of the chromosome that affects envelope growth and cell division?

These possibilities have led to two opposing views on the extent of interplay between envelope growth and DNA replication. In one view the envelope has considerable morphogenetic autonomy; in the other there is explicit interplay between envelope and nucleoid. The first view is based on the visualization of so-called periseptal annuli (PSA) in plasmolysed *E. coli* cells (Cook, MacAlister & Rothfield, 1986). These represent sites of adhesion between the cytoplasmic membrane and the PG layer. Measurements of their localization in various cell length classes has led to a model in which the division site is selected from predetermined sites developed in the growing envelope. In this model, the processes of chromosome duplication and cell division are proposed to be independently triggered by factors related to progression through the cell cycle (Cook *et al.*, 1987).

The second view is based on the placement of constriction sites in filaments of temperature-sensitive DNA replication mutants in which division was allowed to resume after a period of inhibition of DNA synthesis (Woldringh *et al.*, 1985; Woldringh *et al.*, 1990; Mulder & Woldringh, 1989). Combined with autoradiographic measurements of PG synthesis in such filaments (see below), the observations led to the so-called nucleoid-occlusion concept (see also Donachie & Begg, 1989*a*). In this model a direct coupling or dependence of envelope synthesis and cell division on the process of DNA replication and segregation is proposed. Below we will discuss the two views in more detail.

THE LOCALIZATION OF PLASMOLYSIS BAYS ACCORDING TO THE PSA MODEL

Periseptal annuli are zones of adhesion between cytoplasmic membrane and cell wall that become visible upon plasmolysis of Gram-negative cells. Polar plasmolysis bays, which can easily be identified on phase-contrast micrographs, mark the position of annular attachments at old cell division sites. In addition, other sites of membrane-wall attachments are marked by 'partial' plasmolysis bays along the length of the cell. Because these plasmolysis bays were incomplete, they were therefore considered to be nascent. Measurement of their position in a large number of cells led Cook *et al.* (1986, 1987) to a model in which new PSA are generated from those already present in the cell centre of a new-born cell. Subsequently, these nascent annuli grow apart during cell elongation until they have arrived at their final positions at 25% and 75% of the cell length. During this migration, the annuli mature and, as they become extended completely around the cylinder, seal off periplasmic spaces which become the future division sites. In this way each daughter cell is born with a predetermined division site at the middle of the cell (Cook *et al.*, 1987).

Recently, the effect of mutations in the cell division genes *ftsZ* and *ftsA* on the proposed maturation and displacement of periseptal annuli was determined (Cook & Rothfield, 1991). Unlike the annuli formed at the non-permissive temperature in *ftsA* filaments, those in *fts*Z failed to mature. This was indicated by the occurrence of 'partial' plasmolysis bays which were randomly dispersed along the filament. As an explanation for this difference the authors suggested that FtsZ blocks the development of the periseptal annular apparatus at an earlier state than the *ftsA* gene product, before the maturation and localization of PSA.

Because the annular as well as the incomplete attachment sites have not yet been isolated, biochemical evidence for a specific function of the annulus is lacking.

AN ALTERNATIVE RULE FOR THE LOCALIZATION OF PLASMOLYSIS BAYS

Olÿhoek *et al.* (1982) have observed that, in an *E. coli* population, long cells are more prone to plasmolysis than short cells. This result was mainly ascribed to the easier compressibility of a larger cell in obtaining a plasmolysis bay of a given size. A second observation was that constricting cells were less frequently plasmolysed than non-constricting cells of the same size class. The authors concluded that the non-separated daughter cells apparently behaved as two smaller independent compartments.

A possible explanation for this last phenomenon is suggested by the fact that plasmolysis bays are not observed at the site of constriction, neither by phase contrast microscopy (e.g. Cook *et al.*, 1986), nor by electron microscopy (Woldringh, unpublished observations). Such exclusion of plasmolysis bays from the constriction site might result from an increase in membrane-wall interactions, due to the local activation of PG synthesis at the leading edge (Wientjes & Nanninga, 1989). In fact, the association of the PG-cytoplasmic membrane attachment at the leading edge appeared rather tight (MacAlister *et al.*, 1987). The above observations led us to consider an alternative explanation for the relatively local sensitivity of envelope sites to plasmolysis.

To test these ideas, we determined the locations of plasmolysis bays, visualized by whole-mount electron microscopy in relation to the constriction sites and in relation to cell poles and nucleoids. In Fig. 18 the distribution of plasmolysis bays is shown in *E. coli ftsZ* (ts) cells grown at the permissive temperature. Most of the cells contained a large plasmolysis bay at one pole and a smaller plasmolysis bay, either in the lateral wall or at the other

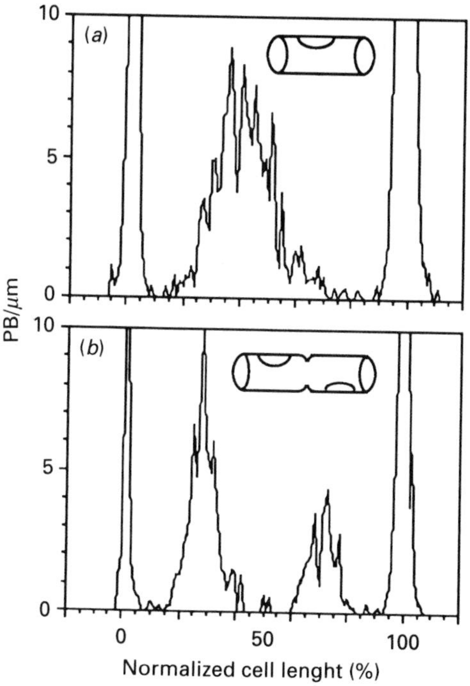

Fig. 18. Distribution of plasmolysis bays in *E. coli ftsZ* (ts) grown at 30 °C, treated for 3 min with 12.5% sucrose. (*a*), non-constricting cells ($N = 685$); (*b*), constricting cells ($N = 388$). For measurement the better plasmolysed cell was placed at 100% of the cell length. (From E. Mulder, Ph D Thesis, University of Amsterdam, 1991).

pole. Orientation of the cells with the larger pole placed at 100% of normalized cell length (Fig. 18(*a*)) showed that the average position of the lateral polar bay is not a mid-cell but at 38% of normalized cell length. In constricting cells (Fig. 18(*b*)), the cell half with the larger polar bay contained less lateral bays. This asymmetric distribution of plasmolysis bays suggests a difference in susceptibility of the cells to plasmolysis from one pole to the other. Because active constrictions may involve a close interaction between plasma membrane and cell wall at the leading edge (MacAlister *et al.*, 1987; Wientjes & Nanninga, 1989), the newest cell pole may be more refractory to plasmolysis than the oldest pole where the interaction may have deteriorated. In accordance with the observations of Cook *et al.* (1987), our observations also show that the placement of lateral plasmolysis bays requires a minimum distance from the polar bays or from the refractory constriction sites (Fig. 18).

Measurement of the positions of plasmolysis bays in *ftsA* (ts) filaments grown at restrictive temperature confirmed the observations by Cook & Rothfield (1991) in that bays occurred at ¼ and ¾ cell lengths, with additional positioning at ⅛, ⅜ and ⅝ in larger filaments. As previously reported (Taschner *et al.*, 1988 and see above), these filaments contained abortive constrictions ('bottle necks'). We suggest that, in contrast to active division sites, such abortive constrictions become more susceptible to plasmolysis, just as the oldest pole in a newborn cell. By contrast in *ftsZ* filaments in which no sign of abortive constrictions is visible (Taschner *et al.*, 1988), no such positioning of bays was observed. The nucleoids in these filaments had nevertheless segregated to regular positions (see Fig. 7 in Woldringh *et al.*, 1991). This observation does not support the suggestion raised by de Boer, Cook & Rothfield (1990) that the periseptal annuli could be used as chromosome attachment sites in the process of nucleoid segregation.

The above results and interpretations can be summarized by the following: (i) plasmolysis bays are randomly positioned along the length of the cell, (ii) old cell poles or abortive constrictions in filaments are most susceptible to plasmolysis, (iii) active constrictions are resistant to plasmolysis, and (iv) zones refractory to plasmolysis extend from polar plasmolysis bays and from active constriction sites (Fig. 18; see also Fig. 4 in Cook *et al.*, 1987). It is this refractory zone extending from the polar plasmolysis bay, which, we believe, causes the asymmetric distribution of bays as shown in Fig. 18(*a*).

In our view, plasmolysis bays are positioned randomly and can either occur as 'complete' or 'partial' bays (terminology of Cook & Rothfield, (1991) depending on the sensitivity of envelope zones to plasmolysis. In other words, we do not consider the plasmolysis bays to be indicative for specific annular structures at their borders. We rather see the bays as envelope domains with relative-low PG synthetic activity and therefore, reduced cytoplasmic membrane–wall interaction.

IS THERE NUCLEOID-DIRECTED ENVELOPE GROWTH?

The explanation for plasmolysis-bay location discussed above ascribes a positioning effect with respect to active or abortive (including polar) constriction sites. Is there a role for the nucleoid(s) in positioning of the division site? Autoradiography of [^{3}H]-Dap incorporation in *E. coli dnaX* (ts) filaments showed that PG is synthesized at a reduced rate in the nucleoid-containing filaments (Fig. 19). It would seem that in this way, the cell prohibits division at the site of an unsegregated nucleoid, because it has been shown that division requires a local increase in the rate of Dap incorporation (see above and Fig. 6).

Inhibition or slowing down of the transcription/translation machinery allows a continuation or a relative increase in the rate of division, leading to so-called residual division (Grossman, Ron & Woldringh, 1982) or smaller cells (Schaechter, Maaloe & Kjeldgaard, 1958), respectively. Is there a differentiation of protein synthesis along the length axis of the cell? Autoradiography of [^{3}H]-leucine incorporation in the above-mentioned *dnaX* (ts) cells

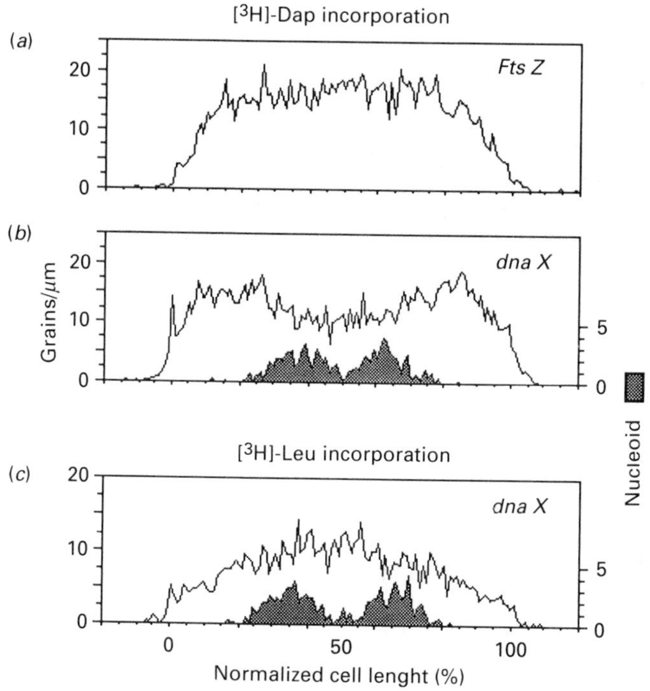

Fig. 19. Autoradiography of [^{3}H]-Dap incorporation (*a*) and (*b*) and [^{3}H]-leucine incorporation (*c*). (*a*) silver grain distribution over whole-mount *ftsZ* filaments at 42 °C. (*b*), the same of *dnaX* filaments; (*c*), *DnaX* filaments pulse-labelled with [^{3}H]-leucine. (cf. text for further details; From: Woldringh *et al.*, 1991).

revealed a higher incorporation in the nucleoid-containing area than in the nucleoid-free area (Woldringh *et al.*, 1991). These observations suggest that transcription/translation activity around the nucleoid coincides with a reduction in PG synthesis in its vicinity.

A seemingly puzzling phenomenon occurs in slowly-growing *E. coli* B/r cells, in which cell constriction has been observed to initiate before complete separation of the nucleoids (Fig. 20; Woldringh *et al.*, 1977). How can this growing cell activate peptidoglycan synthesis at its centre if the repressing nucleoid is still present there? Our working hypothesis is that termination of DNA replication releases a diffusable factor which can stimulate peptidoglyan synthesis. Because during replication the DNA is immediately segregated in *E. coli* (Woldringh, 1976; Donachie & Begg, 1989*b*), the daughter nucleoids are usually separated at the termination of DNA replication. This implies that the last steps in the replication process, termination and decatenation, must occur at the centre of the duplicated nucleoid, which normally coincides with the centre of the cell. Therefore, our hypothetical termination factor is also released at mid-cell.

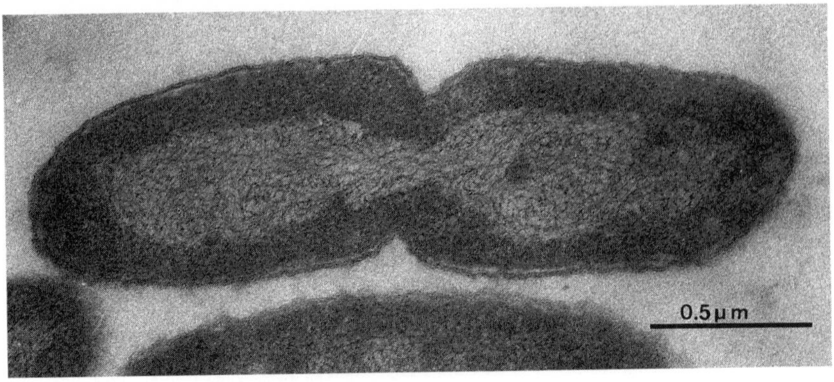

Fig. 20. Thin section of slowly grown *E. coli* B/rA cells (doubling time 90 min). Note that constriction has started before separation of the nucleoids.

In the case of the slow-growing B/r cell described above (Fig. 20), it was estimated by DNA labelling and autoradiography that termination of DNA replication occurred before initiation of constriction in the average cell. It can thus be argued that the termination factor released at the cell centre overrules the repressing effect around the nucleoid and causes the local activation of peptidoglycan synthesis, which is necessary for constriction. In more rapidly growing cells, the repressing effect of the nucleoid, the so-called nucleoid-occlusion (see below), is stronger due to the higher activity of the protein synthesizing machinery. Consequently, the activation of peptidoglycan synthesis has to wait for full separation of the daughter nucleoids. This inhibiting effect of the nucleoid on cell division has been called nucleoid occlusion (Cook, de Boer & Rothfield, 1989; Mulder & Woldringh, 1989; Donachie & Begg, 1989*a*). An alternative explanation according to the PSA model as has been discussed above would be that the PSA in the B/r cell of Fig. 20 have matured at mid-cell and have started the division process independently from DNA replication and nucleoid segregation. A dependence of cell division on termination of DNA replication, however, is supported by the observations of Grossman, Rosner & Ron, (1989). In their system of synchronized *E. coli* K 12 cells, it could be shown that the cells were not able to replicate their terminus. Following resumption of protein synthesis, cell division was observed to be delayed until a pulse of DNA-terminus-replication had occurred.

THE NUCLEOID-OCCLUSION MODEL

The above proposal for the existence of a repressing factor (nucleoid-occlusion) and of an activating factor (released at termination), the balance of which determines the rate of peptidoglycan synthesis, have been summarized in a so-called nucleoid-occlusion model (Mulder & Woldringh, 1989; Wold-

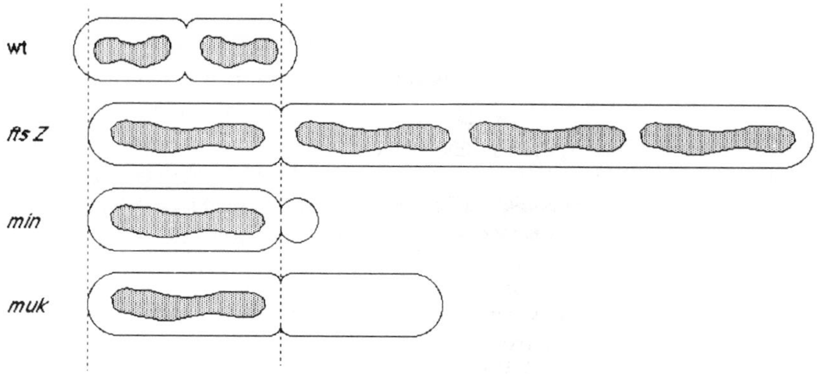

Fig. 21. Schematic representation of a wild type cell showing normal segregation of the nucleoid, in comparison to three mutants showing abnormal cell division. In filaments of *ftsZ*, recovering at 30 °C after growth at 42 °C for two mass doublings, the first divisions probably take place between two pairs of replicated but not yet separated nucleoids. The same holds true for *min* and *muk* mutants. In all three cases the nucleoid-containing cell part is on average longer than a newborn cell but shorter than a normally dividing cell. See text for references.

ringh *et al.*, 1990, 1991). The model can explain a variety of division phenomena from the point of view that envelope synthesis can be locally directed by the nucleoid. These include division recovery in *ftsZ* filaments, the formation of minicells in *min* mutants and the formation of anucleate cells in *muk* mutants. As shown in Fig. 21 these three kinds of cells have one property in common: constriction initiates before nucleoid segregation. The nucleoid occlusion model predicts in all three cases that a cell, almost twice the length of a newborn cell, contains a replicated but unsegregated nucleoid and starts a constriction next to it. In the case of *mukA1* (Hiraga *et al.*, 1989) and *mukB* (Niki *et al.*, 1991) strains, the nucleate cells have been shown (by graphic analysis of fluorescence microscope pictures) to contain two copies of the chromosome. In other words, termination can be expected to have occurred in these cells. Whether this also holds for minicell-forming mutants remains to be established. Minicell formation, like normal division (see above) can be considered from two points of view, first, with respect to a certain autonomy of the cell envelope (De Boer *et al.*, 1990) and secondly, while including a positioning role of the nucleoid. Since the first mechanism has been discussed extensively by Lutkenhaus (this volume) we will only mention that *minB* mutants resemble *gyrB* mutants (Mulder & Woldringh, 1989) and *muk* mutants (see Fig. 21) in the sense that nucleoid segregation is impaired.

An old phenomenon, however, the division pattern in *dnaA* (ts) filaments already described by Hirota *et al.*, (1968), has been brought up to contradict the model (D'Ari *et al.*, 1990). In this mutant, DNA synthesis is blocked without induction of the SOS-response. The cells grow into filaments with one or two nucleoids and continue to divide at a lower rate, by pinching off DNA-less cells at their ends. Both Hirota *et al.* (1968) and Jaffé *et al.* (1986) have claimed that the DNA-less cells were of normal length. If so, this would suggest that the positioning of the division site was not affected by the absence of a nucleoid and that the growing envelope itself directed division at predetermined sites.

Although the uniformity in size of the anucleate cells is indeed remarkable in published micrographs, actual measurement of the DNA-less cell ends of *E. coli dnaA46* (ts) showed that their average cell length (3.1 μm) is significantly higher than that of newborn cells (1.4 to 2.0 μm). In addition, the coefficient of variation (21%) was also much higher than that of newborn cells (11%; Mulder & Woldringh, 1989).

MINICELL FORMATION

Can minicell formation be interpreted in the light of the nucleoid occlusion model? These DNA-less cells are formed in mutants of *E. coli* in which the expression of genes of the *minB* locus is absent or disturbed (De Boer *et al.*, 1990). It has been postulated that the gene products of the *minB* locus (MinC, MinD and MinE) are required for correct selection of the division site. In the selection process MinE activates central division at normal concentration and additional polar division (minicell formation) at higher concentrations. The problem remains, of course, how MinE finds its target.

According to our concept the topological regulation of cell division is primarily determined by the spatial distribution of the nucleoid. Is there any supportive evidence for this view with respect to minicell formation? Cytological observations on a MinE-overproducing strain revealed that: (i) cells with polar constrictions contain predominantly unsegregated nucleoids; (ii) minicell forming cells are on average shorter than normally dividing cells and (iii) that the fraction of constricting cells had doubled. These observations support the idea that MinE is a division activator. However, they also suggest that the polar cell division is guided by the segregation status of the nucleoid.

CONCLUSIONS

In the foregoing we have discussed the spatial and temporal organization of envelope growth in *Escherichia coli*. The spatial organization results in a particular bacterial shape: a cylinder covered with two hemispherical caps. Although no clear-cut difference in peptidoglycan composition between polar caps and lateral wall has so far been found, other differences do occur. These reside in enzyme specificity. A postulated penicillin-insensitive peptidoglycan synthesis (PIPS) system starts division in conjunction with the division protein FtsZ. In a second step this is continued with the activity of PBP3, together with the division protein FtsA and possibly FtsW. Autoradiographic experiments have shown that Dap-incorporation takes place predominantly at the leading edge of the invaginating PG layer. Chemical analysis has indicated that glycan chain insertion is multi-stranded at the leading edge whereas it is single-stranded in the lateral wall. The latter (and presumably that part of the polar cap not belonging to the leading edge) are specifically made by PBP2 in combination with RodA. PBPs 1A and 1B are believed to produce primers during the whole cell cycle for the other PG synthesizing enzymes to act upon. A further distinction with respect to lateral wall synthesis and polar cap formation might be the specificity of acceptor substrates for PG synthesis, the idea being that PBP3 prefers tripeptides as acceptor.

The temporal organization of envelope growth has been discussed in relation to nucleoid position in the cells. Nucleoids, active in transcription/translation, appear to repress PG synthesis in their immediate surroundings (nucleoid occlusion), thus allowing division as soon as termination of DNA replication has occurred and the nucleoids have separated. The switch from lateral wall growth to polar cap formation is thus linked with the DNA replication and separation cycle. In this sense, the latter is a determining factor of cell shaped and division.

ACKNOWLEDGEMENT

This work was supported by the Foundation for Fundamental Biological Research (BION) which is subsidized by The Netherlands Organization for Scientific Research (NWO).

We thanks Mrs E. Lutz-Langezaal and Mr J. D. Leutscher for their help in preparing the text and the figures.

REFERENCES

Adachi, H., Ohta, T. & Matsuzawa, H. (1987). A water-soluble form of penicillin-binding protein 2 of *Escherichia coli* constructed by site-directed mutagenesis. *FEBS Letters*, **226**, 150–4.

Beck, B. D. & Park, J. T. (1976). Activity of three murein hydrolases during the cell division cycle of *Escherichia coli* K-12 as measured in toluene-treated cells. *Journal of Bacteriology*, **126**, 1250–60.

Beck, B. D. & Park, J. T. (1977). Basis for the observed fluctuation of carboxypeptidase II activity during the cell cycle in BUG6, a temperature-sensitive division mutant of *Escherichia coli*. *Journal of Bacteriology*, **130**, 1292–302.

Begg, K. J. & Donachie, W. D. (1985). Cell shape and division in *Escherichia coli*: experiments with shape and division mutants. *Journal of Bacteriology*, **163**, 615–22.

Begg, K. J., Takasuga, A., Edwards, D. H., Dewar, S. J., Spratt, B. G., Adachi, H., Otha, T. Matsuzawa, H. & Donachie, W. D. (1990). The balance between different peptidoglycan precursors determines whether *Escherichia coli* cells will elongate or divide. *Journal of Bacteriology*, **172**, 6697–703.

Botta, G. A. & Park, J. T. (1981). Evidence for involvement of penicillin-binding protein 3 in murein synthesis during septation but not during cell elongation. *Journal of Bacteriology*, **145**, 333–40.

Bowler, L. D. & Spratt, B. G. (1989). Membrane topologyof penicillin-binding protein 3 of *Escherichia coli. Molecular Microbiology*, **3**, 1277–86.

Braun, V., Gnirke, H. Henning, U. & Rehn, K. (1973). Model for the structure of the shape-maintaining layer of the *Escherichia coli* cell envelope. *Journal of Bacteriology*, **114**, 1264–70 and **116**, 1089 (correction).

Broome-Smith, J. K. & Spratt, B. G. (1982). Deletion of the penicillin-binding protein 6 gene of *Escherichia coli. Journal of Bacteriology*, **152**, 904–6.

Buchanan, C. E. (1981). Topographical distribution of penicillin-binding proteins in the *Escherichia coli* membrane. *Journal of Bacteriology*, **145**, 1293–8.

Burdett, I. D. J. & Murray, R. G. E. (1974). Septum formation in *Escherichia coli*: characterization of septal structure and the effects of antibiotics on cell division. *Journal of Bacteriology*, **119**, 303–24.

Burman, L. & Park, J. T. (1984). Molecular model for elongation of the murein sacculus of *Escherichia coli. Proceedings National Academy of Sciences, USA*, **84**, 7144–8.

Chatterjee, A. P., Dasgupta, A. & Chatterjee, A. N. (1988). Spatial dependence of stress distribution for rod-shaped bacteria. *Journal of Theoretical Biology*, **135**, 309–21.

Cook, W. R. & Rothfield, L. I. (1991). Biogenesis of cell division sites in *ftsA* and *ftsZ* filaments. *Research in Microbiology*, **142**, 321–4.

Cook, W. R., De Boer, P. A. J. & Rothfield, L. I. (1989). Differentiation of the bacterial cell division site. *International Review of Cytology*, **118**, 1–31.

Cook, W. R. MacAlister, T. J. & Rothfield, L. I. (1986). Compartmentalization of the periplasmic space at division sites in Gram-negative bacteria. *Journal of Bacteriology*, **168**, 1430–8.

Cook, W. R., Joseleau-Petit, F., MacAlister, T. J. & Rothfield, L. I. (1987). Proposed mechanism for generation and localization of new division sites during the division cycle of *Escherichia coli. Proceedings National Academy of Sciences, USA*, **84**, 7144–8.

Cooper, S. (1988). Rate and topography of cell wall synthesis during the division cycle of *Salmonella typhimurium. Journal of Bacteriology*, **170**, 422–30.

Cooper, S. (1990). Relationship between the acceptor/donor radioactivity ratio and cross-linking in bacterial peptidoglycan: application to surface synthesis during the division cycle. *Journal of Bacteriology*, **172**, 5506–10.

Cooper, S., Hsieh, M.-L & Guenther, B. (1988). Mode of peptidoglycan synthesis in *Salmonella typhimurium*: single-strand insertion. *Journal of Bacteriology*, **170**, 3509–12.

Dai, R. & Ishiguro, E. E. (1988). *MurH*, a new genetic locus in *Escherichia coli* involved in cell wall peptidoglycan biosynthesis. *Journal of Bacteriology*, **170**, 2197–201.

D'Ari, R., Magnin, E., Bouloc, P., Jaffé, A., Robin, A., Liébart, J.-C & Joselaeu-Petit, D. (1990). Aspects of cell cycle regulation. *Research in Microbiology*, **141**, 9–16.

De Boer, P. A. J., Cook, W. R. & Rothfield, L. I. (1990). Bacterial cell division. *Annual Review of Genetics*, **24**, 249–74.

De Jonge, B. L. M., Wientjes, F. B., Jurida, I., Driehuis, F., Wouters, J. T. M. & Nanninga, N. (1989). Peptidoglycan synthesis during the cell cycle of *Escherichia coli*: composition and mode of insertion. *Journal of Bacteriology*, **171**, 5783–94.

Donachie, W. D. & Begg, K. J. (1989*a*). Cell length, nucleoid separation and cell division of rod-shaped and spherical cells of *Escherichia coli. Journal of Bacteriology*, **171**, 4633–9.

Donachie, W. D. & Begg, K. J. (1989*b*). Chromosome partitioning in *Escherichia coli* requires postreplication protein synthesis. *Journal of Bacteriology*, **171**, 5404–9.

Donachie, W. D. & Begg, K. (1990). Genes and the replication cycle of *Escherichia coli. Research in Microbiology*, **141**, 64–75.

Driehuis, F. & Wouters, J. F. M. (1987). Effects of growth rate and cell shape on the peptidoglycan composition in *Escherichia coli. Journal of Bacteriology*, **169**, 97–101.

Edelman, A., Bowler, L., Broome-Smith, J. K. & Spratt, B. G. (1987). Use of a β-lactamase function vector to investigate the organization of penicillin-binding protein 1B in the cytoplasmic membrane of *Escherichia coli. Molecular Microbiology*, **1**, 101–6.

Glauner, B. (1982). Ph D Thesis. University of Tübingen.

Glauner, B. (1988). Separation and quantification of muropeptides with high-performance liquid chromatography. *Analytical Biochemistry*, **172**, 451–64.

Glauner, B. & Schwarz, U. (1987). Investigation of murein structure and metabolism by high-pressure liquid chromatography. In *Bacterial Cell Surface Techniques – Modern Microbiological Methods*, ed. I.C. Hancock & I. R. Poxton, pp. 158–74. New York: John Wiley & Sons, Inc.

Goodell, E. W. (1985). Recycling of murein by *Escherichia coli. Journal of Bacteriology*, **63**, 305–10.

Goodell, E. W. & Schwarz, U. (1985). Release of cell wall peptides into culture medium by exponentially growing *Escherichia coli. Journal of Bacteriology*, **162**, 391–7.

Greenway, D. L. A. & Perkins, H. R. (1985). Turnover of cell wall peptidoglycan during growth of *Neisseria gonorrhoeae* and *Escherichia coli*. Relative stability of newly synthesized material. *Journal of General Microbiology*, **131**, 253–63.

Grossman, N., Ron, E. Z. & Woldringh, C. L. (1982). Changes in cell dimensions during amino acid starvation of *Escherichia coli. Journal of Bacteriology*, **152**, 35–41.

Grossman, N., Rosner, E. Z. & Ron, E. Z. (1989). Termination of DNA replication is required for cell division in *Escherichia coli. Journal of Bacteriology*, **171**, 74–9.

Hara, H. & Suzuki, H. (1984). A novel transglycosylase that synthesizes uncross-linked peptidoglycan in *Escherichia coli. FEBS Letters*, **168**, 155–60.

Hartmann, R., Bock-Hennig, S. B. & Schwarz, U. (1974). Murein hydrolases in the envelope of *Escherichia coli. European Journal of Biochemistry*, **41**, 203–8.

Henning, U. (1975). Determination of cell shape in bacteria. *Annual Review of Microbiology*, **29**, 45–9.

Hiemstra, H., Nanninga, N., Woldringh, C. L., Inouye, M. & Witholt, B. (1987). Distribution of newly synthesized lipoprotein over the outer membrane and the peptidoglycan sacculus of an *Escherichia coli lac-lpp* strain. *Journal of Bacteriology*, **169**, 5434–44.

Higgins, M. L. & Shockmann, G. D. (1976). Study of a cycle of cell wall assembly in *Streptococcus faecalis* by three-dimensional reconstruction of thin sections of cells. *Journal of Bacteriology*, **127**, 1346–58.

Hiraga, S., Niki, H., Ogura, T., Ichinosec, C., Mori, H., Ezaki, B. & Jaffé, A. (1989). Chromosome partitioning in *Escherichia coli:* novel mutants producing anucleate cells. *Journal of Bacteriology*, **171**, 1496–505.

Hirota, Y., Jacob, F., Ryter, A., Buttin, G. & Nakai, T. (1968). On the process of cellular division in *E. coli*. – 1. Asymmetrical cell division and production of DNA-less bacteria. *Journal of Molecular Biology*, **35**, 175–92.

Hobot, J. A., Carlemalm, E., Villiger, W. & Kellenberger, E. (1984). Periplasmic gel: new concept resulting from the reinvestigation of bacterial cell envelope ultrastructure by new methods. *Journal of Bacteriology*, **160**, 143–52.

Höltje, J.-V. & Glauner, B. (1990). Structure and metabolism of murein sacculus. *Research in Microbiology*, **141**, 75–89.

Höltje, J.-V. & Schwarz, U. (1985). Biosynthesis and growth of the murein sacculus. In *Molecular Cytology of* Escherichia coli, ed. N. Nanninga, pp. 77–119. New York: Academic Press, Inc.

Ikeda, M., Sato, T., Wachi, M., Jung, H. K., Ishino, F., Kobayashi, Y. & Matsuhashi, M. (1989). Structural similarity among *Escherichia coli* FtsW and RodA proteins and Bacillus Spo VE protein functioning in cell division, cell elongation and spore formation. *Journal of Bacteriology*, **171**, 6375–8.

Ingraham, J. L., Maaløoe, O. & Neidhardt, F. C. (1983). *Growth of the Bacterial Cell.* Sunderland, Massachusetts.

Ishino, F., Park, W., Tomioka, S., Tamaki, S., Takase, I. Kunugita, K., Matsuzawa, H., Asoh, S., Ohta, T., Spratt, B. G. & Matsuhashi, M. (1986). Peptidoglycan synthetic activities in membranes of *Escherichia coli* caused by overproduction of penicillin-binding protein 2 and RodA protein. *Journal of Biological Chemistry*, **261**, 7024–31.

Iwaya, M. & Strominger, J. L. (1977). Simultaneous deletion of D-alanine carboxypeptidase IB-C and penicillin-binding component 4 in a mutant of *Escherichia coli. Proceedings National Academy Sciences, USA*, **74**, 2980–4.

Jacob, F., Brenner, S. & Cuzin, F. (1963). On the regulation of DNA replication in bacteria. *Cold Spring Harbor Symposium on Quantitative Biology*, **28**, 329–48.

Jacoby, G. H. & Young, K. D. (1991). Cell cycle-independent lysis of *Escherichia coli* cefsulodin, an inhibitor of penicillin-binding proteins 1a and 1b. *Journal of Bacteriology*, **173**, 1–5.

Jaffé, A., D'Ari, R. & Hiraga, S. (1988). Minicell-forming mutants of *Escherichia coli:* Production of minicells and anucleate Rods. *Journal of Bacteriology*, **170**, 3094–101.

Jones, C. & Holland, I. B. (1985). Role of the SulB (FtsZ) protein in division inhibition during the SOS response in *Escherichia coli:* FtsZ stabilizes the inhibitor SulA in maxicells. *Proceedings National Academy of Sciences, USA*, **82**, 6045–9.

Keck, W., van Leeuwen, A. M., Huber, M. & Goodell, E. W. (1990). Cloning and characterization of *mepA*, the structural gene of the penicillin-insensitive murein endopeptidase from *Escherichia coli. Molecular Microbiology*, **4**, 209–19.

Kennedy, E. P. (1987). Membrane-derived oligosaccharides. In Escherichia coli *and* Salmonella typhimurium: *Cellular and Molecular Biology*, ed. F. C. Neidhardt, J. L. Ingraham, B. Magasanik, K. B. Low, M. Schaechter & H. E. Umbarger, pp. 672–9. Washington, DC: American Society for Microbiology.

Koch, A. L. (1985). How bacteria grow and divide in spite of internal hydrostatic pressure. *Canadian Journal of Microbiology*, **31**, 1071–84.

Koch, A. L. (1988). Biophysics of bacterial walls viewed as stress-bearing fabric. *Microbiological Reviews*, **52**, 337–53.

Koch, A. L. (1990). The surface stress theory for the case of *Escherichia coli:* the paradoxes of Gram-negative growth. *Research in Microbiology*, **141**, 119–30.

Labischinsky, H. Goodell, E. W., Goodell, A. & Hochberg, M. L. (1991). Direct proof of a 'more-than-single-layered' peptidoglycan architecture of *Escherichia coli* W7: a neutron small-angle scattering study. *Journal of Bacteriology*, **173**, 751–6.

Leduc, M., Frehel, C., Siegel, E. & van Heijenoort, J. (1989). Multilayered distribution of peptidoglycan in the periplasmic space of *Escherichia coli. Journal of General Microbiology*, **135**, 1243–54.

Lleo, M. M., Canepari, P. & Satta, G. (1990). Bacterial cell shape regulation. *Journal of Bacteriology*. **172**, 3758–71.

MacAlister, T. J., Cook, W. R., Weigand, R. & Rothfield, L. I. (1987). Membrane-murein attachment at the leading edge of the division septum: a second membrane-murein structure associated with morphogenesis of the gram-negative bacterial division septum. *Journal of Bacteriology*, **169**, 3945–51.

Markiewicz, Z., Broome-Smith, J. K., Schwarz, U. & Spratt, B. G. (1982). Spherical *E. coli* due to elevated levels of D-alanine carboxypeptidase. *Nature*, London, **297**, 702–4.

Matsuhashi, M., Wachi, M. & Ishino, F. (1990). Machinery for cell growth and division: penicillin-binding proteins and other proteins. *Research in Microbiology*, **141**, 89–103.

Mirelman, D., Yashouv-Gan, Y. & Schwarz, U. (1977). Regulation of murein biosynthesis and septum formation in filamentous cells of *Escherichia coli* PAT84. *Journal of Bacteriology*, **129**, 1593–600.

Mizushima, S. (1985). Structure, assembly, and biogenesis of the outer membrane. In *Molecular Cytology of* Escherichia coli, ed. N. Nanninga, pp. 39–75. London: Academic Press Inc.

Mulder, E. & Woldringh, C. L. (1989). Actively replicating nucleoids influence the positioning of division sites in DNA-less cell forming filaments of *Escherichia coli. Journal of Bacteriology*, **171**, 4303–14.

Mühlradt, P. F. (1976). Topography of outer membrane assembly in Salmonella. *Journal of Supramolecular Structure*, 5, 103–8.

Nanninga, N. (1988). Growth and form in microorganisms: morphogenesis of *Escherichia coli. Canadian Journal of Microbiology*, **34**, 381–9.

Nanninga, N. (1991). Cell division and peptidoglycan assembly in *Escherichia coli. Molecular Microbiology*, **5**, 791–5.

Nanninga, N. & Woldringh, C. L. (1985). Cell growth, genome duplication and cell division in *Escherichia coli*. In *Molecular Cytology of Escherichia coli*, ed. N. Nanninga, pp. 259–318. London: Academic Press, Inc.

Nanninga, N., Wientjes, F. B., De Jonge, B. L. M. & Woldringh, C. L. (1990). Polar cap formation during cell division in *Escherichia coli. Research in Microbiology*, **141**, 103–18.

Niki, H., Jaffé, A., Inamura, R., Ogura, T & Hiraga, S. (1991). The new gene *mukB* codes for a 177 kd protein with coiled domains involved in chromosome partitioning of *E. coli. The EMBO Journal*, **10**, 183–93.

Olÿhoek, A. J. M., van Eden, C. G., Trueba, F. J., Pas, E. & Nanninga, N. (1982). Plasmolysis during the division of *Escherichia coli. Journal of Bacteriology*, **152**, 479–84.

Park, J. T. & Burman, L. (1973). Fl-1060: a new penicillin with a unique mode of action. *Biochemical and Biophysical Research Communications*, **51**, 863-8.

Pisabarro, A. G., Prats, R., Vázquez, D. & Rodríguez-Tébar, A. (1986). Activity of penicillin-binding protein 3 from *Escherichia coli. Journal of Bacteriology*, **168**, 199–206.

Pratt, J. M., Jackson, M. E. & Holland, I. B. (1986). The C-terminus of penicillin-binding protein 5 is essential for localisation to the *E. coli* inner membrane. *EMBO Journal*, **5**, 2399–405.

Preusser, H.-J. (1969). Strukturveränderungen am 'Murein-Sacculus' von *Spirillum serpens* nach Einwirkung von Uranylacetat. *Archiv für Mikrobiologie*, **68**, 150–64.

Schaechter, M., Maaløe, O. & Kjeldgaard, N. O. (1958). Dependency on medium and temperature of cell size and chemical composition during balanced growth of *Salmonella typhimurium. Journal of General Microbiology*, **19**, 592–606.

Smit, J. & Nikaido, H. (1978). Outer membrane of Gram-negative bacteria. XVIII Electron microscopic studies on porin insertion sites and growth of cell surface of *Salmonella typhimurium*. *Journal of Bacteriology*, **135**, 687–702.

Spratt, B. G. (1975). Distinct penicillin-binding proteins involved in the division, elongation and shape of *Escherichia coli* K 12. *Proceedings of the National Academy of Sciences, USA*, **72**, 2999–3003.

Spratt, B. G. (1977). Properties of the penicillin-binding proteins of *Escherichia coli* K12. *European Journal of Biochemistry*, **72**, 341–52.

Spratt, B. G. (1980). Deletion of the penicillin-binding protein 5 gene of *Escherichia coli*. *Journal of Bacteriology*, **144**, 1190–2.

Stoker, N. G., Pratt, J. M. & Spratt, B. G. (1983). Identification of the *rodA* gene products of *Escherichia coli*. *Journal of Bacteriology*, **155**, 854–9.

Taschner, P. E. M., Huls, P., Pas, E. & Woldringh, C. L. (1988). Division behaviour and shape changes in isogenic *ftsZ, ftsQ, ftsA, pbpB*, and *ftsE* cell division mutants of *Escherichia coli* during temperature shift experiments. *Journal of Bacteriology*, **170**, 1533–40.

Tormo, A., Ayala, J. A., De Pedro, M. A., Aldea, M. & Vincente, M. (1986). Interaction of FtsA and PBP3 proteins in the *Escherichia coli* septum. *Journal of Bacteriology*, **166**, 985–92.

Tuomanen, E. & Cozens, R. (1987). Changes in peptidoglycan composition and penicillin-binding proteins in slowly growing *Escherichia coli*. *Journal of Bacteriology*, **169**, 5308–10.

Verwer, R. W. H., Nanninga, N., Keck, W. & Schwarz, U. (1978). Arrangement of glycan chains in the sacculus of *Escherichia coli*. *Journal of Bacteriology*, **136**, 723–9.

Verwer, R. W. H., Beachy, E. H., Keck, W., Stoub, A. M. & Poldermans, J. E. (1980). Oriented fragmentation of *Escherichia coli* sacculi by sonication. *Journal of Bacteriology*, **141**, 327–32.

Vos-Scheperkeuter, G. H., Pas, E. Brakenhoff, G. J., Nanninga, N. & Witholt, B. (1984). Topography of the insertion of LamB protein into the outer membrane of *Escherichia coli* wild-type and *lac–labB* cells. *Journal of Bacteriology*, **159**, 440–7.

Walker, J. R., Kovarik, A., Allen, J. S. & Gustafson, R. A. (1975). Regulation of bacterial cell division: temperature-sensitive mutants of *Escherichia coli* that are defective in septum formation. *Journal of Bacteriology*, **123**, 693–703.

Waxman, D. J. & Strominger, J. L. (1983). Penicillin-binding proteins and the mechanism of action of β-lactam antibiotics. *Annual Review of Biochemistry*, **52**, 825–69.

Wientjes, F. B. & Nanninga, N. (1989). Rate and topography of peptidoglycan synthesis during cell division in *Escherichia coli*: the concept of a leading edge. *Journal of Bacteriology*, **171**, 3412–9.

Wientjes, F. B. & Nanninga, N. (1991). On the role of the high molecular weight penicillin-binding proteins in the cell cycle of *Escherichia coli*. *Research in Microbiology*, **142**, 333–44.

Wientjes, F. B., Olijhoek, A. J. M., Schwarz, U. &Nanninga, N. (1983). Labelling pattern of major penicillin-binding proteins of *Escherichia coli* during the division cycle. *Journal of Bacteriology*, **153**, 1287–93.

Woldringh, C. L. (1976). Morphological analysis of nucleoid and cell division during the life cycle of *Escherichia coli*. *Journal of Bacteriology*, **125**, 248–57.

Woldringh, C. L., De Jong, M. A., van den Berg, W. & Koppes, L. J. H. (1977). Morphological analysis of the cell division cycle of two *Escherichia coli* substrains during slow growth. *Journal of Bacteriology*, **131**, 270–9.

Woldringh, C. L., Mulder, E., Huls, P. G. & Vischer, N. (1991). Toporegulation of bacterial division according to the nucleoid occlusion model. *Research in Microbiology*, **142,** 309–20.

Woldringh, C. L., Huls, P. G., Pas, E., Brakenhoff, G. J. & Nanninga, N. (1987). Topography of peptidoglycan synthesis during elongation and polar cap formation in a cell division mutant of *Escherichia coli* MC4100. *Journal of General Microbiology*, **133**, 575–86.

Woldringh, C. L., Huls, P. G. Nanninga, N., Pas, E., Taschner, P. E. M. & Wientjes, F. B. (1988). Autoradiographic analysis of peptidoglycan synthesis in shape and division mutants of *Escherichia coli* MC4100. In *Antibiotic Inhibition of Bacterial Cell Surface Assembly and Function*, ed. P. Actor *et al.*, pp. 66–78. Washington DC: American Society of Microbiology.

Woldringh, C. L., Mulder, E., Valkenburg, J. A. C., Wientjes, F. B. Zaritsky, A. & Nanninga, N. (1990). Role of nucleoid in toporegulation of division. *Research in Microbiology*, **141**, 39–49.

Woldringh, C. L., Valkenburg, J. A. C., Pas, E., Taschner, P. E. M., Huls, P. G. & Wientjes, F. B. (1985). Physiological and geometrical conditions for cell division in *Escherichia coli. Annales Institut Pasteur/Microbiologie*, **136A**, 131–8.

Yousif, S. Y., Broome-Smith, J. K. & Spratt, B. G. (1985). Lysis of *Escherichia coli* by B-lactam antibiotics: Deletion analysis of the role of penicillin-binding proteins 1A and 1B. *Journal of General Microbiology*, **131**, 2839–45.

SUPERFAMILIES OF BACTERIAL TRANSPORT SYSTEMS WITH NUCLEOTIDE BINDING COMPONENTS

ANTHONY P. PUGSLEY

Unité de Génétique Moléculaire, Institut Pasteur, 25, rue du Dr Roux, Paris 75724 Cedex 15, France

INTRODUCTION

A wide variety of molecules ranging in size from small ions, simple carbohydrates, amino acids and nucleotides through to huge macromolecules such as proteins, polysaccharides and nucleic acids are actively transported across proteolipid membranes in all living organisms. This transport is almost invariably mediated by proteins that span the membrane, and in some cases may be further facilitated by soluble accessory factors located on one or both sides of the membrane.

In bacteria, most active transport occurs across the cytoplasmic membrane. Nutrients are actively accumulated against a concentration gradient, and secondary metabolites and toxic metal ions are excreted, as are certain antibiotics that penetrate into the cytoplasm from the medium. In Gram-negative bacteria, small solutes such as vitamin B-12 and siderophore–iron complexes may be actively transported across the outer membrane, although this membrane is permeable to small hydrophilic solutes (ranging in size from 600–2000 D depending on the bacterium) which diffuse across it through proteinaceous pores (porins) with varying degrees of specificity or selectivity. Bacteria also actively export and secrete proteins into the cytoplasmic membrane, the external media and, in the case of Gram-negative bacteria, the outer membrane and the periplasm between the two membranes. Large polysaccharides are also actively transported to the outside of the cells to form capsules. In addition, many species of bacteria exhibit a phenomenon known as natural competence that allows them to take up DNA actively from the surrounding medium.

Almost all examples of active transport in bacteria share two basic characteristics: they involve at least one transmembrane protein with several segments composed of 12 to 20 almost entirely hydrophobic or apolar amino acids that come into direct contact with lipid fatty acyl chains in the membrane, and transport requires energy, usually as either a transmembrane chemical or electrical potential or as hydrolysable nucleotide triphosphates.

An increasing number of components of various bacterial transport systems are being identified and characterized. These studies show that components of transport systems for structurally dissimilar substrates often share extensive sequence similarities suggesting the existence of common structures extending beyond those of the membrane-spanning segments. The aim of this chapter is to consider whether some bacterial transport systems belong to a relatively small number of distinct (super)families identifiable by the extensive sequence similarities of some of their components rather than by similarities in the chemical nature, size and complexity of the substance they transport. Systems for transporting proteins to and through membranes studied in the author's laboratory work will be described first. This work was partially instrumental in the recent discovery of one such transport superfamily.

THE UNIVERSAL EXTENSION TO THE GENERAL EXPORT PATHWAY IN GRAM-NEGATIVE BACTERIA

The pullulanase secretion pathway

Gram-negative bacteria export a wide variety of proteins by the signal sequence-dependent General Export Pathway (GEP) which, in *Escherichia coli*, essentially comprises the *sec* gene products and two signal peptidases (for review see Schatz & Beckwith, 1990). The author, in his laboratory, is interested in how GEP has been extended to allow Gram-negative bacteria to secrete proteins across the entire cell envelope and into the medium.

For various reasons (Pugsley, *et al.*, 1990) we chose the cell surface-anchored enzyme pullulanase of *Klebsiella* as a model system. Pullulanase is one of a small but nevertheless important group of bacterial proteins that carry a diacylglyceride and an amine-linked fatty acid on the cysteine that becomes the N-terminal residue when the signal peptide is removed. The fatty acids anchor the protein to the cell surface such that the entire polypeptide chain (118 kD) is fully exposed in the medium. The protein can be produced by *E. coli* strains carrying the cloned structural gene (*pulA*). Not surprisingly, the protein is exported by GEP in this bacterium (Pugsley *et al.*, 1991*a*), but it does not reach the cell surface unless 14 other genes (*pulC-O* and *pulS*) that are close to *pulA* in the *Klebsiella* chromosome are also expressed in the recombinant *E. coli* strain (Fig. 1).

Each of these genes has now been sequenced (Genbank/EMBL bank references: X52181, M32702, M32613, M24118 and X14945) and the locations of the polypeptides determined (d'Enfert & Pugsley, 1989; d'Enfert *et al.*, 1989; Pugsley & Reyss, 1990; Reyss & Pugsley, 1990; Pugsley, Reyss, d'Enfert & Possot, unpublished observations; for review see Pugsley *et al.*, 1990). All but one of them are only expressed when *pulA* is also expressed, and are therefore probably specific for pullulanase secretion. The products of these 14 genes are only involved in the distinct, second step of the pullulanase secretion pathway (Pugsley *et al.*, 1991*b*), the first step, leading to signal peptide processing, fatty acylation and at least partial translocation across

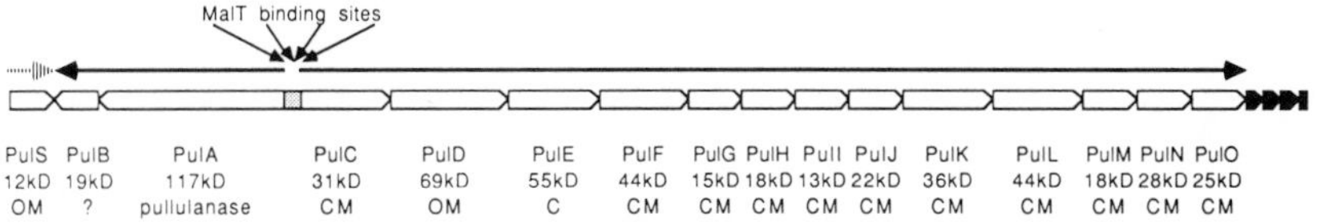

Fig. 1. Schematic representation of the genetic organization of the *pul* genes cloned from the chromosome of *Klebsiella oxytoca* strain UNF5023. Genes are represented by open boxed arrows. The names of the gene products are indicated beneath the genes, together with their sizes and their determined subcellular locations (OM, outer membrane; CM, cytoplasmic membrane; C, cytoplasm). PulA is pullulanase; all of the other proteins except PulB are required for pullulanase secretion. The thin dark arrows indicate transcription units that are induced by maltose-(maltotriose-) activated MalT protein that binds at four different sites between *pulA* and *pulC*. The short shaded arrow represents the constitutively expressed *pulS* transcript, and the small black boxes the short DNA repeats located immediately after *pulO* (see Pugsley *et al.*, 1990 and references therein for further details).

the cytoplasmic membrane, being exclusively and entirely dependent on components of the GEP (Pugsley *et al.*, 1991*a*) (Fig. 2).

Pullulanase secretion pathway homologues in other bacteria

Pullulanase is by no means the only extracellular protein produced as a signal peptide-bearing precursor by Gram-negative bacteria. As more of these systems are characterized genetically, it is becoming abundantly clear that many of their components are similar to those required for pullulanase secretion. In particular, their gene organization is highly conserved and the gene products often display extremely high levels of sequence identity (Filloux *et al.*, 1990; Bally *et al.*, 1991; Dums, Dow & Daniels, 1991; He *et al.*, 1991, Nunn, Bergmann & Lory, 1990 and unpublished observations summarized in Fig. 3). The examples shown in Fig. 3 are taken from the best-characterized systems; others are continually being identified and characterized, and it seems likely that almost all protein-secreting Gram-negative bacteria use such a system for at least some of their extracellular proteins. It is therefore proposed that this pathway represents the universal extension of GEP in Gram-negative bacteria, and that it should be called the General Secretion Pathway (GSP). GEP is thus part of GSP.

In hindsight, it is not too surprising that different species of Gram-negative bacteria should use the same pathway for secreting extracellular proteins, although it was initially thought that the highly complex, pullulanase-specific pathway might be unique. Indeed, the finding that distantly related bacteria have similar protein secretion pathways might help understand the roles of certain components of the pullulanase secretion pathway. For example, a gene corresponding to *pulS* has yet to be found in other bacteria, which leads to speculation that it might be required only for fatty acylated proteins such as pullulanase (most of the extracellular proteins secreted by other Gram-negative bacteria by GSP are not fatty acylated).

The reader should not be mislead into thinking that Gram-negative bac-

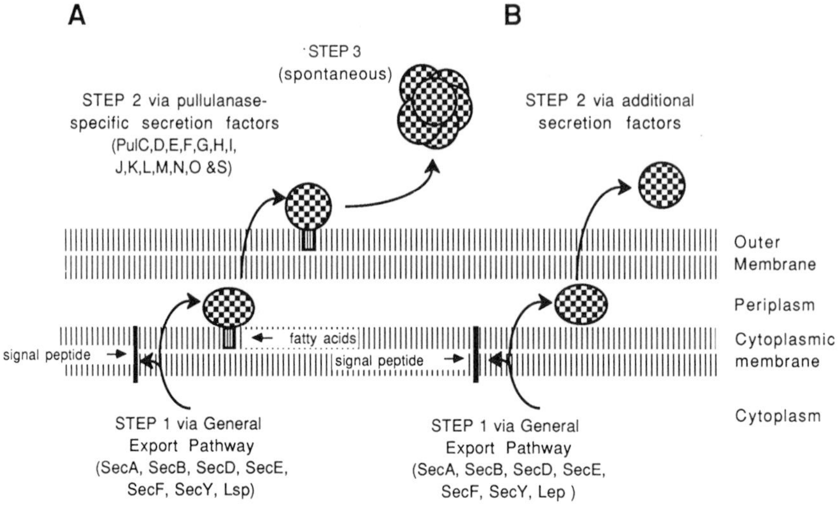

Fig. 2. Models for extracellular protein secretion by GSP. Part A, representing the pathway used by the fatty acylated enzyme pullulanase, is divided into three steps, the first two requiring respectively components of the General Export Pathway and the pullulanase-specific secretion factors encoded by the *pul* genes. The third step (pullulanase release into the medium) is probably spontaneous. The transitory secretion intermediate may not be in the cytoplasmic membrane as shown, but is clearly not in the inner leaflet of the outer membrane (Pugsley *et al.*, 1991b; Pugsley & Kornacker, 1991). Part B represents the more common variant of this pathway found in *Erwinia* sp, *Pseudomonas* sp and *Xanthomonas* sp, which secrete proteins that are not fatty acylated via a transitory periplasmic intermediate. Lep is leader peptidase, Lsp is lipoprotein signal peptidase.

teria secrete all of their extracellular proteins by the GSP. There are, in fact, at least two signal peptide-*independent* pathways for extracellular secretion in Gram-negative bacteria (one of these will be discussed below). Furthermore, certain Gram-negative bacteria secrete extracellular proteins by extended general export pathways that do not involve proteins similar to those of GSP (Alm & Manning, 1990; Miyazaki *et al.*, 1989; Pohlner *et al.*, 1987; Schiebel, Schwartz & Braun, 1989; Uphoff & Welch, 1990; for review see Pugsley *et al.*, 1990).

Prepilin peptidase as part of the General Secretion Pathway

In as much as pili (fimbriae) are appendages formed by the polymerization of extracellular (and therefore secreted) proteins, it is not too surprising that at least some of them (the type IV pili in *Pseudomonas aeruginosa;*

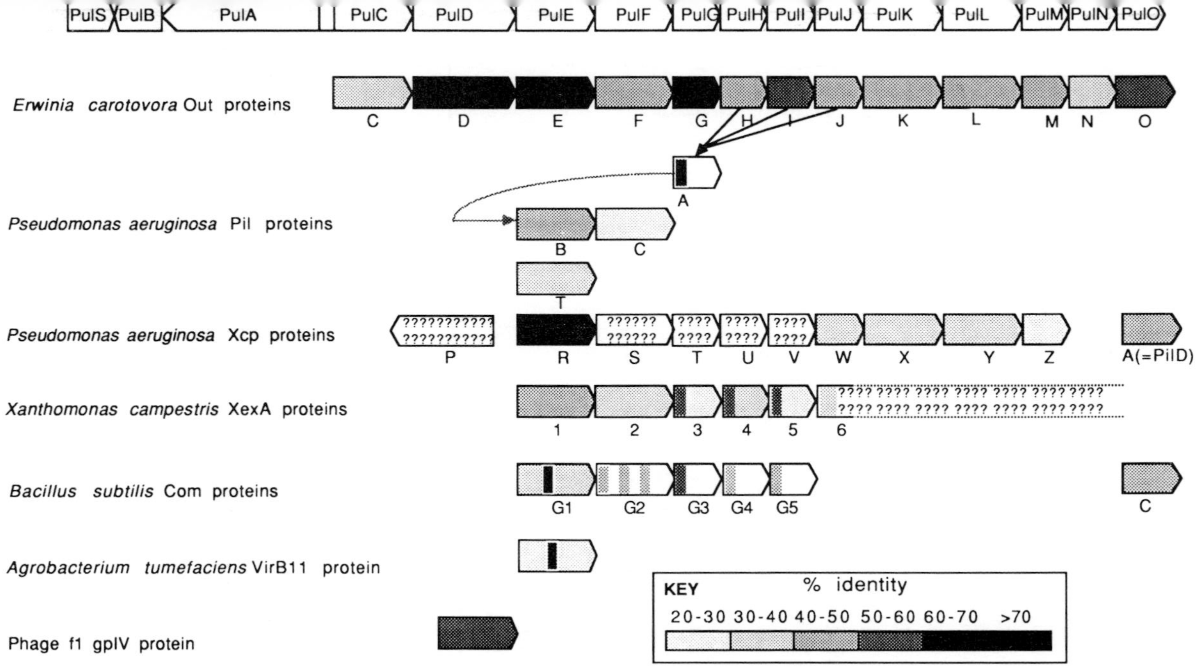

Fig. 3. Alignment of *pul* genes from *Klebsiella oxytoca* with corresponding genes from other bacteria. The genes for the homologous proteins are aligned with the corresponding *pul* gene to indicate their approximate organization. The approximate homologies of the gene products with corresponding Pul proteins are indicated by shading. Special features to note are that *xcpA* is also called *pilD* and is located adjacent to *pilC* in a different region of the chromosome to other *xcp* genes; that *pilA*, the pilus structural gene, is adjacent to *pilB* and oriented in the opposite direction; that the N-terminus of the *pilA* gene product is similar to that of the *pulG*, *pulH*, *pulI* and *pulJ* genes; that the *xcpP,R,S,T,U* and *V* genes have been identified but only partially sequenced, as has *xexA6*; that *xcpP*, coding for a PulD homologue, is close to but not contiguous with *xcpR* and is oriented in the opposite direction (J. Tommassen, personal communication); that other *xex* genes are located in a 5 kb fragment downstream from *xexA6*, and that *outN* may not exist in *Erwinia chrysanthemi*. Data presented are taken from published sequences of the genes (Albano *et al.*, 1989; Bally *et al.*, 1991; Brisette & Russell, 1990; d'Enfert & Pugsley, 1989; d'Enfert *et al.*, 1989; Dubnau, 1991; Dums *et al.*, 1991; Filloux, *et al.*, 1990; Mohan *et al.* 1989; Nunn *et al.*, 1990; Pugsley, d'Enfert, Reyss & Possot, unpublished observations; Pugsley & Reyss, 1990; Reyss & Pugsley, 1990; Shirasu *et al.*, 1990; Whitchurch *et al.*, 1991 and references therein) as analysed from personal communications by Andrée Lazdunski, Corrine Dorel, Jan Tommassen, George Salmond, Alain Kotoujansky and Mike Daniels.

Fig. 3) require *pul*-like genes for their secretion and/or assembly. The PilD protein required for type IV pilus assembly is the peptidase which cleaves the first six amino acids from the precursor pilin subunit, the product of the *pilA* gene (Nunn & Lory, unpublished observations). The same gene was independently characterized as the *xcpA* locus required for extracellular protein secretion (Bally *et al.*, 1991). The sequences of type IV pilin subunits are highly conserved, and are unusual among secreted or exported proteins in that their signal sequences are cleaved six or seven residues from their N-termini, rather than at a site recognized by signal (leader) peptidase. The cleavage site is readily identifiable (Fig. 4). Sequence analyses of the *pulG, pulH, pulI* and *pulJ* genes reveal that they too could encode precursors with the same consensus cleavage site (Fig. 4), as do the corresponding *xcp, out* and *xexA* genes (not shown; He *et al.*, 1991; A. Lazdunski, G. Salmond, A. Kotoujansky & M. Daniels, personal communications). Thus, PulG, H, I and J proteins and their homologues could be pilus subunits, but this seems unlikely since the rest of the sequences of these proteins are quite different from those of the type IV pili (Fig. 4 and not shown) and because *Klebsiella* and other Enterobacteriacese are not known to produce type IV pili. An interesting alternative possibility is that these proteins share features which allow type IV pilins to assemble into complex structures which, in this case, might form a channel or other structure required for protein secretion. We are currently testing the idea that PulO is necessary for pullulanase secretion because it processes, and therefore activates, the PulG, PulH, PulI and PulJ precursors. Preliminary results indicate that PulO can process type IV prepilin *in vivo* (B. Dupuy, unpublished observations).

Protein secretion and DNA uptake by a common pathway

The observation that certain of the Com proteins of *Bacillus subtilis* are also highly homologous to the Pul secretion proteins (Fig. 3) is of considerable importance because these proteins are not involved in protein secretion (*B. subtilis* does not have an outer membrane anyway), but in the uptake of (transforming) DNA. The Com–Pul protein pair showing the highest degree of similarity is ComC–PulO, and three other Com proteins (ComG.3, 4 and 5) all have the consensus type IV pilin cleavage site (Fig. 3; see Fig. 4 for examples of sequences at cleavage sites). Thus, ComC is probably the peptidase which cleaves the precursor forms of the latter proteins. ComG.3 has been shown to be required for DNA binding to competent *B. subtilis* cells (Breitling & Dubnau, 1990), raising the possibility that the similar proteins in the General Secretion Pathway might bind to the extracellular proteins during their secretion.

```
Neisseria     MNTLQKG*FTLIELMIVIAIVGILAAVALPAYQDYTARAQV---
Bacteroides   MKSLQKG*FTLIELMIVVAIIGILAAFAIPAYNDYIARSQA---
Pseudomonas   MKA-QKG*FTLIELMIVVAIIGILAAIAIPQYQNYVARSEG---
Moraxella     MNA-QKG*FTLIELMIVIAIIGILAAIALPAYQDYISKSQT---

PulG          MQR-QRG*FTLLEIMVVIVILGVLASLVVPNLMGNKEKADR---
PulH          MR--QRG*FTLLEMMLILLLMGVSAGMVLLAFPASRDDSAA---
PulI          MKK-QSG*MTLIEVMVALVVFALAGLAVMQATLQQTRQLGR---
PulJ          MIRRSSG*FTLVEMLLALAILAALSVAAVTVLQNVMRADTL---

ComG.3        MNEK--G*FTLVEMLIVLFIIGILLLITIPNVYKHNQTIQK---

Consensus     ----Q-G*FTLVE-MVVVaViG-La--V-pa-q-----s--
                      M  L   LLAL L A S   A
                         I   III I   A
```

Fig. 4 Homologies between the extreme N-termini of type IV prepilins of different bacteria (lines 1–4) and the putative precursor PulG, PulH, PulI, PulJ proteins of *K. oxytoca* (lines 5–8) and preComG.3 of *B. subtilis* (line 9) showing the peptide bond that is known or thought to be cleaved and consensus sequences that might be recognized by the specific *pilD-xcpA/pulO/comC*-encoded peptidase. Highly conserved residues are in shaded areas. The phenylalanine (F) at the N-terminus of mature pilin subunits is methylated. Data taken from Albano *et al.* (1989), Dubnau (1991) Marrs *et al.* (1985) and Reyss & Pugsley (1990) and from references therein.

ABC proteins in the General Secretion Pathway

The only other Com–Pul protein pair with a relatively high degree of homology is ComG.1–PulE. The latter is a cytoplasmic protein (Pugsley, Reyss, d'Enfert & Possot, unpublished observations) that is predicted to have two short sequences (G-P-T-G-S-G-K-S and R-G-R-T-G-I-H-E-L-L-L-V-D) which are characteristically present in proteins with ATPase, kinase or ATP synthase activities (ABC (*A*TP *b*inding *c*assette) proteins), the so-called Walker motifs A (G-X-S-G-X-G-G-K-T/S, where X is virtually any amino acid) and B [R-(1-3)-G-(3)-L-ø-ø-ø-(0-2)-D, where ø indicates any hydrophobic residue and numbers (0–3) indicate the spacing between the conserved residues] (Walker *et al.*, 1982). The region around motif A is highly conserved in the PulE homologues, including ComG.1 and VirB.11 (Fig. 5), but the region around the Walker motif B is less well conserved (not shown; see Dubnau (1991) for alignment of these regions). Walker motifs are also found in a variety of other integral and peripheral membrane proteins involved in the transport of various solutes across membranes (the so-called traffic ATPases or transport ABC proteins; Table 1). Based on studies with other ABC proteins (see below), it seems likely that ATP is directly involved in the reactions mediated by PulE and its homologues.

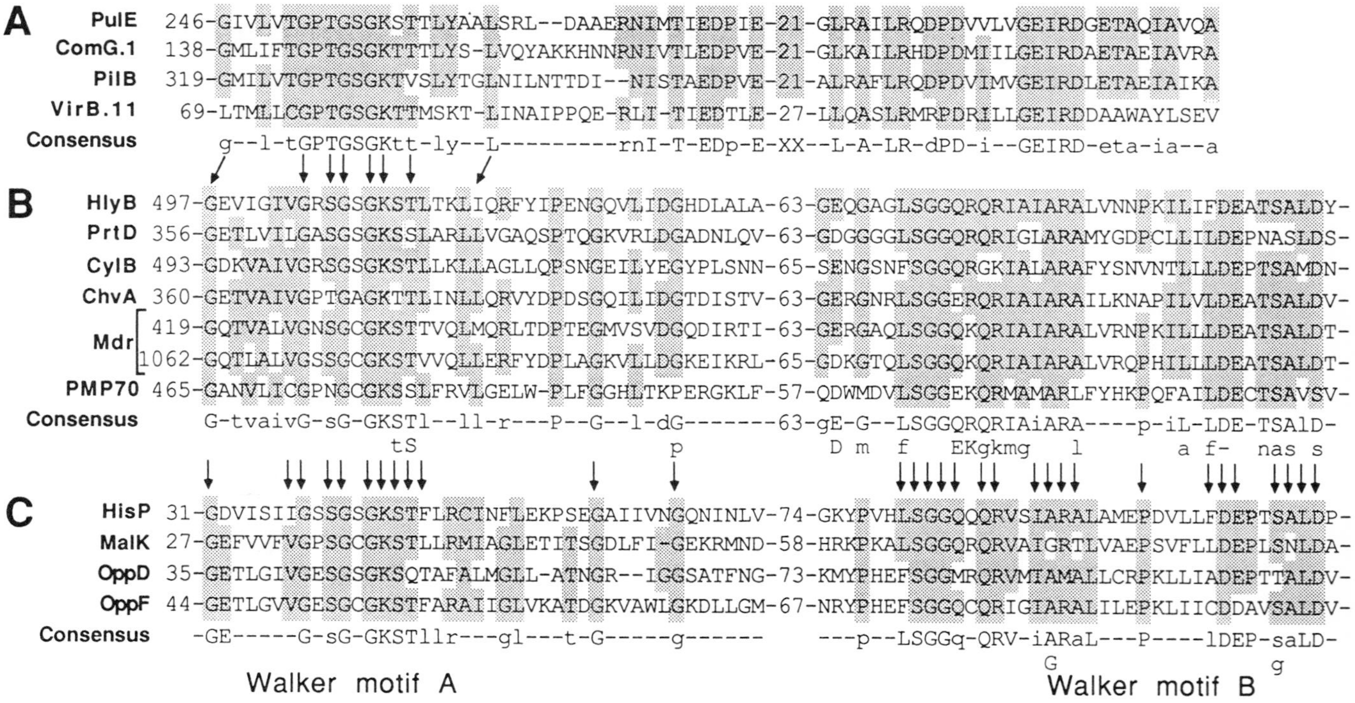

Fig. 5. Alignment of highly conserved sequences in representative proteins of three groups of ABC proteins described in the text and listed in Table 1. Highly conserved residues are in shaded areas and arrows indicate residues that are conserved between the three groups. Consensus sequences include pairs of residues that may be present at a given site, and indicate whether these residues are very frequently (upper case letters) or less frequently (lower case) found. The Mdr protein is the product of the human Mdr-1 gene. Data are taken from our own unpublished sequences and from Albano *et al.* (1989), Cangelosi *et al.* (1989), d'Enfert *et al.* (1989), Felmlee, *et al.* (1985), Gilmore *et al.* (1990), Hyde, *et al.* (1990), Kamijo, *et al.* (1990), Létoffé, *et al.* (1990), Mimura, *et al.* (1991), Nunn *et al.* (1990), Shirasu *et al.* (1990), and references therein. The consensus sequence for group C proteins is derived from the larger collection of sequences compiled in Mimura *et al.* (1991) but is not the same as that given by these authors. The approximate positions of the Walker motifs are indicated. The functions and origins of the proteins are listed in Table 1.

Filamentous phage secretion and the General Secretion Pathway

The homology between phage f1-encoded gpIV and PulD (Fig. 3) was noted in the original report on the sequence of the *pulD* gene (d'Enfert *et al.*, 1989), but the role of gpIV in phage morphogenesis and its location in the outer membrane (Brisette & Russel, 1990) were poorly defined or unknown at that time. The stage at which f1 phage development is arrested in the absence of gpIV has not been determined, but it is important to recall that this phage is *secreted* without cell lysis (Brisette & Russel, 1990). The idea that the phage particles are released across the outer membrane via a pore formed by gpIV, and that phage assembly cannot occur unless secretion can occur, is particularly attractive in view of our own preliminary evidence that pullulanase can adopt considerable, permanent tertiary structure before it is transported across the outer membrane (Pugsley *et al.*, 1991*b*).

Substrate specificity in the General Secretion Pathway

Despite the overall high degree of similarity between the components of the different representatives of GSP, each system has its own particular features, not least of which is its substrate specificity. There are as yet no documented cases in which the individual components of different pathways can replace a defective component in another pathway (although for reasons discussed above, we predict that PulO-ComC-XcpA might be interchangeable), and attempts to secrete proteins from one bacterium via the related pathway even in a closely related bacterium have so far failed (He *et al.*, 1991 and unpublished observations).

Specificity of protein secretion is presumably ensured by the recognition of the protein to be transported by one or more of its cognate secretion factors (e.g. the PulG, H, I and J homologues; see above). This implies the existence of a secretion or sorting signal in the various extracellular proteins. According to our own computer analyses, proteins that are secreted by the same or similar pathways (even proteins secreted by the same bacterium) do not appear to have any conserved sequence motifs that might represent such sorting signals. The fact that these proteins may adopt considerable tertiary structure before they enter the second step in the secretion pathway (Pugsley *et al.*, 1991*b*) suggests that the secretion machinery probably recognize a patch signal (Pugsley, 1989) rather than a linear stretch of amino acids.

A SIGNAL PEPTIDE-INDEPENDENT PROTEIN SECRETION PATHWAY IN BACTERIA

In this section, what is known about another pathway, for extracellular protein secretion by Gram-negative bacteria will be summarized, together

Table 1. *Examples of 'Traffic ATPases' or ABC proteins with 'Walker-type' ATP binding sites that are involved in various aspects of transport across membranes.*

Type	Name	Source	**Function**
A	PulE	*Klebsiella pneumoniae*	protein secretion
	OutE	*Erwinia carotovora*	protein secretion
	XcpR	*Pseudomonas aeruginosa*	protein secretion
	PilD	*Pseudomonas aeruginosa*	pilus assembly
	PilT	*Pseudomonas aeruginosa*	pilus twitching mobility
	ComG.1	*Bacillus subtilis*	DNA uptake
	VirB.11	*Agrobacterium tumefaciens*	conjugation
B	HlyB	*Escherichia coli*	haemolysin secretion
	CyaB	*Bordetella pertussis*	cyclolysin secretion
	PrtD	*Erwinia chrysanthemi*	metalloprotease secretion
	CvaB	*Escherichia coli*	colicin V secretion
	CylB	*Enterococcus faecalis*	bacteriocin secretion
	ComA	*Streptococcus pneumoniae*	competence factor secretion
	ChvA	*Rhizobium meliloti*	*B*-glucan excretion
	NdvA	*Agrobacterium tumefaciens*	*β*-glucan excretion
	PMP70	rat liver peroxisomes	protein import?
	STE6	*Saccharomyces cerevisiae*	pheromone **a** secretion
	Mdr	man & mouse	drug excretion
	Pfmdr	*Plasmodium falciparum*	chloroquinone excretion?
	CFTR	man (cystic fibrosis)	unknown
	Cim/PSF/Ring4	mouse & man	peptide transport in ER
C	HisP	*Salmonella typhimurium*	histidine uptake
	MalK	*Escherichia coli*	maltose uptake
	OppD&F	*Escherichia coli*	peptide uptake
Others	McbF	*Escherichia coli*	microcin B17 secretion
	ArsA	*Escherichia coli*	arsenate excretion
	SecA	*Escherichia coli*	protein export ATPase
	Pas1	*Saccharomyces cerevisiae*	peroxisomal protein import
	Sec18	*Saccharomyces cerevisiae*	secretory vesicle fusion

Data collected from Blight & Holland (1990), Cangelosi *et al.* (1989), Christie *et al.* (1989), Dums *et al.* (1991), Garrido *et al.* (1988), Deverson *et al.* (1990), Dubnau (1991), Erdman *et al.* (1991), Gilmore *et al.* (1990), Glaser *et al.* (1988), Higgins *et al.*, (1986), Hui & Morrison (1991), Kamijo *et al.* (1990), Létoffé *et al.* (1990), Mimura *et al.* (1991), Nunn *et al.* (1990), Rosen (1990), Schmidt *et al.* (1988), Stanfield *et al.* (1988), Whitchurch *et al.* (1991), Wilson *et al.* (1989) and references therein, from personal communications by Andrée Lazdunski and George Salmond, and from our own unpublished results. All of the proteins in this list contain at least the Walker motif A, and the vast majority also have motif B.

with recent results showing that it is even more widespread and has an even wider range of substrates than GSP.

Components of a signal peptide-independent secretion pathway

The archetypal representative of this pathway, at least in bacteria, is the α-haemolysin (HlyA) secretion pathway found in some uropathogenic strains of *E. coli*. It was the first to be identified and reconstituted in *E. coli* K12.

Two integral cytoplasmic membrane proteins (HlyB and HlyD) were found to be required for HlyA secretion. Like the pullulanase secretion genes described above, their genes are co-transcribed with the HlyA structural gene (Fig. 6). Subsequent studies revealed that the secretion of several other extracellular proteins (all lacking amino-terminal signal sequences and including some which are structurally and functionally quite different to HlyA) by a variety of Gram-negative bacteria requires membrane proteins closely-related to HlyB and HlyD. Figure 6 shows the genetic organization of four members of this family, other members of which are either so far incompletely characterized, or are highly similar to that of the *E. coli* Hly system. The similarity between the various secretion factors is high (60–90%) only in the cases of the secretion factors for proteins related to HlyA (the cytolytic RTX toxin family, as defined by Welch (1991)) and is relatively low (20–30%) for components of pathways for the secretion of unrelated proteins (Létoffé Delepelaire & Wandersman, 1990; Gilson, Mahatry & Kolter, 1990).

More recent studies on one of these secretion pathways, the metalloprotease secretion pathway from *Erwinia chrysanthemi*, revealed the requirement for a third membrane protein that had not been identified in the HlyA secretion pathway (Létoffé *et al.*, 1990). This protein, PrtF, was found to be similar to the outer membrane TolC protein of *E. coli* K-12 (Fig. 6). Subsequent studies revealed that TolC is indeed absolutely required for HlyA secretion by *E. coli* K-12 (Wandersman & Delepelaire, 1990), and that TolC-like proteins are required in the related secretion pathways discovered in other bacteria (Fig. 6).

Mutations in *tolC* have several other effects besides preventing HlyA secretion, including defective chromosomal DNA partitioning (Hiraga *et al.*, 1989), tolerance to colicin E1, increased dye and detergent sensitivity (Davies & Reeves, 1975) and reduced translation of mRNA for the major outer membrane OmpF porin (Morona & Reeves, 1982). Many of these features might be explained by general perturbation of membrane function, leading one to question whether TolC is directly or indirectly involved in HlyA secretion. TolC could be required, for example, for the correct assembly of the HlyB-D complex, of OmpF porin, of a component of the colicin E1 import machinery and of envelope components involved in chromosome partitioning between daughter cells.

The effect of *tolC* mutations on colicin E1 import is particularly intriguing in the context of membrane transport. Colicins, the only proteins imported into *E. coli* cells, first adsorb to specific cell surface receptors. Subsequent import of all colicins except E1 across the outer membrane requires either the TonB protein or a major outer membrane porin (Pugsley, 1984). Colicin E1 is the only one to require TolC protein. One way of rationalizing this would be for TolC to be directly involved both in HlyA secretion and colicin E1 import across the outer membrane. However, vesicles prepared from a TolC mutant are unaffected by colicin E1, whereas vesicles prepared from

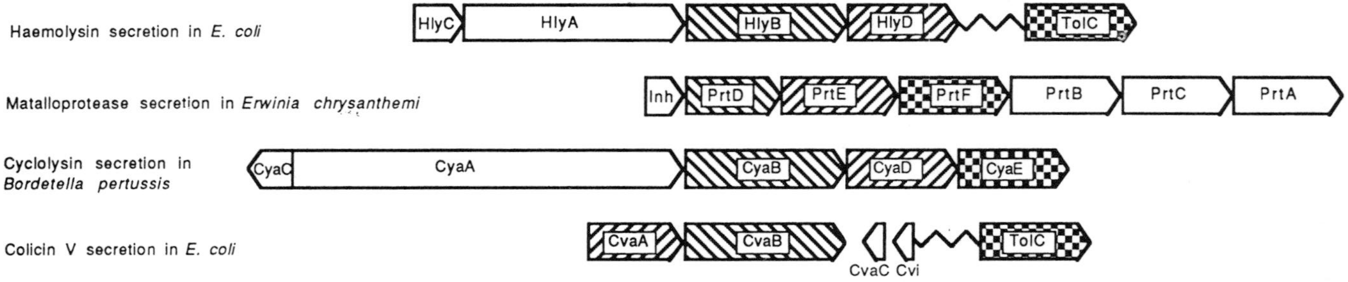

Fig. 6. Alignment of genes involved in three protein secretion pathways from Gram-negative bacteria that are related to the α-haemolysin secretion pathway from *E. coli*. The genes are represented by boxes and identified by the name of the protein they encode. Arrowheads indicate the direction of transcription. Secretion genes are shaded according to their relationship to the *hlyB, hlyD* and *tolC* genes. The wavy line indicates that the genes are not genetically linked. Unshaded boxes represent structural genes for secreted proteins or genes coding for proteins that modify their action. Data taken from Felmlee *et al.* (1985), Gilson *et al.* (1990), Glaser *et al.* (1988), Létoffé *et al.* (1990) and Wandersman & Delepelaire (1990).

a BtuB mutant lacking the colicin E1 receptor are sensitive to the pore-forming action of the colicin (Bhattachacharyya *et al.*, 1970). This implies that the TolC protein is required for colicin E1 action at a stage *after* the colicin has crossed the outer membrane.

Direct involvement of TolC in HlyA secretion has not been tested experimentally, but it is supported by at least one piece of circumstantial evidence, namely that the *tolC* homologue *prtF* is in the same operon as the other (HlyB- and the HlyD-related) metalloprotease secretion genes (Létoffé *et al.*, 1990; Fig. 6). It would be informative to study *prtF* in *E. chrysanthemi* to see if it is constitutively expressed, whether *prtF* mutations in *E. chrysanthemi* provoke the same effects as *tolC* mutations in *E. coli* or, indeed, whether *E. chrysanthemi* has a second TolC analogue that performs the other functions of TolC in *E. coli.*

Secretion signals for extracellular secretion

The secretion signals in HlyA and in metalloprotease B are located at the carboxy-terminus (Mackman *et al.*, 1987; Delepelaire & Wandersman, 1990; Hess *et al.*, 1990). Sequence analyses of these parts of proteins secreted by this pathway reveal primary and secondary structure similarities only in closely related proteins (Koronakis, Koronakis & Hughes, 1989; Delepelaire & Wandersman, 1990 and P. Delepelaire, personal communication). One would therefore predict that each system would be specific for its own particular substrate, but this is not the case since some (but not all) attempts to secrete members of this family of extracellular proteins by heterologous pathways have been at least partially successful (Guzzo *et al.*, 1991; Masure *et al.*, 1990; Welch, 1991). Perhaps this is one example of a situation in which the computer and human eye are unable to detect sequence or structural motifs that are deciphered by components of a secretion machinery.

The idea that the secretion signal is located at the extreme C-terminus would appear to be contradicted by recent studies on the secretion of colicin V (Cva) and in particular of colicin V-alkaline phosphatase (Cva–PhoA) hybrid proteins (Gilson *et al.*, 1990). These studies showed that when the extreme *amino* terminal 39 residues of Cva are placed at the amino terminus of PhoA, the enzyme is released into the *periplasm* by a process that is strongly stimulated by the HlyB-D analogues CvaB and CvaA (see Fig. 6). This is the first time that any protein secreted by a complete Hly-type pathway has been shown to appear in the periplasm under any conditions, and also shows that the extreme C-terminus of Cva is not required for exit from the cytoplasm. These data are difficult to reconcile with those showing that a PhoA–HlyA hybrid with a C-terminal extension composed of the last 60 residues of HlyA is secreted directly into the medium (Hess *et al.*, 1990). Possible non-trivial explanations for this discrepancy are (i) that the secretion signal need not be located at the C-terminus; (ii) that Cva, being relatively small (11 kD), may not be organized in the same way as HlyA (107 kD) or metalloprotease B (52 kD); (iii) that the N-terminus represents one part of the secretion signal, with a second signal (for outer membrane translocation) being located elsewhere; and (iv) that the Cva–PhoA hybrid cannot be transported through the complete secretion pathway because of structural incompatibilities caused by the PhoA moiety, and therefore enters the periplasm by default.

Signal peptide-independent protein secretion in Gram-positive bacteria

The demonstration that secretion of a haemolysin/bacteriocin by the Gram-positive bacterium *Enterococcus faecalis* requires an HlyB-type protein (Gilmore *et al.* 1990) shows that this pathway is not restricted to Gram-negative bacteria. Likewise, the ComA protein required for transformation competence of *Streptococcus pneumoniae* is also a homologue of HlyB (Hui & Morrison, 1991). Although the exact function of ComA is unknown, it is proposed to facilitate the secretion of a competence-inducing protein. It may be important that an HlyD-homologue was not identified in either of these bacteria.

Other substrates for HlyB-type transporters

The extracellular excretion of β-1,2 glucan polysaccharide by the Gram-negative bacteria *Agrobacterium tumefaciens* and *Rhizobium meliloti* requires an HlyB-like protein to transport the glucan into the periplasm (Stanfield *et al.*, 1988; Cangelosi *et al.*, 1989). This shows that bacterial HlyB homologues are not restricted to proteinaceous substrates. Once again, there was no evidence for the involvement of HlyD (or TolC)-like proteins in β-glucan excretion.

Models for HlyB-type transport activity in bacteria

On the basis of these data, it is possible to propose tentative models for the transport of macromolecules across bacterial membranes by this family of transport pathway components (Fig. 7). A HlyB-type protein is required for transport across the cytoplasmic membrane, and is the only component required in Gram-positive bacteria (Fig. 7A). β-glucans may not require other components of the pathway because they diffuse across the outer membrane from the periplasmic pool (Fig. 7B). HlyD- and TolC-like components of the pathway may be required to couple protein transport across the cytoplasmic and outer membranes (Fig. 7C). TolC-like proteins are firmly embedded in the outer membrane, where they could form a channel, while HlyD-like proteins have a single N-terminal hydrophobic segment that could anchor them in the cytoplasmic membrane in such a way that their long C-termini span the periplasm to contact the outer membrane, possibly via TolC. HlyD

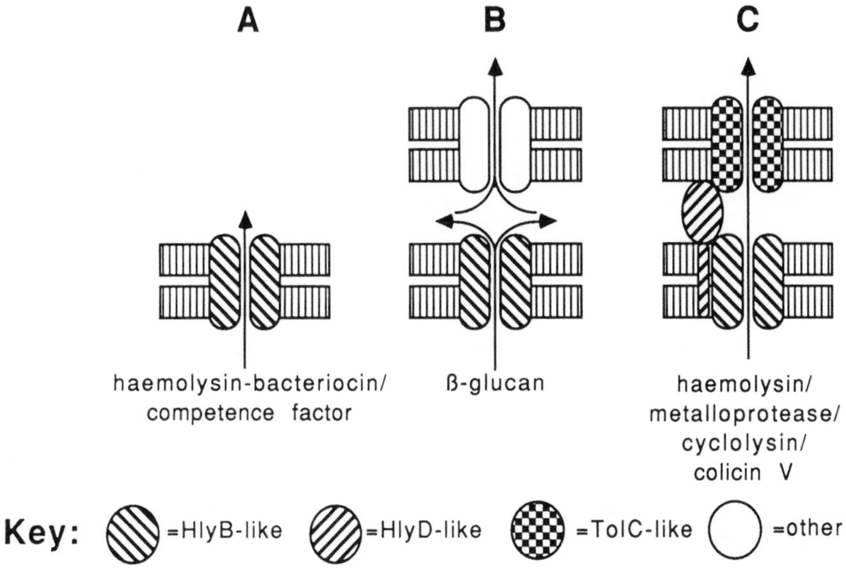

Fig. 7. Models for macromolecular transport reactions mediated by HlyB homologues in bacteria. See text for details.

homologues may be required to open the channel formed by the HlyB-like proteins, which could explain why $HlyB^{+}$, $HlyD^{-}$ strains do not release HlyA into the periplasm. As already proposed, TolC protein may be required for the correct assembly of the HlyB–D complex.

HlyB homologues in eukaryotes

HlyB-like proteins, the key feature of all of the systems described so far, are by no means restricted to bacteria. In fact, HlyB homologues in eukaryotic cells have recently received considerable attention because many of them are clinically important. They include the Mdr family of proteins (P-glycoproteins) that are often overproduced by drug-resistant tumour cells, the related Pfmdr protein involved in chloroquinone resistance in *Plasmodium falciparum*, and the CFTR protein that affects the activity of a chloride channel that is defective in patients with cystic fibrosis (for review see Blight & Holland, 1990). Further examples of HlyB-like proteins in eukaryotes are listed, together with their proposed or known functions, in Table 1 (section B).

HlyB homologues are integral membrane ABC proteins

All of the HlyB-type proteins are integral membrane proteins. The Mdr and CFTR proteins have two domains that appear to anchor the protein in the plasma membrane (see example in Currier, *et al.*, 1989). Each of

	CyaB	LktB	PrtD	Mdr	CylB	ChvA	NdvA	CvaB
HlyB	62	90	36	48	41	44	44	45
CyaB		63	40	53	36	44	45	39
LktB			40	47	39	45	45	39
PrtD				36	29	39	38	34
Mdr					38	48	47	40
CylB						37	37	32
ChvA							80	36
NdvA								36

Fig. 8. Per cent identities between residues in the hydrophylic regions of HlyB-like proteins and the first hydrophylic domain of human Mdr-1 protein. Calculations based on data compiled from published sequences by P. Delepelaire (personal communication). See Table 1 for further details on these proteins. LktB is the HylB homologue in the leukotoxin secretion pathway of *Pasteurella haemolytica* (see Welch, 1991, for details).

these domains includes six readily identifiable hydrophobic segments which could span the membrane to form a channel through which the transported molecules pass. The region between these two domains, and the region C-terminal to the second membrane-spanning domain, include sequences with good homology to the Walker motifs A and B found in ATPases (see above; Walker *et al.*, 1982). They are therefore typical ABC proteins. Both halves of the protein appear to be required for activity (Currier *et al.*, 1989), which contrasts with other members of the group (e.g. HlyB) which have only one integral membrane domain and one pair of Walker motifs in the C-terminal hydrophilic domain (Fig. 5). Although most of the bacterial HlyB homologues have six readily-identifiable transmembrane segments in their N-terminal domains (not shown), studies on HlyB protein (Wang *et al.*, 1991; Blight & Holland, 1990) suggest that its membrane topology might be somewhat different. Irrespective of this, all of these proteins have long hydrophilic C-terminal tails that include the Walker motifs. By analogy with Mdr and CFTR, and for other reasons discussed below, it has been proposed that the functional form of HlyB and its homologues of bacterial origin is a dimer (Hyde *et al.*, 1990; Higgins, 1990). In all cases, the hydrophilic, Walker motif-bearing domains face the cytosol. Apart from the very closely related proteins in the family (e.g. those involved in the secretion of related RTX toxins, the β-glucan transporters in *Agrobacterium* and *Rhizobium*, or Mdr proteins from different mammals), the sequence homologies are most striking in the regions surrounding the Walker motifs (Figs 5 and 8).

DISTINCT GROUPS OF ABC PROTEINS INVOLVED IN MEMBRANE TRANSPORT

I have so far described two groups of ABC proteins (the PulE group and the HlyB–Mdr group) that are involved in the transport of macromolecules across membranes. Before discussing whether and how ATP might be involved in these transport processes, it is instructive to study the extent to which these and other groups of ABC proteins are related.

Comparison of the sequences of the PulE group of proteins with those of the hydrophilic domains of the HlyB–Mdr group (A and B in Table 1 and Fig. 5) reveals several important differences:

(i) the sequences around motif A are highly conserved within, but not between, the two groups (Fig. 5).
(ii) The Walker B motif itself is readily identifiable but poorly conserved in the HlyB-Mdr group of ABC proteins (Fig. 5) and in the PulE group. Sequences around the B motif are highly conserved in the HlyB–Mdr group (Fig. 5) but not in the PulE group (not shown). Furthermore, the Walker B motif in PulE is approximately 50 residues further away from the A motif towards the C-terminus than in the HlyB-Mdr group.
(iii) The most highly conserved region of the PulE group of proteins is located approximately 40 residues after the Walker A motif, a region which is relatively poorly conserved in the HlyB–Mdr group and which is anyway completely different from that in the PulE group.

It is striking that the individual members of the groups are considerably more homologous to each other than to most other, transport-unrelated ABC proteins. As noted by Higgins *et al.* (1986) and Higgins (1990), however, there are examples of bacterial proteins which are not known to be involved in membrane transport functions and yet are highly similar to the hydrophilic domains of the HlyB–Mdr group proteins, particularly in the regions shown in Fig. 5(*b*). Therefore, one cannot conclude that the presence of these regions of extensive homology indicates membrane transport-related activity.

Nevertheless, a group of ABC proteins (referred to here as the periplasmic permease ABC proteins; group C in Table 1 and Fig 5) involved in the uptake of amino acids, peptides and small saccharides in *E. coli* and other bacteria are highly homologous to the hydrophilic regions of the HlyB–Mdr group proteins. Periplasmic permease ABC proteins are cytoplasmic proteins that are part of multi-component permeases made up of one or more (usually two) integral cytoplasmic membrane proteins (corresponding to but not homologous to the N-terminal domain of the HlyB–Mdr group) and a soluble periplasmic protein that delivers the substrate to the permease-proper in the membrane. Their properties will be discussed in the following sections; for now it is important to note that there is considerable variation in the stoichiometry of the different components of periplasmic permeases with different substrate specificities. For example, the ABC proteins of the histidine (HisP in Fig. 5), maltose (MalK) and most other permeases have a single pair of ABC motifs (Fig. 5 and not shown), while there are two ABC proteins for the oligopeptide permease of *Salmonella typhimurium* (OppD and OppF in Fig. 5), and a single ABC protein with two pairs of ABC motifs in the case of the *E. coli* ribose permease (RbsA; Hyde *et al.*, 1990, Mimura *et al.*, 1991). Thus, periplasmic permease ABC proteins except RbsA are probably homo- or heterodimers.

Finally, it is important to note that several other membrane transport-related phenomena involve ABC proteins which are quite distinct from each other and from those already mentioned (Table 1). For example, Microcin B17 (also called colicin X) is a small peptide antibiotic secreted by *E. coli*. The secretion pathway comprises two proteins (McbE and McbF) whose structual genes are co-regulated with that of the antibiotic, and the OmpF porin of the outer membrane. McbE is an apparently cytoplasmic protein with a Walker A motif (but not a readily-identifiable B motif) that is otherwise quite unrelated to any of the other ABC transport protein (Garrido *et al.*, 1988). Thus, although this antibiotic is similar in size to colicin V, it is secreted by an ABC protein-dependent pathway that is not related to the HlyB-Mdr pathway. Similarly, the excretion of arsenate and other toxic anions in *E. coli* is an ATP-dependent process involving a cytoplasmic protein (ArsA; Rosen, 1990) which, despite having two Walker A motifs (but again no recognizable B motifs) is otherwise unrelated to the periplasmic permease ABC proteins (data not shown). Another example of an *E. coli* ABC protein with a single pair of Walker A and B motifs and involved in protein traffic is SecA, an essential component of the General Export Pathway (Schmidt *et al.*, 1988). SecA seems to have multiple activities including the ability to bind and hydrolyse ATP, the latter of which is stimulated by signal peptide-bearing precursor polypeptides, acidic phospholipids and the membrane components of the protein export machinery (Lill, Dowhan & Wickner, 1990). Finally, it is worth noting two other examples of proteins with Walker A proteins, the PAS1 protein involved in protein import into peroxisomes in the yeast *Saccharomyces cerevisiae* (Erdmann *et al.*, 1991) and the SEC 18/NSF protein involved in secretory transport vesicle fusion in animal and yeast cells (Wilson *et al.*, 1989). These two proteins are quite closely related (Erdmann *et al.*, 1991) and have well-conserved A motifs and reasonable B motifs, but are not membrane-anchored and, at least in the case of SEC 18/NSF, are not involved in transport of molecules across membranes. However, both proteins are involved in reactions that require ATP hydrolysis. These examples serve to illustrate the high degree of conservation of the Walker motifs, and provide a basis from which to explore the role of ATP in protein secretion in bacteria.

ATP BINDING, ATP HYDROLYSIS AND KINASE ACTIVITIES OF BACTERIAL AND OTHER TRANSPORT ABC PROTEINS

What is the role, if any, of ATP in the transport reactions which involve ABC proteins? The following paragraphs discuss data from studies on all three major classes of transport ABC proteins described in this review, and deal with the demonstrated requirement for ATP in these reactions, the role of the Walker motifs in ATP binding activity and in ATP hydrolysis, and evidence for kinase activities.

There is no doubt that a wide variety of (transport-unrelated) ABC proteins are involved in ATP-dependent reactions (Walker *et al.*, 1982, Higgins *et al.*, 1986), but is this sufficient to conclude that all reactions involving such proteins are ATP-dependent? There is currently no doubt that the periplasmic permeases are ATP-dependent, although earlier results suggested that this might not be the case (for review see Ferro-Luzzi Ames & Joshi, 1990). Indeed, there is now abundant evidence that periplasmic permease ABC proteins possess ATP-binding activity (Higgins *et al.*, 1985), that permeases reconstituted *in vitro* require ATP, and that the ABC proteins hydrolyse ATP (for review see Ferro-Luzzi Ames & Joshi, 1990). Can one extrapolate from this to the HlyB-Mdr group of ABC proteins? The predicted structures of the highly conserved regions of periplasmic permease ABC proteins can be superimposed on the known structure of adenylate kinase (which also has Walker motifs A and B) (Hyde *et al.*, 1990; Mimura *et al.*, 1991), as can the corresponding region of HlyB homologues (Hyde *et al.*, 1990) and data not shown). Accordingly, these regions of the HlyB-Mdr proteins are often referred to as the nucleotide-binding folds.

Mutations that destroy the consensus sequence around the Walker B motif in the nucleotide-binding fold of HlyB protein drastically reduce α-haemolysin secretion (Koronakis, Koronakis and Hughes, 1988). The best evidence for the ATP requirement for reactions mediated by this group of ABC proteins comes, however, from studies on Mdr and CFTR. Multiple drug resistance is strictly dependent on a supply of cytoplasmic ATP. Mdr proteins have been shown to bind an ATP analogue (Schurr, *et al.*, 1989) and immunoaffinity-purified human Mdr1 has ATPase activity, although this activity was measured independently of drug efflux (Hamada & Tsuruo, 1988). Mutations that replace the strictly conserved Gly or Lys residues in either of the Walker A motifs of mouse Mdr1 abolish its ability to confer multiple drug resistance (Azzaria, Schurr & Gross, 1989). However, even destruction of both motifs did not block the ability of this protein to bind an ATP analogue, implying that these residues are not directly involved in nucleotide binding but rather in a subsequent step, possibly ATP hydrolysis, or that the Walker A motifs are not the only sites at which ATP binds to these proteins.

Likewise, mutations in patients with cystic fibrosis have been shown to affect residues in the regions of both Walker motifs in CFTR (Cutting *et al.*, 1990; Kerem *et al.*, 1990). A synthetic 67 amino acid peptide corresponding to part of the first nucleotide binding fold of CFTR, including the Walker A motif but lacking the Walker B motif, has been shown to have nucleotide binding activity (Thomas *et al.*, 1991). Thus, neither the most highly conserved residues of the A motif nor the B region seem to be required for ATP binding. The unique features of these motifs and the surrounding regions (high glycine content, opposing charge distribution due to conserved lysine and aspartate residues, and high hydrophobicity in the B motif) may favour their direct interaction and the formation of an ATP-binding site without them being directly involved in it.

As noted by Higgins (1990), there is no evidence that any of the periplasmic permease ABC proteins or the HlyB–Mdr proteins have kinase activity, although both the Walker A motif and the region around the B motif (particularly the spacing of the Gly, Arg and Asp residues) are conserved in proteins with known kinase activities (Koronakis *et al.*, 1988). There is thus no reasonable doubt that ATP binding and probably also hydrolysis are required for HlyB-Mdr-type transport activity, although it should be pointed out that CFTR protein, on which this conclusion is partially based, is thought to regulate the activity of a conductance channel (it has a putative regulatory domain that is absent from Mdr) and is not known to have intrinsic transport activity. Clearly, the coupling of ATP hydrolysis to transport will only be definitively demonstrated when these transport pathways are reconstituted *in vitro* from purified components, as was the case with the periplasmic permease (Ferro-Luzzi Ames & Joshi, 1990). This might be easier to achieve with the CylB protein from *Enterococcus faecalis* than with the more complex, multicomponent pathways from Gram-negative bacteria.

The situation with regards to the PulE–ComG.1–VirB.11 proteins is far less clear, partly because the proteins are poorly characterized and partly because there is evidence that one of them (VirB.11) has autophosphorylation as well as ATP binding and ATPase activities (Christie *et al.*, 1989). Furthermore, ComG.1-facilitated DNA uptake has been reported to operate in the absence of cellular ATP (for review see Dubnau, 1991). The situation is further complicated by the presence in this group of the PilT protein that is not required for the transport of pili subunits through the *Ps. aeruginosa* cell envelope, but rather to confer upon these pilin the ability to move or twitch (Whitchurch *et al.*, 1991). It seems unlikely that twitching is an ATP hydrolysis-dependent reaction because the pili are anchored to the cell surface while PilT seems to be a cytoplasmic protein. One can, however, envisage the transmission of a signal generated by PilT-mediated ATP hydrolysis through membrane proteins (e.g. PilC?) to the pilus. Another attractive possibility is that PilT is a kinase that uses ATP to phosphorylate (and therefore activate) an envelope component that causes the pili to twitch. If either of these scenarios is correct, then it could be extended to VirB.11, PulE, ComG.1, PilB and XcpR which might energize or phosphorylate envelope components required for conjugal DNA transfer, protein translocation and DNA uptake. It is worth remembering that the PulE-type secretion factors are required for the secretion of proteins that have probably completely crossed the cytoplasmic membrane (Pugsley *et al.*, 1990, 1991*b*). It will be interesting to determine whether these proteins have ATP binding, ATPase or kinase activities, and to study the importance of the highly-conserved region that immediately follows the Walker A motif (Fig. 5).

CONCLUSION

Irrespective of how ATP is used to drive the various transport reactions described here, it is important to note that transport can occur *in either direction* across the membrane. For example the HlyB–Mdr group of ABC proteins, VirB.11 and the PulE secretion factor homologues are involved in transporting molecules of varying sizes and complexities towards the outside of the cell, while the periplasmic permease ABC proteins and ComG.1 are involved in the import of small solutes and DNA. If all of these pathways are indeed operationally similar, then this degree of substrate and directional flexibility is quite remarkable. The recent discovery of a putative peptide-transporting ABC protein with remarkable homology to the HlyB–Mdr class of proteins in the membrane of the endoplasmic reticulum (Table 1; Deverson *et al.*, 1990) implies that further examples of this class of proteins await discovery. Where else should we look for such proteins? Possible candidates include proteins involved in the signal peptide-independent secretion of interleukin B, in ATP-dependent nuclear protein import and in the delivery of ubiquitinated proteins to the lysosome (Pugsley, 1989). It is clear, however, that ABC protein-dependent pathways do not represent the sole alternative to the signal peptide-dependent pathway for transport of proteins through membranes, since dependent protein import into mitochondria requires ATP for events occurring before and after translocation, but is itself ATP-independent (Pugsley, 1989).

ACKNOWLEDGEMENTS

This chapter would not have been possible without the valuable help of many friends, colleagues and collaborators, especially Andrée Lazdunski, Corrine Dorel, Jan Tommassen, George Salmond, Alain Kotoujansky and Mike Daniels who supplied unpublished sequence data and comparisons shown in Fig. 3, Philippe Delepelaire who provided most of the basic information on the homologies between members of the HlyB–Mdr group of ABC proteins, Odile Possot and Philippe Delepelaire who willingly listened to and corrected my ideas on this group of proteins, Bruno Dupuy and Christian Marchal who provided unpublished data and original ideas on pilin assembly, David Nunn and Stephen Lory for discussions on Pseudomonas pilin assembly and unpublished information on the role of PilD/XcpA, Dave Dubnau for being the first to point out the existence of a family of proteins related to PulE and for helpful discussions, Marjorie Russel and Peter Model for discussions on filamentous phage assembly and secretion, John Ward for discussions concerning the *vir* operon, Caroline Day and Odile Possot for carefully correcting the manuscript, and last but by no means least, Isabelle Reyss and Christophe d'Enfert who worked on the pullulanase secretion genes.

REFERENCES

Albano, M., Breitling, R. & Dubnau, D. A. (1989). Nucleotide sequence and genetic organization of the *Bacillus subtilis comG* operon. *Journal of Bacteriology*, **171**, 5386–404.

Alm, R. A. & Manning, P. M. (1990). Characterization of the *hlyB* gene and its role in the production of the El Tor haemolysin of *Vibrio cholerae* O1. *Molecular Microbiology*, **4**, 413–25.

Azzaria, M., Schurr, E. & Gros, P. (1989). Discrete mutations introduced in the predicted nucleotide binding sites of the *mdr1* gene abolish its ability to confer multidrug resistance. *Molecular and Cellular Biology*, **9**, 5289–97.

Bally, M., Ball, G., Badere, A. & Lazdunski, A. (1991). Protein secretion in *Pseudomonas aeruginosa*; the *xcpA* gene encodes an integral membrane protein homologous to *Klebsiella pneumoniae* secretion function PulO. *Journal of Bacteriology*, **173**, 479–86.

Bhattacharyya, P., Wendt, L., Whitney, E. & Silver, S. (1970). Colicin-tolerant mutants of *Escherichia coli*: resistance of membranes to colicin E1. *Science*, **168**, 998–1000.

Blight, M. A. & Holland, I. B. (1990). Structure and function of haemolysin B, P-glycoprotein and other members of a novel family of membrane translocators. *Molecular Microbiology*, **4**, 873–80.

Breitling, R. & Dubnau, D. (1990). A membrane protein with similarity to *N*-methyphenylalanine pilins is essential for DNA binding by competent *Bacillus subtilis*. *Journal of Bacteriology*, **172**, 1499–508.

Brisette, J. L. & Russel, M. (1990). Secretion and membrane integration of a filamentous phage-encoded morphogenic protein. *Journal of Molecular Biology*, **211**, 565–80.

Cangelosi, N., Martinetti, G., Leigh, J. A., Lee, C. C., Theines, C. & Nester, E. W. (1989). Role of *Agrobacterium tumefaciens* ChvA protein in export of β-1,2-glucan. *Journal of Bacteriology*, **171**, 1609–15.

Christie, P. J., Ward, J. E., Gordon, M. P. & Nester, E. W. (1989). A gene required for transfer of T-DNA to plants encodes an ATPase with autophosphorylating activity. *Proceedings of the National Academy of Sciences, USA*, **86**, 9677–81.

Currier, S. J., Ueda, K., Willingham, M. C., Pastan, I. & Gottesman, M. M. (1989). Deletion and insertion mutants of the multidrug transporter. *Journal of Biological Chemistry*, **264**, 14376–81.

Cutting, G. R., Kasch, L. M., Rosenstein, B. J., Zielenski, J., Tsui, L-C., Antonarakis, S. E. & Kazazian, H. H. (1990). A cluster of cystic fibrosis mutations in the first nucleotide-binding fold of the cystic fibrosis conductance regulator protein. *Nature*, London, **346**, 366-9.

Davies, J. K. & Reeves, P. (1975). Genetics of resistance among colicins in *Escherichia coli* K-12: cross resistance among colicins of group A. *Journal of Bacteriology*, **123**, 102–17.

Delepelaire, P. & Wandersman, C. (1990). Protein secretion in Gram-negative bacteria. The extracellular metalloprotease B from *Erwinia chrysanthemi* contains a C-terminal secretion signal analogous to that of *Escherichia coli* α-hemolysin. *Journal of Biological Chemistry*, **265**, 17118–25.

Deverson, E. V., Gow, I. R., Coadwell, W. J., Monaco, J. J., Butcher, G. W. & Howard, J. C. (1990). MHC class II region encoding proteins related to the multidrug resistance family of transmembrane transporters. *Nature,* London, **348**, 738–41.

Dubnau, D. (1991). Genetic competence in *Bacillus subtilis. Microbiological Reviews* (in press).

Dums, F., Dow, J. M. & Daniels, M. J. (1991). Structural characterization of protein export genes of the bacterial phytopathogen *Xanthomonas campestris* pathovar *campestris*: relatedness to export systems of other Gram-negative bacteria. *Molecular and General Genetics* (in press).

d'Enfert, C & Pugsley, A. P. (1989). *Klebsiella pneumoniae pulS* gene encodes an outer membrane lipoprotein required for pullulanase secretion. *Journal of Bacteriology*, **171**, 3673–9.

d'Enfert, C., Reyss, I., Wandersman, C. & Pugsley, A. P. (1989). Protein secretion by Gram-negative bacteria. Characterization of two membrane proteins required for pullulanase secretion by *Escherichia coli* K-12. *Journal of Biological Chemistry*, **264**, 17462-8.

Erdman, R., Wiebel, F. F., Flessau, A., Rykta, J., Beyer, A., Frölich, K-U. & Kunau, W-H (1991). PAS1, a yeast gene required for peroxisome biogenesis, encodes a member of a novel family of putative ATPases. *Cell*, **64**, 499–510.

Felmlee, T., Pellett, S. & Welch, R. A. (1985). Nucleotide sequence of an *Escherichia coli* chromosomal hemolysin. *Journal of Bacteriology*, **163**, 94–105.

Ferro-Luzzi Ames, G. & Joshi, A. K. (1990). Energy coupling in bacterial periplasmic permeases. *Journal of Bacteriology*, **172**, 4133-7.

Filloux, A., Bally, M., Ball, G., Akrim, M., Tommassen, J. & Lazdunski, A. (1990). Protein secretion in Gram-negative bacteria: transport across the outer membrane involves common mechanisms in different bacteria. *EMBO Journal*, **9**, 4323–39.

Garrido, M Del C., Herrero, M., Kolter, R. & Moreno, F. (1988). The export of the DNA replication inhibitor Microcin B17 provides immunity for the host cell. *EMBO Journal*, **7**, 1853–62.

Gilmore, M. S., Segarra, R. A. & Booth, M. R. (1990). A HylB-type function is required for expression of the *Enterococcus faecalis* hemolysin/bacteriocin. *Infection and Immunity*, **58**, 3914–23.

Gilson, L., Mahatny, H. K. & Kolter, R. (1990). Genetic analysis of an MDR-like export system: the secretion of colicin V. *EMBO Journal*, **9**, 3875–84.

Glaser, P., Sakamoto, H., Bellalou, J., Ullmann, A. & Danchin, A. (1988). Secretion of cyclolysin, the calmodulin-sensitive adenylate cyclase-haemolysin bifunctional protein of *Bordetella pertussis*. *EMBO Journal*, **7**, 3997–4004.

Guzzo, J., Duong, F., Wandersman, C., Murgier, M. & Lazdunski, A. (1991). The secretion genes of *Pseudomonas aeruginosa* alkaline protease are functionally related to those of *Erwinia chrysanthemi* proteases and *Escherichia coli* α-haemolysin. *Molecular Microbiology*, **5**, 447–53.

Hamada, H. & Tsuro, T. (1988). Purification of the 170- to 180 kilodalton membrane glycoprotein associated with multidrug resistance. 170–180-kilodalton membrane glycoprotein as an ATPase. *Journal of Biological Chemistry*, **263**, 1454–58.

He, S. Y., Lindeberg, M., Chatterjee, A. K. & Collmer, A. (1991). Cloned *Erwinia chrysanthemi out* genes enable *Escherichia coli* to selectively secrete a diverse family of heterologous proteins into its milieu. *Proceedings of the National Academy of Sciences, USA*, **88**, 1079–83.

Hess, J., Gentschev, I., Goebel, W. & Jarchau, T. (1990). Analysis of the haemolysin secretion system by PhoA–HlyA fusion proteins. *Molecular and General Genetics*, **224**, 201-8.

Higgins, C. F. (1990). The role of ATP in binding-protein-dependent transport system. *Research in Microbiology*, **141**, 353–60.

Higgins, C. F., Hiles, I. D., Salmond, G. P. C., Gill, D. R., Downie, J. A., Evans, I. J., Holland, I. B., Gray, L., Buckel, S. D., Bell, A. W. & Hermondson, M. A. (1986). A family of related ATP-binding subunits coupled to many distinct biological processes in bacteria. *Nature*, London, **323**, 448–50.

Higgins, C. F., Hiles, I. D., Whalley, K. & Jamieson, D. J. (1985). Nucleotide binding by membrane components of bacterial periplasmic binding protein-dependent transport systems. *EMBO Journal*, **4**, 1033–40.

Hiraga, S., Niki, H., Ogura, T., Ichinose, C., Mori, H., Ezaki, B. & Jaffe, A. (1989). Chromosome partitioning in *Escherichia coli*: novel mutants producing anucleate cells. *Journal of Bacteriology*, **171**, 1496–505.

Hui, F. M. & Morrison, D. A. (1991). Genetic transformation in *Streptococcus pneumoniae*: nucleotide sequence analysis shows *comA*, a gene required for competence induction, to be a member of the bacterial ATP-dependent transport protein family. *Journal of Bacteriology*, **173**, 372–81.

Hyde, S. C., Emsley, P., Hartshorn, M. J., Mimmack, M. M., Gileadi, U., Pearce, S. R., Gallagher, M. P., Gill, D. R., Hubbard, R. E. & Higgins, C. F. (1990). Structural model of ATP-binding proteins associated with cystic fibrosis, multidrug resistance and bacterial transport. *Nature*, London, **346**, 362-5.

Kamijo, K., Taketani, S., Yokota, S., Osumi, T. & Hashimoto, T. (1990). The 70-kDa peroxisomal membrane protein is a member of the Mdr (P-glycoprotein)-related ATP-binding protein superfamily. *Journal of Biological Chemistry*, **265**, 4534–40.

Kerem, B-S., Zielenski, J., Markiewicz, D., Bozon, D., Gazit, E., Yahav, J., Kennedy, D., Riordan, J. R., Collins, F. C., Rommens, J. M. & Tsui, L-C. (1990). Identification of mutations in regions corresponding to the two putative nucleotide (ATP)-binding folds of the cystic fibrosis gene. *Proceedings of the National Academy of Sciences, USA*, **87**, 8447–51.

Koronakis, V., Koronakis, E. & Hughes, C. (1988). Comparison of the haemolysin secretion protein HlyB from *Proteus vulgaris* and *Escherichia coli*; site-specific mutagesis causing impairment of export function. *Molecular and General Genetics*, **213**, 551-5.

Koronakis, V., Koronakis, E. & Hughes, C. (1989). Isolation and analysis of the C-terminal signal directing export of *Escherichia coli* hemolysin protein across both bacterial membranes. *EMBO Journal*, **8**, 595–605.

Létoffé, S., Delepelaire, P. & Wandersman, C. (1990). Protease secretion by *Erwinia chrysanthemi*: the specific secretion functions are analogous to those of *Escherichia coli* α-haemolysin. *EMBO Journal*, **9**, 1375–82.

Lill, R., Dowhan, W. & Wickner, W. (1990). The ATPase activity of SecA is regulated by acidic phospholipids, SecY, and the leader and mature domains of precursor proteins. *Cell*, **60**, 271–80.

Mackman, N., Baker, K., Gray, L., Haigh, R., Nicaud, J-M & Holland, I. B. (1987). Release of a chimeric protein into the medium from *Escherichia coli* using the C-terminal secretion signal of haemolysin. *EMBO Journal*, **6**, 2835–41.

Marrs, C. F., Schoolnick, G., Koomey, J. M., Hardy, J.,Rothbard, J., Falkow, S. (1985). Cloning and sequencing of the *Moraxella bovis* pilin gene. *Journal of Bacteriology*, **163**, 132-9.

Masure, H. R., Au, D. C., Gross, M. K., Donovan, M. G. & Storm, D. R. (1990). Secretion of the *Bordetella pertussis* adenylate cyclase from *Escherichia coli* containing the hemolysin operon. *Biochemistry*, **29**, 140-5.

Mimura, C. S., Holbrook, S. R. & Ferro-Luzzi Ames, G. (1991). Structural model of the nucleotide-binding conserved component of periplasmic permeases. *Proceedings of the National Academy of Sciences, USA*, **88**, 84-8.

Miyazaki, H., Yaninga, H., Horinouchi, S. & Beppu, T. (1989). Characterization of the precursor of *Serratia marcescens* serine protease and COOH-terminal processing during excretion through the outer membrane. *Journal of Bacteriology*, **171**, 6566–72.

Mohan, S., Aghion, J., Guillen, N. & Dubnau, D. (1989). Molecular cloning and characterization of *comC*, a late competence gene in *Bacillus subtilis*. *Journal of Bacteriology*, **171**, 6043–6051.

Morona, R. & Reeves, P. (1982). The *tolC* locus of *Escherichia coli* affects the expression of three major outer membrane proteins. *Journal of Bacteriology*, **150**, 1016–23.

Nunn, D., Bergman, S. & Lory, S. (1990). Products of three accessory genes, *pilB*, *pilC* and *pilD* are required for biogenesis of *Pseudomonas aeruginosa* pili. *Journal of Bacteriology*, **172**, 2911–19.

Pohlner, J., Halter, R., Beyreuther, K. & Meyer, T. F. (1987). Gene structure and extracellular secretion of *N. gonnorhoeae* IgA protease. *Nature*, London, **325**, 458–62.

Pugsley, A. P. (1984). The ins and outs of colicins. *Microbiological Sciences*, **1**, 168–75.

Pugsley, A. P. (1989). *Protein Targeting*. Academic Press, San Diego, USA.

Pugsley, A. P., d'Enfert, C., Reyss, I., & Kornacker, M. G. (1990). Genetics of extracellular protein secretion by Gram-negative bacteria. *Annual Review of Genetics*, **24**, 67–90.

Pugsley, A. P. & Kornacker, M. G. (1991). Secretion of the cell-surface lipoprotein pullulanase in *Escherichia coli*: collaboration or competition between the specific secretion pathway and the lipoprotein sorting pathway? *Journal of Biological Chemistry*, **266**, in press.

Pugsley, A. P., Kornacker, M. G. & Poquet, I. (1991*a*). The general protein-export pathway is directly required for extracellular pullulanase secretion in *Escherichia coli* K12. *Molecular Microbiology*, **5**, 343–52.

Pugsley, A. P., Poquet, I. & Kornacker, M. G. (1991*b*). Two distinct steps in pullulanase secretion by *Escherichia coli* K12. *Molecular Microbiology*, **5**, in press.

Pugsley, A. P. & Reyss, I. (1990). Five genes at the 3′ end of the *Klebsiella pneumoniae pulC* operon are required for pullulanase secretion. *Molecular Microbiology*, **4**, 365–79.

Reyss, I. & Pugsley, A. P. (1990). Five additional genes in the *pulC-O* operon of the Gram-negative bacterium *Klebsiella oxytoca* UNF5023 that are required for pullulanase secretion. *Molecular and General Genetics*, **222**, 176–84.

Rosen, B. P. (1990). The plasmid-encoded arsenical resistance pump: an anion-translocating ATPase. *Research in Microbiology*, **141**, 336–41.

Schatz, P. & Beckwith, J. (1990). Genetic analysis of protein export in *Escherichia coli*. *Annual Review of Genetics*, **24**, 215–48.

Schiebel, E., Schwartz, H. & Braun, V. (1989). Subcellular location and unique secretion of the hemolysin of *Serratia marcescens*. *Journal of Biological Chemistry*, **264**, 16311–20.

Schmidt, M. G., Rollo, E. E., Grodberg, J. & Oliver, D. B. (1988). Nucleotide sequence of the *secA* gene and *secA*(Ts) mutations preventing protein export in *Escherichia coli*. *Journal of Bacteriology*, **170**, 3404–14.

Schurr, E., Raymond, M., Bell, J. C. & Gros, P. (1989). Characterization of the multidrug resistance protein expressed in cell clones stably transfected with the mouse *mdr-1* cDNA. *Cancer Research*, **49**, 2729–34.

Shirasu, K., Morel, P. & Kado, C. I. (1990). Characterization of the *virB* operon of *Agrobacterium tumefaciens* Ti plasmid: nucleotide sequence and protein analysis. *Molecular Microbiology*, **4**, 1153–63.

Stanfield, S. W., Ielpi, L., O'Brochta, D., Helinski, D. R. & Ditta, G. S. (1988).

The *ndvA* gene product of *Rhizobium meliloti* is required for β-(1→2) glucan production and has homology to the ATP-binding export protein HlyB. *Journal of Bacteriology*, **170**, 3523–30.

Thomas, P. J., Shenbagamurthi, P., Ysern, X. & Pederson, P. L. (1991). Cystic fibrosis transmembrane conductance regulator: nucleotide binding to a synthetic peptide. *Science*, **251**, 555-7.

Uphoff, T. S. & Welch, R. A. (1990). Nucleotide sequence of the *Proteus mirabilis* calcium-independent hemolysin genes (*hmpA* and *hpmB*) reveals sequence similarity with the *Serratia marcescens* hemolysin genes (*shlA* and *shlB*). *Journal ofBacteriology*, **172**, 1206–16.

Walker, J. E., Saraste, M., Runswick, M. J. & Gay, N. J. (1982). Distantly related sequences in the α- and β- subunits of ATP synthase, myosin, kinases and other ATP-requiring enzymes and a common nucleotide binding fold. *EMBO Journal*, **1**, 945–51.

Wandersman, C & Delepelaire, P. (1990). TolC, an *Escherichia coli* outer membrane protein required for hemolysin secretion. *Proceedings of the National Academy of Sciences, USA*, **87**, 4776–80.

Wang, R., Seror, S. J., Blight, M., Pratt, J. M., Broome-Smith, J. K. & Holland, I. B. (1991). Analysis of the membrane organization of an *Escherichia coli* protein translocator, HlyB, a member of a large family of prokaryote and eukaryote surface transport proteins. *Journal of Molecular Biology*, **217**, 441–54.

Welch, R. (1991). Pore-forming cytolysins of Gram-negative bacteria. *Molecular Microbiology*, **5**, 521-8.

Whitchurch, C. B., Hobbs, M., Livingston, S. P., Krishnapillai, V., and Mattick, J. S. (1991). Characterization of a *Pseudomonas aeruginosa* twitching motility gene and evidence for a specialised protein export system widespread in bacteria. *Gene*, in press.

Wilbur, W. J. & Lipman, D. J. (1983). Rapid similarity searches of nuclei acid and protein sequence data banks. *Proceedings of the National Academy of Sciences, USA*, **80**, 726–30.

Wilson, D. W., Wilcox, C. A., Flynn, G. C., Chen, E., Kuang, W.-J., Henzel, W. J., Block, M. R., Ullrich, A. & Rothman, J. E. (1989). A fusion protein required for vesicle-mediated transport in both mammalian cells and yeast. *Nature*, London, **369**, 335-9.

INTRACYTOPLASMIC MEMBRANES IN BACTERIAL CELLS: ORGANIZATION, FUNCTION AND BIOSYNTHESIS

GERHART DREWS

Institute of Biology II, Microbiology, Albert-Lüdwigs-University 7800 Freiburg, Germany

INTRODUCTION

During evolution, a large number of physiological groups of prokaryotes has been selected which have been adapted to different environments. Each ecological niche is characterized by physical and chemical parameters which, however, can vary considerably. Thus prokaryotes have developed numerous mechanisms to keep alive under various conditions and to respond to sudden or long-term variations of pH, temperature, ion and substrate concentrations in the environment. The single bacterial cells are much more in contact with the environment than cells in eukaryotic tissues.

The border between cytoplasm and environment is the cell envelope. The envelope consists of the cytoplasmic membrane which is the main physiological boundary of the protoplast and the cell wall of Gram-positive bacteria or the periplasmic space, the peptidoglycan layer and the outer membrane of Gram-negative bacteria. The cytoplasmic membrane (CM) of bacteria is a multifunctional system which has been specialized to mediate different types of communication between protoplasts and the environment.

CYTOPLASMIC MEMBRANE

The multiple functions of the cytoplasmic membrane

The oxidation of substrates in aerobic and anaerobic respiring bacteria leads finally to an electron transfer along the respiratory chain to oxygen or another electron acceptor at the CM and a proton translocation across the membrane. The energy is stored in a ΔpH and the membrane potential, $\Delta\psi$, and is proportional to the proton electrochemical potential which is used for ATP synthesis by the H^+-ATPase or other energy consuming processes. In some phototrophic bacteria the photosynthetic and the respiratory apparatus and other redox systems are located in the CM. In anaerobic bacteria vectorial redox processes can be coupled with the generation of a proton motive force in the cytoplasmic membrane. Examples are: reduction of methylcoenzyme M ($CoM\text{-}S\text{-}CH_3$) to methane in methanogenic bacteria, reduction of sulphate to sulphide by hydrogenase and sulphite reductase during dissimilatory sulphate reduction and the generation of an electrochemical gradient of sodium ions at the membrane-associated oxalacetate decarboxylase in *Enterobacter aerogenes* which can be transformed into a proton gradient (Gottschalk, 1986).

Gates and pumps in the CM regulate the flow of ions and protons across the CM. There is need for specific ions which are essential constituents of enzymes and the redox system. Iron, for example, is under most conditions limited in nature and in host organisms and is taken up by highly efficient, low-molecular weight chelators (Crosa, 1989). Other ions are not covalently bound to protein but are necessary for ion balance, activation of enzymes and for homeostasis of pH and ions in the cytoplasm (Krulwich & Guffanti, 1989; Matin, 1990). Many of the transport systems, including periplasmic permeases, are energy coupled (Ferro-Luzzi Ames & Joshi 1990; Ferro-Luzzi Ames, Mimura & Shyamala, 1990).

Many prokaryotes accumulate solutes against a concentration gradient using $\Delta\tilde{\mu}$ H^+, or one of its components, i.e. the membrane potential $\Delta\psi$ or the pH gradient ΔH^+, as the immediate driving force. One of many prototypes is the β-galactoside (*lac*) transport system of *Escherichia coli* (Kabak, 1990). The *lac* permease is a large, integral membrane protein with twelve hydrophobic α helical domains each crossing the CM. Some amino acid positions are important for lactose-coupled H^+ translocation (Kabak, 1990). The electron transfer chain not only functions as the proton motive force-generating system but also plays a direct role in the regulation of secondary transport systems. Furthermore, in addition to the electron transfer chain, K^+ plays an important role in solute transport in the photo-synthetic bacterium *Rhodobacter sphaeroides* (Abee, Hellingwerf & Konings, 1988; Abee *et al.*, 1989). Another type of transport system, which has been studied thoroughly, is the phosphoenolpyruvate:sugar phosphotransferase system (PTS). The uptake and vectorial phosphorylation of sugars is driven by phosphoenol pyruvate (PEP). PEP as the phosphoryl donor and three protein fractions, the enzymes I, HPr and enzyme II, are required. The potentials of $P \sim EI$, $P \sim HPr$ and $P \sim III$ are close to that of PEP, so the phosphotransfer reactions are reversible. The PTS is strongly controlled (Meadow, Fox & Roseman, 1990).

Many bacteria are able to sense chemical or physical gradients in the environment and direct the movement upwards or downwards in these gradients or to adapt to different osmotic pressures in the environment. Components of the chemotaxis or phototaxis apparatuses are located in the membrane. The transduction of signals from the outside is mediated by the CM (Mizuno & Mizushima, 1990; Stock, Stock & Mottonen, 1990; Gross, Arico & Rappuoli, 1989).

All systems contain membrane-bound components for uptake of substrates and export of molecules to the periplasmic space, into the outer membrane or into the extracellular space. The CM contains enzymes which modify molecules vectorially during export or import or synthesize membrane-bound constituents. Adaptation of bacteria to fluctuations in the osmolarity of their surroundings is another function which is located mainly in the CM. A primary response of *E. coli* to an increase in osmolarity is the uptake of K^+ via the Trk or the Kdp system (Epstein, 1986). The long-range response to increase in osmolarity is the accumulation or synthesis of so-called compatible solutes. These are compounds which can be accumulated to high intracellular concentrations without causing harm to vital functions of cells (Higgins *et al.* 1987). These osmo-protectants include amino acids like glutamate and proline, betaines and sugars (Abee *et al.* 1990). DNA of bacteriophages or plasmids is transported across the CM, possibly at specific sites. Flagella and pili are anchored in the CM and are located in some bacteria at the cell pole.

Structure, composition and physical properties of the CM

Biomembranes consist of lipids and proteins. Phospholipids are the major class of bacterial CM lipids which are completed by glyco-lipids, hopanoid triterpenes (Rohmer, Bouvier-Nave & Ourisson, 1984), quinones, carotenoids and other lipids. Structure, flexibility, stability, semipermeability and fluidity of biomembranes are determined by the lipids. The amphiphilic membrane lipids aggregate above the critical micellar concentration at defined physical conditions to lipid double layers of about 4nm diameter. Hydrophobic and other non-covalent interactions stabilize the lipid bilayer.

Membranes are asymmetric due to the structure and organization of peripheral or integral membrane proteins and in some cases by differences in lipid composition of the cytoplasmic and periplasmic leaflet of the lipid bilayer. Membranes are fluid structures. Lipids and proteins diffuse rapidly in the plane of the membrane but do not rotate across the membrane. Membrane proteins tower above the plane of the lipid bilayer and increase the overall diameter of biomembranes to about 6–8 nm. Many membrane proteins aggregate non-covalently to form regular supramolecular structures such as the photochemical reaction centre, the b/c_1 oxidoreductase or the H^+- ATP-ase, which may, for their part, interact with other functional complexes. These supramolecular protein complexes and the high protein to lipid ratio of the CM may reduce the lateral mobility of proteins in the lipid bilayer. The fluidity of prokaryotic membranes is regulated by the number of double bonds, the length of fatty acyl chains and the presence of hopanoids and other stiffening components.

It is well known that biomembranes in eukaryotic cells are highly specialized and therefore differ considerably in their protein to lipid ratio (0.2 to 3) as well as in their protein and lipid composition. The CM of bacteria has a protein to lipid ratio of 2–4 (Salton, 1978; Rogers, Perkins & Ward, 1980).

Lateral diversity of the CM

Although most bacteria have only one type of plasma membrane in which the components are equilibrated by lateral diffusion, there exists, at least temporarily, a lateral diversity, caused by local processes of differentiation. Most cells divide symmetrically by formation of a cross septum and concomitantly centripetal invagination of the CM in the middle of the cell. Formation of the cross septum and division of the chromosome seem to be coordinated, membrane-bound processes.

The differentiation of a swarmer cell and a stalk cell after division of *Caulobacter* cells is finally a membrane-directed polar process (Shapiro, 1985).

The purple membrane is a patch-like differentiation of the cytoplasmic membrane of *Halobacterium halobium*. It is a large asymmetric crystal of bacteriorhodopsin which functions as a light-drive proton pump. The bean-shaped cells of the lithoautotrophic bacterium *Gallionella ferruginea* excrete in a distinct area at the concave side of the cell, where the CM is invaginated, a twisted ferric hydroxide stalk (Lütters & Hanert, 1989; Lütters-Czekalla, 1990). Buds and prostheka originate under participation of the CM by local evagination of the cell envelope, very often non-randomly at specific sites of the cell surface or of the prostheka. It is unknown how polarity and lateral diversity are determined in a fluid membrane.

INTRACYTOPLASMIC MEMBRANES

Intracytoplasmic membranes (ICM) are located in the cytoplasm and separated from the CM. It is believed that they are formed by invagination of the CM. The thylakoids of cyanobacteria, however, ICMs which seem to be independent from the CM (Golecki & Drews, 1982).

Structure and distribution of intracytoplasmic membranes in bacteria

Intracytoplasmic membranes (ICMs) are regular constituents of photosynthetic purple bacteria and cyanobacteria. In these organisms, the energy-transducing processes of the photosynthetic apparatus are localized on ICMs. ICMs are organized as loose or ordered aggregates of vesicles, or as tubular or flat vesicular membranes (Table 1). Lithoautotrophic bacteria also contain ICMs, but less regularly. Chemotrophic bacteria sporadically contain small ICMs but rarely larger ICM aggregates. The appearance of ICM in a relative small number of physiologically defined groups of bacteria suggests that ICMs are examples of a primitive form of compartmentalization of prokaryotic cells. Unfortunately, our knowledge of ICM function and differentiation is very restricted to a few groups of bacteria (Mayer, 1986).

ICM in chemotrophic bacteria

Methanotrophs develop extensive ICM when growing on methane (Table 1; Higgins *et al.*, 1981). They are not present or much less extended in facultative methylotrophs growing on methanol. The structure and extension of ICM seem to be dependent upon culture conditions. It was reported that low oxygen tension and the presence of methane induce the formation of ICM. In late stationary phase the ICM seem to degenerate. Broad-specificity methane monooxygenase and $NAD(P)^+$-independent methanol dehydrogenase seem to be bound to the ICM. The energy transduction and electron transport of these enzymes are dependent upon the type of enzyme (Higgins *et al.*, 1981).

In dinitrogen-fixing cells of *Azotobacter vinelandii* the area of ICM per cell increased with increasing aeration of the culture. A steep increase was observed between 1% and 25% air saturation. In cells growing with ammonium as sole nitrogen source the number of ICM vesicles stayed almost constant per unit of cell volume and increased 2.9-fold per cell. It was concluded that the oxygen tension controls the formation of ICM while the nitrogen source controls the amounts of ICM at a given partial pressure (Post, Golecki & Oelze, 1982). The cellular respiration increased with dissolved oxygen concentrations increasing from about 1% to 30% air saturation. Only a single membrane fraction with a buoyant density of $\rho^{20} = 1{,}176\ g\,cm^{-2}$ was detected in all cells adapted to various oxygen tensions. Maximum activities of NADH-H, NADPH-H and malate dehydrogenases increased in parallel as the oxygen tension was increased from 1% to 30% air saturation. It was concluded that CM and ICM do not differ with respect to respiratory activities (Post, Vakalopoulou & Oelze, 1983).

Gallionella ferruginea cells contain ICM invaginating in the form of long vesicotubular channels at the concave cell side. It is unknown whether these ICM are concerned with excretion of the ferric hydroxide stalk (Lütters & Hanert, 1989; Lütters-Czekalla, 1990).

ICM in nitrifying bacteria

Many, but not all, nitrifying bacteria have ICMs. In *Nitrosomonas europaea* peripheral, lamellar ICM are arranged in parallel layers. Other ammonia-oxidizing bacteria contain centrally located parallel bundles of lamellar membranes or single branched lamellae. *Nitrobacter winogradskyi* cells contain flattened vesicles in the peripheral region of the cell; in contrast, *Nitrococcus*

Table 1. *Presence, structure and function of intracytoplasmic membranes (ICM) in bacteria.*

Organisms	Structure and organization of ICM	Function of ICM	Growth conditions determining formation of ICM
(*a*) photosynthetically active			
Rhodospirillum rubrum, R. salinareum, Rhodopila, Rhodobacter, Heliobacterium, Thiocystis, Thiocapsa roseo-persicina, Lamprocystis, Chromatium, Thiospirillum, Lamprobacter, Amoebobacter, Thiopedia.	Network of *vesicular structures* 35–70 nm in diameter	Photosynthetic apparatus, anoxygenic photosynthesis	Formation induced by lowering of oxygen tension and lowering of light intensity under anaerobic conditions
Thiocapsa pfennigii	Bundled tubes	Anoxygenic photosynthesis	
Rhodospirillum tenue *Rhodocyclus gelatinosa*	Single, small tubular invaginations of CM		
Rhodospirillum photometricum R. molischianum, R. fulvum, R. salexigens, Ectothiorhodospira	Short lamellar stacks	Anoxygenic photosynthesis	Oxygen tension and light intensity dependent modulation
Cyanobacteria	Single *thylakoids* (flat sacs) in parallel concentric layers or irregular arrangements; phycobilisomes are attached	Oxygenic photosynthesis	Light intensity and nitrogen metabolism modulate the thylakoid differentiation
Prochlorales	Thylakoids like cyanobacteria without phycobilisomes	Oxygenic photosynthesis	
Green photosynthetic bacteria *Chlorobium, Prosthecochloris, Pelodiction, Ancalochloris, Chloroflexus*	No ICM, chlorosomes (light-harvesting chlorophyll c–containing structures)	Photosynthetic apparatus in the CM; anoxygenic photosynthesis	Modulation by light intensity and O_2-tension (*Chloroflexus*)
(*b*) non-photosynthetic chemotrophic bacteria			
Nitrifying bacteria *Nitrobacter*	Polar cap of flattened vesicles in the peripheral region of the cell, cytoplasmic surface of ICM studded with 7-9 nm particles arranged in paired rows under lithotrophic growth conditions. Nitrite oxidase	Nitrite oxidase respiratory system	
Nitrococcus	Tubular ICM, randomly arranged		

Table 1 continued opposite

Table 1 continued from previous page

Organisms	Structure and organization of ICM	Function of ICM	Growth conditions determining formation of ICM
Ammonia oxidizing bacteria *Nitrosomonas* *Nitrosococcus*	ICM in some species flattened lamellae arranged centrally, peripherally or randomly branched		
Nitrosospira	no ICM		
Chemotrophic bacteria *Gallionella*	In the central region of the cell at the concave side vesico-tubular channels by invagination from the ICM; membranes only 5nm thick	Excretion of iron hydroxide?	
Crenothrix polyspora	Membrane stacks perpendicular to the CM		
Azotobacter	Flat membranes	NADH-, NADPH-, malate DH[2] activities increased when air saturation increased from 1 to 30%	Number and area of ICM increased with aeration under N_2-fixing conditions.
	ICM and CM do not differ with respect to respiratory activities		
Many bacteria	Mesosomes seen as a cluster of tubular membranes	Unknown	
Staphylococcus aureus	Murosomes	Doubled invaginations of CM; cell division; lysis of peptidoglycan	
Gluconobacter	Polar complexes of loosely coiled ICM, organized in lamellae	Increase of glycerol oxidation in cells containing ICM	True membranes?
Methylococcus[1]	Bundles of vesicular discs or paired membranes aligned to periphery.	Respiration; methane oxidation	Present when grown on methane
Methylomonas[1]	I Uniform array of membranes, stacked series of flattened discoidal vesicles, continuous with CM, branching II Paired membranes enclosing variable-sized lumens.	Methane oxidase	Oxygen-tension regulated.

[1] Whittenbury & Krieg 1984 [2] DH = dehydrogenase

mobilis contains tubular ICM which are randomly arranged. The morphology, fine structure and biochemistry of these lithoautrophic bacteria have been reviewed recently (Watson, Valois & Aterbury, 1981; Bock *et al.*, 1991). The enzymes for the oxidation of NH_3 to HNO_2 seem to be located on the periplasmic side of the CM or in the periplasmic space. It has not been reported whether these enzymes and the components of these respiratory chains are equally distributed on CM and ICM. Nitrite oxidoreductase and nitrite reductase are integral membrane enzymes. Nitrite oxidoreductase forms large multicomponent ICM particles (9.5 nm in diameter) composed of three polypeptide subunits (116k, 65k, 32k M_r, iron, molybdenum, sulphur and copper. The particles are densely packed and equally distributed on CM and ICM (Sundermeyer-Klinger *et al.*, 1984).

Other chemotrophic bacteria such as *Crenothrix polyspora* and *Gluconobacter* have been reported to form ICM. In *Gluconobacter* the ICM content increases when glycerol is oxidized.

Mesosomes are aggregates of tubular membrane structures close to the sites of cell division described on the basis of electron microscopical investigations of both Gram-positive and Gram-negative bacteria. Functions in energy metabolism, respiration, cell and chromosome division and enzyme secretion have been proposed. The discussion about the existence and function of mesosomes is very controversial. The literature has been reviewed (Drews & Giesbrecht, 1981; Greenawalt & Whiteside, 1975).

ICM in phototrophic bacteria

Apart from a few exceptions (*Gloeobacter, Rhodospirillum tenue, Heliobacterium chlorum*), all purple and cyanobacteria contain ICMs on which the photosynthetic apparatus is located (Drews & Imhoff, 1991). The composition and development of ICM has been extensively studied in a few facultative phototrophic bacteria in which membrane formation and synthesis of the photosynthetic apparatus are correlated. Both processes can be induced or repressed by the external factor, oxygen tension, and be modulated by light intensity and other growth factors. In green sulphur and non-sulphur photosynthetic bacteria the photosynthetic apparatus is located in the CM and the attached chlorosomes (Table 1). The synthesis of the photosynthetic apparatus of *Chloroflexus aurantiacus*, a facultative green non-sulphur bacterium, is induced by lowering of oxygen partial pressure. The differentiation of CM in this bacterium is under investigation in several laboratories and will not be discussed in this article (review: Olson *et al.*, 1988).

Since the literature on the composition, function and formation of ICM of facultative phototrophic purple bacteria has been reviewed recently (Drews & Oelze, 1981; Donohue & Kaplan, 1986; Sprague & Varga, 1986; Kiley & Kaplan, 1988; Drews, 1991) this chapter concentrates on the problems of ICM differentiation in facultative purple bacteria.

Is the ICM in facultative photosynthetic bacteria a specific compartment or an enlarged CM?

CM and one or more fractions of ICM can be separated from each other by equilibrium sucrose density centrifugation (Niederman & Gibson, 1978). They differ from each other in their density, composition and functional equipment (Niederman & Gibson, 1978; Oelze & Drews, 1981). Enzymes and redox components of the respiratory chain are present in different concentrations in all membrane fractions (Takemoto & Bachmann, 1979; Cox, Beatty & Favinger, 1983; Ferguson, Jackson & McEwan, 1987; Kaufmann *et al.*, 1982; Oelze & Drews, 1981). The pigment proteins of the photosynthetic apparatus, however, are enriched in the ICM (Niederman, Mallon & Parks, 1979; Bowyer *et al.*, 1985; Kaufmann *et al.*, 1982; Lampe & Drews, 1972). The different densities and compositions of the two or three membrane fractions isolated from bacteria grown phototrophically or chemotrophically at low oxygen tension correlate with different particle populations in membrane fractions visualized by electron microscopy of freeze–fractured cells (Fig. 1). The number of intra-membrane particles per unit area, their mean diameter and their orientation relative to the cytoplasmic or external fracture face have been determined for several types of bacteria that had been grown under chemotrophic or phototrophic conditions (Golecki & Oelze, 1980; Golecki, Schumacher & Drews, 1980; Varga & Staehelin, 1983; Fig. 1). High intramembrane-particle densities were observed in photosynthetic intracytoplasmic membrane stacks and in those areas of the CM which adhere to the underlying ICM of *Rhodopseudomonas palustris*. Lowering of light-intensity effects an increase of particle diameter (Varga & Staehelin, 1983). A comparison of isolated pigment–protein complexes and complexes embedded in liposomes with the intramembrane particles suggested that the different size classes of particles were identical with the 7.5 nm LHII (B800–850), the 10 nm reaction centre (RC) + LHI complex (B880) and the 12.5 nm RC + LHI + LHII, particles (Varga & Staehelin, 1985*a*). The LHI complex was proposed to be the adhesion factor for membrane stacking (Varga & Staehelin, 1985*b*). Other macromolecular membrane-bound enzyme complexes which should be visible as particles, like ATPase or b/c_1 complex, have not been identified.

ICMs are formed by invagination of the CM and remain connected with the CM (Drews & Giesbrecht, 1963; Tauschel & Drews, 1967; Golecki & Oelze, 1975; Varga & Staehelin, 1983). CM and ICM form a large membrane continuum (Hurlbert, Golecki & Drews, 1974; Dierstein, Schumacher & Drews, 1981). Newly synthesized pigment proteins are inserted into a membrane fraction that is distinct from the CM but still less dense than mature ICM. It was proposed that this newly synthesized ICM fraction is derived from specific invagination sites of the CM (Niederman, Mallon & Parks, 1979; Dierstein *et al.*, 1981; Bowyer *et al.*, 1985). These and numerous other

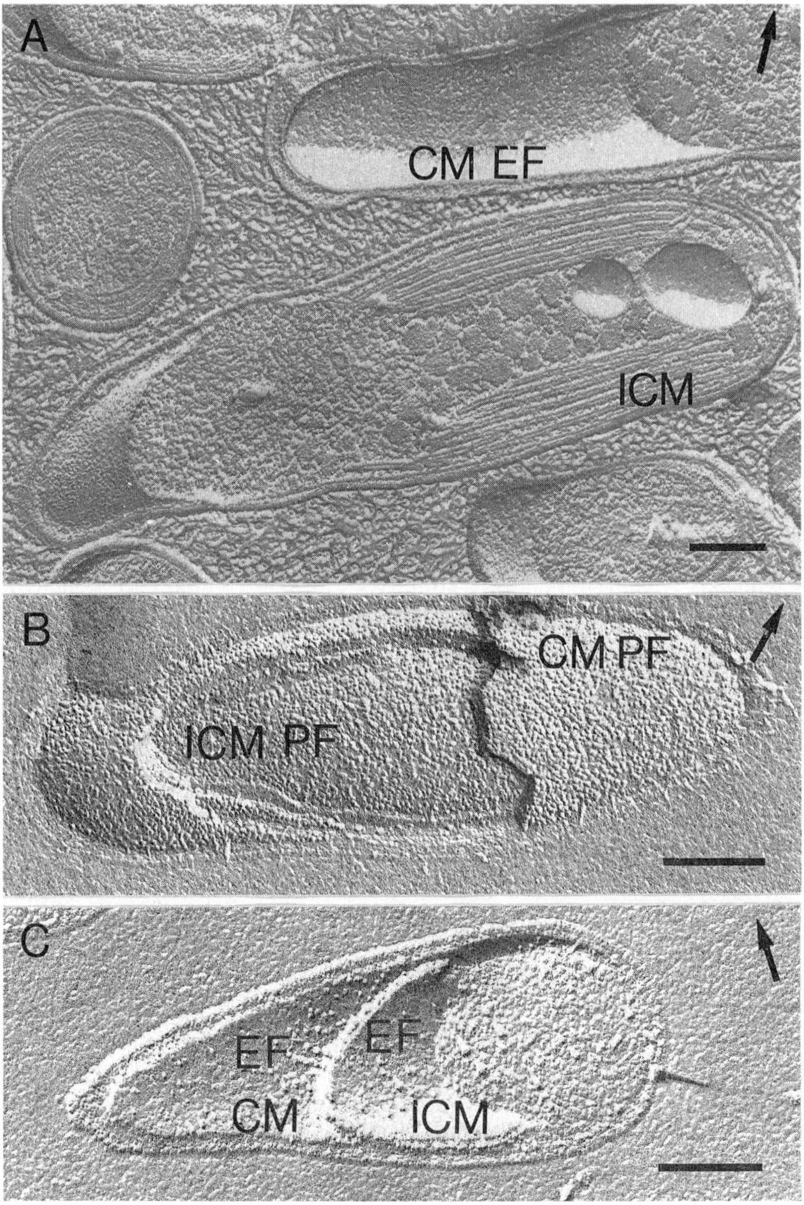

Fig. 1. Freeze–fracture electron micrographs of cytoplasmic (CM) and intracytoplasmic membranes (IC) of *Rhodopseudomonas palustris*. EF: exoplasmic fracture face, PF: protoplasmic fracture face. A: cross fracture shows the stacked lamellar, double ICM. B,C: The PF of CM and ICM are densely studded with integral membrane particles, while the EFs showed much less particles. EM-micrograph by J. R. Golecki.

studies have shown that CM and ICM form a continuous membrane system. The ratio of ICM to CM may vary considerably. It is roughly correlated to the number and size of the photosynthetic units (mol light-harvesting (LH) bacteriochlorophyll per reaction centre). A shift from high to low light intensity resulted in a five-fold increase of pmol reaction centre per mg cell protein, a doubling of the size of the photosynthetic unit and a six-fold increase of the number of ICM vesicles per cell of *Rhodobacter capsulatus* (Reidl, Golecki & Drews, 1985). The ICM is not a separate compartment of photosynthetic bacteria but an enlarged and specialized portion of the membrane system. It is remarkable that, in spite of a lateral diffusion in membranes, the ICM/CM system is not homogenous in its composition. It is differentiated in functionally specialized zones. The main function of ICM is photosynthesis, while the CM is specialized in anaerobic and aerobic respiration, transport and phospholipid bio-synthesis (Kaplan 1978; Tai & Kaplan, 1985).

The ICM of anoxigenic purple bacteria is organized in species-specific patterns of vesicles, tubules, stacked lamellar aggregates and irregular branched non-stacked membranes (Oelze & Drews, 1981; Remsen, 1978; Sprague & Varga, 1986; Table 1). It has been proposed that the integral membrane proteins have morphogenetic functions. The incorporation of the α and β polypeptides of the B800–850 (LHII) complex into ICM of *Rhodobacter sphaeroides* was supposed to be responsible for molding of ICM (Hunter *et al.*, 1988; Kiley, Varga & Kaplan, 1988). However, *Rhodospirillum rubrum, Rhodobacter sphaeroides* and *Rhodobacter capsulatus* form vesicular ICM without incorporation of LHII polypeptides (Golecki *et al.*, 1989). The function of the LHI complex as adhesion factor for stacking (Varga & Staehelin, 1985*b*) has to be confirmed.

Although no distinct membrane protein has yet been shown to be a morphogenetic factor for molding the ICM, the composition and organization of integral membrane proteins should direct morphogenetic events during invagination and pattern formation, as shown for budding of enveloped viruses at the plasma membrane and formation of cell organelles in eukaryotic cells.

The influence of oxygen tension and light intensity on composition and formation of the ICM in facultative phototrophic bacteria.

The concentration of dissolved oxygen in the growth medium is the major external factor which determines via a signal chain the formation of the ICM in facultative phototrophic bacteria. It is unknown how the bacteria sense variations of oxygen partial pressure. Aerotaxis, i.e. movement towards an increasing oxygen gradient, is determined by the increase of proton motive force (Armitage, Ingham & Evans, 1985). Since, however, the facultative phototrophic bacteria can respire and produce an electrochemical gradient of protons across the CM from high to very low oxygen tensions, the Δ p does not seem to be the signal for ICM formation. The redox state of a constituent of the electron transport chain was proposed to be a link in the signal chain from the sensing system to the regulation of membrane differentiation (Cohen-Bazire, Sistrom & Stanier, 1957; Marrs & Gest, 1973). However, how a shift of oxygen tension is transduced into a transcriptional or post-transcriptional regulation of membrane differentiation has not been established. The threshold value of oxygen tension at which the formation of ICM is induced or inhibited is species-specific. *Erythrobacter spec.* OCH 114 and *Rhodobacter sulfidophilus* barely respond to variations of oxygen tension while *Rhodospirillum rubrum* or *Rhodobacter capsulatus* initiate the formation of ICM and of the photosynthetic apparatus below 200–500 Pa (=1.5–3.8 mm Hg) dissolved oxygen (Dierstein & Drews, 1974). Lowering the oxygen tension increased the rate of transcription initiation and the level of mRNA for some enzymes of bacteriochlorophyll and carotenoid synthesis about two- to three-fold (Yang & Bauer, 1990; Biel & Marrs, 1983; Hornberger, Wieseler & Drews, 1991; Guiliano *et al.*, 1988). The steady-state level of mRNA coding for pigment-binding proteins of the reaction centre and light-harvesting complexes increased much more than the level of mRNA of genes for pigment synthesis when the oxygen tension was lowered (Kiley & Kaplan, 1988). The operons for reaction centre and light-harvesting complex I (B870, LHI) proteins (*puf*QBAMLX), for the H-subunit of the reaction centre (*puh*H) and for LHII (B800–850) proteins (*puc*BA) are coordinately but not synchronously regulated (Klug, Kaufmann & Drews, 1985; Bauer, Young & Marrs, 1988, Berand, Bélanger & Gingras, 1989). The oxygen-regulated promoter of the *puf* operon was located about 700 bp upstream from the *puf B* structural gene (Bauer *et al.*, 1988; Adams *et al.*, 1989).

The coordination of syntheses of pigments and pigment-binding proteins is partially caused by clustering of genes encoding various components of the photosynthetic apparatus on the chromosome of *Rhodobacter capsulatus* and *Rhodobacter sphaeroides*. It has been shown that the three adjacent operons *crtEF*, *bchCA* and *pufQBAMLX* are co-transcribable (Young *et al.*, 1989). An open reading frame (*pufQ*) was detected coding for a polypeptide which is proposed to bind bacteriochlorophyll (Bchl) and functions as a carrier polypeptide. The *pufQ* gene is co-transcribed with other genes of the *puf* operon and may be another factor coordinating synthesis of Bchl and Bchl-binding proteins and assembly of the complexes (Bauer & Marrs, 1988). Other transcription factors as well as the rates of decay of the various segments of the polycistronic mRNA determine the steady state levels of mRNA for the genes of the photosynthetic apparatus (Adams *et al.*, 1989). A general control of gene expression after switching from aerobic to anaerobic growth conditions was proposed to involve DNA topology (Yamamoto & Droffner, 1985). The model predicted that DNA gyrase introduces into the chromosome negative supercoils which are essential for anaerobic gene expression. Experiments with *Rhodobacter capsulatus* could not detect changes in the superhelicity during metabolic transition (Cook *et al.*, 1989).

Besides a transcriptional regulation of photosynthetic genes, post-transcriptional events are involved in regulation of ICM differentiation. The fine regulation of key enzymes of the tetrapyrrole biosynthetic pathway by a feed-back mechanism has been concluded from results of *in vitro* experiments using inhibitors and activators of δ-aminolevulinate synthase (Lascelles, 1978). Unfortunately, the experiments were hampered by the lack of stable, purified enzyme preparations.

Variation of light intensity strongly effects formation of ICM. Lowering of light intensity under anaerobic conditions results in an increase of the amount of light-harvesting antenna Bchl per reaction centre, increasing the size of the photosynthetic unit. Bacteria like *Rhodobacter capsulatus*, which have two antenna complexes (LHI-B870 and LHII B800–850), increase the amount of LHII relative to the core complex consisting of reaction centre and LHI. The ratio of LHI to reaction centre remains relatively stable at about 30 mol LH Bchl per mol reaction centre. Besides an enlargement of the photosynthetic unit, the number of photosynthetic units per cell and the area of ICM per cell increase when the light intensity is decreased (Reidl *et al.*, 1985; Kiley & Kaplan, 1988). The concentration of cytochromes *b* and *c* and of ubiquinone also decreased at a lower light intensity (Garcia *et al.*, 1987). Cells of *Rhodobacter capsulatus*, grown anaerobically at a low intensity of light, contain about four times as much LHII complexes as cells grown under high light intensity. After the shift from high light to low light the level of mRNA for the *puc* operon increased transiently about 15-fold with a maximum after 50 min (Klug *et al.*, 1985). The steady state level of *pucBA* mRNA in high-light grown cells was about four times as great as in cells grown under low light intensity. However, the half-lives of *pucBA* mRNA were about 10 min in high light grown cells and approximately 19 min in cultures grown with low intensity light. It was concluded that the relative amount of LHII complexes is controlled by post-transcriptional processes (Zucconi & Beatty, 1988). A single base pair exchange within a sequence of dyad symmetry upstream of the *puf* promoter affects both the oxygen and the light regulation of the formation of reaction centre and LHI complexes in *Rhodobacter capsulatus*. It was concluded that oxygen and light variations act via different signal chains on the same regulatory DNA sequence (Klug *et al., 1991*).

The co-regulation of pigment proteins and ICM formation

The coordination of pigment protein and ICM synthesis have been observed very often but the molecular basis of coregulation remains unsolved (Kiley & Kaplan, 1988). In synchronized cultures of *Rhodobacter sphaeroides*, the phospholipid to protein ratio of ICM varies throughout the cell due to the continuous insertion of pigments and proteins into the ICM but incorporation of previously synthesized phospholipid into the membrane just prior to cell division (Kaplan *et al.*, 1983). The enzymes involved in phospholipid biosynthesis are localized in the CM (Cain *et al.*, 1984). The cell cycle-specific accumulation of phospholipid in the ICM was proposed to be regulated by controlling the movement of phospholipid into the ICM. ICM vesicles with a high protein to phospholipid ratio (isolated from cells prior to cell division) are better substrates for assaying phospholipid transfer *in vitro* than those isolated from cells just after cell division (Tai, Hoger & Kaplan, 1986; Kiley & Kaplan, 1988). Low oxygen tension and low light intensity increase whole cell phospholipid synthesis when compared with high-oxygen and high-light intensity grown cells. The incorporation of phospholipid into the membrane system seems to be uncoupled from phospholipid synthesis. The former process takes place during the course of cell division; the latter occurs throughout the cell cycle but is interrupted at the time of cell division (Cain *et al.*, 1984; Kiley & Kaplan, 1988).

The regulation of electron transport systems

The cyclic, light-driven electron transport system and the dark, linear respiratory chain share common constituents. In facultative phototrophic purple bacteria these are the cytochrome bc_1 oxidoreductase, ubiquinone as mobile electron carrier in the membrane and the free, mobile cytochrome c_2 which shuttles electrons from the bc_1-complex to the reaction centre and to cytochrome oxidase, respectively. The proton motive force generated by photosynthetic or respiratory electron transport is used by the same proton ATPase to generate ATP. Respiration is suppressed by strong illumination. Thus cells having both the photosynthetic and the respiratory system developed under low oxygen tension use only one of them for the production of an electrochemical gradient of protons. Optimum energy transduction needs a suitable redox balance of the redox carriers. Facultative phototrophic bacteria, living in ecological niches of frequently changing concentrations of substrates, oxygen tension and light intensity, have developed a finely tuned redox control of electron transport systems which allows electron transport under anaerobic light, aerobic dark and anaerobic dark conditions (Ferguson *et al.*, 1987). They have a branched electron transport system with several acceptor oxidoreductases. Some of the functional complexes are active under most of the conditions described above and they are constitutively synthesized. Others, like pigment proteins of the photosynthetic apparatus and cytochrome oxidase are synthesized in variable amounts. The concentrations of quinones and cytochromes are also regulated. High concentrations of cytochrome oxidase and potentially high rates of respiration have been observed in cells grown at low oxygen tension in the dark (Lampe & Drews, 1972; King & Drews, 1975; Cox *et al.*, 1983; Hüdig, Stark & Drews, 1987).

The differentiation of the membrane system

It has been mentioned above that two to three membrane fractions have been isolated from cells of facultative phototrophic bacteria which differ in specific density, composition and enzymic activities. ICM is formed in growing and non-growing cells. New material is incorporated at specific 'growing sites' of the membrane (Niederman *et al.*, 1979; Bowyer *et al.*, 1985). Although the molecular mechanism of membrane differentiation is unknown, numerous observations suggest that variations in oxygen tension, light intensity or substrate concentrations and other factors (Dierstein & Drews, 1974, 1975; Kiley & Kaplan, 1988; Oelze, 1988) induce via different signal chains transcriptional and post-transcriptional processes of gene expression or enzyme regulation resulting in a coordinated but differential synthesis of functional membrane-bound complexes. CM and ICM can incorporate different components at specific sites with variable rates (Bowyer *et al.*, 1985; Kaufman *et al.*, 1982). This might be regulated by the protein to lipid ratio of the membrane and specific targeting and translocation proteins at specific sites (Kiley & Kaplan, 1988). Nucleotide transport carriers are incorporated into CM and ICM (Carmeli & Lifshitz, 1989). The direction of transport of protons, ions and substrates was measured and the results confirmed the cytological observation that ICM vesicles (chromatophores) are inside-out vesicles (Konings & Michels, 1980). The incorporation of phospholipids into the membrane is correlated with the cell cycle. The relatively high protein to lipid ratio of CM and ICM and the possibly transient connections between ICM vesicles and CM may inhibit equilibration of all membrane constituents of the membrane system by lateral diffusion. Furthermore targeting of different constituents at specific membrane sites may cause diversity within the membrane system. Although the ICM is not a cell compartment, it can be considered as an early step in the evolution of compartmentalization.

The assembly process of membrane-bound functional complexes

Functional studies have shown that the supramolecular intramembrane particles, visualized by freeze–fracture electron microscopy (Golecki & Oelze, 1980; Varga & Staehelin, 1983, 1985*a*), are asymmetrical along the vertical axis to the plane of the membrane. This has been confirmed by high resolution X-ray studies on the photo-chemical reaction centre. The special pair of Bchl is localized in the periplasmic half of the membrane while the quinones A and B are in the cytoplasmic half of the membrane (Deisenhofer & Michel, 1989; Feher *et al.*, 1989). The strict orientation of polypeptide chains and cofactors in the membrane implies that the process of incorporation and assembly of constituents is highly ordered. This has been shown by studies of the formation of light-harvesting complexes in *Rhodobacter capsulatus*. Each subunit of LHI and LHII contains two of the low molecular weight polypeptides α and β (M_r αI 6455, βI 5341, αII 7322, βII 5087; Drews, 1985). The polypeptides α and β bind two or three, respectively, molecules of Bchl and one molecule of carotenoid. The basic subunits form oligomeric structures of about 80,000 to 100,000 (M_r (Shiozowa *et. al.*, 1982; Kramer *et al.*, 1984; Hunter *et al.*, 1988). The polypeptides have three domains. The amphiphilic N-terminal domain is exposed on the cytoplasmic surface of the membrane, but the C-termini are exposed on or point to the periplasmic space (Tadros *et al.*, 1987). A central hydrophobic and α-helical structured domain of about 20 amino acyl residues penetrates the lipid bilayer. The axes of the α-helices are slanting to the axis vertical to the plane of the membrane (Nabedryk & Breton, 1981). Bacteriochlorophyll molecules fit exactly between α-helical chains of the two Bchl-binding polypeptides. The central magnesium atom of Bchl is coordinately ligated to a conserved histidine residue (Deisenhofer & Michel, 1989; Lutz & Mäntele, 1990). Carbonyl functions of the tetrapyrrol ring interact with amino acyl residues in specific positions (Lutz & Mäntele, 1990). The exact adjustment of Bchl and polypeptides has been demonstrated by X-ray high resolution spectroscopy (Deisenhofer & Michel, 1989; Feher *et al.*, 1989), probing the Bchl binding site by reconstitution of LH complexes with Bchl analogues (Parkes-Loach *et al.*, 1990), exchange of Bchl by Bchl analogues in reaction centres (Struck *et al.*, 1990) or by replacement of amino acyl residues of the putative Bchl-binding site within the hydrophobic domain by site specific mutagenesis (Bylina, Robles & Youvan (1988). The formation of LH complexes depends on the coordinated synthesis and assembly of all constituents (Klug, Liebetanz & Drews, 1986; Dörge *et al.*, 1990). The precursor pools of free Bchl (Beck & Drews, 1982) and of free α, β proteins in the membrane and the cytoplasm, respectively, are very small (P. Richter, H. Stiehle, N. Cortez, G. Drews, unpublished observations). The Bchl-binding polypeptide β of LHI is inserted first into the membrane followed by the α polypeptide after about 2 min. No cleavable signal peptides are present. The N-terminal sequence of α polypeptides has a structure similar to signal peptides (Dalbey, 1990). It contains positively charged amino acyl residues near the N-terminus and close to the membrane-spanning α-helical central region interspersed by a stretch of hydrophobic amino acids. The N-terminus of β polypeptides is unlike a signal sequence. It is negatively charged and contains no hydrophobic domain of amino acyl residues (Dörge *et al.*, 1990; Stiehle *et al.*, 1990; Tadros & Drews, 1990). An exchange of four positively charged amino acids by negatively charged residues in α, or a substitution of the conserved Trp-8 of α by Ala resulted in a complete loss of the LHI from the membrane (Stiehle *et al.*, 1990; Richter, Cortez & Drews, 1991). Little or no α

polypeptide was incorporated into the membrane. The *β* polypeptide of these mutants was incorporated but did not remain stable. It disappeared from the membrane in about 10 min. If the gene for the *α* polypeptide was deleted, *β*, was inserted into the membrane but disappeared after about 10 min. It was then detected in the supernatant. Deletion of the gene for the *β* polypeptide inhibited incorporation of *α* into the membrane. (P. Richter & G. Drews, unpublished observations). The results show that the stable incorporation of *α* and *β* polypeptides into the membrane is dependent on both proteins and their intact N-terminal domains. Mutation of the *α* N-terminal domain inhibits the assembly process.

Similar N-terminal mutations in *β* have a less dramatic effect. An exchange of four charged amino acyl residues in the *β* N-terminal region for oppositely charged residues decreased the stability and the size of the LHI antenna but did not inhibit formation of LHI (Stiehle *et al.*, 1990). It is thought that the pigments are bound as soon as the polypeptides are translocated across the lipid layer. These results and other reports support the hypothesis that the *α* and *β* polypeptides interact during targeting and stable insertion. The process is proposed to be assisted by a membrane-bound translocation protein and possibly by soluble proteins which stabilize *β* in an unfolded or loosely folded state. The *α* polypeptide binds to the membrane at the same target site as *β* binds. Interactions of N-termini of *α* and *β* and the proton motive force are necessary for a stable translocation of the polypeptides *α* and *β* (Dierstein & Drews, 1986). Finally, binding of Bchl stabilizes the complex (Fig. 2).

Non-pigment binding proteins contribute to the assembly process of the LHII complex. Three open reading frames have been detected in the *puc* operon downstream of *puc BA* (coding for the pigment-binding polypeptides *α* and *β*; Tichy *et al.*, 1989). The gene products of *pucD* and *pucE* stabilize the B800–850 LHII complex. Deletion of *pucD* and *pucE* resulted in a fast decrease of the B800 absorption peak. Deletion of *pucC* inhibited completely the formation of LHII (Tichy *et al.*, 1991; in press).

The steady-state level of *puc* operon mRNA is about six times lower if *pucC* is deleted. The *pucC* product is a polypeptide of 403 amino acids with a high proportion of hydrophobic and charged amino acyl residues. It is assumed that PucC is involved in signal transduction (Tichy *et. al.*, 1991; in press). Post-transcriptional regulation of the LHII complex has also been observed in *R. sphaeroides* (Lee, Kiley & Kaplan, 1989). Experiments with a cell free transcription- translation system of *R. capsulatus* indicate that other membrane-bound and soluble polypeptides support the assembly process (Troschel & Müller, 1990 and unpublished observations).

In *Rhodobacter sphaeroides* an *R* gene product seems to control the synthesis or assembly of the LHI (B875) complex. The *R* gene has been mapped upstream of the *pufQ* gene in *R. sphaeroides* (Kiley & Kaplan, 1988). The *puhH* gene product is present in the CM of *R. sphaeroides* even when grown

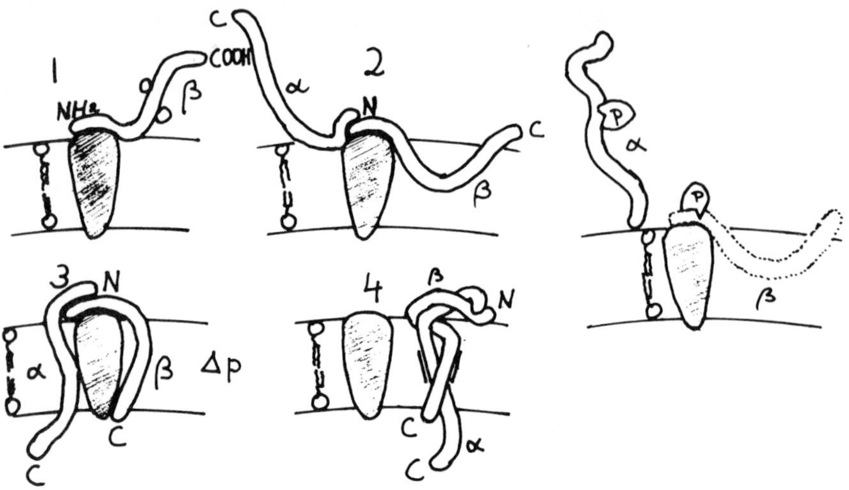

Fig. 2. Insertion and assembly of the pigment-binding proteins α and *β* of the LHI (B870) complex. Hypothetical scheme. (1) The *β* polypeptide binds by its N-terminus to a translocator protein in the membrane. Attached to *β* are chaperones inhibiting folding of *β*. (2) α polypeptide is docked to the same translocator at a different site with the N-terminal signal peptide. At the same time *a* and *β* interact. (3) The interaction between *a* and *β*, binding to the translocator and proton motive force are the prerequisites for translocation of *a* and *β* across the lipid layer. Lack of *a* or *β* inhibits the process of stable translocation. The single or mutated proteins are degraded by proteases. 4) Stable assembly of the LHI complex is established after translocation in the presence of bacteriochlorophyll. Mutated polypeptides are degraded by proteases (P).

under high oxygen conditions. It was proposed to anchor or to target the reaction centre polypeptides to the membrane (Donohue, Hoger & Kaplan, 1986).

Cytoplasmic and intracytoplasmic membranes of cyanobacteria

Cyanobacteria are characterized by a plant-like oxygenic photo-synthesis (Table 1). The photosynthetic apparatus seems to be present exclusively on thylakoids which are flat, sack-like lamellae containing both photosystems and phycobilisomes. One exception is *Gloeobacter* which does not possess thylakoids. Cyanobacteria perform aerobic respiration in the dark. There are contradictory results on the localization of the respiratory activity. High cytochrome oxidase activity was found in the thylakoid membrane (Omata & Murata, 1985) or in cytoplasmic and in intracytoplasmic membranes (Wastyn *et al.*, 1988; Pescheck, Kurz & Erber, 1988). The cytochrome oxidase is of the aa_3-type (Häfele, Scherer & Böger, 1988, 1989; Wastyn *et al.*, 1988). The cytoplasmic membrane was attributed as the major site of interaction and adaptation to stress signals from the surroundings (Khomutov *et al.*, 1990) and light-induced H^+ efflux and Na^+ (Kaplan, Scherer & Lerner, 1989).

It is unknown whether the thylakoids (ICM) are a membrane system which is independently synthesized from CM.

ACKNOWLEDGEMENT

Experiments of our laboratory described in this chapter were supported by grants of the Deutsche Forschungsgemeinschaft and Fond der Chemischen Industrie.

REFERENCES

Abee, T. Hellingwerf, K. J. & Konings, W. N. (1988). Effects of potassium ions on proton motive force in *Rhodobacter sphaeroides*. *Journal of Bacteriology*, **170**, 5647–53.

Abee, T., Palmen, R., Hellingwerf, K. S. & Konings, W. N. (1990). Osmoseregulation in *Rhodobacter sphaeroides*. *Journal of Bacteriology*, **172**, 149–54.

Abee, T., Van der Wal, F. J., Hellingwerf, K. J. & Konings, W. N., (1989). Binding-protein dependent alanine transport in *Rhodobacter sphaeroides* regulated by the internal pH. *Journal of Bacteriology*, **171**, 5148–54.

Adams, C. W. Forrest, M. E., Cohen, S. N. & Beatty, J. T. (1989). Structural and functional analysis of transcriptional control of the *Rhodobacter capsulatus puf* operon. *Journal of Bacteriology*, **171**, 473–82.

Armitage, J. P., Ingham, C. & Evans, M. C. W. (1985). Role of proton motive force in phototactic and aerotactic responses in *R. sphaeroides*. *Journal of Bacteriology*, **161**, 967–72.

Bauer, C. E., Young, D. A. & Marrs, B. L. (1988). Analysis of the *Rhodobacter capsulatus puf* operon. *Journal of Biological Chemistry*, **263**, 4820-7.

Bauer, C. E. & Marrs, B. L. (1988). *Rhodobacter capsulatus puf* operon encodes a regulatory protein (PufQ) for bacteriochlorophyll synthesis. *Proceedings of the National Academy of Sciences, USA*, **85**, 7074-8.

Beck, J. & Drews, G. (1982). Tetrapyrrol derivatives shown by fluorescence emission and excitation spectroscopy in cells of *Rhodopseudomonas capsulata* adapting to phototrophic conditions. *Zeitschrift für Naturforschung*, **37c,** 199–204.

Berand, J., Bélanger, G. & Gingras, G. (1989). Mapping of the *puh* messenger RNAs from *Rhodospirillum rubrum*. *Journal of Biological Chemistry*, **264**, 10897–903.

Biel, A. J. & Marrs, B. L. (1983). Transcriptional regulation of several genes for bacteriochlorophyll biosynthesis in *Rhodopseudomonas capsulata* in response to oxygen. *Journal of Bacteriology*, **156,** 686–94.

Bock, E., Koops, H.-P., Harms, H. & Ahlers, B. (1991). The biochemistry of nitrifying organisms. In: *Variation in Autotrophic Life*, eds. J. M. Shively, L. L. Barton, pp. 171–200, *Academic Press*, New York.

Bowyer, J. R., Hunter, C. N., Ohnishi T. & Niederman, R. A. (1985). Photosynthetic membrane development in *Rhodopseudomonas sphaeroides*. *Journal of Biological Chemistry*, **260**, 3295–304.

Bylina, E. J., Robles, S. J. & Youvan, D. C. (1988). Directed mutations affecting the putative bacteriochlorophyll-binding sites in the light-harvesting antenna of *Rhodobacter capsulatus*. *Israel Journal of Chemistry*, **28**, 73-8.

Cain, B. D., Donohoe, T. J., Shephard, W. D. & Kaplan, S. (1984). Localization of phospholipid biosynthetic enzyme activities in cell-free fractions derived from *Rhodopseudomonas sphaeroides*. *Journal of Biological Chemistry*, **259**, 942-8.

Carmeli, C. & Lifshitz; Y. (1989). Nucleotide transport in *Rhodobacter capsulatus*. *Journal of Bacteriology*, **171**, 6521–6525.

Cohen-Bazire, G., Sistrom, W. R. & Stanier, R. Y. (1957). Kinetic studies of pigment synthesis by non-sulfur purple bacteria. *Journal of Cellular Comparative Physiology*, **49**, 25–35.

Cook, D. N., Armstrong, G. A. & Hearst, J. E. (1989). Induction of anaerobic gene expression in *Rhodobacter capsulatus* is not accompanied by a local change in chromosomal supercoiling as measured by a novel assay. *Journal of Bacteriology*, **171**, 4836–43.

Cox, J. C., Beatty, J. T. & Favinger, J. L. (1983). Increased activity of respiratory enzymes from photosynthetically grown *Rhodopseudomonas capsulata* in response to small amounts of oxygen. *Archives of Microbiology*, **134**, 324-8.

Crosa, J. H. (1989). Genetics and molecular biology of siderophore-mediated iron transport in bacteria. *Microbial Review*, **53**, 517–30.

Dalbey, R. E. (1990). Positively charged residues are important determinants of membrane protein topology. *Trends in Biochemical Sciences*, **15**, 253-7.

Deisenhofer, J. & Michel, H. (1989). The photosynthetic reaction center from the purple bacterium *Rhodopseudomonas viridis*. *Science*, **245**, 1463–73.

Dierstein, R. & Drews, G. (1974). Nitrogen-limited continuous culture of *Rhodopseudomonas capsulata* growing photosynthetically or heterotrophically under low oxygen tensions. *Archives of Microbiology*, **99**, 117–28.

Dierstein, R. & Drews, G. (1975). Control of composition and activity of the photosynthetic apparatus of *Rhodopseudomonas capsulata* grown in ammonium-limited continuous culture. *Archives of Microbiology*, **106**, 227–35.

Dierstein, R. & Drews, G. (1986). Effect of uncoupler on assembly pathway for pigment-binding protein of bacterial photosynthetic membranes. *Journal of Bacteriology*, **168**, 167–72.

Dierstein, R. Schuhmacher, A. & Drews, G. (1981). On insertion of pigment-associated polypeptides during membrane biogenesis in *Rhodopseudomonas capsulata*. *Archives of Microbiology*, **128**, 376–83.

Donohue, T. J., Hoger, J. H. & Kaplan, S. (1986). Cloning and expression of the *Rhodobacter sphaeroides* reaction center H gene. *Journal of Bacteriology*, **168**, 953–61.

Donohue, T. J. & Kaplan, S. (1986). Synthesis and assembly of bacterial photosynthetic membranes. In *Photosynthesis III* eds. L. A. Staehelin, C. J. Arntzen, *Encyclopedia of Plant Physiology*. **vol 19**, pp. 632-9.

Dörge, B., Klug, G. Gadon, N., Cohen, S. & Drews, G. (1990). Effects on the formation of antenna complex B870 of *Rhodobacter capsulatus* by exchange of charged amino acids in the N-terminal domain of the α and β pigment-binding proteins. *Biochemistry*, **29**, 7754-8.

Drews, G. (1985). Structure and functional Organization of light-harvesting complexes and photochemical reaction centers in membranes of phototrophic bacteria. *Microbiological Reviews*, **49**, 59–70.

Drews, G. (1991). Regulated development of the photosynthetic apparatus in anoxigenic bacteria. In *The Molecular Biology of Plastids and Mitochondria*, eds. L. Bogorad, I. K. Vasil, pp. 114–48, *Academic Press*, New York.

Drews, G. & Giesbrecht, P. (1963). Zur Morphogenese der Bakterien Chromatophoren und zur Synthese des Bakteriochlorophylls bei *Rhodopseudomonas sphaeroides* und *Rhodospirillum rubrum*. *Zentralblatt für Bakteriologie, Parasitenkunde, Infektionskrankheiten und Hygiene*, 1 Original **190**, 508–36.

Drews, G. & Giesbrecht, P. (1981). Die Strukturelemente der Prokaryotenzelle. In *Die Zelle*, ed. H. Metzner, 3rd edn., pp. 471–541, Stuttgart, *Wissenschaftliche Verlagsgesellschaft*.

Drews, G. & Imhoff, J. F. (1991). Phototrophic purple bacteria. In *Variations in*

Autotrophic Life, eds. J. M. Shively & L. L. Barton, pp. 51–97, *Academic Press*, New York.

Drews, G. & Oelze, J. (1981). Organization and differentiation of membranes of phototrophic bacteria. *Advances in Microbial Physiology*, **22**, 1-92.

Epstein, W. (1986). Osmoregulation by potassium transport in *Eschericia coli. FEMS Microbiological Review*, **39**, 73-8.

Feher, G., Allen, J. P. Okamura, M. Y. & Rees, D. C. (1989). Structure and function of bacterial photosynthetic reaction centers. *Nature*, London. **339**, 111–16.

Ferguson, S. J., Jackson, J. B. & McEwan, A. G. (1987). Anaerobic respiration in the *Rhodospirillacaeae*: characterization of pathways and evaluation of roles in redox balancing during photosynthesis. *FEMS Microbiology Reviews*, **46**, 117–43.

Ferro-Luzzi Ames, G. & Joshi, A. K. (1990). Energy coupling in bacterial periplasmic permeases. *Journal of Bacteriology*, **172**, 4133–7.

Ferro-Luzzi Ames, G., Mimura; C. S. & Shyamala, V. (1990). Bacterial periplasmic permeases belong to a family of transport proteins operating from *Escherichia coli* to human: traffic ATPases. *FEMS Microbiology Reviews*, **75**, 429–46.

Garcia, A. F., Venturoli, G., Gadon, N., Fernandez-Velasco, J. G., Melandri, B. A. & Drews, G. (1987). The adaptation of the electron transfer chain of *Rhodopseudomonas capsulata* to different light intensities. *Biochimica et Biopysica Acta*, **890**, 335–45.

Golecki, J. R. & Oelze, J. (1975). Quantitative determination of cytoplasmic membrane invaginations in phototrophically growing *Rhodospirillum rubrum. Journal of General Microbiology*, **88**, 253-8.

Golecki, J. R. & Oelze, J. (1980). Differences in the architecture of cytoplasmic and intracytoplasmic membranes of three chemotrophically and phototrophically grown species of the*Rhodospirillaceae. Journal of Bacteriology*, **144**, 781-8.

Golecki, J. R., Tadros, M. H., Ventura, A. S. & Oelze, J. (1989). Intracytoplasmic membrane vesiculation in light-harvesting mutants of *Rhodobacter sphaeroides* and *Rhodobacter capsulatus. FEMS Microbiology Letters*, **65**, 315–18.

Golecki, J. R. & Drews, G. (1982). Supramolecular organization and composition of membranes. In *The Biology of Cyanobacteria*, eds. N. G. Carr & B. A. Whitton, pp. 125–41, Oxford, London, Edinburgh, Boston, Melbourne, Blackwell Scientific Publication.

Golecki, J. R., Schumacher, A. & Drews, G. (1980). The differentiation of the photosynthetic apparatus and the intracytoplasmic membrane in cells of *Rhodopseudomonas capsulata* upon variation of light intensity. *European Journal of Cell Biology*, **23**, 1-5.

Gottschalk, G. (1986). *Bacterial Metabolism*. Springer Publication, New York, Berlin, Heidelberg.

Greenawalt, J. W. & Whiteside, T. L. (1975). Mesosomes, membranes bacterial organelles. *Bacteriological Review*, **39**, 405–63.

Gross, R., Arico, B. & Rappuoli, Q. (1989). Families of bacterial signal-transducing proteins. *Molecular Microbiology*, **3**, 1661–7.

Guiliano, G., Pollock, D., Stapp, H. & Scolnik, P. (1988). A genetic-physical map of the *Rhodobacter capsulatus* carotenoid biosynthesis gene cluster. *Molecular General Genetics*, **213**, 78–83.

Häfele, U., Scherer, S. & Böger, P. (1988). Cytochrome aa_3 from heterocysts of the cyanobacterium *Anabaena variabilis*: isolation and spectral characterization. *Biochimica et Biophysica Acta*, **934**, 186–90.

Häfele, U., Scherer, S. & Böger, P. (1989). Cytochrome c oxidase of the cyanobacterium *Phormidium foveolarum. Zeitschrift für Naturforschung*, **44c**, 378–82.

Higgins, I. J., Best, D. J., Hammond, R. C. & Scott, D. (1981). Methanoxidizing microorganisms. *Microbiological Reviews* **45**, 556–90.

Higgins, C. F., Cairney, J., Stirling, D. A., Sutherland, L. & Booth, I. K. (1987). Osmotic regulation of gene expression: ionic strength as an intracellular signal? *Trends in Biochemical Sciences*, **12**, 339–44.

Hornberger, U., Wieseler, B. & Drews, G. (1991). Oxygen-tension regulated expression of the *hem*A gene of *Rhodobacter capsulatus. Archives of Microbiology*, in press.

Hüdig, H., Stark, G. & Drews, G. (1987). The regulationof cytochrome oxidase of *Rhodobacter capsulatus* by light and oxygen. *Archives of Microbiology*, **149**, 12–18.

Hunter, C. N., Pennoyer, J. D., Sturgis, J. N., Farrelly, D. & Niedermann, R. A. (1988). Oligomerization states and associations of light-harvesting pigment-protein complexes of *Rhodobacter sphaeroides* as analysed by lithium dodecylsulfate-polyacrylamide gel electrophoresis. *Biochemistry*, **27**, 3459–67.

Hurlbert, R. E., Golecki, J. R. & Drews, G. (1974). Isolation and characterization of *Chromatium vinosum* membranes. *Archives of Microbiology*, **101**, 169–86.

Kabak, H. R. (1990). The *lac* permease of *Escherichia coli*: a prototypic energy-transducing membrane protein. *Biochimica et Biophysica Acta*, **1018**, 160-2.

Kaplan, S. (1978). Control and kinetics of photosynthetic membrane development. In *The Photosynthetic Bacteria*, pp. 809–39, eds. R. K. Clayton, W. R. Sistrom, *Plenum Press*, New York.

Kaplan, S., Cain, B. D., Donohue, T. J., Shephard, W. D. & Yen, G. S. L. (1983). Biosynthesis of the photosynthetic membranes of *Rhodopseudomonas sphaeroides. Journal of Cellular Biochemistry*, **22**, 15–29.

Kaplan, A., Scherer, S., & Lerner, M. (1989). Nature of the light-induced H^+ efflux and Na^+ uptake in cyanobacteria. *Plant Physiology*, **89**, 1220-5.

Kaufmann, N. Reidl, H.-H., Golecki, J. R., Garcia, A. F., & Drews, G. (1982). Differentiation of the membrane system in cells of *Rhodopseudomonas capsulata* after transition from chemotrophic to phototrophic growth conditions. *Archives of Microbiology*, **131**, 313–22.

Khomutov, G., Fry, I. V., Huflejt, M. E. & Packer, L. (1990). Membrane lipid composition, fluidity and surface charge changes in response to growth of the fresh water cyanobacterium *Synechococcus* 6311 under high salinity. *Archives of Biochemistry and Biophysics*, **277**, 263-7.

Kiley, P. J. & Kaplan, S. (1988). Molecular genetics of photosynthetic membrane biosynthesis in *Rhodobacter sphaeroides. Microbiological Reviews*, **52**, 50–69.

Kiley, P. J., Varga, A. & Kaplan, S. (1988). Physiological and structural analysis of light-harvesting mutants of *Rhodobacter sphaeroides. Journal of Bacteriology*, **170**, 1103–15.

King, M. T. & Drews, G. (1975). The respiratory electron transport system of heterotrophically-grown *Rhodopseudomonas palustris. Archives of Microbiology*, **102**, 219–31.

Klug, G., Gad'on, N., Jock, S. & Narro, M. L. (1991). Light and oxygen effects share a common regulatory DNA sequence in *Rhodobacter capsulatus. Molecular Microbiology*, in press.

Klug, G., Kaufmann, N. & Drews, G. (1985). Gene expression of pigment-binding proteins of the bacterial photosynthetic apparatus. Transcription and assembly in the membrane of *Rhodopseudomonas capsulata. Proceedings National Academy of Sciences, USA*, **82**, 6485–9.

Klug, G., Liebetanz, R. & Drews; G. (1986). The influence of bacteriochlorophyll biosynthesis on formation of pigment-binding proteins and assembly of pigment protein complexes in *Rhodopseudomonas capsulata*. *Archives of Microbiology*, **146**, 284–91.

Konings, W. N. & Michels, P. A. M. (1980). Electron transfer-driven solute translocation across bacterial membranes. In *Diversity of Bacterial respiratory Systems*, ed. C. J. Knowles, Vol. I, pp. 33–86, *CRC Press*, Boca Raton, FL.

Kramer, H. J. M., Van Grondelle, R., Hunter, C. N., Westerhuis, W. H. & Amesz, J. (1984). Pigment organization of the B800–850 antenna complex of *Rhodopseudomonas sphaeroides*. *Biochimica et Biophysica Acta*, **765**, 156–65.

Krulwich, T. A. & Guffanti, A. A. (1989). The Na^+ cycle of extreme alkalophiles: a secondary Na^+/H^+ antiporter and Na^+/solute symporters. *Journal of Bioenergetics and Biomembranes*, **21**, 663–77.

Lampe, H.-H & Drews, G. (1972). Die Differenzierung des Membransystems von *Rhodopseudomonas capsulata* hinsichtlich seiner photosynthetischen und respiratorischen Funktionen. *Archives of Microbiology*, **84**, 1–19.

Lascelles, J. (1978). Regulation of pyrrole synthesis. In *The Photosynthetic Bacteria*, eds. R. K. Clayton, W. R. Sistrom, pp. 795–808. *Plenum Press*, New York.

Lee, J. K. Kiley, P. J. & Kaplan, S. (1989). Post-transcriptional control of *puc* operon expression of B800–850 light-harvesting complex formation in *Rhodobacter sphaeroides*. *Journal of Bacteriology*, **171**, 3391–405.

Lütters-Czekalla, S. (1990). Lithoautotrophic growth of the iron bacterium *Gallionella ferruginea* with thiosulfate or sulfide as energy source. *Archives of Microbiology*, **154**, 417–21.

Lütters, S. & Hanert, H. H. (1989). The ultrastructure of chemo-lithoautotrophic *Gallionella ferruginea* and *Thiobacillus ferroxidans* as revealed by chemical fixation and freeze-etching. *Archives of Microbiology*, **151**, 245–251.

Lutz, M. & Mäntele, W. (1990). Vibrational spectroscopy of chlorophyll. In *Chlorophylls*, ed. H. Scheer, pp. 855–902. *CRC Press, Boca Raton*, FL.

Marrs, B. L. & Gest, H. (1973). Regulation of bacteriochlorophyll synthesis by oxygen in respiratory mutants of *Rhodopseudomonas capsulata*. *Journal of Bacteriology*, **114**, 1052–7.

Matin, A. (1990). Bioenergetics parameters and transport in obligate acidophiles. *Biochimica et Biophysica Acta*, **1018**, 267–70.

Mayer, F. (1986). Cytology and morphogenesis of bacteria. vol VI, 2, *Encyclopedia of Plant Anatomy*, eds. H. J. Braun, S. Carlquist, P. Ozenda, I. Roth, *Gebrüder Bornträger*, Berlin, Stuttgart.

Meadow, N. D., Fox, D. K., & Roseman, S. (1990). The bacterial phosphoenol–pyruvate–glucose phosphotransferase system. *Annual Review of Biochemistry*. **59**, 497–542.

Mizuno, T. & Mizushima, S. (1990). Signal transduction and gene regulation through the phosphorylation of two regulatory components: the molecular basis for the osmotic regulation of the porin genes. *Molecular Microbiology*, **4**, 1077–82.

Nabedryk, E. & Breton, J. (1981). Orientation of intrinsic proteins in photosynthetic membranes. Polarized infrared spectroscopy of chloroplasts and chromatrophores. *Biochimica et Biophysica Acta*, **635**, 5115–24.

Niedermann, P. A. & Gibson, K. D. (1978). Isolation and physico-chemical properties of membranes from purple photosynthetic bacteria. In *The Photosynthetic Bacteria*, eds. R. K. Clayton, W. R. Sistrom, pp. 79–118. *Plenum Press*. New York.

Niedermann, P. A., Mallon, D. E. & Parks, L. C. (1979). Isolation of a fraction enriched in newly synthesized bacteriochlorophyll a-protein complexes. *Biochimica et Biophysica Acta*, **555**, 210–20.

Oelze, J. (1988). Regulation of tetrapyrrole synthesis by light in chemostat cultures of *Rhodobacter sphaeroides*. *Journal of Bacteriology*, **170**, 4652–7.

Oelze, J. & Drews, G. (1981). Membranes of phototrophic bacteria. In *Organization of prokaryotic cell membranes*, ed. B. K. Ghosh, vol. II, pp. 131–95, *CRC Press*, Boca Raton, FL.

Olson, J. M. Ormerod, J. G., Amesz, J., Stackebrandt, E. & Trüper, H. G. eds. (1988). *Green Photosynthetic Bacteria*, *Plenum Press*, New York.

Omata, T. & Murata, N. (1985). Electron-transport reactions in cytoplasmic and thylakoid membranes prepared from the cyanobacteria *Anacystis nidulans* and *Synechocystis* PCC6714. *Biochimica et Biophysica Acta*, **810**, 354–61.

Parkes-Loach, P. S., Michalski, T. J., Bass, W. J., Smith, U. & Loach, P. A. (1990). Probing the bacteriochlorophyll binding site by reconstitution of the light-harvesting complex of *Rhodospirillum rubrum* with bacteriochlorophyll a analogues. *Biochemistry*, **29**, 2951–60.

Peschek, G. A., Kurz, M. A. & Erber, W. W. A. (1988). Impermeant electron acceptors and donors to the plasma membrane-bound respiratory chain of intact cyanobacterium *Anacystis nidulans*. *Physiologia Plantarum*, **73**, 175–81.

Post, E., Golecki, J. R. & Oelze, J. (1982). Morphological and ultrastructural variations in *Azotobacter vinelandii* growing in oxygen-controlled continuous culture. *Archives of Microbiology*, **133**, 75–82.

Post, E., Vakalopoulou, E. & Oelze, J. (1983). On the relationship of intracytoplasmic to cytoplasmic membranes in nitrogen-fixing *Azotobacter vinelandii*. *Archives of Microbiology*, **134**, 265-9.

Reidl, H.-H., Golecki, J. R., & Drews, G. (1985). Composition and activity of the photosynthetic system of *Rhodopseudomonas capsulatus*. *Biochimica et Biophysica Acta*, **808**, 328–33.

Remsen, C. C. (1978). Comparative subcellular architecture of photosynthetic bacteria. In *The Photosynthetic Bacteria*, R. K. Clayton, W. R. Sistrom, eds., pp. 31–60, *Plenum Press*, New York.

Richter, P., Cortez, N. & Drews, G. (1991). Possible role of the highly conserved amino acids Trp-8 and Pro-13 in the N-terminal segment of the pigment-binding polypeptide LHIα of *Rh. capsulatus*. *FEBS Letters*, **285**, 80–4.

Rogers, H. J., Perkins, H. R. & Ward, J. B. (1980). *Microbial Cell Walls and Membranes*, *Chapman & Hall*, London, New York.

Rohmer, M., Bouvier-Nave, P. & Ourisson, G. (1984). Distribution of hopanoid triterpenes in prokaryotes. *Journal of General Microbiology*, **130**, 1137–50.

Salton, M. (1978). Structure and function of bacterial plasma membranes, In: *Relations between Structures and Function in the Prokaryotic Cell*, eds R. Y. Stanier, H. J. Rogers & B. J. Ward, pp. 201–23, *Cambridge University Press*, Cambridge.

Shapiro, L. (1985). Generation of polarity during *Caulobacter* cell differentiation. *Annual Reviews of Cell Biology*, **1**, 175–207.

Shiozowa, J. A., Welte, W., Hodapp & N., Drews, G. (1982). Studies on the size and composition of the isolated light-harvesting B800–850 pigment-protein complex of *Rhodopseudomonas capsulata*. *Archives of Biochemistry and Biophysics*, **213**, 473–85.

Sprague, S. G. & Varga, A. R. (1986). Membrane architecture of anoxygenic photosynthetic bacteria. In *Photosynthesis* III, eds L. A. Staehelin, C. J. Arntzen, pp. 603–19. *Springer Verlag*, Berlin, Heidelberg.

Stiehle, H., Cortez, N., Klug, C. & Drews, G. (1990). A negative charged N-terminus in the α polypeptide inhibits formation of light-harvesting complex I in *Rhodobacter capsulatus*. *Journal of Bacteriology*, **172**, 7131–7.

Stock, J. B., Stock, A. M. & Mottonen, J. M. (1990). Signal transduction in bacteria. *Nature*, London, **344**, 395–400.

Struck, A., Cmiel, E., Katheder, I. & Scheer, H. (1990). Modified reaction centers from *Rhodobacter sphaeroides* R 26:2: Bacteriochlorophylls with modified C-3 substituents at sites B_A and B_B. *FEBS Letters*, **268**, 180–184.

Sundermeyer-Klinger, H., Meyer, W., Warninghoff, B. & Bock, E. (1984). Membrane-bound nitrite oxidoreductase of *Nitrobacter*. *Archives of Microbiology*, **140**, 153-8.

Tadros, M. H. & Drews, G. (1990). Pigment-proteins of antenna complexes from purple non-sulfur bacteria: localization in the membrane, alignments of primary structure and structural predictions. In *Molecular Biology of Membrane Bound Complexes in Phototrophic Bacteria*, eds. G. Drews & E. A. Dawes, pp. 181–192, *Plenum Press*, New York.

Tadros, M. H., Frank, R., Dörge, B., Gad'on, N., Takemoto, J. Y. & Drews, G. (1987). Orientation of the B800–850, B870 and reaction center polypeptides on the cytoplasmic and periplasmic surfaces of *Rhodobacter capsulatus* membranes. *Biochemistry*, **26**, 7680–7.

Tai, S. P., Hoger, J. H. & Kaplan, S. (1986). Phospholipid transfer activity in synchronous populations of *Rhodobacter sphaeroides*. *Biochimica et Biophysica Acta*, **859**, 198–208.

Tai, S. P. & Kaplan, S. (1985). Intracellular localization of phospholipid transfer activity in *Rhodopseudomonas sphaeroides* and a possible role in membrane biogenesis. *Journal of Bacteriology*, **164**, 181-6.

Takemoto, J. & Bachmann, R. C. (1979). Orientation of chromatophores and spheroplast derived membrane vesicles of *Rhodopseudomonas sphaeroides*. Analysis by localization of enzyme activities. *Archives of Biochemistry and Biophysics*, **195**, 526–34.

Tauschel, H.-D. & Drews, G. (1967). Thylakoidmorphogenese bei *Rhodopseudomonas palustris*. *Archiv für Mikrobiologie*, **59**, 381–404.

Tichy, H. V., Oberlé, B., Stiehle, H., Schiltz, E. & Drews, G. (1989). Genes downstream from *pucB* and *puc*A are essential for formation of the B800–850 complex of *Rhodobacter capsulatus*. *Journal of Bacteriology*, **171**, 4914–22.

Troschel, D. & Müller, M. (1990). Development of a cell-free system to study the membrane assembly of photosynthetic proteins of *Rhodobacter capsulatus*. *Journal of Cell Biology*, **111**, 87–94.

Varga, A. R., & Staehelin, A. L. (1983). Spatial differentiation in photosynthetic and non-photosynthetic membranes of *Rhodopseudomonas palustris*. *Journal of Bacteriology*, **154**, 1414–30.

Varga, A. R. & Staehelin, L. A. (1985a). Pigment-protein complexes from *Rhodopseudomonas palustris*. Isolation, characterization and reconstitution into liposomes. *Journal of Bacteriology*, **161**, 921–7.

Varga, A. R. & Staehelin, L. A. (1985*b*). Membrane adhesion in photosynthetic bacterial membranes. Light-harvesting complex I appears to be the main adhesion factor. *Archives of Microbiology*, **141**, 290–6.

Wastyn, M. Achatz, A. Molitor, V. & Peschek, G. A. (1988). Respiratory activities and aa_3-type cytochrome oxidase in plasma and thylakoid membranes from vegetative cells and heterocysts of the cyanobacterium *Anabaena* ATCC29413. *Biochimica et Biophysica Acta*, **935**, 217–24.

Watson, S. W., Valois, F. W. & Aterbury, J. B. (1981). The family nitrobacteraceae. In *The Prokaryotes*, eds. M. P. Starr, H. Stolp, H. G. Trüper, A. Balows, H. G. Schlegel, vol. 1, pp. 1005–22. *Springer Verlag*, Berlin, Heidelberg, New York.

Whittenbury, R. & Krieg, N. R. (1984). Methylococcaceae. In *Bergey's Manual of Systematic Bacteriology*, eds. N. R. Krieg & J. G. Holt, vol. 1, pp. 256–261, *Williams & Wilkins*, Baltimore, London.

Yamamoto, N. & Droffner, M. L. (1985). Mechanisms determining aerobic or anaerobic growth in the facultative anaerobe *Salmonella typhimurium. Proceedings of the National Academy of Sciences, USA*, **82**, 2077–81.

Yang, Z., & Bauer, C. E. (1990). *Rhodobacter capsulatus* genes involved in early steps of bacteriochlorphyll biosynthetic pathway. *Journal of Bacteriology*, **172**, 5001–10.

Young, D. A., Bauer, C. E. Williams, J. A. C. & Marrs, B. L. (1989). Genetic evidence for superoperonal organization of genes for photosynthetic pigments and pigment-binding proteins in *Rhodobacter capsulatus. Molecular General Genetics*, **218**, 1-12.

Zucconi, A. P. & Beatty, J. T. (1988). Post-transcriptional regulation by light of the steady-state levels of mature B800–850 light-harvesting complexes in *Rhodobacter capsulatus. Journal of Bacteriology*, **170**, 877–82.

COMPARTMENTALIZED GENE EXPRESSION DURING *BACILLUS SUBTILIS* SPORULATION

ANNE MOIR

Department of Molecular Biology and Biotechnology, University of Sheffield, Sheffield S10 2UH, UK

THE SPORULATION CYCLE

The formation of the *Bacillus* endospore is the best characterised of bacterial developmental processes. When the supply of an essential nutrient is exhausted, in post-exponential phase, the cell activates a number of stationary phase responses, the most extreme of which is to initiate sporulation (Sonenshein, 1989). This is a complex process, involving extensive *de novo* protein synthesis and taking 7–8 hours at 37 °C. As sporulation can be induced relatively synchronously, it has been possible to correlate the timing of biochemical and morphological changes in the sporulating cell (Ryter, Schaeffer & Ionesco, 1966). The subject has been extensively reviewed (for example, Young & Mandelstam, 1979; Losick, Youngman & Piggot, 1986, Mandelstam & Errington, 1987; Losick & Kroos, 1989). This chapter reviews the basic process in brief, but concentrates on aspects of the process related to forespore-specific expression.

The sporulation process has been classified, for convenience, into a series of morphological stages (stages 0–VII), each of which takes approximately 1 hour at 37 °C. These are represented diagrammatically in Fig. 1, whereas Fig. 2 shows electron micrographs of cells at various stages of sporulation. A large number of mutants blocked in the sporulation process (*spo* mutants) have been characterised in terms of their morphology, physiology and genetics (classified by Piggot & Coote, 1976). The mutations in these strains lie in genes whose products are required for the successful formation of a heat-resistant spore, and the genes are named on the basis of the morphological stage at which sporulation is blocked (*spo0*, *spoII*, *spoIII* genes, etc). As genetic mapping was pursued, the nomenclature was extended; as a number of *spoII* mutations, for example, were located at different positions on the genetic map of *B. subtilis*, the loci were distinguished by an additional letter – the first mapped was called *spoIIA* the next *spoIIB*, etc. (Piggot & Coote, 1976). With the advent of gene cloning it became easier to distinguish individual genes; for example the *spoIIA* locus proved to be an operon of three genes – these are now called *spoIIAA*, *spoIIAB* and *spoIIAC*, in order along the operon.

Stage 0 is defined as the initiation of sporulation. Mutations blocking the process at this stage have identified genes whose products are essential

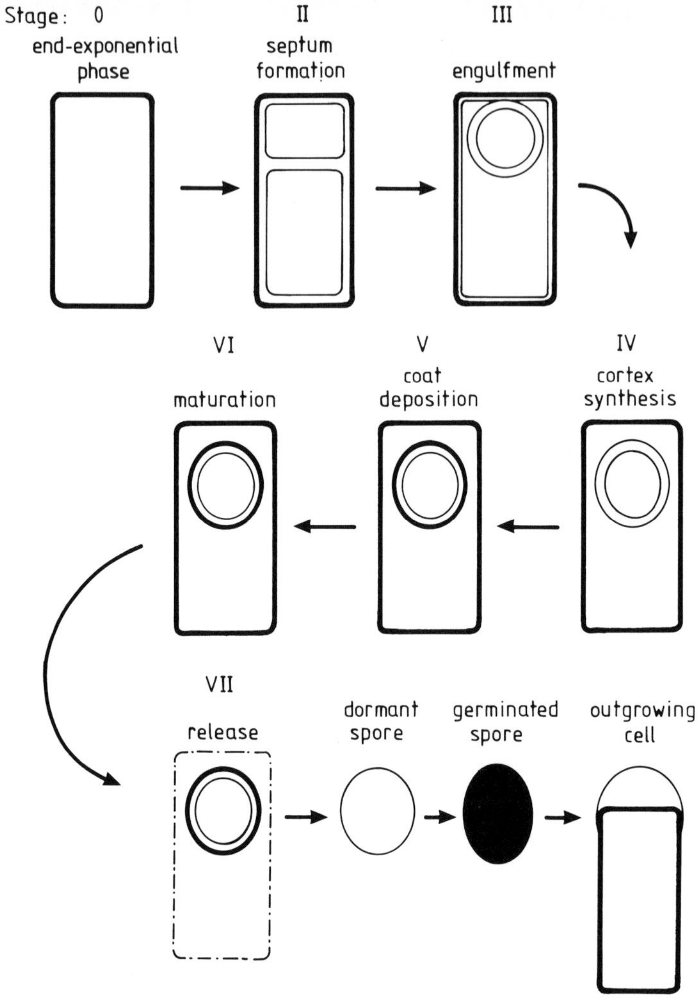

Fig. 1. Morphological stages of the sporulation and germination cycle of *Bacillus* species.

for entry into the sporulation pathway. Stage I, defined from electron microscopy as an apparent condensation of the replicated chromosome to form an axial filament, has not been recognised as a distinct stage identifiable in blocked mutants. On entry to sporulation, the normal cell cycle is replaced by a specialised cell division, in which a septum forms, not at the normal central position, but towards one pole of the cell (Stage II). A variety of Stage II mutants, blocked in formation of a normal polar septum, have been described (Piggot & Coote, 1976), although their morphological phenotypes have recently been reassessed (Illing & Errington, 1991). The products of the asymmetric cell division are two cellular compartments of unequal volume, each containing one copy of the genome. The smaller forespore (also called prespore) compartment is destined to form the cellular core of the spore, whereas the larger mother cell compartment is mortal; it contributes to spore development but is lysed at the end of the sporulation process. The sporulation septum is distinct from the normal cell division septum in more than location; at first it contains a small amount of peptidoglycan, but this is rapidly digested away (Holt, Gauthier & Tipper, 1975; Fig. 2). The mother cell septal membrane continues to grow, migrating round the prespore compartment and eventually engulfing it (Stage III). The *spoIII* mutants generally complete engulfment but do not progress further (Piggot & Coote, 1976). As discussed in a previous volume of this series, the result of engulfment is a forespore compartment separated from the mother cell cytoplasm by two membranes of opposite polarity (Ellar, 1978). This has major implications for transmembrane processes such as energy generation and transport; the mechanism of calcium uptake into the spore, for example, has been characterised as a two-stage process in which Ca^{2+} is accumulated in the mother cell by active transport from the surrounding medium, followed by uptake into the forespore by facilitated diffusion (Ellar, 1978). The biochemistry of the forespore has been reviewed more recently by Setlow (1981). After engulfment, peptidoglycan is synthesised across the forespore membranes: vegetative type forms the germ cell wall, which will form the cell wall of the germinated spore, whereas cortex peptidoglycan, less extensively crosslinked and containing muramic lactam residues, is essential to dehydration of the spore and the development of heat resistance (Warth, 1978) – it is degraded during germination (Foster & Johnstone, 1990). There is evidence that cortex-type peptidoglycan is synthesised across the outer forespore membrane and the germ cell wall across the inner forespore membrane (Tipper & Linnett, 1976, discussed in Ellar, 1978). Formation of the electron-transparent cortex is considered morphological Stage IV. The *spoIV* mutants form cortex but do not proceed further. At the same time as cortex synthesis, dipicolinic acid (DPA) is synthesised in the mother cell and transported into the spore (Andreoli *et al.*, 1975). In the mother cell cytoplasm, coat proteins are synthesised and are assembled into layers on the outer surface of the forespore in a controlled fashion; the gradual assembly of coat layers has been followed by [^{125}I]-labelling of proteins exposed on the surface of the developing spore (Jenkinson, Sawyer & Mandelstam, 1981). The formation of spore coats is designated Stage V. By this stage spores are resistant to toluene, octanol and to mild heating. The *spoV* mutants generally produce coats, but are to a variable extent cortex-defective (Piggot & Coote, 1976). Development of the spore's full heat resistance properties (maturation; Milhaud & Balassa, 1973) follows as Stage VI. Some mutants that make abnormal coats but apparently normal cortex have been designated *spoVIA*, *B* and *C* mutants; these structurally defective spores are also partially germi-

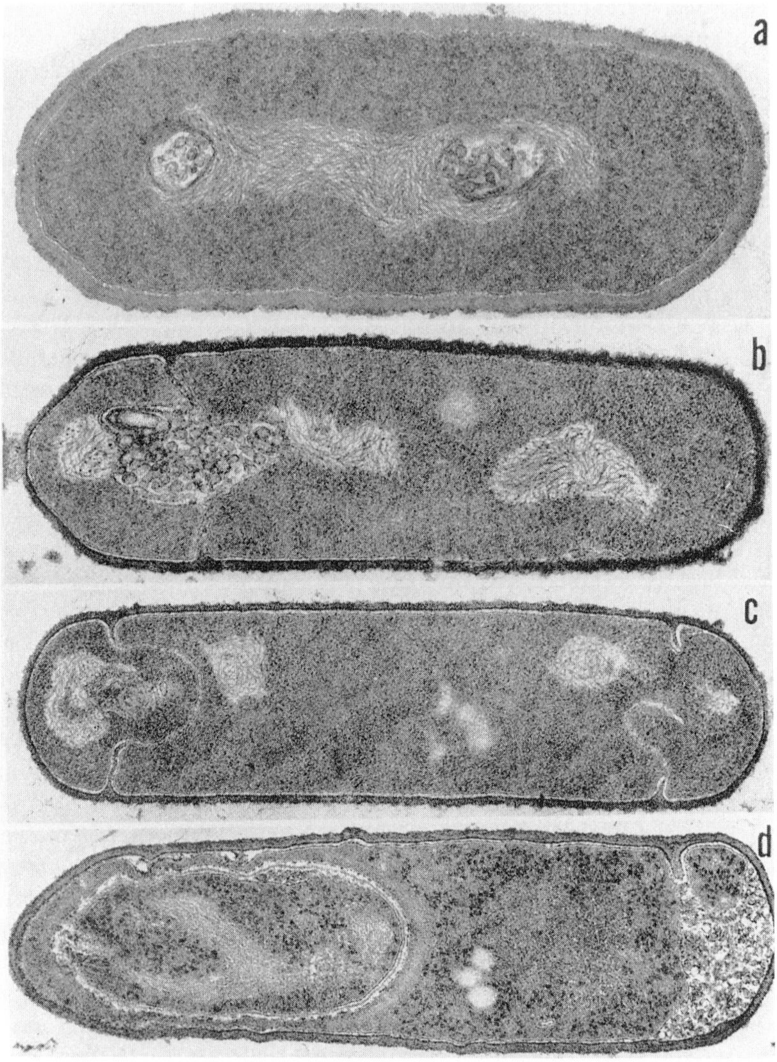

nation defective (for review see Moir & Smith, 1990). The final morphological change, lysis of the mother cell to release the mature spore, is classified as Stage VII.

The spores will then remain dormant until stimulated to germinate; germi-

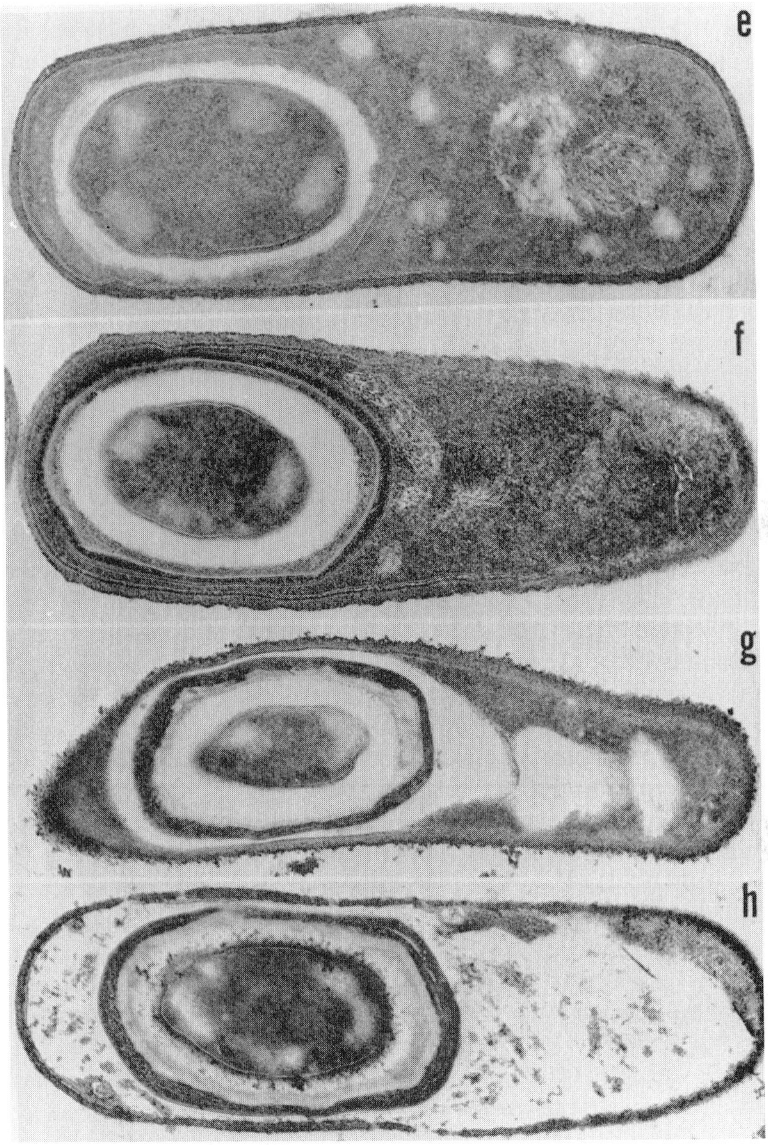

Fig. 2. Morphological stages of development in sporulating *B. subtilis*. (*a*) Axial filament formation; (*b*) Stage II: forespore septum formation; (*c*) Early stage III: commencement of engulfment; (*d*) Late stage III: engulfment of forespore is complete; (*e*) Stage IV: cortex formation; (*f*) Stage V: coat deposition; (*g*) Stage VI: maturation; (*h*) Stage VII: mother cell lysis. (Reproduced, with permission, from Kaneko, Doi & Santo, 1974.)

nation is essentially a hydrolytic process in which dormancy and resistance properties are lost as the cortex is hydrolysed and the core rehydrates (Setlow, 1981; Foster & Johnstone, 1990; Moir & Smith, 1990). The subsequent period in which RNA, protein and DNA synthesis are resumed in sequence, is termed outgrowth.

Perhaps 100 regulatory *spo* genes have now been identified, but relatively few of the genes that are controlled, other than additional elements of the regulatory cascades, have been unearthed. Several of the genes encoding major spore-specific proteins have been characterised – such genes are named according to the characteristics of the product, for example glucose dehydrogenase (*gdh*; Lampel *et al.*, 1986), small acid-soluble proteins (*ssp*; Connors, Mason & Setlow, 1986*a,b*) or coat proteins (*cot*; Donovan *et al.*, 1987). The germination apparatus is present in the mature spore – germination proceeds in the absence of *de novo* protein synthesis and most of the *ger* loci identified by mutation encode spore-specific proteins (Moir & Smith, 1990).

REGULATION OF SPORULATION

The pleiotropic effects of *spo* mutations suggested that the sporulation process involves a dependent sequence of events (Mandelstam, 1976). With the advent of recombinant DNA technology, it has been possible to analyse the function and interdependence of the products of many of the *spo* genes (Mandelstam & Errington, 1987; Stragier & Losick, 1990; Kunkel, 1991). The initiation of sporulation is induced by a combination of signals, the transduction of which is becoming less obscure. The *spo0A* gene, for example, is of pivotal significance to the initiation of sporulation (Grossman, 1991), and is involved in activation, either directly or indirectly, of a number of sporulation genes. It encodes a transcription factor (a member of the regulator class of the two-component sensor-regulator family) whose activity is regulated through a phosphorelay system responsive to environmental and intracellular signals; several kinases and additional Spo0 proteins have been implicated, some in the phosphate transfer process and others in its control (Perego, Spiegelman & Hoch, 1988; Burbulys *et al.*, 1991; Grossman, 1991; Perego *et al.*, 1991).

The period from initiation of sporulation through the establishment of balanced compartment-specific gene expression is currently the subject of extremely active research. A complex series of controls, involving the expression and activation of a series of sigma factors, is gradually being elucidated; this is beyond the scope of this article, but has been recently reviewed by Stragier & Losick (1990).

During sporulation an asymmetric and specialised cell division event results in the formation of two non-equivalent daughter cells. The compartments are engaged in parallel but closely coordinated processes of differential gene expression; the regulatory mechanisms concerned in establishing compartment-specific expression are discussed by P. Stragier in the next chapter. The general outlines of the regulatory sigma factor cascade have been established, but the nature of signals that act as 'checkpoints' between development in the two cell compartments (such as that mediated by *spoIVB*, as discussed below) is still unclear.

Compartment-specific sigma factors G and K are responsible for directing gene expression during forespore and late mother cell development respectively; they highlight two different modes by which sigma activity is modulated during sporulation. The sigma G protein is encoded in an active form, but other Spo proteins are implicated in controlling its activity (as discussed by P. Stragier). In contrast, the sigma K protein, expressed from a gene formed by a mother-cell specific DNA rearrangement, is synthesised as a pre-protein, and its processing is subject to further controls (Cutting *et al.*, 1991; Kunkel, 1991).

Once a functional sigma G-containing RNA polymerase is present in the forespore, expression of a number of forespore specific genes is set in train. As well as the functions required for the establishment and maintenance of dormancy, the mature spore must contain the apparatus required for spore germination and earliest stages of outgrowth. Amongst the genes expressed in the forespore are a number with a recognized role in reestablishing a growing organism; the following section describes a number of forespore-regulating genes that have been characterised, and discusses their contribution to the properties of the mature endospore.

FORESPORE-EXPRESSED GENES

Procedures for obtaining purified forespores (Andreoli *et al.*, 1973, 1981; Ellar *et al.*, 1975; Singh, Setlow & Setlow, 1977; Nicholson & Setlow, 1990*a*) have been crucial to an overview of the activities and proteins present in the forespore, and to the assessment of whether a particular protein is present exclusively in one compartment. Andreoli and coworkers estimated from a comparison of Coomassie-stained 2D polyacrylamide gels that very early after engulfment, 25% of the proteins differed in quantitative terms; within one further hour, 15% of the total detected protein species in the developing forespores were 'new' in the sense that they were not detectable in mother cell or pre-engulfment extracts. Although some of these may represent proteins that are modified, so that their mobility is altered in either of the two dimensions, many of them will represent proteins synthesised *de novo*. Of the very large number of new proteins that contribute to the mature spore, only a small proportion – some derived from the mother cell, (such as the coat proteins) and some from the forespore – have been recognized. Enzyme activities, too, can be measured; glucose dehydrogenase (Fujita, Ramaley & Freese, 1977) and the small acid-soluble spore proteins (Setlow,

1988) are found only in the forespore; the former serves as a particularly useful marker enzyme for the forespore fraction.

Identification of forespore-expressed genes

Fractionation experiments are technically difficult; cross-contamination of fractions must be closely monitored, and it is very difficult to prove biochemically that a particular protein is *exclusively* synthesised in one compartment. Immunoelectron microscopy can help to define the location of a protein, provided it is expressed at a relatively high level; the technique can also be generalised to identify the compartmentalisation of gene expression by using promoter fusions to *lacZ* and detecting by immunogold-labelling the cellular compartment in which β-galactosidase is expressed (Francesconi *et al.*, 1988).

Genetic approaches can be used to distinguish between *spo* genes whose products are *required* in different spore compartments. These methods depend on the formation of genetic mosaics, in which only one cell compartment has the wild type gene. The classic experiment is that of DeLencastre and Piggot (1979), in which *spo*$^-$ cells are transformed with *spo*$^+$ DNA at the end of growth; recombination of the incoming *spo*$^+$ allele with one of the two chromosomes can generate a genetically *spo*$^+$ forespore and a *spo*$^-$ mother cell or vice versa (Fig. 3). Selection for heat-resistant organisms at the end of sporulation allows recovery and characterisation of colonies derived from mature spores. If the *spo*$^+$ gene were required in the forespore compartment, and the forespore copy of the chromosome is exclusively responsible for expression of that gene, then it would be impossible for a genetic mosaic carrying a *spo*$^-$ forespore genome to generate a mature spore. If, in contrast, the gene product does not need to be expressed in the forespore compartment (either because the protein is not required in that compartment, or because the protein is expressed before septation) then *spo*$^-$ forespore compartments could, in a genetic mosaic, develop into mature spores. The proportion of *spo*$^-$ spores approaches zero for forespore expressed genes and can be up to a theoretical maximum of 50% for genes that are not forespore-expressed. Such experiments were the first to demonstrate that the product(s) of a *spo* locus (*spoVA*) must be expressed in the forespore; eight other *spo* loci tested gave the opposite result. A recent modification of this approach exploits *spo* mutants in which the lesion results from integration of a plasmid; in this case an excision frequency of 10^{-4} per cell results in restoration of *spo*$^+$ and loss of Cm^R. Simply measuring the relative proportions of Cm^R (*spo*$^-$) and Cm^S (*spo*$^+$) mature spores in a culture can establish whether forespore-specific expression is required (Illing, Young & Errington, 1990).

If a *spo* gene is expressed before septation, the encoded protein will be present in both compartments and the result of the above genetic test will

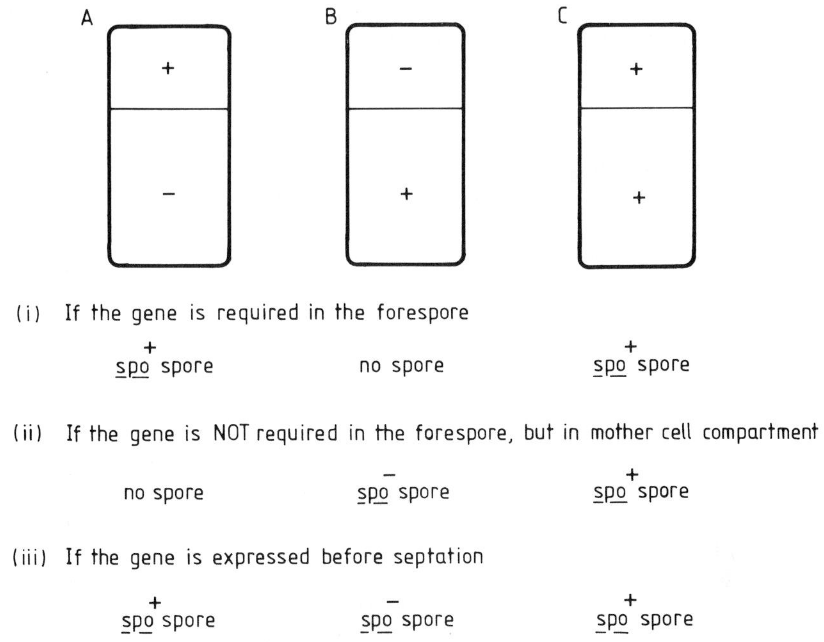

Fig. 3. Genetic mosaics during sporulation. Sporangia A and B arise if one chromosome is restored to *spo*$^+$; C arises if both are *spo*$^+$.

be indistinguishable from that of a *spo* gene whose expression is confined to the mother cell; other information, such as the time of expression, dependence of gene expression on other *spo* genes or immunological localisation of the gene product can be used to distinguish these two classes of expression (Gholamhoseinan & Piggot, 1989).

Protoplast fusion between *spo*$^-$ and *spo*$^+$ sporulating cultures was exploited to test whether providing a *spo*$^+$ mother cell environment would allow an engulfed *spo*$^-$ forespore to develop to maturity – entirely analogous to, and generating results consistent with, the genetic mosaic experiments. The approach could also be used for complementation analysis (Dancer & Mandelstam, 1981).

As *spo* genes do not encode easily assayable products, the construction of *lac* fusions (as in Errington & Mandelstam, 1986) has been crucial to the analysis of *spo* gene expression – its timing during sporulation, its dependence on other *spo* gene products and its compartmentalisation. The ease of introducing DNA into the *Bacillus* chromosome has allowed the integration of *lac* fusions by a single crossover with homologous sequences in a Campbell-type integration event (as in Fig. 4(*a*)) or by substitution of DNA, via two crossover events, as in introduction of DNA at the *amy*

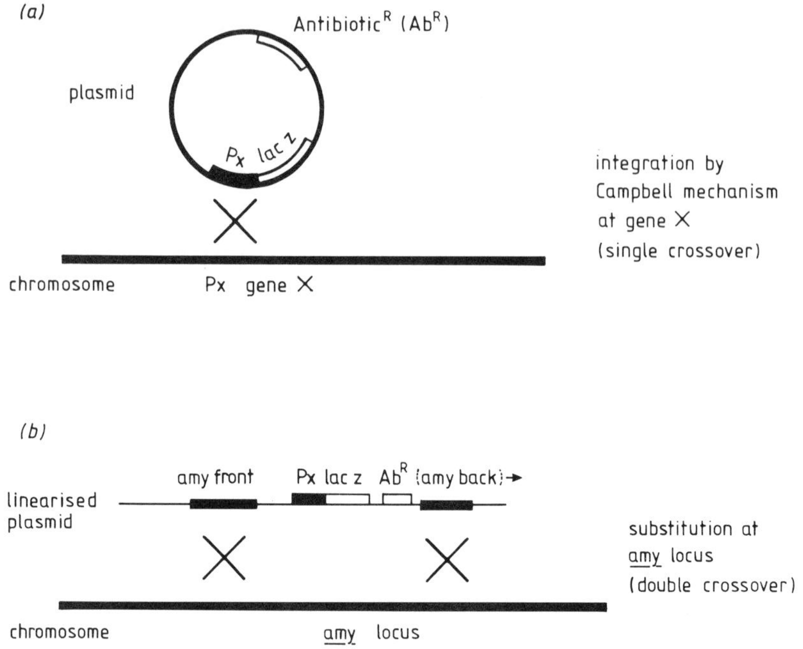

Fig. 4. Two types of chromosomal *lacZ* fusion. Px represents the promoter of gene X.

locus (Fig. 4(*b*); Shimotsu & Henner, 1986). The *lac* fusions can be transferred into different *spo* mutant backgrounds by classical transformation, so that the dependence of expression on other *spo* gene products can be explored.

Regulation of forespore-expressed genes

The regulation of forespore-expressed genes has been clarified by the discovery of a forespore-specific sigma factor, sigma G (Sun, Stragier & Setlow, 1989). The small acid-soluble proteins, discussed in more detail later, are major constituents of the spore and are exclusively expressed, at a relatively high level, in the developing forespore (Setlow, 1988). Fractionated extracts were tested for their ability to transcribe *in vitro* an *sspE* template – two separate holoenzyme forms of RNA polymerase were identified; the more effective, containing a sigma factor named sigma G, is encoded by the *spoIIIG* gene. The other holoenzyme, containing sigma F, product of the *spoIIAC* gene, was less effective *in vitro* and not at all effective *in vivo* on the *sspE* promoter; it appears to recognise a promoter sequence very similar to the sigma G consensus (Sussman & Setlow, 1991). As discussed in the next chapter by P. Stragier, an active sigma F-containing RNA polymerase is essential for *spoIIIG* (sigma G) expression, although the mode of action is controversial.

Table 1. *Forespore-expressed genes and their function.*

spoIIIG	(sigmaG)	Activator of forespore genes
spoIVB		Control of sigmaK processing
spoVAA-AE		Some membrane association?
sspA,B	(SASP sα and β	UV resistance: aa source
sspE	(SASP γ)	aa source in outgrowth
gpr	(germination protease)	Initiates SASP cleavage
gdh	(glucose dehydrogenase)	Glucose metabolism
gerAA-AC	(ala germination)	Alanine receptor in spore
gerD	(germination)	Common pathway component
0.3 kb gene		Possible aa source

For most of the known forespore-specific genes, sigma G is necessary for transcription (Nicholson *et al.*, 1989). Fusion of the *spoIIIG* gene downstream of the IPTG-inducible *spac* promoter (Yansura & Henner, 1984) allows induction of sigma G in vegetative cells; this is an effective means of identifying those genes whose expression is dependent solely on sigma G – or, in some cases, on other regulators that are sigma G-induced (Sun *et al.*, 1989). The rapid expression of most of the forespore genes on IPTG induction of sigma G suggests that it acts directly and is sufficient to induce expression (an exception to this is the *0.3kb* gene, discussed later); the ability to delete sequences upstream of the promoter region without affecting transcription of *sspE* and *B* (Fajardo-Cavazos, Tovar-Rojo & Setlow, 1991) is consistent with this interpretation. Thus expression or activation of functional sigma G is the crucial inducing element in forespore-specific gene expression.

Genes known to be expressed in the forespore are listed in Table 1. Only a minute fraction of the total number of forespore-expressed genes have been identified; many of those identified are important to the spore for the later developmental stages of germination and outgrowth. Some were identified genetically as genes required for sporulation (*spoIVB*, *spoVA*), others were first identified biochemically as protein or mRNA (*gdh*, *ssp*, *0.3kb* genes), and a third class were recognized as affecting spore germination (*gerA*, *gerD*). The transcriptional startpoints of these genes have been determined and a promoter consensus derived from 23 sigma G-dependent promoter regions (Fajardo-Cavazos *et al.*, 1991).

Forespore expressed genes essential for formation of a heat-resistant spore

spoIIIG

The regulation of transcription of this forespore-specific gene is discussed at length in Patrick Stragier's article, to which the reader is referred. The sigma G protein is synthesised some time before expression of sigma G-dependent genes; genetic evidence argues that the SpoIIAB protein inhibits sigma G activity (Rather *et al.*, 1990), delaying its activation until after engulfment.

spoIVB

A *lac* fusion to this gene (Van Hoy & Hoch, 1990) is expressed in a sigma G-dependent and forespore-specific manner (Cutting *et al.*, 1991). Mutations of *spoIVB* block sporulation; strikingly, the gene product appears to activate the *spoIVF*-dependent proteolysis of the mother cell pro-sigma K protein. This coordination of activity between forespore and mother cell expression is important to proper development – mutants bypassing this processing-mediated checkpoint control produce structurally defective spores (Cutting *et al.*, 1990). The SpoIVB protein contains a potential signal or membrane anchor sequence, suggesting that it may be located, either free or inner membrane-bound, in the region between the two forespore membranes.

spoVA

The *spoVA* locus contains an operon of at least five genes, mutations in any of which block sporulation in a similar fashion. In the mutants, the spore protoplast is surrounded by an almost complete cortex and coat structure, but the spore core does not dehydrate; the spores are partially resistant to toluene and lysozyme, but do not become heat or chloroform resistant (Errington & Mandelstam, 1984). The spores do not accumulate dipicolinate (DPA), although the mutant can produce this compound, as evidenced by its coculture with a DPA-dependent *spoVF* mutant (Errington, Cutting & Mandelstam, 1988). All five *spoVA* genes contain stretches of hydrophobic amino acids sufficiently long to represent membrane-traversing helices; the two smallest gene products (those of *spoVAB* and *spoVAC*) are strongly hydrophobic (Fort & Errington, 1985) and probably are integral membrane proteins. A sixth open reading frame, only part of which has been cloned, is not required for sporulation (J. Errington, personal communication), but the sequence of the first 155 amino acids of this ORF is 30% identical to the N-terminal sequence of germination protein GerAA, which is also a forespore protein. Inactivation of this ORF does not block either of the two major germination pathways of *B. subtilis* (E. H. Kemp, unpublished observations).

Forespore genes with a function during dormancy, germination and outgrowth

Small acid soluble proteins (SASPs): The three major SASP proteins in *B. subtilis* (SASP α, β and γ, encoded by *sspA*, *B* and *E* genes respectively) constitute approx. 15% of the total spore core protein (Johnson & Tipper, 1981). Of these, SASP γ is the most abundant, representing 8% of the total protein. The SASP α/β (or A/B) group of proteins are homologous, and four members of this gene family (*sspA-D*) have been cloned from *B. subtilis*

(Connors *et al.*, 1986*b*). The genes all map at different locations and are monocistronic (Connors *et al.*, 1986*a*). The SASP γ type represents a different family of proteins, whose sequence is unrelated to that of SASP α/β, except at the site of endoproteolytic cleavage by a SASP-specific protease. All of the genes are transcribed *in vivo* and *in vitro* by sigma G-containing RNA polymerase; there are suggestions of autoregulation, probably at the translational level, at least for *sspB* (Mason, Hackett & Setlow, 1988). The SASP proteins are found in the core of the spore and are degraded to free amino-acids early after triggering of germination (Setlow, 1988). They serve as a source of amino acids for the resumption of protein synthesis in outgrowth. As dormant spores are depleted in free amino acids and lack some of the amino acid biosynthetic enzymes, the SASP proteins are a crucial source of amino acids for rapid outgrowth. Cross-linking studies suggested that in *B. megaterium* the α/β type SASP was associated with spore DNA (Setlow & Setlow, 1979). Immunoelectron microscopy (Francesconi *et al.*, 1988) has confirmed that SASP proteins are located in the spore core, with α/β type present in the region of the nucleoid and γ distributed throughout the core.

Reverse genetics has revealed a role for SASP α/β in UV resistance. Deletion of both *sspA* and *B* genes does not affect spore formation, but the resulting spores are extremely UV sensitive – even more so than vegetative cells (Mason & Setlow, 1986, 1987). UV light interacts with spore DNA to form a different type of thymine dimer, the so-called 'spore photoproduct', which is efficiently repaired during germination. In the α/β double mutant a significant quantity of vegetative type thymine dimer is formed; this is less effectively repaired in germinating spores, and the UV resistance is hence low (Setlow, 1988). Spore heat resistance is also reduced, but to a much lesser extent (Mason & Setlow, 1987). The formation of a different thymine adduct suggests that the DNA is in a different conformation in the spore; this is confirmed on isolation of plasmid DNA from spores – it is particularly strongly negatively supercoiled; this topology is relaxed back to the vegetative level on germination, coincident with the degradation of SASP proteins (Nicholson & Setlow, 1990*b*). The overexpression of SspC protein has allowed a similar phenomenon to be demonstrated *in vitro*; the altered supercoiling is dependent on binding of SASP (to a level of one molecule per 5bp) and the addition of topoisomerase I. SASP binding causes a shift in the spore DNA from the B to the A conformation (Mohr *et al.*, 1991); on release from this state by extraction of SASP proteins, the DNA would return to B form and the plasmid DNA, in maintaining a constant torsional strain, will appear more negatively supercoiled (Nicholson, Setlow & Setlow, 1990).

The effect of expressing large concentrations of SASPs has been graphically demonstrated in *E. coli*. Within 2.5 h after induction of SspC protein, the chromosome became condensed and no longer transcriptionally active (Setlow, Hand & Setlow, 1991). The other major class, SASP γ, has no effect on DNA conformation and does not bind DNA; this is consistent with observations that it does not contribute to spore UV resistance (Hackett & Setlow, 1988).

The role of SASPs in supplying amino acids for outgrowth is demonstrated in mutants progressively deleted for more SASP genes, which outgrow at progressively slower rates. If spores are plated on minimal agar, the loss of the genes encoding the three major SASP proteins can reduce the spore's ability to regenerate to form colonies by at least 100-fold (Hackett & Setlow, 1988).

The gpr *gene*

This encodes a pre-form of the sequence specific SASP protease; it is expressed from approximately 1 h earlier than the *ssp* genes, on the basis of analysis of *lac* fusion kinetics (Sussman & Setlow, 1991). The pre-protease, originally 46 kD, is processed to 41 kD around the time of *ssp* gene expression; this is further reduced to 40 kD on germination. Both 41 and 40 kD forms are enzymically active (Hackett & Setlow, 1983); nothing is known of the mechanisms responsible for processing the protease, or for maintaining it in an inactive form in the spore. The *gpr* gene is transcribed from the same promoter region by both sigma F- and sigma G-containing polymerase (Sussman & Setlow, 1991).

The 0.3kb *gene*

The so-called *0.3kb* gene was cloned as the gene encoding a sporulation-specific and relatively stable 300 base mRNA (Stephens *et al.*, 1984). The encoded protein is small and relatively basic, and it is possible that it may have a function analogous to a SASP, although it is not homologous to either of the major classes of these proteins. Unlike the other forespore-expressed genes tested, it is not transcribed *in vitro* by sigma G-containing holoenzyme (Setlow, 1989), and does not share the classical consensus promoter sequence; its expression in vegetative cells lags 60 min behind that of the other sigma G-induced products, suggesting either that transcription of this gene requires another auxiliary regulator in addition to sigma G, or that it uses a different, and as yet undiscovered, forespore sigma factor that follows later in the developmental cascade (Panzer *et al.*, 1989).

*The glucose dehydrogenase (*gdh*) gene*

Glucose dehydrogenase activity is not found in vegetative cells of *B. subtilis*, but is present in spores; up to 70% of glucose added to germinated spores is metabolised by this enzyme (Otani *et al.*, 1986). The gene has been cloned and is the second ORF in an operon; inactivation of either of these ORFs has no effect on the rate of spore germination (Rather & Moran, 1988; R. F. Ramaley, Abstracts of the 10th International Spores Meeting).

Spore germination genes

Forespore-expressed genes include several required for the spore germination response. *B. subtilis* spores will germinate in response to two different types of germination stimulus – alanine and its analogues will induce germination, but a range of other amino acids, such as asparagine, that are inactive on their own, are germinative in combination with glucose, fructose and potassium salts (AGFK). A number of *ger* genes have been identified by mutation (for review see Moir & Smith, 1990); the *gerA* mutants are specifically defective in alanine-triggered germination, whereas *gerB* and *gerK* mutants are defective in the AGFK combination but germinate normally in alanine. A second group of mutants are defective in both germination pathways, germinating more slowly in alanine but not at all in the asparagine combination; these are *gerD* and *gerF*, and define proteins that are accessory for alanine germination but absolutely essential for the AGFK response. Finally, members of the remaining class (*gerE*, *gerJ* and *gerM*) are pleiotropically defective in the structural development of the spore and are blocked at late stages of germination in either type of germinant – heat resistance is lost normally, but the late stages of cortex hydrolysis are blocked.

Only one member of each of the first two groups, blocked at the initiation of germination, has yet been characterised in detail; these, the *gerA* and *gerD* loci are both forespore-expressed (Feavers *et al.*, 1990; Kemp *et al.*, 1991). In contrast, the mutations that affect spore structure and block germination at an intermediate stage are not in forespore expressed genes – both *gerJ* and *gerM* are expressed before or around septation (Warburg *et al.*, 1986; Sammons, Slynn & Smith, 1987) and *gerE* is a DNA binding protein modulating sigma K-dependent late coat gene expression in the mother cell (Zheng & Losick, 1990).

Mutations in the *gerA* locus block the germination response to alanine and its analogues; this locus is likely to encode a receptor for alanine in the spore. The *gerA* region contains an operon of three genes (*gerAA*, *AB* and *AC*; Fig. 5); mutations in any of these block germination in alanine but leave the other germination pathway intact (Zuberi, Feavers & Moir, 1987). The deduced amino acid sequence of the gene products suggests that all are in some way membrane associated; *GerAA* has a hydrophilic 250 residue N-terminal domain, followed by a 200 residue hydrophobic domain containing approximately six blocks of hydrophobic amino acids that are likely to represent membrane-spanning helices; the C-terminal 30 residues are highly charged and could lie on the outer or inner face of the membrane, depending on whether the protein contains an odd or even number of membrane traversing helices. The GerAB protein would form an integral

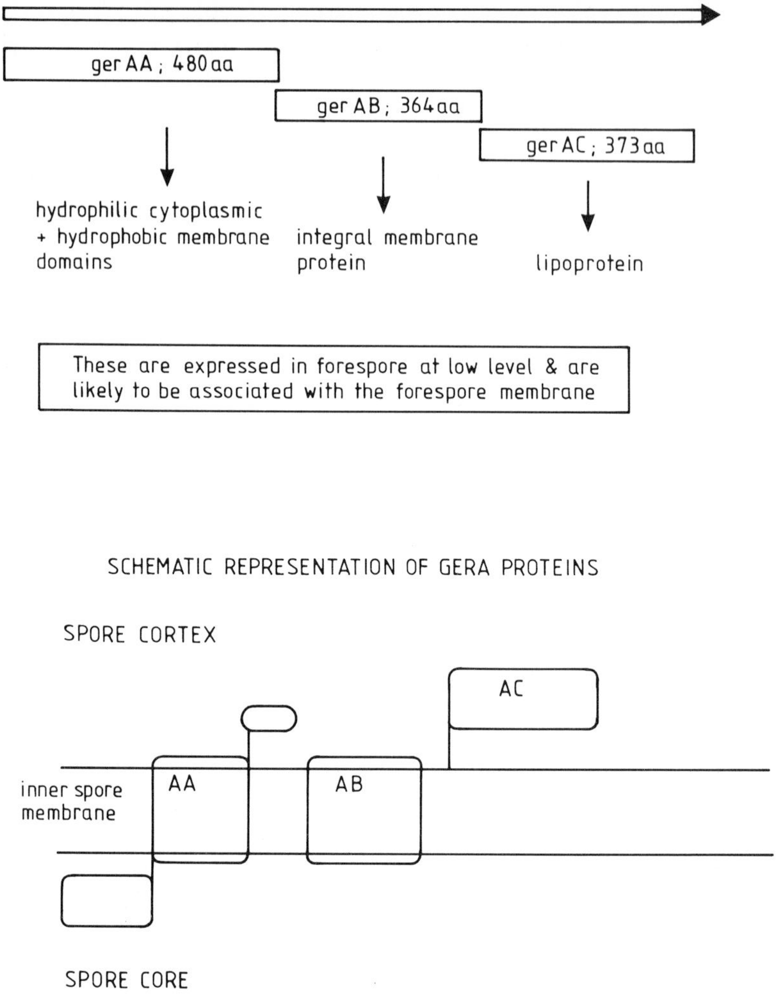

Fig. 5. Spore germination proteins – a receptor for alanine in the inner spore membrane?

membrane-spanning protein, with about 10 membrane spans; two *gerA* mutants that show an altered alanine concentration dependence map in this gene (Sammons, Moir & Smith, 1981; Zuberi, Feavers & Moir, 1985). The GerAC protein is hydrophilic, but has a signal sequence characteristic of pre-lipoproteins, so that it is likely to be transferred across, and anchored to the outer surface of, the membrane. As the genes are expressed in the forespore compartment, their likely location is in the inner spore membrane; this would imply that alanine, in order to act as germinant, must penetrate the spore to the inner membrane. A diagrammatic representation of their possible organisation is shown in Fig. 5. Interestingly, sequence information for the cloned *gerB* locus, required for AGFK germination (B. A. Corfe, R. L. Sammons & D. A. Smith, personal communication) suggests that it encodes proteins homologous to the GerA proteins; the proteins share around 30–35% identical residues, except in the hydrophobic region of GerAA (65% identical), suggesting that the sequence here is particularly important to the germination function of both complexes. The GerA and GerB proteins represent members of a family of transducers whose detailed molecular functions are not yet understood.

The *gerD* gene, which is monocistronic (Yon, Sammons & Smith, 1989), is also expressed in the forespore, as demonstrated from *lac* fusion data and sigma G-dependent transcription. The GerD protein is hydrophilic, but has a signal sequence or membrane anchor; like GerAC, it too will either be located on the outer face of the inner spore membrane or be free in the cortex. Transcribed at about a 15-fold higher level than the *gerA* genes, the *gerD* gene product may help to transduce the germination signal from the initial complexes, before events that lead to loss of heat resistance.

CONCLUSIONS

Sporulation is an extremely complex and highly coordinated process; it cannot really be viewed as a 'primitive' developmental system (Mandelstam, 1976). We could with justification argue that its culmination is not only the formation of a dormant, mature spore; the aim of this sophisticated developmental pathway is to produce a structure that although resistant and dormant, is able to respond to changes in the environment, germinating and outgrowing to found a new colony. Many, but by no means all, of the genes expressed specifically in the forespore are committed to these later stages of the cycle; SASP proteins, for example, have a dual function, binding DNA in the spore and serving as a storage polymer, providing amino acids needed in outgrowth. The germination proteins, although not well characterised, represent a new class of transducer, with a role only in the breakage of dormancy. Whether these types of proteins are confined solely to the bacterial endospore or have homologues with functions related to other, vegetative, cellular processes remains an open question.

REFERENCES

Andreoli, A. J., Kao, M., Chui, R., Cabrera, J. & Wong, S. K. S. (1981). Two dimensional polyacrylamide gel electrophoresis analyses of cytoplasmic and membrane proteins from sporulating cells and forespore and mother cell compartments of *Bacillus cereus*. In *Sporulation and Germination*, ed. H. S. Levinson, A. L. Sonenshein & D. J. Tripper, pp. 168–73. Washington DC: American Society for Microbiology.

Andreoli, A. J., Saranto, J., Baecker, P. A., Suehiro, S., Escamilla, E. & Steiner, A. (1975) Biochemical properties of forespores isolated from *Bacillus cereus*. In

Spores VI, ed. P. Gerhardt, R. N. Costilow & H. L. Sadoff, pp. 418–24. Washington: American Society for Microbiology.

Andreoli, A. J., Suehiro, S., Sakiyama, D., Takemoto, J., Vivanco, E., Lara, J. C. & Klute, M. C. (1973). Release and recovery of forespores from *Bacillus cereus*. *Journal of Bacteriology*, **115**, 1159–66.

Burbulys, D., Trach, K. A. & Hoch, J. A. (1991). Initiation of sporulation in *Bacillus subtilis* is controlled by a multicomponent phosphorelay. *Cell*, **64**, 545–52.

Connors, M. J., Howard, S., Hoch, J. & Setlow, P. (1986*a*). Determination of the chromosomal locations of four *Bacillus subtilis* genes which code for a family of small, acid-soluble proteins, *Journal of Bacteriology*, **166**, 412–16.

Connors, M. J., Mason, J. M. & Setlow, P. (1986*b*). Cloning and nucleotide sequencing of genes for three small, acid-soluble proteins from *Bacillus subtilis* spores. *Journal of Bacteriology*, **166**, 417–25.

Cutting, S., Driks, A., Schmidt, R., Kunkel, B. & Losick, R. (1991). Forespore-specific transcription of a gene in the signal transduction pathway that governs pro-σ^K processing in *Bacillus subtilis*. *Genes and Development*, **5**, 456–66.

Cutting, S., Oke, V., Driks, A., Losick, R., Lu, S. & Kroos, L. (1990). A forespore checkpoint for mother cell gene expression during development in *Bacillus subtilis*. *Cell*, **62**, 239–50.

Dancer, B. N. & Mandelstam, J. (1981). Complementation of sporulation mutations in fused protoplasts of *Bacillus subtilis*. *Journal of General Microbiology*, **123**, 17–26.

DeLencastre, H. & Piggot, P. J. (1979). Identification of different sites of expression for *spo* loci by transformation of *Bacillus subtilis*. *Journal of General Microbiology*, **114**, 377–89.

Donovan, W., Zheng, L., Sandman, K. & Losick, R. (1987). Genes encoding spore coat polypeptides from *Bacillus subtilis*. *Journal of Molecular Biology*, **196**, 1–10.

Ellar, D. J. (1978). Spore specific structures and their function. In *Relations between Structure and Function in the Prokaryotic Cell: Symposium of the Society for General Microbiology*, **28**, ed. R. Y. Stanier, H. J. Rogers & B. J. Ward, pp. 295–325. Cambridge: Cambridge University Press.

Ellar, D. J., Eaton, M. W., Hogarth, C., Wilkinson, B. J., Deans, J. & La Nauze, J. (1975). Comparative biochemistry and function of forespore and mother cell compartments during sporulation of *Bacillus megaterium* cells. In *Spores VI*, ed. P. Gerhardt, R. N. Costilow & H. L. Sadoff, pp. 425–33. Washington: American Society for Microbiology.

Errington, J., Cutting, S. & Mandelstam, J. (1988). Branched pattern of regulatory interactions between late sporulation genes in *Bacillus subtilis*. *Journal of Bacteriology*, **170**, 796–801.

Errington, J. & Mandelstam, J. (1984). Genetic and phenotypic characterisation of a cluster of mutations in the *spoVA* locus of *Bacillus subtilis*. *Journal of General Microbiology*, **130**, 2115–21.

Errington, J. & Mandelstam, J. (1986). Use of a *lacZ* fusion to determine the dependence pattern and the spore compartment expression of sporulation operon *spoVA* in *spo* mutants of *Bacillus subtilis*. *Journal of General Microbiology*, **132**, 2977–85.

Fajardo-Cavazos, P., Tovar-Rojo, F. & Setlow, P. (1991). Effect of promoter mutations and upstream deletions on the expression of genes coding for small, acid-soluble proteins of *Bacillus subtilis*. *Journal of Bacteriology*, **173**, 2011–16.

Feavers, I. M., Foulkes, J., Setlow, B., Sun, D., Nicholson, W., Setlow, P. & Moir, A. (1990). The regulation of transcription of the *gerA* spore germination operon of *Bacillus subtilis*. *Molecular Microbiology*, **4**, 275–82.

Fort, P. & Errington, J. (1985). Nucleotide sequence and complementation analysis of a polycistronic sporulation operon, *spoVA*, in *Bacillus subtilis*. *Journal of General Microbiology*, **131**, 1091–105.

Foster, S. & Johnstone, K. (1990). Pulling the trigger: the mechanism of spore germination. *Molecular Microbiology*, **4**, 137–41.

Francesconi, S. C., McAlister, T. J., Setlow, B. & Setlow, P. (1988). Immunoelectron microscopic localization of small, acid-soluble spore proteins in sporulating cells of *Bacillus subtilis*. *Journal of Bacteriology*, **170**, 5963–7.

Fujita, Y., Ramaley, R. & Freese, E. (1977). Location and properties of glucose dehydrogenase in sporulating cells and spores of *Bacillus subtilis*. *Journal of Bacteriology*, **132**, 282–93.

Gholamhoseinan, A. & Piggot, P. J. (1989). Timing of *spoII* gene expression relative to septum formation during sporulation of *Bacillus subtilis*. *Journal of Bacteriology*, **171**, 5747–9.

Grossman, A. (1991). Integration of developmental signals and the initiation of sporulation in *B. subtilis*. *Cell*, **65**, 5–8.

Hackett, R. H. & Setlow, P. (1983). Determination of the enzymatic activity of the precursor form of the *Bacillus megaterium* spore protease. *Journal of Bacteriology*, **153**, 375–8.

Hackett, R. H. & Setlow, P. (1988). Properties of spores of *Bacillus subtilis* which lack the major small acid-soluble protein. *Journal of Bacteriology*, **170**, 1403–4.

Holt, S. C., Gauthier, J. J. & Tipper, D. J. (1975). Ultrastructural studies of sporulation in *Bacillus sphaericus*. *Journal of Bacteriology*, **122**, 1322–38.

Illing, N., & Errington, J. (1991). Genetic regulation of morphogenesis in *Bacillus subtilis*: roles of σ^E and σ^F in prespore engulfment. *Journal of Bacteriology*, **173**, 3159–69.

Illing, N., Young, M. & Errington, J. (1990). Use of integrational plasmid excision to identify cellular localization of gene expression during sporulation in *Bacillus subtilis*. *Journal of Bacteriology*, **172**, 6937–41.

Jenkinson, H. F., Sawyer, W. D. & Mandelstam, J. (1981). Synthesis and order of assembly of spore coat proteins in *Bacillus subtilis*. *Journal of General Microbiology*, **123**, 1–16.

Johnson, W. C. & Tipper, D. J. (1981). Acid-soluble spore proteins of *Bacillus subtilis*. *Journal of Bacteriology*, **146**, 972–82.

Kaneko, I., Doi, R. H. & Santo, L. Y. (1974). Bacterial sporulation and germination. *The Cell*, **6**, 154–77.

Kemp, E. H., Sammons, R. L., Moir, A., Sun, D. & Setlow, P. (1991). Analysis of transcriptional control of the *gerD* spore germination gene of *Bacillus subtilis* 168. *Journal of Bacteriology*, **173**, 4646–52.

Kunkel, B. (1991). Regulation of compartmentalised gene expression during sporulation in *Bacillus subtilis*. *Trends in Genetics*, **7**, in press.

Lampel, K. A., Uratani, B., Chaudry, G. R., Ramaley, R. F. & Rudikoff, S. (1986). Characterization of the developmentally regulated *Bacillus subtilis* glucose dehydrogenase gene. *Journal of Bacteriology*, **166**, 238–43.

Losick, R. & Kroos, L. (1989). Dependence pathways for the expression of genes involved in endospore formation in *Bacillus subtilis*. In *Regulation of Prokaryotic Development*, ed. I. Smith, R. A. Slepecky & P. Setlow, pp. 223–54. Washington: American Society for Microbiology.

Losick, R., Youngman, P. & Piggot, P. J. (1986). Genetics of endospore formation in *Bacillus subtilis*. *Annual Review of Genetics*, **20**, 625–69.

Mandelstam, J. (1976). Bacterial sporulation: a problem in the biochemistry and genetics of a primitive developmental system. *Proceedings of the Royal Society of London, B*, **193**, 89–106.

Mandelstam, J. & Errington, J. (1987). Dependent sequences of gene expression controlling spore formation in *Bacillus subtilis*. *Microbiological Sciences*, **4**, 238–44.

Mason, J. M. & Setlow, P. (1986). Essential role of small, acid-soluble spore proteins in resistance of *Bacillus subtilis* spores to UV light. *Journal of Bacteriology*, **167**, 174–8.

Mason, J. M. & Setlow, P. (1987). Different small, acid-soluble proteins of the α/β type have interchangeable roles in the heat and UV radiation resistance of *Bacillus subtilis* spores. *Journal of Bacteriology*, **169**, 3633-7.

Mason, J. M., Hackett, R. H. & Setlow, P. (1988). Regulation of expression of genes coding for small, acid-soluble proteins of *Bacillus subtilis* spores: studies using *lacZ* gene fusions. *Journal of Bacteriology*, **170**, 239–44.

Milhaud, P. & Balassa, G. (1973). Biochemical genetics of bacterial sporulation. IV. Sequential development of resistances to chemical and physical agents during sporulation of *Bacillus subtilis*. *Molecular and General Genetics*, **125**, 241–50.

Mohr, S. C., Sokolov, N. V. H. A., He, C. & Setlow, P. (1991). Binding of small, acid-soluble spore proteins from *Bacillus subtilis* change the conformation of DNA from B to A. *Proceedings of the National Academy of Sciences, USA*, **88**, 77–81.

Moir, A. & Smith, D. A. (1990). The genetics of bacterial spore germination. *Annual Reviews of Microbiology*, **44**, 531–53.

Nicholson, W. L. & Setlow, P. (1990*a*). Sporulation, germination and outgrowth. In *Molecular Biology Methods for* Bacillus, ed. C. R. Harwood & S. M. Cutting. pp. 391–429. Chichester, UK: John Wiley & Sons.

Nicholson, W. L. & Setlow, P. (1990*b*). Dramatic increase in negative superhelicity of plasmid DNA in the forespore compartment of sporulating cells of *Bacillus subtilis*. *Journal of Bacteriology*, **172**, 7–14.

Nicholson, W. L., Setlow, B. & Setlow, P. (1990). Binding of DNA *in vitro* by a small acid-soluble spore protein from *Bacillus subtilis* and the effect of this binding on DNA topology. *Journal of Bacteriology*, **172**, 6900–6.

Nicholson, W. L., Sun, D., Setlow, B. & Setlow, P. (1989). Promoter specificity of σ^G-containing RNA polymerase from sporulating cells of *Bacillus subtilis*: identification of a group of forespore-specific promoters. *Journal of Bacteriology*, **171**, 2708–18.

Otani, M., Ihara, N., Umezawa, C. & Sano, K. (1986). Predominance of gluconate formation during germination of *Bacillus megaterium QMB1551* spores. *Journal of Bacteriology*, **167**, 148–52.

Panzer, S., Losick, R., Sun, D. & Setlow, P. (1989). Evidence for an additional temporal class of gene expression in the forespore compartment of sporulating *Bacillus subtilis*. *Journal of Bacteriology*, **171**, 561–4.

Perego, M., Spiegelman, G. B. & Hoch, J. A. (1988). Structure of the gene for the transition state regulator *abrB*: regulator synthesis is controlled by the *spoOA* sporulation gene in *Bacillus subtilis*. *Molecular Microbiology*, **2**, 689–99.

Perego, M., Higgins, C. F., Pearce, S. R., Gallaher, M. P. & Hoch, J. A. (1991). The oligopeptide transport system of *Bacillus subtilis* plays a role in the initiation of sporulation. *Molecular Microbiology*, **5**, 173–85.

Piggot, P. J. & Coote, J. G. (1976). Genetic aspects of bacterial endospore formation. *Bacteriological Reviews*, **40**, 908–62.

Rather, P. N., Copolecchia, R., DeGrazia, H. & Moran, Jr., C. P. (1990). Negative regulator of σ^G-controlled gene expression in stationary-phase *Bacillus subtilis*. *Journal of Bacteriology*, **172**, 709–15.

Rather, P. N. & Moran, Jr., C. P. (1988). Compartment-specific transcription in *Bacillus subtilis*: identification of the promoter for *gdh*. *Journal of Bacteriology*, **170**, 5086–92.

Ryter, A., Schaeffer, P. & Ionesco, H. (1966). Classification cytologique pour leur étude de blocage des mutants de sporulation de *Bacillus subtilis Marburg*. *Annales de l'Institut Pasteur*, **100**, 305.

Sammons, R. L., Moir, A. & Smith, D. A. (1981). Isolation and properties of spore germination mutants deficient in the initiation of germination. *Journal of General Microbiology*, **124**, 229–41.

Sammons, R. L., Slynn, G. M. & Smith, D. A. (1987). Genetical and molecular studies on *gerM*, a new developmental locus of *Bacillus subtilis*. *Journal of General Microbiology*, **133**, 3299–312.

Setlow, P. (1981). Biochemistry of forespore development and spore germination. In *Sporulation and Germination*, ed. H. S. Levinson, A. L. Sonenshein & D. J. Tipper, pp. 13–28. Washington DC: American Society for Microbiology.

Setlow, P. (1988). Small, acid-soluble proteins of *Bacillus* species: structure, synthesis, genetics, function and degradation. *Annual Review of Microbiology*, **42**, 19–38.

Setlow, P. (1989). Forespore-specific genes of *Bacillus subtilis*: function and regulation of expression. In *Regulation of Procaryotic Development*, ed. I. Smith, R. A. Slepecky & P. Setlow. pp. 211–21. Washington: American Society for Microbiology.

Setlow, B., Hand, A. R. & Setlow, P. (1991). Synthesis of a *Bacillus subtilis* small, acid-soluble spore protein in *Escherichia coli* causes cell DNA to assume some characteristics of spore DNA. *Journal of Bacteriology*, **173**, 1642–53.

Setlow, B. & Setlow, P. (1979). Localisation of low molecular weight basic proteins in *Bacillus megaterium* spores by irradiation with UV light. *Journal of Bacteriology*, **139**, 486–94.

Shimotsu, H. & Henner, D. J. (1986). Construction of a single-copy integration vector and its use in analysis of regulation of the *trp* operon of *Bacillus subtilis*. *Gene*, **43**, 85–94.

Singh, R. P., Setlow, B. & Setlow, P. (1977). Levels of small molecules and enzymes in the mother cell compartment and the forespore of sporulating *Bacillus megaterium*. *Journal of Bacteriology*, **130**, 1130–8.

Sonenshein, A. L. (1989). Metabolic regulation of sporulation and other stationary-phase phenomena. In *Regulation of Procaryotic Development*, ed. I. Smith, R. A. Slepecky & P. Setlow, pp. 109–30. Washington, DC: American Society for Microbiology.

Stephens, M. A., Lang, N., Sandman, K. & Losick, R. (1984). A promoter whose utilisation is temporally regulated during sporulation in *Bacillus subtilis*. *Journal of Molecular Biology*, **176**, 333–48.

Stragier, P. & Losick, R. (1990). Cascades of sigma factors revisited. *Molecular Microbiology*, **4**, 1801–6.

Sun, D., Stragier, P. & Setlow, P. (1989). Identification of a new factor involved in compartmentalised gene expression during sporulation of *Bacillus subtilis*. *Genes and Development*, **3**, 141–9.

Sussman, M. D. & Setlow, P. (1991). Cloning, nucleotide sequence and regulation of the *Bacillus subtilis gpr* gene, which codes for the protease that initiates degradation of small, acid-soluble proteins during spore germination. *Journal of Bacteriology*, **173**, 291–300.

Tipper, D. J. & Linnett, P. E. (1976). Distribution of peptidoglycan synthetase activities between sporangia and forespores in sporulating cells of *Bacillus sphaericus*. *Journal of Bacteriology*, **126**, 213–21.

Van Hoy, B. E. & Hoch, J. A. (1990). Characterization of the *spoIVB* and *recN* loci of *Bacillus subtilis*. *Journal of Bacteriology*, **172**, 1306–11.

Warburg, R. J., Buchanan, C. E., Parent, K. & Halvorson, H. O. (1986). A detailed study of *gerJ* mutants of *Bacillus subtilis*. *Journal of General Microbiology*, **132**, 2309–19.

Warth, A. D. (1978). Molecular structure of the bacterial spore. *Advances in Microbial Physiology*, **17**, 1–47.

Yansura, D. G. & Henner, D. J. (1984). Use of the *Escherichia coli lac* repressor and operator to control gene expression in *Bacillus subtilis*. *Proceedings of the National Academy of Sciences, USA*, **81**, 439–43.

Yon, J. R., Sammons, R. L. & Smith, D. A. (1989). Cloning and sequencing of the *gerD* gene of *Bacillus subtilis*. *Journal of General Microbiology*, **135**, 3431–45.

Young, M. & Mandelstam, J. (1979). Early events during bacterial endospore formation. *Advances in Microbial Physiology*, **20**, 103–62.

Zheng, L. & Losick, R. (1990). Cascade regulation of spore coat gene expression in *Bacillus subtilis*. *Journal of Molecular Biology*, **212**, 645–60.

Zuberi, A. R., Feavers, I. M. & Moir, A. (1985). Identification of three complementation units in the *gerA* spore germination locus of *Bacillus subtilis*. *Journal of Bacteriology*, **162**, 756–62.

Zuberi, A. R., Feavers, I. M. & Moir, A. (1987). The nucleotide sequence and gene organisation of the *gerA* spore germination operon of *Bacillus subtilis* 168. *Gene*, **51**, 1–11.

ESTABLISHMENT OF FORESPORE-SPECIFIC GENE EXPRESSION DURING SPORULATION OF *BACILLUS SUBTILIS*

PATRICK STRAGIER

Institut de Biologie Physico-Chimique, 13 rue Pierre et Marie Curie, 75005 Paris, France

INTRODUCTION

Sporulation of *Bacillus subtilis* is often considered as a paradigm of cellular differentiation because of its 'simplicity'. The metamorphosis of a growing *B. subtilis* cell into a dormant spore (described by A. Moir in this volume and shown in Fig. 1) involves only two cell types, the forespore and the mother cell, and is controlled by about 100 gene products. The absence of one of these products (due to a mutation in the corresponding gene) either arrests sporulation at a specific developmental stage, or leads to a spore with altered resistance or germination properties (Losick, Youngman & Piggot, 1986). Each cell compartment contains one copy of the bacterial genome but different genes are expressed in the two cell types (Losick & Kroos, 1989): for instance, the *ssp* genes, encoding the small, acid-soluble proteins that bind to the spore chromosome and provide UV resistance to the maturing spore, are transcribed only in the forespore (Mason, Hackett & Setlow, 1988). Conversely, the *cot* genes, encoding the coat proteins that encase the spore in a tough outer shell, are transcribed only in the mother-cell (Zheng & Losick, 1990). As summarized in Fig. 1, differential gene expression during sporulation is mainly controlled by the sequential activity of developmental sigma factors that channel RNA polymerase towards new classes of promoters (Stragier & Losick, 1990). Therefore, it is of major importance to understand the mechanisms that regulate the synthesis and the activity of these sporulation sigma factors.

At the time of asymmetric septation, the first recognizable morphological sporulation event ('stage II'), several sigma factors are present in the bacteria. Some, such as σ^A and σ^H, which are synthesized throughout vegetative growth and are required for transcription of some 'early' sporulation genes, ultimately become inactive (Moran, 1989). Others, such as σ^E and σ^F, which are synthesized before septation (and therefore expected to be randomly segregated in the two compartments), are required for the subsequent engulfment of the smaller compartment by the larger one to form an isolated

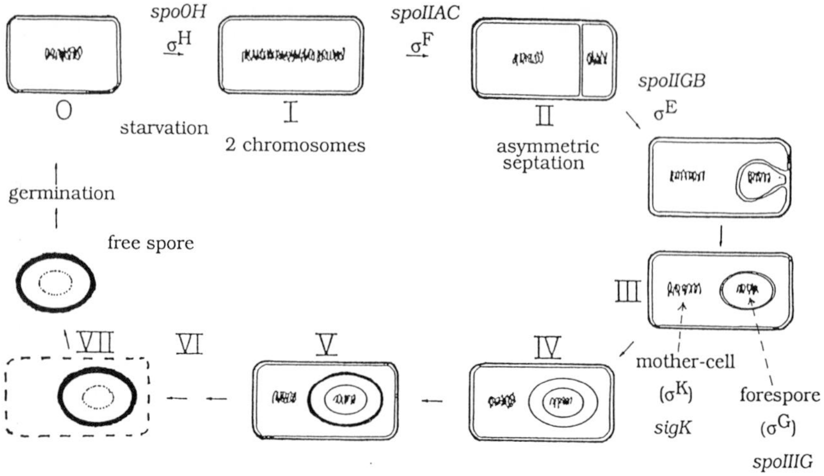

Fig. 1. Stages of sporulation (adapted from Losick *et al.*, 1986). The sigma factors involved in sporulation are indicated with the names of the genes coding for them. Mutations in *spoIIAC* and *spoIIGB* give the same phenotype, blockage at stage II; the precise time of action of σ^F and σ^E with regard to asymmetric septation is not known and their placement in the figure is arbitrary. Mutations in *sigK* block sporulation at stage IV.

forespore within the mother-cell (stage III). Interestingly, both σ^E (the product of the *spoIIGB* gene, the second gene of the *spoIIG* operon) and σ^F (the product of the *spoIIAC* gene, the third gene of the *spoIIA* operon) appear to be synthesized in inactive form. σ^E is made as a precursor, pro-σ^E, whose amino-terminus has to be proteolytically removed (presumably by the *spoIIGA* product) to become active (LaBell, Trempy & Haldenwang, 1987; Stragier, Bonamy & Karmazyn-Campelli, 1988), while σ^F is inhibited by the product of the *spoIIAB* gene through an unknown mechanism which is counteracted by the *spoIIAA* product (Schmidt *et al.*, 1990). Since activation of σ^E and σ^F coincides with the time of asymmetric septation it is formally possible that one (or both) of these processes takes place in only one of the two compartments, which would generate differential gene expression at a very early stage of sporulation.

Bona fide compartment-specific sigma factors have been identified that act at a later stage. Most of the transcription in the forespore depends on σ^G, the *spoIIIG* product (Sun, Stragier & Setlow, 1989; Karmazyn-Campelli *et al.*, 1989; Nicholson *et al.*, 1989), while later transcription in the mother cell depends on σ^K (Kroos, Kunkel & Losick, 1989), the product of a gene created by a chromosomal rearrangement splicing the *spoIVCB* and *spoIIIC* reading frames (Stragier *et al.*, 1989; Sato, Samori & Kobayashi, 1990; Kunkel, Losick & Stragier, 1990). Several redundant controls appear to restrict σ^K synthesis to the mother-cell (reviewed by Kunkel, 1991) but only one (forespore-specific transcription of *spoIIIG*) has been reported to partition

Table 1. *Some characteristics of sporulation sigma factors.*

Sigma factor	encoding gene	regulator[1]	target genes
σ^E	*spoIIGB*	SpoIIGA (+)	*spoIID* *spoIIIA* *spoIIID*
σ^F	*spoIIAC*	SpoIIAA (+) SpoIIAB (−) SpoIIIE (−)	*spoIIIG* *gpr*
σ^G	*spoIIIG*	SpoIIAB (−) SpoIIIA (+)	*spoIIIG* *ssp* genes *gpr*
σ^K	*spoIVCB::spoIIIC*[2]	SpoIVFA (−) SpoIVFB (+)	*spoIVCB::spoIIIC* *cot* genes

[1]The positive or negative effect of each regulator on sigma factor activity is indicated.
[2]The product of the rearranged *spoIVCB::spoIIIC* gene is an inactive precursor, pro-σ^K, where processing is controlled by the products of the *spoIVF* operon (S. Cutting & R. Losick, personal communication).

σ^G activity to the forespore (Karmazyn-Campelli *et al.*, 1989). This latter observation has stimulated further experimentation in several laboratories and has led to the emergence of two very different models for establishment of forespore-specific gene expression. In order to clarify what has become a confusing situation the various data which support the two models will be summarized in this chapter and some hints about apparent discrepancies will be provided. For an easier understanding, an abridged lexicon is given in Table 1.

spoIIIE RESTRICTS σ^F ACTIVITY TO THE FORESPORE WHERE IT TRANSCRIBES *spoIIIG*

The model

In this first model *spoIIIG*, the gene encoding the forespore sigma factor σ^G, is mainly transcribed by σ^F from a promoter located immediately upstream of the *spoIIIG* coding sequence. Transcription of *spoIIIG* occurs only in the forespore because σ^F activity is itself restricted to the forespore compartment (Partridge, Foulger & Errington, 1991). As stated above σ^F, the *spoIIAC* product, is synthesized before septation and its activity is modulated by the antagonistic proteins SpoIIAA and SpoIIAB. It is proposed that the product of the *spoIIIE* locus, a vegetative protein which is absolutely and exclusively required for expression of the forespore genes, boosts SpoIIAA activity (and relieves the inhibitory effect of SpoIIAB on σ^F) shortly after asymmetric septation and only in the small compartment (Foulger & Errington, 1989). In this model the synthesis of σ^G in the forespore is a simple and direct consequence of the already existing compartmentalization of σ^F-directed transcription which is established by the *spoIIIE* product.

The experimental data

The *spoIIIG* locus is located immediately downstream of the *spoIIG* operon. Insertion in the *spoIIG–spoIIIG* intergenic region of a whole circular plasmid by a single homologous recombination event taking place in the cloned DNA fragment ('Campbell-type' recombination) does not interfere with sporulation (Foulger & Errington, 1989). This indicates that *spoIIIG* does not belong to the *spoIIG* operon and is transcribed from its own promoter. This promoter has been identified by primer extension experiments and its sequence appears to be closely related to the sequence of σ^G-recognized promoters (Nicholson *et al.*, 1989). However, although *spoIIIG* can actually be transcribed both *in vitro* and *in vivo* by σ^G-RNA polymerase, it is also transcribed by σ^F-RNA polymerase and with much more efficiency (Schmidt *et al.*, 1990; Partridge *et al.*, 1991; Sun *et al.*, 1991). Due to the strong similarity between the amino-acid sequences of σ^F and σ^G some overlap of their promoter recognition specificity is expected (Masuda *et al.*, 1988; Karmazyn-Campelli *et al.* 1989). Therefore additional genetic evidence is required to discriminate between a σ^F- and a σ^G-controlled promoter.

A *spoIIIG–lacZ* translational fusion was constructed and inserted by a 'Campbell-type' recombination event at the *spoIIIG* locus (Foulger & Errington, 1989). Synthesis of β-galactosidase from this fusion starts around t_2 (2 hours after the end of exponential growth) and is abolished by mutations in several *spo* genes, *spoIIAA, spoIIAC, spoIIE* and *spoIIIE*, but not by mutations in *spoIIIA* and *spoIIIG*. Another translational *spoIIIG–lacZ* fusion was subsequently constructed that inactivates the *spoIIIG* locus after recombination into the chromosome (Partridge *et al.*, 1991). Synthesis of β-galactosidase from this fusion is similar to what is observed with the previous one and the 5′ end of the *spoIIIG–lacZ* hybrid mRNA has been characterized by primer extension analysis. This fusion was used to select for secondary mutations allowing β-galactosidase synthesis during vegetative growth. One such mutation was identified and found to decrease the efficiency of translation of the *spoIIAA* and *spoIIAB* cistrons (Partridge *et al.*, 1991). Since an intact *spoIIAC* gene is still required in this mutant to activate *spoIIIG–lacZ* expression during vegetative growth (but not any of the other genes usually required during sporulation) it is likely that the decreased level of SpoIIAB due to this mutation leads to the presence during growth of some active σ^F molecules that in turn allow transcription of *spoIIIG*. The *spoIIIG–lacZ* mRNA synthesized during growth in this mutant starts at the same position as in the wild type strain during sporulation but appears to be much less abundant.

Another forespore-specific gene, *gpr*, encoding the protease which degrades the small, acid-soluble proteins in the germinating spore, is – at least partially – controlled by σ^F (Sussman & Setlow, 1991). Expression of a *gpr–lacZ* translational fusion (integrated by a double recombination event at the *gpr* locus) is completely abolished by mutations in *spoIIAC* and *spoIIIE* but is reduced only about 50% in a *spoIIIG* mutant. This is in sharp contrast with all other forespore-specific genes (excluding the special case of *spoIIIG* itself) which are completely dependent on σ^G, the *spoIIIG* product (Sun *et al.*, 1989). Moreover, expression of the *gpr–lacZ* fusion appears to precede expression of other forespore-specific genes. Altogether these results suggest that *gpr* is first transcribed by σ^F-RNA polymerase, and later by σ^G-RNA polymerase. It is not known if the σ^F-dependent mode of transcription of *gpr* is compartmentalized to the forespore.

Putting the pieces together

All the above data fit nicely with the proposed model in which *spoIIIG* is transcribed by σ^F-RNA polymerase whose activity is restricted to the smaller compartment by the action of the *spoIIIE* product. The role of *spoIIE* is more obscure since mutations in *spoIIE* block transcription of *spoIIIG* but also block mother-cell specific gene expression, presumably because they prevent processing of pro-σ^E to its active form (Trempy, Morrison-Plummer & Haldenwang, 1985). The situation is made even more complex by taking into account the fact that the absence of σ^E (due to a nonsense mutation in the proximal part of *spoIIGB*) blocks expression of *spoIIIG* and of all σ^G-dependent genes (Mason *et al.*, 1988; Errington *et al.*, 1990). The effect of the same mutation on transcription of *gpr* has not been reported. Within the frame of the current model the immediate interpretation would be that σ^E (or the product of a gene controlled by σ^E) is required for σ^F activity. And here comes the paradox: null mutations in *spoIIAC* (as well as missense mutations in *spoIIAA*) prevent processing of pro-σ^E (Stragier *et al.*, 1988; Jonas & Haldenwang, 1989) and block the whole σ^E-dependent regulon. In other words σ^E and σ^F are dependent upon each other for their activation. How could it work?

An explanation has been provided by the discovery that some missense mutations in *spoIIGB* (affecting σ^E activity) do not interfere with *spoIIIG–lacZ* expression (which presumably reflects σ^F activity) and that some missense mutations in *spoIIAC* (affecting σ^F activity) allow expression of a *spoIIID–lacZ* fusion (*spoIIID* is transcribed by σ^E-RNA polymerase). Therefore it is proposed that the reciprocal activation of σ^E and σ^F is not mediated by their sigma factor function but merely by some protein-protein interaction (Errington *et al.*, 1990).

Attempts have been made to formalize these results and interpretations (without betraying the authors' intentions!) in Fig. 2. Several proteins (which are all synthesized prior to asymmetric septation) are supposed to interact with each other. Since the *spoIIGA* (Stragier *et al.*, 1988) and *spoIIIE* (Foulger & Errington, 1989) products are predicted from their sequence to be membrane bound the whole complex could be formed at the immediate vicinity of the membrane. The inactive form of σ^F (inhibited by SpoIIAB) would interact with SpoIIGA (itself in contact with its normal substrate pro-σ^E) and would block its processing activity. Such a dual interaction would at the same time protect both inactive σ^F and pro-σ^E from the housekeeping proteases which remove inactive enzymes from the cell. This protection could be the basis for the reciprocal relationship existing between σ^E and σ^F activity.

The interaction of inactive σ^F with SpoIIGA would be disrupted in response to some temporal and/or spatial signal. SpoIIGA could then process pro-σ^E and SpoIIAA could convert σ^F to its active form. A further refinement of the model involves the SpoIIIE protein that would act after that step (since it is not required for pro-σ^E processing) and somehow would channel either σ^F activation or the active σ^F molecules towards the smaller compartment. It has also been proposed that pro-σ^E processing occurs only in the larger compartment since all σ^E-dependent genes tested to date appear to be required only in the mother-cell (Errington *et al.*, 1990). Since no specific regulator has been identified that could act at that step the simplest interpretation is to suggest that the disruption of the σ^F-SpoIIGA interaction is biased in such a way that SpoIIGA becomes active in the larger compartment while σ^F would be available to SpoIIAA and/or SpoIIIE in the smaller one.

SPOIIIA ACTIVATES σ^G IN THE FORESPORE WHERE IT TRANSCRIBES *SPOIIIG*

The model

In this second model (shown in Fig. 3) *spoIIIG* is initially transcribed as the third gene of the *spoIIG* operon (Masuda *et al.*, 1988; Karmazyn-Campelli *et al.*, 1989). However, translation of the *spoIIIG* reading frame is prevented by a stem-loop structure in this large mRNA which makes its biological relevance questionable (C. Karmazyn-Campelli & P. Stragier, unpublished observations). Actual synthesis of σ^G starts only about 1 hour later from a smaller mRNA initiated by σ^F-RNA polymerase. Despite the presence of σ^G, the transcription of its target genes cannot be detected until a stage III forespore has been formed. This delay is mainly due to the inhibition of σ^G by the SpoIIAB protein. Reactivation of σ^G is mediated by the products of the *spoIIIA* operon and occurs only in the forespore (C. Karmazyn-Campelli & P. Stragier, unpublished observations). *spoIIIG* is then transcribed from a proximal promoter by σ^G-RNA polymerase which leads to a burst of active σ^G in the forespore (Karmazyn-Campelli *et al.*, 1989). No previous

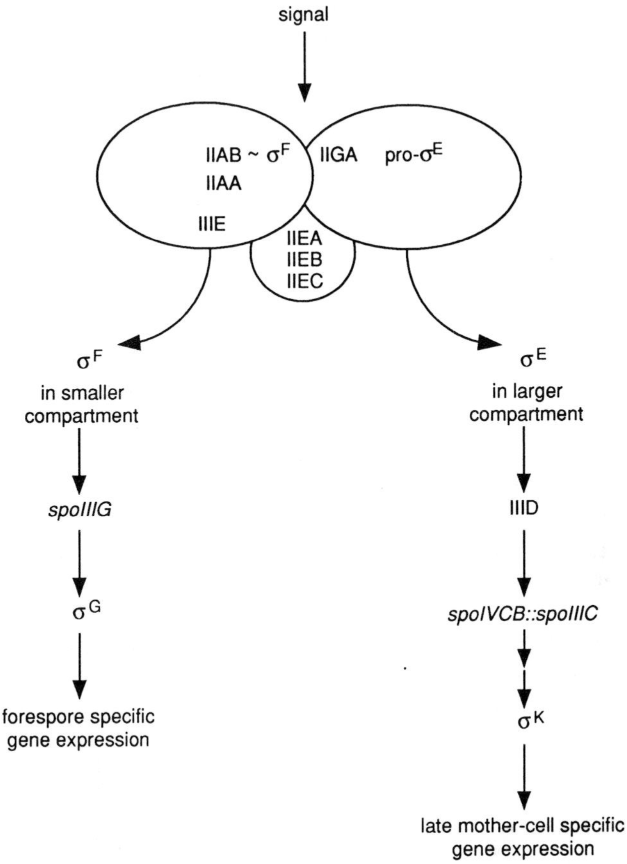

Fig. 2. A model of establishment of compartmentalized gene expression immediately after asymmetric septation. A multiprotein complex (presumably membrane bound) is shown on the top. After receiving some temporal or spatial signal it segregates σ^E and σ^F activities in each of the two compartments of the sporulating cell where they induce the synthesis of σ^K and σ^G respectively. Arrows indicate an activatory effect or a gene–product relationship.

bias in σ^F and/or σ^E compartmentalization is required for achieving forespore-specific transcription of *spoIIIG*.

The experimental data

Various DNA fragments containing the beginning of *spoIIIG* and adjacent sequences were cloned upstream of *lacZ*, either as transcriptional or translational fusions. These *spoIIIG-lacZ* hybrid genes were integrated at the non-essential *amy* locus of the *B. subtilis* chromosome by a double recombination event involving *amy* sequences flanking the fusions on both sides and *β*-

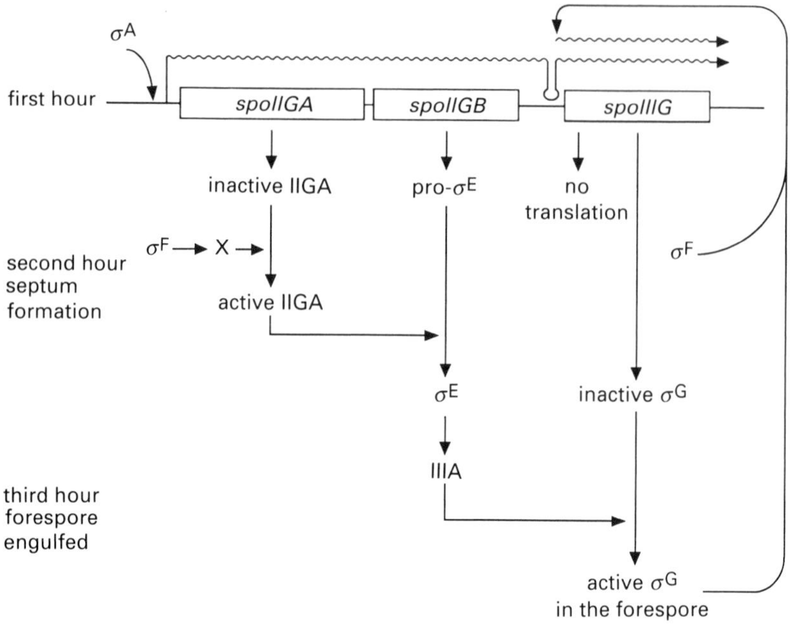

Fig. 3. Cascade of regulations leading to forespore-specific transcription of *spoIIIG* by σ^G. Gene coding sequences are shown as boxes and mRNA are indicated by wavy arrows. A putative stem-loop structure is shown in the larger mRNA. An approximate time scale is given on the left part of the figure with relevant morphological stages. Straight arrows indicate an activatory effect or a gene–product (or precursor–mature product) relationship.

galactosidase synthesis was monitored during sporulation (Karmazyn-Campelli *et al.*, 1989; C. Karmazyn-Campelli & P. Stragier, unpublished data). Strikingly different results were obtained with both types of fusions.

When a DNA fragment carrying the beginning of *spoIIIG* as well as the whole upstream *spoIIG* operon is cloned upstream of *lacZ* as a transcriptional fusion β-galactosidase synthesis starts during the first hour of sporulation with a second peak of activity around t_3. The early mode of transcription of *spoIIIG* requires an intact *spoIIG* promoter but is independent of σ^E, σ^F and σ^G. Conversely, the late mode of transcription of *spoIIIG* is abolished by mutations in *spoIIAC, spoIIGB, spoIID* (a well-known σ^E-dependent gene), *spoIIIA* and *spoIIIG* itself. Only this second mode of transcription is observed when a smaller DNA fragment containing the end of *spoIIGB* and the beginning of *spoIIIG* is cloned upstream of *lacZ* as a transcriptional fusion.

When the 'larger' or the 'smaller' fragments are used to construct translational *lacZ* fusions at the *amy* locus only one peak of β-galactosidase synthesis is observed which starts in both cases during the second hour of sporulation and is reduced to about 30% of the wild-type level in *spoIIGB, spoIID, spoIIIA* and *spoIIIG* mutants. It is completely abolished by mutations in *spoIIE* and *spoIIAC* (C. Karmazyn-Campelli & P. Stragier, unpublished data). Both fusions give indistinguishable results.

The delay observed between synthesis of σ^G (monitored by the expression of the translational *spoIIIG–lacZ* fusions) and expression of σ^G-dependent genes (monitored for instance by the expression of a *sepB–lacZ* fusion) suggests that σ^G is initially made in inactive form. Such a regulatory control was already postulated from previous experiments in which induction of σ^G synthesis during growth led to efficient transcription of σ^G-dependent genes while only poor expression could be obtained when induction was achieved after t_0 (Stragier, 1989). A potential inhibitor has been identified with the isolation of mutations in *spoIIAB* which increase σ^G-dependent transcription (Rather *et al.*, 1990).

Putting the pieces together

The above data fit with a model in which *spoIIIG* is first transcribed by readthrough from the *spoIIG* operon, a mode of transcription which does not depend on any stage II gene. The difference in the kinetics of expression of the 'larger' transcriptional and translational *spoIIIG–lacZ* fusions suggests that translation of *spoIIIG* from the large mRNA is prevented, presumably by a stable stem–loop structure sequestering the *spoIIIG* ribosome binding site (Masuda *et al.*, 1988). Since the 'smaller' and 'larger' translational *spoIIIG–lacZ* fusions show identical patterns of β-galactosidase synthesis it appears that transcription from the upstream *spoIIG* operon plays no significant role in σ^G synthesis which depends exclusively on mRNAs initiated in the *spoIIGB–spoIIIG* intergenic region.

The loss of about 70% of β-galactosidase synthesis from translational *spoIIIG–lacZ* fusions in a *spoIIIG* background indicates that the *spoIIIG* promoter is recognised *in vivo* by σ^G itself. Therefore another sigma factor has to prime σ^G synthesis and σ^F appears as a likely candidate since mutations in *spoIIAC* (and *spoIIE*) completely abolish β-galactosidase synthesis from translational *spoIIIG–lacZ* fusions. Transcription from the *spoIIIG* promoter can actually be induced during vegetative growth by the presence of either σ^F or σ^G (Schmidt *et al.*, 1990). Other experiments have shown that σ^F activity is abolished in a *spoIIE* background (P. Margolis & R. Losick, personal communication). Therefore, it seems likely that the *spoIIE* products control σ^F activity, presumably by modulating its interactions with SpoIIAA and SpoIIAB, which explains the absolute dependence of *spoIIIG–lacZ* translational fusions on *spoIIE*.

A puzzling result is the complete dependence of the 'smaller' transcriptional *spoIIIG–lacZ* fusion on *spoIIIG* itself (Karmazyn-Campelli *et al.*, 1989). Since the same DNA fragment allows 30% of the activity of a translational *spoIIIG–lacZ* fusion in a *spoIIIG* strain (and therefore is able to promote some transcription in the absence of σ^G) we have to assume that the absence of β-galactosidase synthesis from the 'smaller' transcriptional *spoIIIG–lacZ* fusion in a *spoIIIG* strain reflects some unusual inhibition of translation of the *lacZ* gene.

The delay between translation of *spoIIIG* and activation of forespore-specific genes implies that σ^G activity is inhibited, presumably by SpoIIAB. Such a cross reaction with the σ^F inhibitor is not too surprising because of the strong similarities between these two sigma factors. However the relief of SpoIIAB-mediated inhibition seems to follow a very different pathway for σ^F and σ^G. The first molecules of σ^G are synthesized after reactivation of σ^F by SpoIIAA, while activation of σ^G occurs about 1 hour later and requires the product of multiple genes. Only one class of mutations leads to the same morphological block as *spoIIIG* and prevents σ^G activity; they define the *spoIIIA* locus. Mutations in this locus have been shown in several laboratories to abolish transcription of forespore-specific genes (Mason *et al.*, 1988; Rather & Moran, 1988; Karmazyn-Campelli *et al.*, 1989; Cutting *et al.*, 1991). Because of their effect on σ^G activity it appears that the *spoIIIA* products are involved in the switch of σ^G to its active form. The *spoIIIA* operon is transcribed by σ^E-RNA polymerase as is another gene, *spoIID*, in which mutations also block σ^G activity although they do not affect *spoIIIA* expression (Rong, Rosenkrantz & Sonenshein, 1986; A.-M. Guérout-Fleury, G. Gonzy-Tréboul & P. Stragier, unpublished observations). This complex dependence pattern can be explained by assuming that reactivation of σ^G requires an isolated forespore and, more specifically, that the *spoIIIA* products modify some physico-chemical parameter of the forespore cytoplasm that is involved in σ^G inactivation. Any mutation interfering with engulfment of the forespore (as, for example, *spoIID*) will have the same negative effect on σ^G activity as mutations in *spoIIIA*.

RIGHT YOU ARE, IF YOU THINK YOU ARE

A striking discrepancy between the two models described above is about the *spoIIIA* operon which plays a major role in the second model and no role at all in the first one. It has been claimed that mutations in *spoIIIA* have no effect on expression of some forespore-specific genes (Illing & Errington, 1990). This is in sharp contrast with results from several other laboratories that agree on the dramatic effect of such mutations on a variety of forespore-expressed genes (Mason *et al.*, 1988; Rather & Moran, 1988; Karmazyn-Campelli *et al.*, 1989; Cutting *et al.*, 1991). Obviously, there must be some basic difference in the strains or in the experimental procedures used in the various laboratories to explain such an inconsistency.

Another subject of contention is the role of the *spoIIIE* locus. Although not stated in the description of the data leading to the second model a *spoIIIE* mutation is found to block expression of a translational *spoIIIG–lacZ* fusion at the *spoIIIG* locus (C. Karmazyn-Campelli & P. Stragier, unpublished observations), in agreement with results from others (Foulger & Errington, 1989). This result fits with the idea that the *spoIIIE* effect is mediated by the loss of σ^F activity (and therefore the complete absence of *spoIIIG* transcription). However, strikingly opposite results are obtained when expression of *spoIIIG* (and other forespore-specific genes) is monitored through *lacZ* fusions integrated at the *amy* locus where it is observed that a *spoIIIE* mutation actually enhances expression of these genes (C. Karmazyn-Campelli & P. Stragier, unpublished observations). A similar unusual dependency upon the location of the *lacZ* fusion on the chromosome is reported for expression of other genes recognized by variants of σ^F (P. Margolis & R. Losick, personal communication). Since the SpoIIIE protein is made during vegetative growth (Foulger & Errington, 1989) it seems odd that its *raison d'être* could be in establishing compartmentalized gene expression after asymmetric septation. It appears merely that the absence of this membrane-bound protein might modify the global structure of the chromosome and grossly perturbate transcription. Some genes could be silenced and others activated depending on their chromosomal location. It is conceivable that abnormal expression of some genes during early sporulation in a *spoIIIE* mutant ultimately blocks the whole process with the lysis of the forespore (Stragier, 1989). According to this interpretation the interference of the *spoIIIE* product with sporulation could be very indirect.

The model shown in Fig. 3 conveys the idea that there is a cascade of activation leading from σ^F to σ^E and from both σ^F and σ^E to σ^G. The relationship between σ^E and σ^F is not symmetric as it is presented in Fig. 2 because null mutations in *spoIIGB* (destroying σ^E) have been found not to affect σ^F activity (monitored by the expression of a translational *spoIIIG–lacZ* fusion) (C. Karmazyn-Campelli & P. Stragier, unpublished observations). Conversely, it is widely accepted that mutations in *spoIIAC* block pro-σ^E processing (Jonas & Haldenwang, 1989). Since this is also the case for mutations in *spoIIAA* (Stragier *et al.*, 1988; Jonas & Haldenwang, 1989) which encodes the activator of σ^F (Schmidt *et al.*, 1990), it is hard to understand how such an effect could not be mediated by the sigma factor activity of σ^F. Therefore, as suggested some years ago (Stragier *et al.*, 1988), it appears more logical to propose that σ^F activates some unknown gene whose product (here called X) modulates, directly or indirectly, SpoIIGA activity (see Fig. 3). Such a gene could be controlled by a strong σ^F-dependent promoter which would still be activated in missense mutants of *spoIIAC* where pro-σ^E processing takes place (Errington *et al.*, 1990).

The essential difference between the two models deals with the timing and devices of compartmentalization of gene expression. In the first model transcription factors are compartmentalized as early as asymmetric septation, with active σ^F in the smaller compartment and σ^E in the larger one. Some undefined physical interaction between the inactive forms of the two sigma factors has to be postulated and no clue has yet been provided for the molecular basis of the complex process leading to the simultaneous activation of σ^E and σ^F in opposite subcellular compartments. Such an early compartmentalization makes the role of σ^G redundant, especially since it shares similar promoter-recognition specificity with σ^F. In the second model, the potential compartmentalization of σ^E and/or σ^F is irrelevant. It postulates that the whole control of forespore-specific transcription relies on the activation of σ^G through modification of some physico-chemical parameter of the forespore cytoplasm. It does not matter where the *spoIIIA* operon is expressed since its products could extrude some inhibitory component (for instance) from the inside as well as from the outside of the forespore. Although some pieces are still missing (there are no candidates for gene X) that model explains how *spoIID* mutations which block engulfment also prevent forespore-specific gene expression. The whole purpose of such a cascade of controls exerted on σ^G would be to couple transcription of forespore genes to the mere presence of a forespore completely surrounded by its typical double membrane. Similar morphological checkpoints have already been suggested for earlier and later stages of sporulation (Stragier *et al.*, 1988; Cutting *et al.*, 1990) and an additional one would be fulfilling from both an esthetical and a regulatory point of view.

ACKNOWLEDGEMENTS

I am deeply grateful to Céline Karmazyn-Campelli who did all the experiments on *spoIIIG* expression that led to the 'second model' reported in this chapter. I thank Richard Losick for unpublished information and many challenging discussions. Work from my laboratory was supported by grants from CNRS (URA1139) and INSERM (Contrat de Recherche Extrerne 881016).

REFERENCES

Cutting, S., Driks, A., Schmidt, R., Kunkel, B. & Losick, R. (1991). Forespore-specific transcription of a gene in the signal transduction pathway that governs pro-σ^K processing in *Bacillus subtilis*. *Genes & Development*, **5**, 456–66.

Cutting, S., Oke, V., Driks, A., Losick, R., Lu, S. & Kroos, L. (1990). A forespore checkpoint for mother cell gene expression during development in *B. subtilis*. *Cell*, **62**, 239–50.

Errington, J., Foulger, D., Illing, N., Partridge, S. R. & Stevens, C. M. (1990). Regulation of differential gene expression during sporulation in *Bacillus subtilis*. In *Genetics and Biotechnology of Bacilli*, ed. M. M. Zukowski, A. T. Ganesan & J. A. Hoch, vol. 3, pp. 257–67, Academic Press, Inc., San Diego.

Foulger, D. & Errington, J. (1989). The role of the sporulation gene *spoIIIE* in the regulation of prespore-specific gene expression in *Bacillus subtilis*. *Molecular Microbiology*, **3**, 1247–55.

Illing, N. & Errington, J. (1990). The *spoIIIA* locus is not a major determinant of prespore-specific gene expression during sporulation in *Bacillus subtilis. Journal of Bacteriology*, **172**, 6930–6.

Jonas, R. M. & Haldenwang, W. G. (1989). Influence of *spo* mutations on σ^F synthesis in *Bacillus subtilis. Journal of Bacteriology*, **171**, 5226–8.

Karmazyn-Campelli, C., Bonamy, C., Savelli, B. & Stragier, P. (1989). Tandem genes encoding σ-factors for consecutive steps of development in *Bacillus subtilis. Genes and Development*, **3**, 150–7.

Kroos, L., Kunkel, B. & Losick, R. (1989). Switch protein alters specificity of RNA polymerase containing a compartment-specific sigma factor. *Science*, **243**, 526–9.

Kunkel, B. (1991). Regulation of compartmentalized gene expression during sporulation in *Bacillus subtilis. Trends in Genetics*, **7**, 167–72.

Kunkel, B., Losick, R. & Stragier, P. (1990). The *Bacillus subtilis* gene for the developmental transcription factor σ^K is generated by excision of a dispensable DNA element containing a sporulation recombinase gene. *Genes and Development*, **4**, 525–35.

LaBell, T. L., Trempy, J. E. & Haldenwang, W. G. (1987). Sporulation-specific σ factor σ^{29} of *Bacillus subtilis* is synthesized from a precursor protein, P^{31}. *Proceedings of the National Academy of Sciences, USA*, **84**, 1784–8.

Losick, R. & Kroos, L. (1989). Dependence pathways for the expression of genes involved in endospore formation in *Bacillus subtilis*. In *Regulation of Prokaryotic Development*, ed. I. Smith, R. A. Slepecky & P. Setlow, pp. 223–41, American Society for Microbiology, Washington.

Losick, R., Youngman, P. & Piggot, P. (1986). Genetics of endospore formation in *Bacillus subtilis. Annual Review of Genetics*, **20**, 625–69.

Mason, J. M., Hackett, R. H. & Setlow, P. (1988). Studies on the regulation of expression of genes coding for small, acid-soluble proteins of *Bacillus subtilis* spores using *lacZ* gene fusions. *Journal of Bacteriology*, **170**, 239–44.

Masuda, E. S., Anaguchi, H., Yamada, K. & Kobayashi, Y. (1988). Two developmental genes encoding σ factor homologs are arranged in tandem in *Bacillus subtilis. Proceedings of the National Academy of Sciences, USA*, **85**, 7637–41.

Moran, C. P. Jr (1989). Sigma factors and the regulation of transcription. In *Regulation of Prokaryotic Development*, ed. I. Smith, R. A. Slepecky & P. Setlow, pp. 167–84, American Society for Microbiology, Washington.

Nicholson, W. N., Sun, D., Setlow, B. & Setlow, P. (1989). Promoter specificity of σ^G containing RNA polymerase from sporulating cells of *Bacillus subtilis*: identification of a group of forespore-specific promoters. *Journal of Bacteriology*, **171**, 2708–18.

Partridge, S. R., Foulger, D. & Errington, J. (1991). The role of σ^F in prespore-specific transcription in *Bacillus subtilis. Molecular Microbiology*, **5**, 757–67.

Rather, P. N., Coppolecchia, R., DeGrazia, H. & Moran, C. P. Jr (1990). Negative regulator of σ^G-controlled gene expression in stationary-phase *Bacillus subtilis. Journal of Bacteriology*, **172**, 709–15.

Rather, P. N. & Moran, C. P. Jr (1988). Compartment-specific transcription in *Bacillus subtilis*: identification of the promoter for *gdh. Journal of Bacteriology*, **170**, 5086–92.

Rong, S., Rosenkrantz, M. S. & Sonenshein, A. L. (1986). Transcriptional control of the *Bacillus subtilis spoIID* gene. *Journal of Bacteriology*, **165**, 771–9.

Sato, T., Samori, Y. & Kobayashi, Y. (1990). The *cisA* cistron of *Bacillus subtilis* sporulation gene *spoIVC* encodes a protein homologous to a site-specific recombinase. *Journal of Bacteriology*, **172**, 1092–8.

Schmidt, R., Margolis, P., Duncan, L., Coppolecchia, R., Moran, C. P. Jr. & Losick, R. (1990). Control of developmental transcription factor σ^F by sporulation regulatory proteins SpoIIAA and SpoIIAB in *Bacillus subtilis. Proceedings of the National Academy of Sciences, USA*, **87**, 9221–5.

Stragier, P. (1989). Temporal and spatial control of gene expression during sporulation: from facts to speculations. In *Regulation of Prokaryotic Development*, ed. I. Smith, R. A. Slepecky & P. Setlow, pp. 243–54. American Society for Microbiology, Washington.

Stragier, P., Bonamy, C. & Karmazyn-Campelli, C. (1988). Processing of a sporulation sigma factor in *Bacillus subtilis*: how morphological structure could control gene expression. *Cell*, **52**, 697–704.

Stragier, P., Kunkel, B., Kroos, L. & Losick, R. (1989). Chromosomal rearrangement generating a composite gene for a developmental transcription factor. *Science*, **243**, 507–12.

Stragier, P. & Losick, R. (1990). Cascades of sigma factors revisited. *Molecular Microbiology*, **4**, 1801–6.

Sun, D., Stragier, P. & Setlow, P. (1989). Identification of a new σ-factor involved in compartmentalized gene expression during sporulation of *Bacillus subtilis. Genes and Development*, **3**, 141–9.

Sussman, M. D. & Setlow, P. (1991). Cloning, nucleotide sequence, and regulation of the *Bacillus subtilis gpr* gene, which codes for the protease that initiates degradation of small, acid-soluble proteins during spore germination. *Journal of Bacteriology*, **173**, 291–300.

Trempy, J. E., Morrison-Plummer, J. & Haldenwang, W. G. (1985). Synthesis of σ^{29}, an RNA polymerase specificity determinant, is a developmentally regulated event in *Bacillus subtilis. Journal of Bacteriology*, **161**, 340–6.

Zheng, L. & Losick, R. (1990). Cascade regulation of spore coat gene expression in *Bacillus subtilis. Journal of Molecular Biology*, **212**, 645–60.

THE PERIPLASM

S. J. FERGUSON

Department of Biochemistry, University of Oxford, South Parks Road, Oxford, OX1 3QU, U.K.

INTRODUCTION

The intention in this contribution is to summarise aspects of what is presently known concerning the physical dimensions, molecular composition, functions and biosynthesis of the bacterial periplasm. The latter is a feature of Gram-negative organisms which contributes considerably to the life of such cells, although in different ways depending upon both the organism and the growth conditions. At the end of the chapter, brief consideration will be given to the question of the implications for Gram-positive bacteria of the absence from them of a periplasm, at least in the strict sense of the term. It is with consideration of what is meant by 'periplasm' and the description of a molecule as 'periplasmic' that this chapter commences.

The term periplasm was coined by Mitchell (1961) to describe the region between the cell wall of *Escherichia coli* and the outer surface of the cytoplasmic membrane. The requirement to define this region arose because studies on a water-soluble glucose-6-phosphatase activity showed that the enzyme was located externally to the cytoplasmic membrane. Shortly thereafter several other water-soluble proteins were recognised to be located on the exterior side of the bacterial cytoplasmic membrane and in reviewing this work Heppel (1967) wrote, with reference to Mitchell's paper, of the 'periplasmic space'. Subsequently, the terms periplasm and periplasmic space have both been used, often interchangeably. The terms have, however, been distinguished. Beveridge (1981), for instance, has defined the space as a region that separates the outer membrane (cell wall) of a Gram negative organism from the plasma membrane; the substance of this space was termed the periplasm. It is arguable whether the term periplasmic space is helpful because the word 'space' can give an impression of emptiness. As will be seen there is little evidence for space, in this sense of the word, between the outer and cytoplasmic membranes and in the present article periplasm will be used as in Mitchell's original definition. The term periplasm is then broadly comparable with the usage of 'cytoplasm'; cytoplasmic space is not a term in common use.

The periplasm cannot be thought of as having strict boundaries. At the side adjacent to the cytoplasmic membrane, proteins will project from the membrane into the periplasm. A very approximate boundary could be regarded as the phospholipid head group region. Towards the outer membrane, the boundary of the periplasm is even more ill defined. The murein layer will not be a sharp boundary, and, indeed, as will be seen later there is reason to believe that the murein layer may extend into the whole periplasm, there being a gradient of cross linking from high density at the outer membrane side to low density closer to the cytoplasmic membrane. Most research on the periplasm as an entity has been done on *E. coli*. This should be kept in mind throughout this chapter; one cannot necessarily extrapolate from *E. coli* to all other Gram negative organisms. It should also be noted that in some organisms, for example, many genera of photosynthetic bacteria, the cytoplasmic membrane is invaginated. This means that there will be regions that are contiguous with the periplasm as defined above but in which the periplasm is effectively bounded on all sides by the external surface of the cytoplasmic membrane. In such regions it may be supposed that the penetration of murein, even at very low density of cross-linking, is very limited. This proposition has not, however, been put to experimental test.

SIZE OF THE PERIPLASM

The dimensions of the periplasm are uncertain. Most measurements of the physical size of the periplasm have been done with enteric bacteria. As pointed out by Van Wielink & Duine (1990), many textbooks suggest a width of approximately 7 nm (e.g. Ferguson, 1988*a*). This value, derived from electron micrographs, has become relatively established. Given that the smallest dimension of a periplasmic protein can be of the order of 4 nm and that the estimates of size have not been made with a very wide range of organisms, the idea of a universal width of 7 nm for the bacterial periplasm needs to be treated with caution (and see Van Wielink & Duine, 1990). New results from electron microscopy on *E. coli* (Luduc *et al.*, 1989) indicate an overall width of 11–15 nm, with a 8 nm thick sacculus region comprising cross-linked peptidoglycan (murein) juxtaposed to the outer membrane. Widths in the same range have recently been reported for several other genera including Pseudomonads (Graham *et al.*, 1991).

By measuring the water space that was accessible to medium-sized molecules (which could penetrate the outer but not the cytoplasmic membrane) but not to large ones (to which both membranes were impermeable), Stock, Rauch & Roseman (1977) concluded that the periplasm occupied between 20 and 40% of the aqueous regions of *Salmonella typhimurium* and *E. coli*. It has been asked whether figures in this range are compatible with a periplasmic width of 7 nm (Van Wielink & Duine, 1990). Nikaido and Nakae (1979) have queried the results of Stock *et al.* (1977) and have suggested a value for the volume of approximately 5% for unplasmolysed cells. Brass (1986) has also doubted that the periplasm can contain as much as 20–40% of the cell water and has argued for a contribution of 7%. A complication is that the volume contributed by the periplasm can be variable, with it rising to 13% for starved or stationary phase cells and going as high as 40–50% in cells plasmolysed in 0.5 M NaCl (Nikaido & Nakae, 1979). Van Wielink & Duine (1990) have calculated that, if 20 or 40% of the cell volume in the organism *Thiobacillus versutus* were to be periplasm, then the latter would have a width of 32 or 71 nm, respectively. As discussed above, there is at present no evidence for such a large width. But there is clearly uncertainty about the physical dimensions of the periplasm and it is important to keep in mind that there could be considerable variation between different organisms. The implications of the size of the periplasm for the functioning of the proteins that are located therein will be addressed later.

A final facet to the issue of periplasmic size is that the predatory bacterium *Bdellovibrio bacteriovorus* 109J grows in the periplasm of *E. coli* (Rittenberg & Thomashow, 1979). Although *B. bacteriovorus* is a very small cell when it colonises *E. coli*, and must force the cytoplasmic and outer membranes apart upon entry or shortly thereafter, the initial entry of the predator cell is difficult to reconcile with a very narrow periplasm.

GROSS ORGANISATION OF THE PERIPLASM

Hobot *et al.* (1984) have made the important proposal, at least for *E. coli*, that the periplasm is filled by a gel. Evidence in favour of this proposition included (i) the maintenance of a regular distance between the cytoplasmic and outer membranes even when cryosubstituted samples of cells are in organic solvents in which no biological turgour can exist, and (ii) observation of a low average electron density upon staining which is inconsistent with a high occupancy of proteins. Staining for peptidoglycan with phosphotungstic acid provides the evidence that the peptidoglycan forms an uninterrupted layer approximately 8 nm thick (Leduc *et al.*, 1989). This is called the sacculus and is considered to be sufficiently hydrated to form a gel. Relatively uncross-linked, but hydrated, glycan strands extending from the peptidoglycan, together with hydrated oligosaccharides (see below), are proposed to constitute the gel phase that approaches the cytoplasmic membrane (Kellenberger, 1990). As will be discussed below, there is no doubt that there are many proteins in the periplasm. It is generally assumed that these would be excluded from the peptidoglycan layer. In the gel model, it is contemplated that the mesh size of the sacculus would be sufficient to allow for protein diffusion; the analogy with the diffusion of proteins through highly hydrated agarose and acrylamide gels has been drawn (Kellenberger, 1990). Closer to the cytoplasmic membrane, the periplasmic proteins would diffuse through a less cross-linked gel.

The concept of a gel phase in the periplasm implies that the motion of its proteins might be relatively restricted. Brass *et al.* (1986) investigated this point by introducing into the periplasm proteins that had been labelled with a fluorophore. Such introduction of exogenous proteins was possible because a methodology for selectively permeabilizing the outer membrane with Ca^{2+} was devised (Brass, 1986). Once incorporated into the periplasm of enlarged (by drug treatment) cells, the diffusion of the labelled proteins was studied by the measurement of fluorescence recovery following photobleaching. In this way, lateral diffusion coefficients of the order of 5×10^{-10} $cm^2\ s^{-1}$ were obtained. Such values are 1000-fold lower than expected for diffusion in aqueous solution and 100-fold less than estimates of diffusion in the bacterial cytoplasm. The slow diffusion could not be attributed to specific protein–protein interactions, because comparable diffusion coefficients were obtained irrespective of whether the labelled protein was from either the original bacterium or a eukaryotic cell. For technical reasons elongated cells of *E. coli* were used in these studies and the elongation was achieved through treatment with cephalexin. In a subsequent study (Folley *et al.*, 1989) elongation was achieved either with lower cephalexin concentrations and shorter incubation times or by use of a temperature-sensitive mutant (*ftsA*) of *E. coli* that forms filaments at certain temperatures. In this later study lateral diffusion coefficients of 2.5×10^{-9} $cm^2\ s^{-1}$ were obtained. Although larger by a factor of four than found in the original study, these more recent values still point to a very restricted motion of proteins in the periplasm. Data of these types have been used as support for the concept of a periplasmic gel (Kellenberger, 1990). The extent to which the periplasms of other organisms should be regarded as gels rather than aqueous phases remains to be determined. A very recent study has shown that of ten (excluding *E. coli*) organisms investigated six had a periplasmic gel whilst four others, including *Pseudomonas aeruginosa* and *Vibrio cholerae*, did not (Graham *et al.*, 1991). It is not yet established whether factors such as phase of growth and nutrition might influence the formation of a periplasmic gel (Graham *et al.*, 1991).

A further important point has been investigated with the photobleaching technique. There is morphological evidence (MacAllister, Macdonald & Rothfield, 1983) that the periplasm may be compartmentalised by localised zones of adhesion between the cytoplasmic and outer membranes. Such zones are envisaged to define future sites of cell division. A uniformly low rate of recovery of fluorescence after photobleaching has been found at potential sites of cell division, indicating the presence of some local compartments in the periplasm, primarily located at the sites of past or future cell divisions (Folley *et al.*, 1989).

It has been widely considered that the periplasm is interrupted by many connections between the inner and outer membranes. Such connections, which are distinct from attachments between the cytoplasmic membrane and peptidoglycan proposed to be connected with cell division (see above and MacAllister *et al.*, 1983), are known as Bayer Bridges. The existence of such bridges was deduced for *E. coli* using evidence from the electron microscope. They were proposed to play an important role in the trafficking of proteins from the cytoplasm into the outer membrane. However, recent (Kellenberger, 1990) electron microscopy work and the identification of a continuous sacculus (Leduc *et al.*, 1989) indicates that such connections are not real. It is argued that proteins destined for the outer membrane are released from the cytoplasmic membrane into the periplasm in which they diffuse to the outer membrane. This idea is, supported for example, by the ease of release of the periplasmic intermediates of the Lam B protein (Kellenberger, 1990) and the evidence that enterotoxin of *Vibrio chlorae* passes through the periplasm before export from the cell (Hirst & Holmgren, 1987). Doubtless, the last word has not been said on the issue of whether the periplasm is interrupted by close appositions, presumably involving a lipidic fusion, between the two membranes.

Figure 1 summarises this section, and to some extent anticipates points to be made later about periplasmic proteins, by offering a generalised view of the components and the organisation of the periplasm.

CRITERIA FOR DESIGNATION OF A MOLECULE AS PERIPLASMIC

There are undoubtedly water-soluble proteins located in the periplasm. The criteria by which they can be assigned to this compartment have been reviewed previously (Ferguson, 1988*a*; Beacham, 1979). Suffice it to say here that, increasingly, the identification of an amino terminal signal sequence for export of a protein from the cytoplasm provides evidence for a periplasmic location. Ideally, such evidence should be complemented by biochemical criteria. A recent example of immunochemical methods applied to a nitrite reductase is given by Coyne *et al.* (1990). Non-proteinaceous molecules, e.g. sugars, are subject to a smaller number of experimental approaches and generally their selective release from cells under conditions in which cytoplasmic molecules are retained is the criterion used. Not all the polypeptide chains present in the periplasm are water-soluble proteins. A protein may have a substantial domain exposed in the periplasm but be anchored to the cytoplasmic membrane by one or more stretches of alpha helix (Fig. 1). Such proteins will contribute to the function of the periplasm if their domains in this region carry functionally important parts of the protein, e.g. active site of an enzyme, (see also Ferguson, 1988*a*). Similarly, a protein anchored to the outer membrane, or one that interacts with both membranes, e.g. the TonB protein that is involved in active transport across the outer membrane of *E. coli* (Postle, 1990), can be considered to contribute the function and activity of the periplasm.

Study of the water-soluble periplasmic components is facilitated if these components can be selectively released from cells. There are many ways of seeking to do this and traditionally osmotic shock or treatment of cells

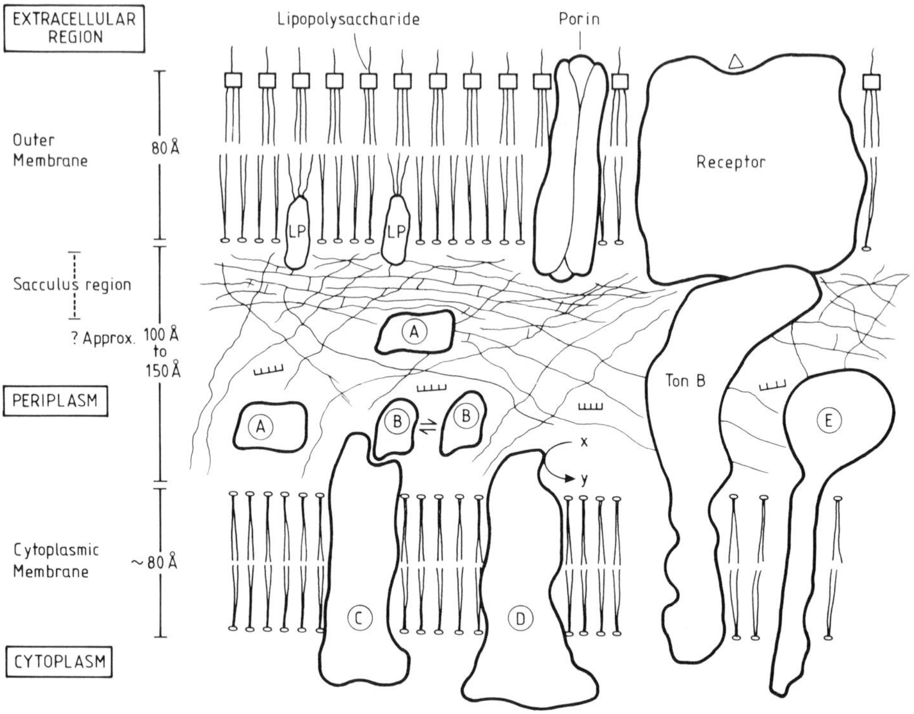

Fig. 1. A schematic representation of the bacterial periplasm. The outer half or sacculus region is show as comprising cross-linked peptidoglycan or murein. The inner region (abutting the cytoplasmic membran is thought to contain glycan strands running from the sacculus together with membrane-derived oligosaccha ides (⊔⊔⊔⊔). The proteins may be largely located in this region which is thought to be viscous. Howeve proteins may be able to reach the outer membrane by passage through regions of the sacculus (Kellenberge 1990). Protein A could be an electron transport protein (cytochrome) than can freely diffuse in the periplas or a water-soluble enzyme such as alkaline phosphatase. Protein B might be a solute-binding protein th can bind to a membrane-bound receptor or pore C. Alternatively B and C could each be electron transpo proteins. Protein D represents an internal membrane protein for which the active site catalysing reactio X→Y (bottom centre) is located in the periplasm whilst protein E is an example of a protein that h its major domain(s) in the periplasm and is anchored to the membrane by an alpha helix. The TonB prote is shown as an example of a system interacting with an outer membrane receptor (see Postle, 1990). A discussed in the test, it is not certain that the features shown here apply to all Gram-negative bacter and the dimensions of the periplasm are uncertain. LP is lipoprotein.

with lysozyme and/or EDTA have been used. Ames, Prody and Kustu (1984) have advocated the use of chloroform-treatment for *Salmonella typhimurium* and subsequently this method has been reported as successful with some other, but not all, genera (Lall, Eribo & Jay, 1989). There is almost certainly no universal recipe, a point that is emphasised by the report that Triton-treatment was the most effective method for release of periplasmic proteins of Rhizobia species (Steeter, 1989).

THE CONTENTS OF THE PERIPLASM AND THEIR FUNCTIONS

Oligosaccharides

The periplasms of *E. coli, Agrobacterium tumefaciens* and *Rhizobium trifolii* contain oligosaccharides (Miller, Kennedy & Rienhold, 1986). Presumably other genera are similarly endowed. In the case of *E. coli*, these saccharides comprise between 6 and 12 glucose units linked by β-(1→2) and β(1→6) bonds. The glucose units can be substituted with phosphoglycerol, phosphoethanolamine and *O*-succinyl ester residues and thus the oligosaccharides are negatively charged. Since the first two of these substituent groups are derived from the turnover of membrane phospholipids these oligosaccharides have been categorised as membrane-derived oligosaccharides (MDO). The molecular weights of these oligosaccharides will be of the order of 2500, and, therefore, they will be retained in the periplasm because the pores of the outer membrane have a molecular weight exclusion limit of approximately 600. In *A. tumefaciens* and *R. trifolii*, the periplasmic oligosaccharides are cyclic (1→2) β-D-glucans.

There is good evidence that the function of these oligosaccharides is to provide a means of contributing to the osmotic pressure of the periplasm. Kennedy and his colleagues have shown for both *E. coli* and *A. tumefaciens* that increasing periplasmic content, to a maximum of approximately 7% of the cell dry weight, of oligosaccharides correlates with decreasing osmolarity of the growth medium. The logical deduction from such observations is that the synthesis of periplasmic oligosaccharides, which requires contributions from between 10 and 15 enzymes and the utilisation of a significant proportion of the cells' metabolic resources, is designed to contribute to the osmolarity of the periplasm. There is evidence that the periplasm is iso-osmotic with the cytoplasm at about 300 mosM. The negatively charged oligosaccharides, together with their attendant cations, would supplement any shortfall in osmolarity provided by the surrounding medium, which would have access to the periplasm through the pores, and the protein content of the periplasm. Colligative properties of matter mean that, on a unit weight basis, proteins will make less contribution to osmolarity than oligosaccharides.

The notion that the periplasmic oligosaccharides supplement the osmolarity of the periplasm obviously requires that their synthesis be regulated. The enzymes for synthesis are apparently constitutive and thus their regulation must be exerted at the level of enzyme activity. Higgins *et al.* (1987) have suggested that variation in cytoplasmic ionic strength may be an intracellular signal and cite Kennedy for having observed that activities of some enzymes that participate in the oligosaccharide synthesis are very sensitive to ionic strength.

The proposed role of the periplasmic oligosaccharides is plausible, although it has not been rigorously established experimentally. Evidence that periplasmic oligosaccharide synthesis was most rapid after growth of a mutant in a medium of high osmolarity (Clark, 1985) has subsequently been rebutted (Kennedy & Rumley, 1988) and thus, at present, all observations are consistent with the osmolarity of the periplasm being supplemented when required by the oligosaccharides.

An interesting facet to the membrane-derived oligosaccharides is that they are released from the periplasm of a prey cell following invasion by the predatory bacterium *Bdellovibrio bacteriovorus* (Ruby & McCabe, 1988). Such loss of the oligosaccharide may contribute to the destabilisation of the prey cell.

Proteins

The functions of most of the periplasmic proteins can be classified in several groups: (a) electron transport proteins; (b) binding proteins that act to capture sugars, amino acids and inorganic ions and which play a role in both active transport and chemotaxis; (c) enzymes involved in degradation of molecules destined for import into the cell; (d) enzymes involved in biosynthesis of the cell wall or components of the periplasm; (e) enzymes that protect the cell by modifying toxic compounds. Several of these are considered in turn, starting with electron transport proteins which have arguably been overlooked as important residents of the periplasm in many genera.

Electron transport proteins

It is ironic that, although an electron transport protein, a *c*-type cytochrome from *E. coli* (cytochrome c_{552} which was later shown to be one of the nitrite reductases of this organism; Fujita & Satoh, 1966; Cole, 1968) was one of the earliest proteins to be assigned to the periplasm, the widespread role of the periplasm as a location for electron transport has not been, at least until fairly recently, widely recognised. Thus, electron transport is scarcely mentioned in the reviews of the periplasm from Heppel (1971) and Beacham (1979). Perhaps this exclusion arose because enteric bacteria do not have extensive periplasmic electron transport reactions. However, in other organisms, for instance denitrifiers (Ferguson, 1987, 1988*a*, *b*) and chemoautolithotrophs in general, periplasmic electron transport reactions are of the utmost importance to the organism. A relatively comprehensive survey of these reactions has been presented recently (Ferguson, 1988*a*). Related points have also been reviewed by Hooper & Dispirito (1985) but a recap of some of the principal implications of periplasmic locations for electron transport proteins is given here, using the organism *Paracoccus denitrificans* for illustration.

P. denitrificans exhibits great versatility in its modes of growth. It will, for example, grow aerobically with either methanol or methylamine as carbon source. The dehydrogenases for these two compounds are in the periplasm. This means that both protons and formaldehyde are released into the periplasm. There might be advantage in forming the toxic formaldehyde in the periplasm rather than in the cytoplasm, whilst the released protons contribute to the generation of a proton electrochemical gradient (Fig. 2). Electrons pass from the dehydrogenases via soluble *c*-type cytochromes or copper proteins to a cytochrome oxidase that is an integral membrane protein. Cytochrome aa_3 can act as the oxidase and Fig. 2 shows that it is currently envisaged to translocate both electrons and protons across the cytoplasmic membrane. Thus the transfer of electrons from the periplasmic site of methanol or methylamine oxidation to the site of oxygen reduction towards the cytoplasmic surface of the membrane generates a proton electrochemical gradient, relatively positive and acidic in the periplasm. This mode of generation is supplemented by the concomitant outward movement of protons catalysed by cytochrome oxidase. The proton electrochemical gradient drives the synthesis of ATP and thus it should be clear from the above description and from Fig. 2 how the water-soluble dehydrogenases for methanol and methylamine can be associated with oxidative phosphorylation system of the bacterium. Mechanistic details of these periplasmic dehydrogenases are not fully understood but the availability of a high resolution structure for a methylamine dehydrogenase will aid progress in this direction (Vellieux *et al.*, 1989).

As its name indicates, *P. denitrificans* can catalyse the denitrification process in which nitrate is reduced in several steps to dinitrogen. Two of these steps involve periplasmic reductases for nitrite and nitrous oxide. It is sometimes not appreciated how such periplasmic reductases which, in contrast to the dehydrogenases for methanol or methylamine (Fig. 2), consume protons in the periplasm, can be associated with the ATP synthesizing system of the cell. Figure 3 shows that electrons destined for the reduction of nitrite or nitrous oxide will, under most conditions, predominantly originate from NADH and thus generation of a proton electrochemical gradient will be associated with reduction of nitrite or nitrous oxide. It is the case that the stoichiometry of charge translocation across the cytoplasmic membrane is lower than would be the case if the electrons passed back across the cytoplasmic membrane, as illustrated in Fig. 3 for the hypothetical electron acceptor XO. Possible rationales for why the reductases for nitrite and nitrous oxide are in the periplasm rather than in the membrane with their active sites at the cytoplasmic surface have been given previously (Ferguson, 1988*a*).

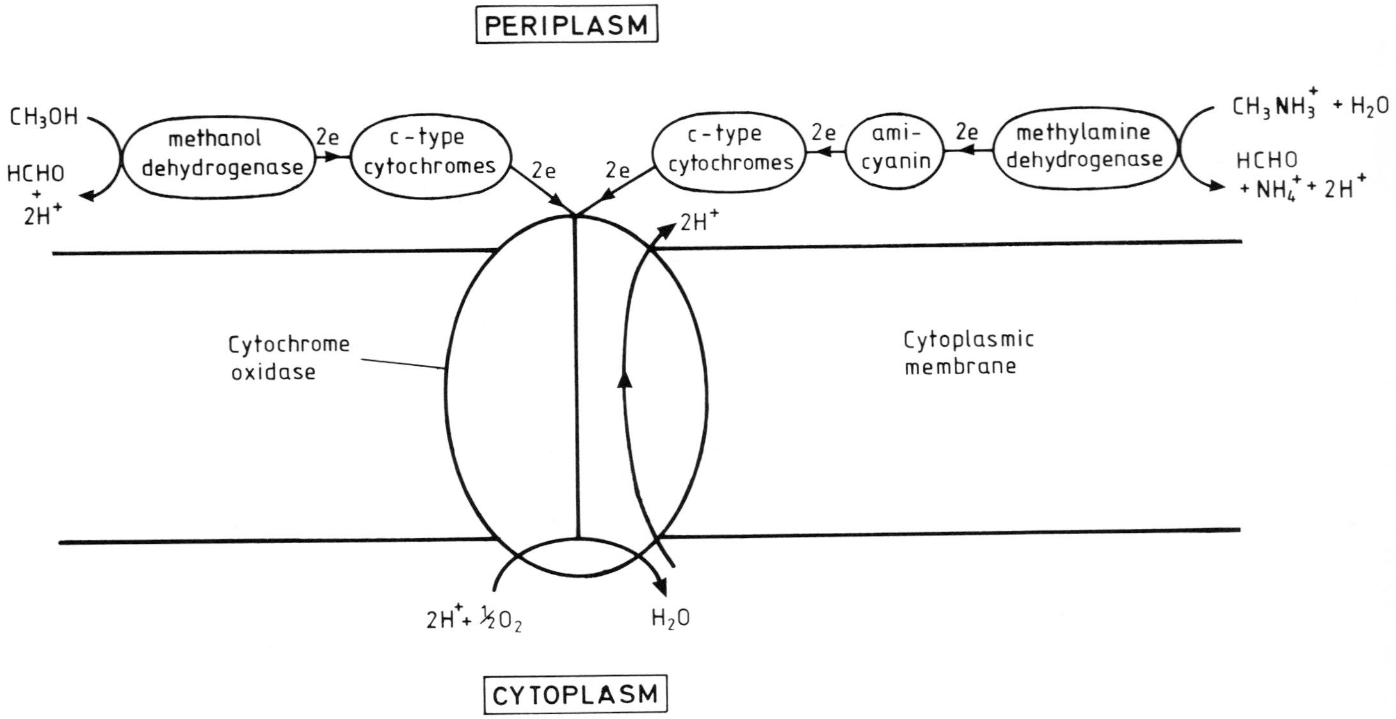

Fig. 2. Schematic representation of periplasmic oxidation of methanol or methylamine in *P. denitrificans*. The figure shows how respective release and uptake of protons in periplasm and cytoplasm, together with transfer of electrons across the cytoplasmic membrane, generates a proton electrochemical gradient, relatively positive and acidic on the periplasmic side. As explained in the text, the generation of the gradient is supplemented by the proton pumping activity of cytochrome aa_3. There is uncertainty about the nature of the c-type cytochromes that participate in electron transfer in the periplasm (see text).

Consideration of Figs. 2 and 3 together shows, however, that electron transfer from methanol or methylamine via periplasmic *c*-type cytochromes to nitrite or nitrous oxide could not be linked to the generation of a proton electrochemical gradient, irrespective of the free energy change for the reaction.

Recently, reduction of nitric oxide has been shown to be a discrete step in the denitrification process catalysed by *P. denitrificans*. The nitric oxide reductase is an integral membrane protein (Carr & Ferguson, 1990) and there are indications that the catalytic site lies at the periplasmic side of the membrane (Shapleigh & Payne, 1985). Thus the same energetic considerations, in terms of proton translocation across the cytoplasmic membrane (Fig. 3), apply to this reaction as to the reduction of nitrite or nitrous oxide by the water-soluble proteins in the periplasm. The reduction of nitrate as the first step of the denitrification process usually occurs on the cytoplasmic side of the membrane in a reaction that is inhibited in the presence of oxygen (Ferguson, 1987). This inhibition is proposed to be exerted on the movement of nitrate across the cytoplasmic membrane (Ferguson, 1987). The recognition that the organism *Thiosphaera pantotropha* could reduce nitrate under aerobic conditions thus raised the question as to how the control on nitrate movement could be avoided. This problem has been resolved by the identification in this organism of a periplasmic nitrate reductase that can function in aerobic conditions (Bell, Richardson & Ferguson, 1990).

The concept that the periplasm is not an aqueous region but rather a gel phase has already been outlined together with the evidence that protein diffusion in this region is relatively restricted. This has implications for electron transport because it is expected that transfer of electrons from one protein to another will occur either upon collision of the two molecules or between two or more molecules that are permanently associated. A narrow periplasm of relatively high viscosity suggests that random three-dimensional diffusion may not account for transfer of electrons within the periplasm. In this context it is striking that two periplasmic reductases of *Rhodobacter capsulatus*, those for nitrate and dimethylsulphoxide each copurify with a specific *c*-type cytochrome (McEwan *et al.*, 1989; Richardson *et al.*, 1990). Thus it may be that the reductase and cytochrome subunits are permanently associated in the periplasm. If this is so, it still remains to be explained how these complexes receive electrons from the membrane bound redox components through which the electrons must pass en route from dehydrogenases for NADH and other reductants. A narrow and viscous periplasm does not seem conducive for efficient transfer of electrons via a mechanism that relies upon collisional interactions between macromolecules (Ferguson, 1988*a*; van Vielink & Duine, 1990).

The specificity of periplasmic electron transfer pathways is not presently fully understood. Two examples illustrate the complexities that have recently arisen. The first of these concerns the acceptance over many years that diffusion of the periplasmic and water-soluble cytochrome c_2 of the

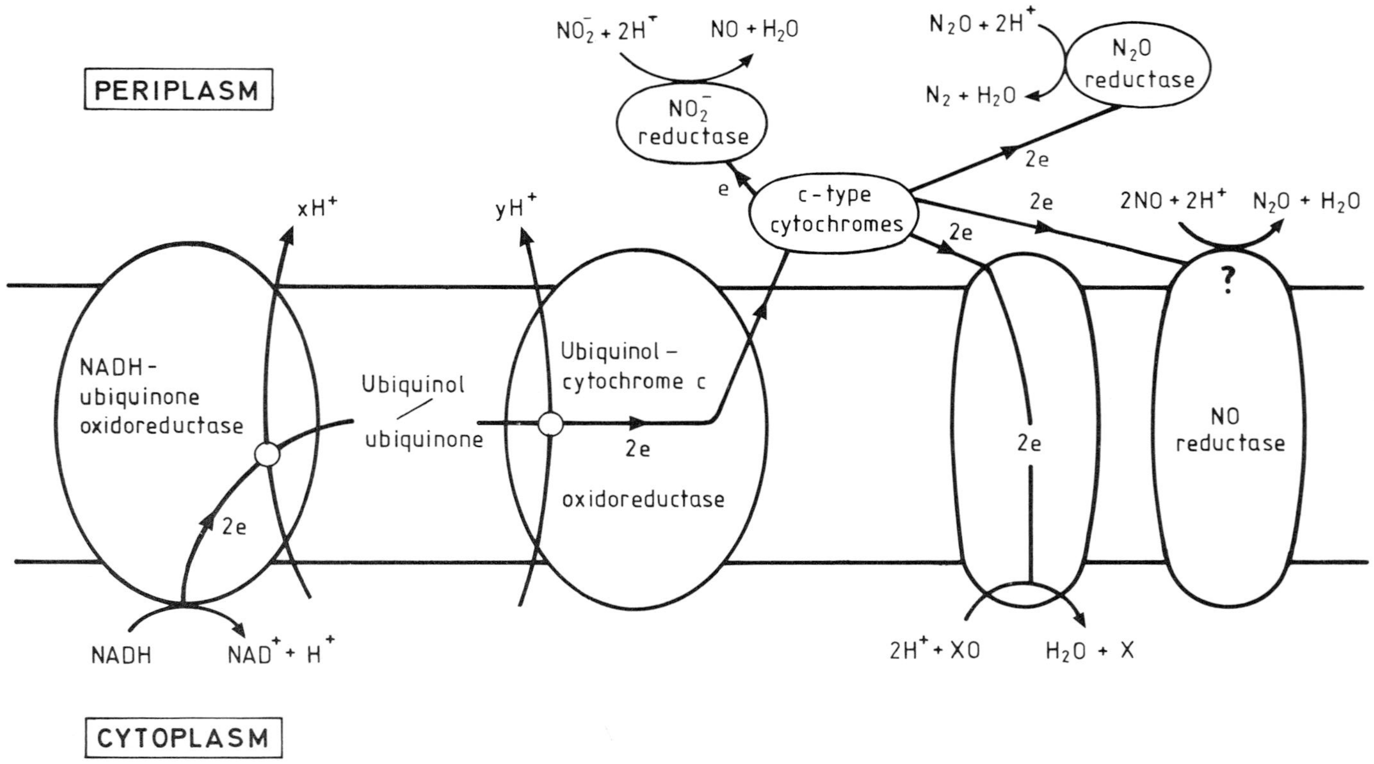

Fig. 3. Electron transport to water-soluble periplasmic reductase enzymes can generate a proton electrochemical gradient. The figure shows electron transport from NADH to periplasmic nitrite and nitrous oxide reductases. x and y are number of protons translocated per two electrons passing through each of the proton translocating segments of the electron transport chain. Provided x + y is greater than 2, the passage of electrons from NADH to nitrite or nitrous oxide will result in the periplasmic phase becoming positive relative to the cytoplasmic phase. The same arguments apply to the transfer of electrons to nitric oxide although confirmation that the site of reduction is on the periplasmic surface is required. The action of the cytochrome bc_1 complex is over-simplified here. Hypothetical electron transport to an acceptor XO on the cytoplasmic surface is also shown for comparison.

photosynthetic bacterium *R. capsulatus* mediates electron flow between the cytochrome bc_1 complex and the photosynthetic reaction centre complex. Deletion of the gene for cytochrome c_2 gave the surprising result that adequate electron transport between the two complexes was possible in the absence of the cytochrome (Daldal *et al.* 1986). Although this was at first interpreted to mean that electron transfer could occur directly between the two complexes, more recent work suggests that an alternative *c*-type cytochrome, probably anchored to the cytoplasmic membrane, functions in the absence of cytochrome c_2 (Jones, McEwan & Jackson, 1990). It is difficult to determine whether the latter membrane bound cytochrome functions when the cytochrome c_2 is present or whether it only substitutes when required as an (inferior?) electron carrier.

The second example concerns *P. denitrificans*. Figures 2 and 3 show unspecified *c*-type cytochromes as connecting the periplasmic dehydrogenases and reductases to the membrane bound electron transport components. One of the periplasmic *c*-type cytochromes in this organism is cytochrome c_{550}, a molecule that is very similar to the cytochrome c_2 of *R. capsulatus* and mitochondrial cytochrome *c*. This cytochrome c_{550} has been considered to participate in electron transfer to and from periplasmic enzymes as well as in aerobic respiration between the cytochrome bc_1 and cytochrome aa_3 complexes. However, van Spanning *et al.* (1990*b*) have found that inactivation of the cytochrome c_{550} gene does not prevent any of these electron transfer processes. As in the case of the cytochrome c_2 of *R. capsulatus*, it appears that the absence of cytochrome c_{550} can be made good by one or more alternative electron transfer proteins. In the case of the link between cytochrome bc_1 and cytochrome aa_3 there is evidence that a membrane-bound redox protein, cytochrome c_{552} is involved. This topic has been further reviewed recently (Ferguson, 1991). In contrast to the proposed degeneracy amongst the *c*-type cytochromes, a copper protein, amicyanin, has proved indispensible for growth on methylamine (Fig. 2; van Spanning *et al.*, 1990*a*).

Periplasmic binding proteins

The periplasms of *E. coli* and *S. typhimurium* and other Gram-negative organisms contain a number of water-soluble proteins that specifically bind a variety of nutrients. The substrates bound by such proteins include sulphate, phosphate, several amino acids, dipeptides, maltose, galactose and ribose. Very high resolution X-ray structures have been obtained for several such proteins. These structures show that, although there is little sequence similarity between the proteins, they share a common structural organisation with the substrate held by an intricate network of hydrogen bonds between two domains. The recent determination at 1.7 Å resolution of the structure for the phosphate binding protein provides an example (Luecke & Quiocho, 1990).

The periplasmic binding proteins are all involved in nutrient transport into the cell. Some of them, those for maltose, galactose, ribose and dipeptides, are also involved in the chemotactic response. The latter involves the interaction of the liganded binding protein with membrane-bound receptors, MCP II (Tar) for maltose and MCP III (Trg) for the other three. Such interaction leads to activation of the methylation reactions and alteration of the tumbling of the bacteria. The reason why a binding protein is required is not clear because chemotactic amino acids act by binding directly to a membrane-bound receptor. This uncertainty is comparable with the problem as to why some transport processes require binding proteins and others do not (see below).

The general feature of periplasmic binding protein-dependent transport is that cells show a very high affinity for substrates that are handled in this way and also that the concentration ratio achieved between the cytoplasm and the external medium is extremely high (up to 10^5). Analysis of genes indicates that these systems are closely related to one another (see, e.g. Ames & Joshi, 1990; Higgins *et al.* 1990). The high affinities and concentration rations associated with these transport systems are characteristics that distinguish them from other bacterial transporters, e.g. that for lactose, which operate by symport with a cation (frequently a proton). Transport systems that are dependent on the periplasmic binding proteins may be more common than previously thought. The alanine uptake system of *Rhodobacter sphaeroides* and that for C4 dicarboxylic acids in *Rhodobacter capsulatus* are examples of systems that are now known to fall into this category having previously been tacitly regarded as proton symporters (Abee *et al.*, 1989; Shaw & Kelly, 1991).

The histidine uptake system of *S. typhimurium* is an archetypal periplasmic transport process. The involvement of four polypeptides has been established. HisJ is the periplasmic binding protein. HisQ and HisM are hydrophobic proteins that are predicted to span the cytoplasmic membrane whilst HisP (which may be present as a stoichiometry of 2:1 with the other components) is predicted to have a substantial domain that is exposed to the cytoplasm and has an ATP binding site. It is not clear whether the HisP protein also might have a transmembrane domain. Intuitively one would expect that the HisJ protein with histidine bound would interact with one or more of the other three polypeptides in the system. Both *in vitro* and *in vivo* cross-linking experiments have provided good evidence that a specific interaction occurs with the HisQ protein, with the interaction being strengthened by the presence of histidine (Prossnitz *et al.*, 1988). In contrast, in cells possessing a mutation such that the HisJ protein still binds histidine but transport is blocked, cross-linking of HisJ to HisQ was not observed. Parenthetically, it is noted that the successful use of the cross-linking strategy might find future application in the study of other protein-protein interactions in the periplasm, for example in electron transport processes (see previous section). Once the complex between HisJ and HisQ has been formed the immediate fate of the histidine molecule is uncertain. Two extreme possibilities can be envisaged. First, the histidine could be discharged into a pore which would be intrinsically non-specific but would be operational only with HisJ bound. Alternatively, one or more of the HisQ, M and P proteins might provide a binding site to which the histidine molecule would be transferred from HisJ. At least two lines of evidence support the latter scheme. First, it is known that maltose transport can function, albeit with considerably lower affinity and rate, in certain mutants that lack the periplasmic maltose binding protein (Shuman, 1982). Second, Speiser & Ames (1991) have recently established that histidine transport by *S. typhimurium* is possible in the absence of HisJ in cells carrying mutations in the HisP protein, provided that the membrane protein of the His system were produced in large amounts.

An important development in the understanding of the periplasmic protein-dependent transport systems is the recent demonstration both with intact cells and reconstituted components, that they are powered by ATP (Higgins *et al.*, 1990; Ames & Joshi, 1990), as had been predicted on the basis of the presence of an ATP binding motif in one of the subunits of the system (e.g. HisP discussed above). Whereas the energy source for transport has finally been identified, the issue of why these systems depend on the periplasmic binding proteins remains unresolved. Quiocho (1990) has argued strongly that these proteins cannot in any sense be regarded as supernumerary. He has pointed out that maltose transport in the absence of a binding protein is a very inefficient process. A crucial role of the binding protein is to dehydrate substrates, especially those, e.g. sulphate, with a very high hydration energy (Quiocho, 1990). A scheme has been proposed in which interaction between the liganded binding protein and the membrane components drives a conformational change, presumably at the expense of ATP hydrolysis, which releases the substrate from the binding protein which is then free to return to the bulk phase of the periplasm (Quiocho, 1990). As the binding proteins are in considerable molar excess (concentration as high as 1 mM) over the membrane components it is reasonable to expect, in line with the results of the cross-linking experiments (above), that the liganded form must have a higher affinity for the membrane proteins than the unliganded form.

The contrast between the binding protein type transport systems, exemplied by the histidine system in *S. typhimurium*, and proton symport, e.g. that for lactose in *E. coli*, has been perplexing for many years. What advantage does the former system have over the latter? Various ideas for advantage conferred by the binding protein have been advanced but most have been rebutted (Brass, 1986). The binding proteins might, because of their high affinity, scavenge for low concentrations of their ligands but the gel-like mature of the periplasm suggests that their three dimensional diffusion in the periplasm would be limited. It has been proposed that the binding proteins may form semi-ordered arrays in the periplasm with the ligand being passed on from one molecule of protein to an adjacent one (Brass *et al.*, 1986). The relatively narrow width of the periplasm might mean that such an array is effectively two dimensional, an arrangement that would correspond to the probable organization of the counterparts to the binding proteins in Gram-positive organisms (see later). At the limit, this mechanism would involve a molecule of binding protein remaining bound to the membrane components of the transport system and acting as the final acceptor in the array. Such relatively permanent residence might be incompatible with the indications that the histidine-loaded binding protein has higher affinity for the membrane components than the unloaded protein (Prossnitz *et al.*, 1988) and would also not be compatible with the model of Quiocho (1990). In the case of sulphate or dicarboxylate transport, the ATP- and binding protein-dependent transport system might have two advantages over cation symports. First, when as is often the case, the proton electrochemical gradient is dominated by the membrane potential, interior of the cell negative, at least three protons would have to be moved in symport with the divalent anions to obtain active transport. With a typical membrane potential of 180 mV symport with four protons would be needed to attain accumulation ratios of 10^5. The second advantage may relate to the effectiveness of the binding proteins in dehydrating the bound ligand. Such dehydration might be useful in transport and more readily achieved by a water-soluble two-domain binding protein than by an integral membrane protein.

The observation of a dual role in both active transport and chemotaxis of some of the binding proteins raises the question of whether such proteins have distinct or overlapping sites for interacting with the two (transport or signal transduction) membrane protein sectors. The properties of a mutant of *E. coli* indicate that these must be separate. This mutant showed no chemotactic response to galactose although it transported galactose normally and its binding protein had similar affinity to the wild type. In the binding protein of the mutant, aspartate had replaced glycine at position 74 (Vyas, Vyas & Quiocho, 1988). The structure of this mutant protein was identical to that of the wild type protein except in the immediate vicinity of the amino acid substitution. Here there were considerably fewer water molecules than in the wild type. Consequently it was suggested (Vyas *et al.*, 1988) that displacement of this water might be important in the interaction of the binding protein with the chemoreceptor protein. The properties of the mutant clearly indicate that a distinct region must be involved in interacting with the transport system.

Degradative enzymes

This category of proteins scavenges for nutrients that are potential sources of carbon, phosphate or nitrogen (Beacham, 1979). In general, unlike the products that result from the action of the enzyme, the substrates of this class of periplasmic enzyme, e.g. alkaline phosphatase or 3′ and 5′ nucleotidases, cannot normally be transported into the cell. A recent example is an amylase that catalyses hydrolysis of amylose and dextrins that are too large to be transported to the cytoplasm (Brass, 1986).

Biosynthetic enzymes

Periplasmic enzymes contribute to the synthesis of peptidoglycan and are generally involved in morphogenesis. This aspect of the role of the periplasm is discussed in several reviews, e.g. Brass (1986) and Nanninga, 1991 and this volume.

Detoxifying enzymes

Most prominent in this category is β-lactamase (Collatz, Labia & Gutmann, 1990) but other proteins are also involved in countering attack by other antibiotics (Beacham, 1979) and the effects of heavy metal ions, e.g. Hg^{2+}, (Brown, 1985).

Role of the periplasm in conjugation

Transfer of the conjugative F plasmid requires the expression of at least 14 transfer operon genes which are associated with the F-pili that extend from the surface of the donor cell. One of these genes, *tra*F, has recently been shown to code for a protein that is located in the periplasm (Wu, Kathir & Ippen-Ihler, 1988). The DNA sequence indicates that the primary translation product has an N-terminal sequence that is typical of periplasmic proteins whilst the mature protein would be a water-soluble rather than an integral membrane protein. The function of the *tra*F gene product is not known for certain, but it has been suggested to be a component of a complex of pilus assembly proteins, especially as other *tra* operon products are also thought to have a periplasmic location (Wu *et al.*, 1988).

Passage across the periplasm

Incoming and outgoing DNA, e.g. see previous section, and proteins destined for export or the outer membrane must effectively pass through the periplasm. The question of whether 'Bayer bridges' play a role in the latter process was discussed in the early part of this article. However, it is not only macromolecules that transit the periplasm but also intermediate sized molecules, for instance iron siderophores and vitamin B_{12}, that are too large (>600 D) to enter the periplasm by the porins of the outer membrane. There are receptors in the outer membrane for these molecules which are taken into the cell in an energy-dependent process such that their concentration in the periplasm is $1000\times$ that outside. How then is the energy supply in the cytoplasm or alternatively across the cytoplasmic membrane in the form of the proton electrochemical gradient connected to the outer membrane? The TonB protein fulfils this role (Postle, 1990). The sequence and other analysis of the protein suggests that it is anchored in the cytoplasmic membrane, but has a substantial domain in the periplasm, including a proline-rich region which is suggested to take up an extended conformation of sufficient length (100 A) to span the periplasm. Both biochemical and genetic evidence indicate that the TonB protein interacts with outer-membrane receptors. This interaction must be subject to alteration such that a conformational change in the outer membrane receptor can be induced with concomitant release of the bound ligand. What powers the change in interaction between TonB and the outer membrane receptors? This is unknown, but under current consideration is the possibility that the amino-terminal transmembrane segment of TonB interacts with a proton channel within the cytoplasmic membrane so that the proton electrochemical gradient across this membrane could be coupled via the TonB protein, and through the periplasm, to the outer membrane. Once a ligand has been released from the outer membrane it interacts with a periplasmic binding protein and then with a cytoplasmic membrane transport component (Postle, 1990). These latter stages of the process are thus analogous to the ATP-dependent transport process discussed earlier.

Biosynthesis of the periplasmic proteins

The proteins of the periplasm are synthesised in the cytoplasm and then translocated across the cytoplasmic membrane. As far as is known, there is not obligatory coupling between translation of the mRNA and translocation across the membrane. What determines whether a protein is translocated? Most, if not all, periplasmic proteins that have been examined are first synthesised with an N-terminal leader or signal sequence that comprises between 18 and 24 amino acids and which is cleaved from the mature protein by a membrane-bound signal peptidase (Randall & Hardy, 1989). Although certain general features of this sequence can be recognised, for example, a region of positively charged amino acids at the N-terminus of the sequence and a core of hydrophobic amino acids, there is essentially no consensus sequence similarity (Randall & Hardy, 1989). A further puzzle is that signal sequences found for periplasmic proteins are similar to those observed for proteins that are located in either the inner or the outer membrane.

Study of mutations within the signal sequence has indicated that insertion of charged amino acids into the hydrophobic region blocks translocation to the periplasm with resulting accumulation of unsecreted full length transcripts in the cytoplasm. Introduction of glycine and/or proline residues, which are generally regarded as destablising for alpha helices, also impairs translocation across the membrane. Ferenci and Silhavy (1987) have drawn attention to the possibility that the absence of a consensus sequence in the leader sequence may mean that there is an interplay between the leader sequence and the mature sequence. The same authors have also emphasised that consequences of mutation in the leader sequence are not always analysed quantitatively. If a mutation allows polypeptide translocation to proceed, but at a drastically reduced rate, it might not be safe to conclude that the amino acid residue that was altered by mutation was not essential for translocation.

If interaction between the leader sequence and the mature polypeptide is not important in translocation of polypeptides to the periplasm, and leader sequences to not clearly differ between periplasmic and externally secreted proteins, one is led to the conclusion that information dictating the location of the mature protein must lie on the mature protein itself. There is no clear experimental evidence to support this proposal. Note, however, that β-galactosidase is not secreted following fusion of a leader sequence to its amino terminus, thus showing that the nature of the mature protein does have an effect on translocation.

A recent investigation sheds some further light on the question of the effect of the mature sequence on translocation. Summers and Knowles (1989) have studied the movement to the periplasm of a hybrid polypeptide comprising chicken triose isomerase and the leader sequence of β-lactamase. The polypeptide was not translocated into the periplasm unless part of the N-terminal sequence of the mature β-lactamase polypeptide was included in the fused protein. Inclusion of the first three N-terminal amino acids was essential for any translocation to occur whilst optimal translocation required the first 12 amino acids of the amino terminus of the mature lactamase. In further studies of this system, it was shown that with a leader sequence but none of the mature β-lactamase sequence present, replacement of the arginine residue at position 3 in the triose phosphate isomerase sequence by serine or proline permitted translocation (Summers, Harris & Knowles, 1989). Subsequent replacement (with proline still at position 3) of the lysine at position 4 in the isomerase sequence by arginine caused complete blockage of translocation. This shows that positive charge alone at a certain position, e.g. position 4, is not a critical factor. However, it does suggest that the higher pKa of arginine might be significant, because there would be less tendency to deprotonate than there would be for a lysine in the hydrophobic milieu of the membrane. It is suggested, on the basis of the block imposed on translocation by arginine at position 3 (or by mutation at position 4) of the mature protein unless this position was separated from the leader signal sequence by fourteen residues, that at least 14 of the N-terminal amino acids of a mature polypeptide may have to be buried in order to facilitate secretion. It is also interesting to note that, whereas activity of the chicken enzyme is observed in the cytoplasm of *E. coli*, it was not observed in the periplasm. This was attributed to the oxidizing environment in the periplasm and possible disulphide bond formation; the question of disulphide bond formation in the periplasm is returned to later.

Many genes, e.g. those of the *sec* group, have been implicated in the translocation of proteins across the cytoplasmic membrane of *E. coli* (Oliver, 1987) but it is not clear whether the N- terminal amino acids discussed above are inserted into either the lipid region of the membrane or a specific protein. There is increasing evidence that proteins destined for the periplasm must be kept in a relatively unfolded state in the cytoplasm. The N-terminal leader sequence contributes to satisfying this requirement, at least for a maltose binding protein (Park *et al.*, 1988) and β-lactamase (Laminet & Pluckthun, 1989), but chaperone proteins, e.g. SecB and GroEL are also important in this regard (Kumamoto, 1991). The translocation appears to require both ATP and the proton electrochemical gradient (Mizushima & Tokuda, 1990). A brief summary of current knowledge and controversies concerning protein export in *E. coli* has recently been presented by Bassford *et al.* (1991).

Many of the periplasmic enzymes have an attached cofactor. Examples include the haem groups of cytochromes, PQQ in methanol dehydrogenase, the zinc ion in alkaline phosphatase and the copper of azurin. How are these groups attached? This question has to be addressed whilst keeping in mind the evidence (see above) that polypeptides are thought to be translocated to the periplasm in an essentially unfolded state. Cofactor or metal binding sites are normally formed only when a protein folds. Cloned genes for an azurin and a related copper protein are expressed in *E. coli* as periplasmic holo-proteins (Karlsson *et al.* 1989; Yamamoto, Uozumi & Beppu, 1987), implying, as *E. coli* does not synthesise this type of protein, that either the copper inserts spontaneously or that there is fortuitously an enzyme in *E. coli* that serves to insert the copper. In contrast, when another copper-containing protein, the periplasmic nitrous oxide reductase of a Pseudomonad, is expressed in *E. coli* only an apo-form of the protein is observed. This reductase has novel cooper centres and their assembly appears to be mediated by the products of several genes. Amongst these is an outer membrane protein (NosA) that is postulated to participate in insertion of copper (Lee, Hancock & Ingraham, 1989; Mokhele *et al.*, 1987) and a set of genes related to the ATP-dependent and binding protein transport systems discussed earlier in this chapter (Zumft *et al.*, 1990).

The periplasmic *c*-type cytochromes present a clear biosynthetic puzzle. By analogy with mitochondrial cytochome *c* the covalent insertion of the haem is expected to be enzyme-catalysed. Indeed the existence of mutants that are deficient in a wide range of *c*-type cytochromes is consistent with the occurrence of one or more haem lyase enzymes. This leads to the question of whether the haem is inserted in either the periplasm or the cytoplasm. The cytochromes are first synthesised with a typical signal sequence which suggests that the translocation is catalysed by the normal export machinery. As discussed earlier, proteins must be in an unfolded state to facilitate translocation. It thus seems improbable that insertion of the haem, which will tend to order the protein, occurs in the cytoplasm. In support of this proposition apo-proteins of two *c*-type cytochromes, with their leader sequences removed, have been identified in the periplasm of a mutant of *P. denitrificans* that is pleiotropically deficient in *c*-type cytochromes (Page & Ferguson, 1989, 1990). In related work it was also shown that blockage of haem synthesis in wild type cells caused accumulation of processed apo-protein in the periplasm. This indicates that neither the gene expression nor the translocation of the polypeptide across the cytoplasmic membrane is strictly coupled to the availability of haem. If haem is attached in the periplasm then it follows that a haem export system may be needed. It will be interesting to see if an ATP-dependent system, analogous to that implicated in copper insertion, emerges. Not all haem groups in the periplasm are covalently attached to proteins. An example of a non-covalent form is the *d*-type haem of the nitrite reductase of *P. denitrificans* (Page & Ferguson, 1990). It is difficult to imagine how such haem could be translocated with an unfolded polypeptide across the cytoplasmic membrane.

In addition to the problem of cofactor acquisition in the periplasm, the successful assembly of periplasmic proteins may also depend upon the activity of periplasmic chaperone proteins comparable with the GroEL protein in the cytoplasm. For *E. coli*, the assembly of the pilus occurs in the periplasm and there is good evidence that the product of the *papD* gene acts as a chaperone to ensure that other *pap* gene products assemble correctly into the pilus (Hultgren *et al.*, 1989). The molecular basis for chaperone action is not known, but the crystal structure of the *papD* protein has been obtained (Holmgren & Branden, 1989); it is striking that it has features associated with the immunoglobulin fold, including the presence of a disulphide bridge.

Disulphide bridges are not only known for the *papD* protein; they are also known to be present in the methylamine dehydrogenase protein for which a crystal structure has been determined (Vellieux *et al.*, 1989). It has also been found that at least some methanol dehydrogenases have disulphide bridges (C. Anthony, personal communication). This raises the question of whether these bridges (bonds) form spontaneously or whether, as in the endoplasmic reticulum of eukaryotic cells, there is a disulphide isomerase. Presumably formation of S–S bridges in the periplasm would reflect the presence of a relatively oxidising environment in the periplasm relative to the cytoplasm. This would account for why the triose phosphate isomerase forms disulphide bridges in the periplasm but not in the cytoplasm (see above). However, it should be taken into account that bacteria can function in the absence of oxygen. What then would be the oxidant for disulphide bridge formation from two cysteines? Nitrous oxide reductase is a protein formed under anaerobic conditions and which is suggested to have a disulphide bridge (Kroneck *et al.*, 1990).

Finally, it is also relevant that a *cis–trans* prolyl peptide isomerase has been recognised in the periplasm (Liu & Walsh, 1990). This activity could be important in ensuring the correct folding of periplasmic proteins.

Gene expression in response to protein–protein interactions in the periplasm

Interaction between polypeptides within the periplasm is proposed in some instances to be of importance for gene activation. This is illustrated by the interaction between the *toxR* and *toxS* gene products in *Vibrio cholerae*. The *toxR* protein is a transmembrane polypeptide that can bind DNA via its cytoplasmic domain and thereby activate transcription of genes that encode cholera toxin (DiRita & Mekalanos, 1991). DNA sequence and alkaline phosphatase fusion studies have shown that *toxR* has a substantial periplasmic domain that might be responsible for detection of specific environmental conditions (e.g. pH, temperature) that influence the coordinate expression of virulence-related genes. The *toxS* gene product enhances the transcriptional activation of *toxR*. Analysis of the *toxS* gene indicates that it codes for a polypeptide which is largely located in the periplasm with a putative N-terminal alpha helix to anchor it to the cytoplasmic membrane. Biochemical and mutational experiments indicate that *toxR* and *toxS* interact within the periplasm to form a complex, possibly including two molecules of *toxR* whose cytoplasmic domains can together create a cytoplasmic DNA binding site (DiRita & Mekalanos, 1991). There is some relationship between this proposed mechanism and models that have been proposed for eukaryotic membrane receptor tyrosine kinases.

The enterotoxin subunits formed by *V. cholerae* are thought to enter the periplasm transiently, where they are assembled en route to the extracellular medium (Hirst & Holmgren, 1987). This type of mechanism clearly avoids the need for connections of the Bayer bridge type between the cytoplasmic and outer membranes.

Gram-positive organisms

In a Gram-positive organism it is believed that the peptidoglycan material of the cell wall is immediately juxtaposed to the cytoplasmic membrane. The corollary is that there cannot be a periplasm in the sense that is described for Gram-negative organisms. However, it has been seen earlier in this chapter that, in at least some Gram-negative organisms, the periplasm is thought to be occupied throughout to a greater or lesser extent by murein and other carbohydrate-derived material. This comparison of Gram-positive and Gram-negative organisms suggests that the difference between them may be only a matter of degree. Such a view is compatible with the recent findings that proteins analogous to those of the periplasmic binding protein dependent systems also occur in Gram-positive organisms (Alloing, Trombe & Claverys, 1990; Gilson *et al.*, 1988; Perego *et al.*, 1991). In one case, an oligopeptide transport system in *Bacillus subtilis*, the involve ment of the proteins in an oligopeptide transport process has been directly demonstrated (Perego *et al.*, 1991). A notable difference between the Gram-negative and -positive organisms is that the counterpart of the periplasmic binding protein of a Gram-negative organism has, in the Gram-positive case, a sequence at the amino terminal end which suggests that the protein is anchored to the cytoplasmic membrane by a post-translationally added lipid. The implication is that the binding protein would be restricted to the immediate region of the surface of the cytoplasmic membrane, rather as it would be in a Gram-negative organism with a very narrow periplasm (Ferguson, 1990; Perego *et al.*, 1991). Such an operational location further argues against the idea (see earlier) that the binding proteins may act by three dimensional diffusion to transfer a nutrient across the periplasm of a Gram-negative cell. A two-dimensional semi-ordered array of binding proteins would, however, be feasible in a Gram-positive organism.

C-type cytochromes have been identified in Gram-positive organisms. They appear to be relatively strongly attached to the cytoplasmic membrane. In one case, the cytochrome c_{550} of *B. subtilis*, sequence data indicate that the protein is attached to the cytoplasmic membrane by a hydrophobic amino terminal sequence (van Wachenfeldt & Hederstedt, 1990). As with the solute binding proteins, one can envisage that cytochrome c_{550} of this organism diffuses on the surface of the membrane. Again the distinction from Gram-negative organisms is blurred because whereas cytochrome c_2 from *R. capsulatus* is water-soluble, it seems that in its absence a *c*-type cytochrome that is more tightly associated with the cytoplasmic membrane can substitute. A similar situation obtains for *P. denitrificans* in respect of cytochromes c_{550} and c_{552} (see earlier). The emerging picture is of proteins in Gram-positive organisms that are equivalent to water-soluble periplasmic proteins being found more firmly attached to the cytoplasmic membrane. This makes sense because we know that water-soluble proteins translocated across the cytoplasmic membranes of Gram-positive organisms are generally secreted from the cell; a retention mechanism is required for proteins that are to function on the external face of the cytoplasmic membrane. However, this might not always be needed because there is a report of a nucleotide diphosphate hydrolase in *B. subtilis* that is water-soluble but retained by the cells on the external surface (Mauck & Glaser, 1970). The molecular weight is reported as 137,000 which might mean that it is impermeable to the cell wall, unlike the smaller *c*-type cytochrome and binding proteins that therefore have to be retained via the mechanism outlined above. It remains to be seen whether other related proteins in Gram-positive and Gram-negative bacteria differ in this way. What does seem valid is that modes of bacterial growth that depend upon extensive periplasmic electron transport activity, e.g. growth dependent upon oxidation of methanol or iron (II) seems to be restricted to Gram-negative organisms (Hooper & Dispirito, 1985;

Ferguson, 1988a). Such growth modes may require the location of proteins to extend some distance from the cytoplasmic surface of the membrane; an organisation that is probably not possible in Gram-positive organisms.

CONCLUSION

This chapter has not comprehensively covered all aspects of the structure and function of the periplasm. In some respects it complements the coverage in earlier reviews of Brass (1986) and Oliver (1987). The periplasm merits only one or two index entries in most textbooks of microbiology. Perhaps this chapter, for all its omissions and over-simplifications, has shown that the importance of the periplasm to an organism is such that greater attention to this region of the cell is warranted.

ACKNOWLEDGEMENTS

I thank all colleagues past and present who have contributed to experimental work from our laboratory that is summarised here, the SERC for grant support and Jeff Cole for his patience and helpful comments during the writing of this chapter.

REFERENCES

Abee, T., van der Wal, F-J., Hellingwerf, K. J. & Konings, W. N. (1989). Binding-protein dependent alanine transport in *Rhodobacter sphaeroides* is regulated by the internal pH. *Journal of Bacteriology*, **171**, 5148–54.

Alloing. G., Trombe, M-C. & Claverys, J-P. (1990). The ami locus of the Gram-positive bacterium *Streptococcus pneumoniae* is similar to binding protein dependent transport operons of Gram-negative bacteria. *Molecular Microbiology*, **4**, 633–44.

Ames, G. F-L. & Joshi, A. (1990). Energy coupling in bacterial periplasmic permeases. *Journal of Bacteriology*, **172**, 4133–7.

Ames, G.F-L., Prody, C. & Kustu, S. (1984). Simple, rapid and quantitative release of periplasmic proteins by chloroform. *Journal of Bacteriology*, **160**, 1181–83.

Bassford, P., Beckwith, J., Ho, K., Kumaoto, C., Mizushima, S., Oliver, D., Randall, L., Silhavy, T., Tai, P. C. & Wickner, W. (1991). The primary pathway of protein export in *E. coli. Cell*, **65**, 367–8.

Beacham, I. R. (1979) Periplasmic enzymes in Gram-negative bacteria. *International Journal of Biochemistry*, **10**, 877–83.

Bell, L. C., Richardson, D. J. & Ferguson, S. J. (1990). Periplasmic and membrane-bound respiratory nitrate reductases in *Thiosphaera pantotropha* – The periplasmic enzyme catalyses the first step in aerobic denitrification. *FEBS Letters*, **265**, 85–7.

Beveridge, T. J. (1981). Ultrastructure, chemistry and function of the bacterial wall. *International Review of Cytology*, **72**, 229–317.

Brass, J. M. (1986) The cell envelope of Gram-negative bacteria: new aspects of its function in transport and chemotaxis. *Current Topics in Microbiology and Immunology*, **129**, 1–96.

Brass, J. M., Higgins, C. F., Folley, M., Rugman, P. A., Birmingham, J. & Garland, P. B. (1986). Lateral diffusion of proteins in the periplasm of *Escherichia coli. Journal of Bacteriology*, **165**, 787–94.

Brown, N. L. (1985) Bacterial resistance to mercury – *reductio ad absurdum*? *Trends in Biochemical Sciences*, **10**, 400–3.

Carr, G. & Ferguson, S. J. (1990) Nitric oxide reductase of *Paracoccus denitrificans*. *Biochemical Journal*, **269**, 423–9.

Clark, D. P. (1985). Mutant of *Escherichia coli* deficient in osmoregulation of periplasmic oligosaccharide synthesis. *Journal of Bacteriology*, **161**, 1049–53.

Cole, J. A. (1968). Cytochrome c_{552} and nitrite reduction in *Escherichia coli Biochimica et Biophysica Acta*, **152**, 356–68.

Collatz, E., Labia, R. & Gutmann, L. (1990). Molecular evolution of ubiquitous β-lactamases towards extended-spectrum enzymes active against new β-lactam antibiotics. *Molecular Microbiology*, **4**, 1615–20.

Coyne, M. S., Arunakumari, A., Pankratz, H. S. & Tiedje, J. M. (1990). Localization of the cytochrome cdl and copper nitrite reductases in denitrifying bacteria. *Journal of Bacteriology*, **172**, 2558–62.

Daldal, F., Cheng, S., Applebaum, J., Davidson, E. & Prince, R. C. (1986). Cytochrome c_2 is not essential for photosynthetic growth of *Rhodopseudomonas capsulata*. *Proceedings of the National Academy of Sciences, USA*, **83**, 2012–16.

DiRita, V. J. & Mekalanos, J. J. (1991). Periplasmic interaction between two membrane regulatory proteins, *toxR* and *toxS*, results in signal transduction and transcriptional activation. *Cell*, **64**, 29–37.

Ferenci, T. & Silhavy, T. J. (1987). Sequence information required for protein translocation from the cytoplasm. *Journal of Bacteriology*, **169**, 5339–42.

Ferguson, S. J. (1987). Denitrification: a question of the control and organization of electron and ion transport. *Trends in Biochemical Sciences*, **12**, 354–7.

Ferguson, S. J. (1988a). Periplasmic electron transport reactions. In *Bacterial Energy Transduction*, ed. C. Anthony, pp. 151–82, Academic Press, London.

Ferguson, S. J. (1988b) The redox reactions of the nitrogen and sulphur cycles. In *The Nitrogen and Sulphur Cycles*. Society for General Microbiology Symposium 42 ed. J. A. Cole and S. J. Ferguson, pp. 1–29, Cambridge University Press, Cambridge.

Ferguson, S. J. (1990). Periplasm underestimated. *Trends in Biochemical Sciences*, **15**, 377.

Ferguson, S. J. (1991). The functions and synthesis of bacterial c-type cytochromes with particular reference to *Paracoccus denitrificans* and *Rhodobacter capsulatus*. *Biochimica et Biophysica Acta*, **1058**, 17–20.

Folley, M., Brass, J. M., Birmingham, J., Cook, W. R., Garland, P. B., Higgins, C. F. & Rothfield, L. I. (1989). Compartmentalisation of the periplasm at cell division sites in *Escherichia coli* as shown by fluorescence photobleaching experiments. *Molecular Microbiology*, **3**, 1329–36.

Fujita, T. & Satoh, R. (1966). Studies on soluble cytochromes in Enterobacteriaceae. III. Localization of cytochrome c_{552} in the surface layer of cells. *Journal of Biochemistry*, **60**, 568–77.

Gilson, E., Alloing, G., Schmidt, T., Claverys, J-P., Dudler, R. & Hofnung, M. (1988). Evidence for high affinity binding-protein dependent transport systems in Gram-positive bacteria and in *Mycoplasma*. *EMBO Journal*, **7**, 3971–4.

Graham, L. L., Harris, R., Villiger, W. & Beveridge, T. J. (1991). Freeze-substitution of Gram-negative eubacteria: general cell morphology and envelope profiles. *Journal of Bacteriology*, **173**, 1623–33.

Heppel, L. A. (1967). Selective release of enzymes from bacteria. *Science*, **156**, 1451–5.

Heppel, L. A. (1971). The concept of periplasmic enzymes. In *Structure and Function of Biological Membranes* ed. L. I. Rothfield, pp. 223–47. Academic Press, New York.

Higgins, C. F., Cairney, J., Stirling, D. A., Sutherland, L. & Booth, I. R. (1987). Osmotic regulation of gene expression: ionic strength as an intracellular signal? *Trends in Biochemical Sciences*, **12**, 339–44.

Higgins, C. F., Hyde, M. M., Mimmack, M. M., Gileadi, U., Gill, D. R. & Gallagher, M. P. (1990). Binding protein-dependent transport systems. *Journal of Bioenergetics and Biomembranes*, **22**, 571–92.

Hirst, T. R. & Holmgren, J. (1987). Transient entry of enterotoxin subunits into the periplasm occurs during their secretion from *Vibrio* cholerae. *Journal of Bacteriology*, **169**, 1037–45.

Hobot, J. A., Carleman, E., Villiger, W. & Kellenberg, E. (1984). Periplasmic gel: new concept resulting from the reinvestigation of bacterial cell envelope ultrastructure by new methods. *Journal of Bacteriology*, **160**, 143–52.

Holmgren, A. and Branden, C-I. (1989). Crystal structure of chaperone protein Pap D reveals an immunoglobulin fold. *Nature*, London, **342**, 248–51.

Hooper, A. B. and DiSpirito, A. A. (1985). In bacteria which grow on simple reductants, generation of a proton gradient involves extracytoplasmic oxidation of substrates. *Microbiological Reviews*, **49**, 140–57.

Hultgren, S. J., Lindberg, F., Magnusson, G., Kihlberg, J., Tennent, J. M. & Normark, S. (1989). The PapG adhesin of uropathogenic *Escherichia coli* contains separate regions for receptor binding and for the incorporation into the pilus. *Proceedings of the National Academy of Sciences, USA*, **86**, 4357–61.

Jones, M. R., McEwan, A. G. & Jackson, J. B. (1990). The role of c-type cytochromes in the photosynthetic electron transport pathway of *Rhodobacter capsulatus. Biochimicaet Biophysica Acta*, **1019**, 59–66.

Karlsson, B. G., Pascher, T., Nordling, M., Arvidsson, R. H. A. & Lundberg, L. G. (1989). Expression of the blue copper protein azurin from *Pseudomonas aeruginosa* in *Escherichia coli. FEBS Letters*, **146**, 211–17.

Kellenberger, E. (1990). The 'Bayer bridges' confronted with results from improved electron microscopy methods. *Molecular Microbiology*, **4**, 697–705.

Kennedy, E. P. and Rumley, M. K. (1988) Osmotic regulation of biosynthesis of membrane-derived oligosaccharides in *Escherichia coli. Journal of Bacteriology*, **170**, 1457–61.

Kroneck, P. M. H., Riester, J., Zumft, W. G. & Antholine, W. E. (1990). The copper site in nitrous oxide reductase. *Biology of Metals*, **3**, 103–9.

Kumamoto, C. A. (1991). Molecular chaperones and protein translocation across the *Escherichia coli* inner membrane. *Molecular Microbiology*, **5**, 19–22.

Lall, S. D., Eribo, B. E. & Jay, J. M. (1989). Comparison of four methods for extracting periplasmic proteins. *Journal of Microbiological Methods*, **9**, 195–9.

Laminet, A. A. & Pluckthun, A. (1989). The precursor of β-lactamase: purification, properties and folding kinetics. *EMBO Journal*, **8**, 1469–77.

Leduc, M., Frehel, C., Siegal, E. & van Heijenoort, J. V. (1989). Multilayered distribution of peptidoglycan in the periplasmic space of *Escherichia coli. Journal of General Microbiology*, **135**, 1243–54.

Lee, H. S. Hancock, R. W. & Ingraham, J. L. (1989). Properties of a *Pseudomonas stutzeri* outer membrane channel forming protein (Nos A) required for production of copper-containing N_20 reductase. *Journal of Bacteriology*, **171**, 2096–100.

Liu, J. & Walsh, C. T. (1990). Peptidyl-propylyl cis–trans-isomerase from *Escherichia coli*: A periplasmic homologue of cyclophilin that is not inhibited by cyclosporin A. *Proceedings of the National Academy of Sciences, USA*, **87**, 4028–32.

Luecke, H. & Quiocho, F. (1990). High specificity of a phosphate transport protein determined by hydrogen bonds. *Nature, London*, **347**, 402–6.

MacAllister, T. J., Macdonald, B. & Rothfield, L. I. (1983). The periseptal annulus.

An organelle associated with cell division in Gram negative bacteria. *Proceedings of the National Academy of Sciences, USA*, **80**, 1372–6.

McEwan, A. G., Richardson, D. J., Hudig, H., Ferguson, S. J. & Jackson, J. B. (1989). Identification of cytochromes involved in electron transport to trimethylamine *N*-oxide /dimethylsulphoxide reductase in *Rhodobacter capsulatus*. *Biochimica et Biophysica Acta*, **973**, 308–14.

Mauck, J. & Glaser, L. (1970). Periplasmic nucleoside diphosphate sugar hydrolase from *Bacillus subtilis*. *Biochemistry*, **9**, 1140–7.

Miller, K. J., Kennedy, E. P. & Reinhold, V. N. (1986). Osmotic adaptation by Gram-negative bacteria: possible role for periplasmic oligosaccharides. *Science*, **231**, 48–51.

Mitchell, P. (1961) Approaches to the analysis of specific membrane transport. In *Biological Structure and Function*. eds. T. W. Goodwin & O. Lundberg, vol 2, pp. 581–603, Academic Press, New York.

Mizushima, S. and Tokuda, H. (1990) *In vitro* translocation of bacterial secretory proteins and energy requirements. *Journal of Bioenergetics and Biomembranes*, **22**, 389–99.

Mokhele, K., Tang, Y. J., Clark, M. A. & Ingraham, J. L. (1987). A *Pseudomonas stutzeri* outer membrane protein inserts copper into nitrous oxide reductase *Journal of Bacteriology*, **169**, 5721–6.

Nanninga, N. (1991). Cell division and peptidoglycan assembly in *Escherichia coli*. *Molecular Microbiology*, **5**, 791–5.

Nikaido, H. & Nakae, T. (1979). The outer membrane of Gram-negative bacteria *Advances in Microbial Physiology*, **20**, 163–250.

Oliver, D. B. (1987) Periplasm and protein secretion. In Escherichia coli *and* Salmonella typhimurium, ed. F. C. Niedhardt, pp. 56–69, ASM Publications, Washington, DC.

Page, M. D. & Ferguson, S. J. (1989). A bacterial c-type cytochrome can be translocated to the periplasm as an apo form; the biosynthesis of cytochrome cd_1 (nitrite reductase) from *Paracoccus denitrificans*. *Molecular Microbiology*, **3**, 653–61.

Page, M. D. & Ferguson, S. J. (1990). Apo forms of cytochrome c_{550} and cytochrome cd_1 are translocated to the periplasm of *Paracoccus denitrificans* in the absence of haem incorporation caused by either mutation or inhibition of haem synthesis. *Molecular Microbiology*, **4**, 1181–92.

Park, S., Liu, G., Topping, T. B., Cover, W. H. & Randall, L. L. (1988). Modulation of folding pathways of exported proteins by the leader sequence. *Science*, **239**, 1033–5.

Perego, M., Higgins, C. F., Pearce, S. R., Gallagher, M. P. & Hoch, J. A. (1991). The oligopeptide transport system of *Bacillus subtilis* plays a role in the initiation of sporulation. *Molecular Microbiology*, **5**, 173–85.

Postle, K. (1990). TonB and the Gram-negative dilemma. *Molecular Microbiology*, **4**, 2019–25.

Prossnitz, E., Nikaido, K., Ulbrich, S. J. & Ames, G.F-L. (1988). Formaldehyde and photoactivatable cross-linking of the periplasmic binding protein to a membrane component of the histidine transport system of *Salmonella typhimurium*. *Journal of Biological Chemistry*, **263**, 17917–20.

Quiocho, F. A. (1990). Atomic structures of periplasmic binding proteins and the high affinity active transport systems in bacteria. *Philosophical Transactions of the Royal Society of London Series B*, **326**, 341–51.

Randall, L. L. & Hardy, S. J. S. (1989). Unity in function in the absence of a consensus in sequence: role of leader peptides in export. *Science*, **243**, 1156–59.

Richardson, D. J., McEwan, A. G., Page, M. D., Jackson, J. B. & Ferguson, S.

J. (1990). Identification of cytochromes involved in the transfer of electrons to the periplasmic nitrate reductase of *Rhodobacter capsulatus* and resolution of a soluble NO_3-reductase–cytochrome-c_{552} redox complex. *European Journal of Biochemistry*, **194**, 263–70.

Rittenberg, S. C. & Thomashow, M. F. (1979). Intraperiplasmic growth – life in a cozy environment. In *Microbiology 1979*, ed. D. Schlessinger, pp. 80–5. American Society for Microbiology, Washington, DC.

Ruby, E. G. & McCabe, J. B., (1988). Metabolism of periplasmic membrane-derived oligosaccharides by the predatory bacterium *Bdellovibrio bacteriovorus* 109J. *Journal of Bacteriology*, **170**, 646–52.

Shapleigh, J. P. & Payne, W. J. (1985). Nitric oxide-dependent proton translocation in various denitrifiers. *Journal of Bacteriology*, **163**, 837–40.

Shaw, J. G. & Kelly, D. J. (1991). Binding protein dependent transport of C_4 dicarboxylates in *Rhodobacter capsulatus. Archives of Microbiology*, **155**, 466–72.

Shuman, H. A. (1982). Active transport of maltose in *Escherichia coli*; role of periplasmic maltose-binding protein and evidence for a substrate recognition site in the cytoplasmic membrane. *Journal of Biological Chemistry*, **257**, 5455–61.

Speiser, D. M. & Ames, G.F-L. (1991). *Salmonella typhimurium* histidine periplasmic permease mutations that allow transport in the absence of histidine-binding proteins. *Journal of Bacteriology*, **173**, 1444–51.

Steeter, J. G. (1989). Analysis of periplasmic enzymes in intact cultured bacteria and bacteroids of *Bradyrhizobium japonicum* and *Rhizobium leguminosarum* biovar *phaseoli. Journal of General Microbiology*, **135**, 34377–84.

Stock, J. B., Rauch, B. & Roseman, S. (1977). Periplasmic space in *Salmonella typhimurium* and *Escherichia coli. Journal of Biological Chemistry*, **252**, 7850–61.

Summers, R. G. & Knowles, J. R. (1989). Illicit secretion of a cytoplasmic protein into the periplasm of *Escherichia coli* requires a signal peptide plus a portion of the cognate secreted protein. Demarcation of the critical region of the mature protein. *Journal of Biological Chemistry*, **264**, 20074–81.

Summers, R. G., Harris, C. R. & Knowles, J. R. (1989). A conservative amino acid substitution arginine for lysine abolishes export of a hybrid protein in *Escherichia coli.* Implications for the mechanisms of protein secretion. *Journal of Biological Chemistry*, **264**, 20082–8.

van Wielink, J. E. & Duine, J. A. (1990). How big is the periplasmic space? *Trends in Biochemical Sciences*, **15**, 136–7.

van Spanning, R. J. M., Wansell, C. W., Reijnders, W. N. M., Oltmann, L. F. & Stouthamer, A. H. (1990*a*). Mutagenesis of the gene encoding amicyanin of *Paracoccus denitrificans* and the resultant effect on methylamine oxidation. *FEBS Letters*, **275**, 217–20.

van Spanning, R. J. M., Wansell, C. W., Harms, N. L., Oltmann, L. F. & Stouthamer, A. H. (1990*b*). Mutagenesis of the gene encoding cytochrome c_{550} of *Paracoccus denitrificans* and analysis of the resultant physiological effects. *Journal of Bacteriology*, **172**, 986–96.

Vellieux, F. M. D., Huitema, F., Groendijk, H., Kalk, K. H., Frank, J. Jzn., Jongejan, J. A., Duine, J. A. Petratos, K. Drenth, J. & Hol, W. G. J. (1989). Structure of quinoprotein methylamine dehydrogenase at 2.25 A resolution. *The EMBO Journal*, **8**, 2172–8.

von Wachenfeldt, C. & Hederstedt, L. (1990). *Bacillus subtilis* 13-kilodalton cytochrome c-550 encoded by cccA consists of a membrane anchor and a heme domain. *Journal of Biological Chemistry*, **265**, 13939–48.

Vyas, N. K., Vyas, M. N. & Quiocho, F. A. (1988). Sugar and signal-transducer sites of the *Escherichia coli* galactose chemoreceptor protein. *Science*, **242**, 1290–5.

Wu, J. H., Kathir, P. & Ippen-Ihler, K. (1988). The product of the F plasmid transfer operon gene, *traF*, is a periplasmic protein. *Journal of Bacteriology*, **170**, 3633–9.
Yamamoto, K., Uozumi, T. & Beppu, T. (1987). The blue copper protein gene of *Alcaligenes faecalis* S-6 directs secretion of blue copper protein from *Escherichia coli* cells. *Journal ofBacteriology*, **169**, 5648–52.
Zumft, W. G., Viebrock-Sambale, A. & Braun, C. (1990). Nitrous oxide reductase from denitrifying *Pseudomonas stutzeri*; genes for copper-processing and properties of the deduced products, including a new member of the family of the ATP/GTP-binding proteins. *European Journal of Biochemistry*, **192**, 591–9.

MULTICELLULARITY IN CYANOBACTERIA

D. G. ADAMS

Department of Microbiology, The University of Leeds, Leeds LS2 9JT, UK

INTRODUCTION

This short chapter will consider the multicellular nature of cyanobacteria and, in particular, the ways in which the various forms of differentiation are controlled. It will concentrate largely on the most recently published work, except where, for reasons of clarity or because of the paucity of recent publications, older work will be cited. The reader who wishes to know more about the earlier literature and the contributions made by earlier workers should refer to the review articles and books given in the various sections and in the supplementary list at the end.

Cyanobacteria are a fascinating group of photosynthetic prokaryotes and the filamentous forms exhibit true multicellularity in their ability to form differentiated cell types which have specific functions and precise spatial relationships to one another (see Figs 1 and 2). However, it will become apparent in reading this chapter that we still know remarkably little about the mechanisms which control the complex differentiation processes of cyanobacteria. The recent application of molecular biological techniques by the groups of Haselkorn and Wolk, among others, has begun to offer the possibility of a clearer understanding of one aspect of cyanobacterial differentiation – the formation of heterocysts. However, there is a great deal that remains to be done and this account will of necessity leave many questions unanswered.

HETEROCYSTS

Cyanobacteria are photosynthetic prokaryotes which resemble eukaryotic algae and higher plants in their ability to photolyse water. However, the photosynthetic production of oxygen creates a problem for the many dinitrogen-fixing forms, which have to find a means of protecting nitrogenase from oxygen inactivation. They achieve this by several mechanisms (Stewart, 1980; Stewart & Rowell, 1986; Gallon & Chaplin, 1987) but the most remarkable is the production of a highly specialized cell, the heterocyst, whose sole purpose is to provide an anaerobic environment for the functioning of nitrogenase within an otherwise aerobic filament. Heterocysts are usually larger than vegetative cells, with less granular cytoplasm, thickened cell walls and polar bodies at either end of the cell at the point of attachment to the vegetative cells (Figs 1 and 2). During its development the heterocyst passes through an intermediate stage, the proheterocyst, which has a similar appearance to the mature cell, but lacks the fully thickened cell walls and polar bodies (Figs 1 and 2(*a*)). Unlike the mature heterocyst, which is terminally differentiated, the proheterocyst can revert to the vegetative form under the appropriate conditions (Fig. 1). In most filamentous cyanobacteria both proheterocysts and heterocysts occur in a very regularly spaced pattern (Figs 1 and 2).

The thickened cell walls characteristic of mature heterocysts are produced by the deposition of three extra layers external to the vegetative cell envelope (Fig. 4). The innermost of these is the laminated layer, consisting of glycolipid, which probably serves to reduce gas diffusion into the heterocyst and so maintain an anaerobic interior (see Wolk, 1982). External to the glycolipid layer are the homogeneous layer which consists largely of polysaccharide, and the outermost fibrous layer, which is not always apparent and which probably consists of uncompacted strands of the same polysaccharide (see Wolk, 1982). Deposition of this loose, fibrous material external to the vegetative cell envelope is one of the earliest signs of heterocyst development (see Fig. 3). During the formation of a heterocyst, each of the junctions with neighbouring vegetative cells becomes characteristically narrowed into a neck and it is in this region that a plug of material is deposited, which probably consists of the nitrogen reserve, cyanophycin (see section on proteolysis). This polar body is readily observed by light and electron microscopy (see Figs 2(*a*) and 4).

The pale and agranular appearance of heterocysts results from a number of changes which occur during their development. They generally lack phycobilisomes which are the complex structures containing the light-harvesting pigments, the phycobiliproteins (see Castenholz & Waterbury, 1989). The photosynthetic membranes become rearranged during heterocyst development, frequently showing a relatively open arrangement in the centre of the cell, but becoming tightly coiled in the regions adjacent to the polar plug (see Adams & Carr, 1981b; Castenholz & Waterbury, 1989). Heterocysts also lack the cyanophycin granules and carboxysomes (composed largely of ribulosebisphosphate carboxylase/oxygenase protein) which can be seen in the vegetative cell cytoplasm (see Adams & Carr, 1981*b*; Wolk, 1982; Castenholz & Waterbury, 1989).

The detailed structure of heterocysts, their metabolism and involvement in dinitrogen fixation have been extensively reviewed (see Stewart, 1980; Adams & Carr, 1981*b*; Wolk, 1982; Neuer, Papen & Bothe, 1983; Stewart & Rowell, 1986) and will not be dealt with here. This chapter will instead consider the recent advances in our understanding of the control of heterocyst development and spacing.

The groups of Haselkorn and Wolk have both applied the techniques of molecular biology to the process of heterocyst development, using related

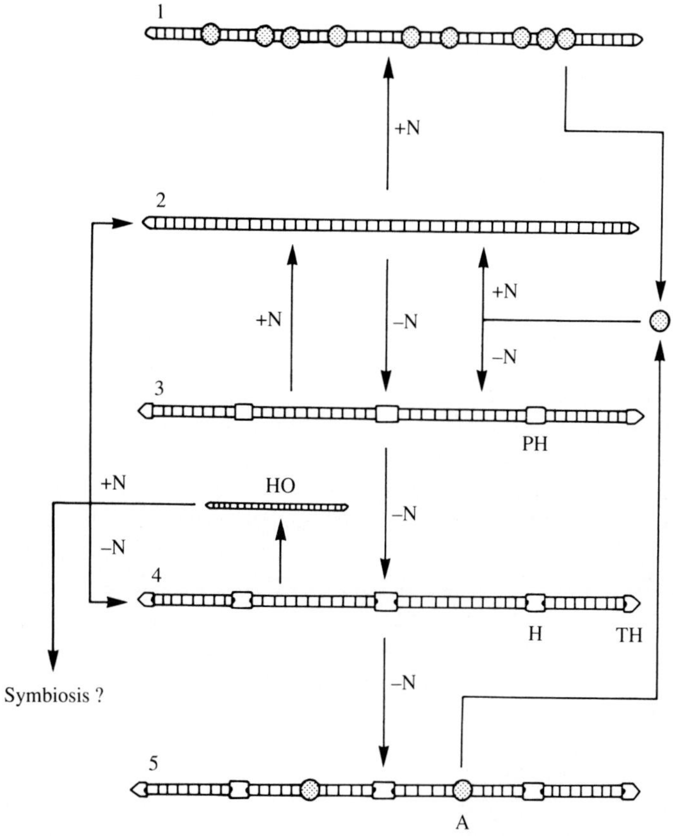

Fig. 1. Schematic diagram illustrating some of the possibilities for morphological development in filamentous cyanobacteria. When grown in the presence of a source of fixed nitrogen the filament consists entirely of undifferentiated vegetative cells (2). At the end of the exponential growth period when light (energy) becomes limiting, some vegetative cells can differentiate into the spore-like cells, akinetes (A) which, in the absence of heterocysts, are randomly situated within the filament (1). In the absence of fixed nitrogen the vegetative filament differentiates the highly specialized dinitrogen-fixing cells, heterocysts, at regular intervals within the filament (H) and in terminal positions (TH). Heterocysts are characterized by their thickened cell walls, relatively agranular cytoplasm and polar bodies at the point of attachment to vegetative cells, there being two in heterocysts within the filament, but only one in terminal heterocysts (4). During their development heterocysts pass through an intermediate stage, the proheterocyst (PH) which, unlike the mature cell, does not have the thickened cell wall and polar bodies and is able to dedifferentiate in the presence of fixed nitrogen (3). When akinetes develop in dinitrogen-fixing cultures they do so at locations with a precise spatial relationship to the heterocyts (see text for details) such as midway between (5). Akinetes can germinate and give rise to filaments with or without heterocysts depending on the availability of fixed nitrogen. Hormogonia (HO) are short, motile, undifferentiated filaments which develop as a result of a variety of stimuli (see text for details). Their formation usually involves the rapid division of vegetative cells without concomitant growth, followed by fragmentation of the filament to release heterocysts and motile hormogonia. The latter can give rise to heterocystous or non-heterocystous filaments depending on fixed nitrogen availability. Hormogonia can also serve as the infective agents in the establishment of symbiotic associations with bryophytes such as *Anthoceros* and *Blasia* (see text for details).

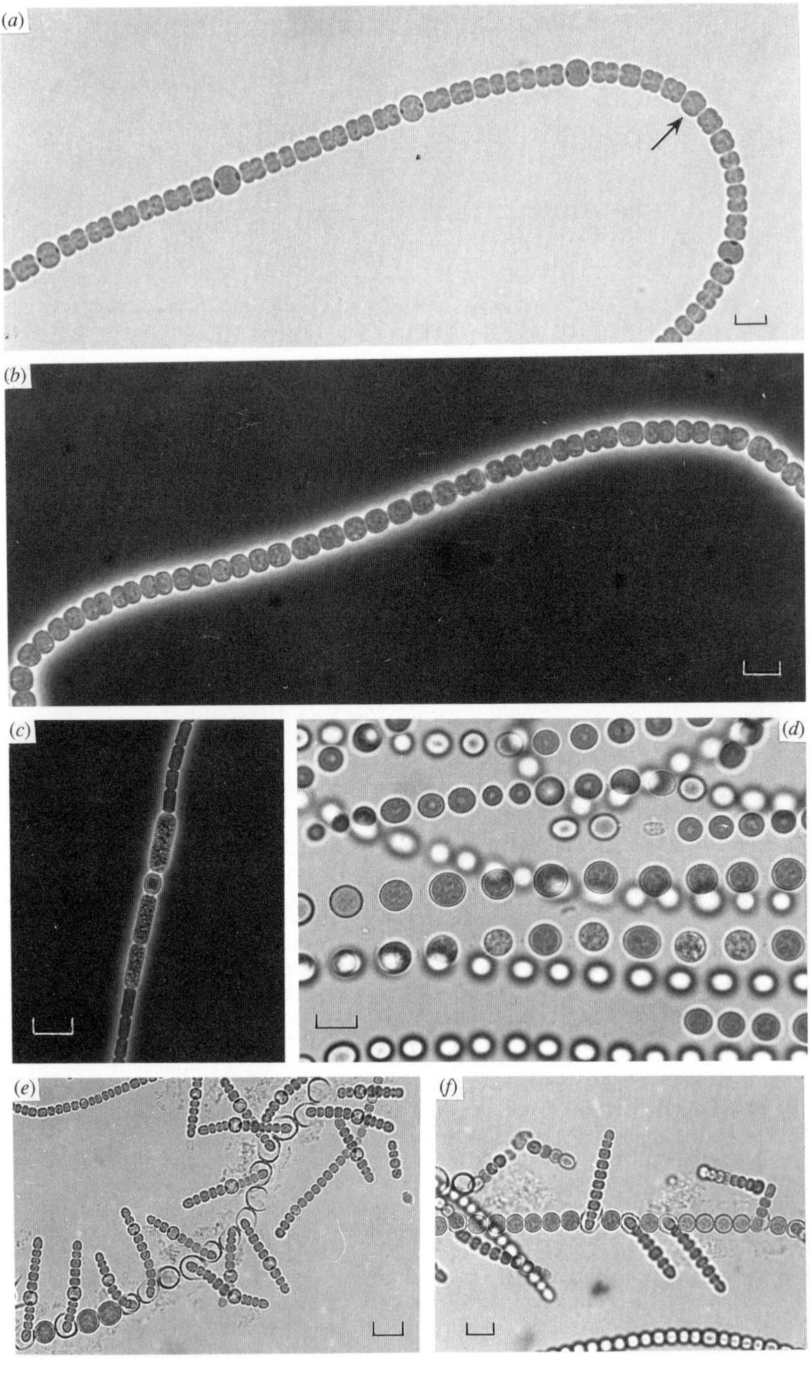
(a)
(b)
(c)
(d)
(e)
(f)

though different approaches. While Haselkorn and colleagues have concentrated mostly on *nif* gene organization, Wolk and co-workers have developed a cloning system for filamentous, heterocystous cyanobacteria with a view to isolating heterocyst specific genes and examining their expression.

Genome rearrangements associated with heterocyst development

To identify the *Anabaena nif* genes Mazur, Rice & Haselkorn (1980) made use of the available *Klebsiella nif* genes which, despite the evolutionary divergence between *Anabaena* and *Klebsiella*, retain sufficient homology to serve as very effective heterologous probes. However, the arrangement of the *nif* genes in *Anabaena* was shown to be different to that in *Klebsiella* (Mazur *et al.*, 1980). Detailed mapping of the genes subsequently revealed that, unlike *Klebsiella*, the *nifK*, *nifD* and *nifH* genes of *Anabaena* are not contiguous, there being an 11 kb segment of DNA separating *nifK* from *nifDH* (Rice, Mazur & Haselkorn, 1982). In addition, the *nifS* gene is to the right of *nifHD* rather than to the left. It was later shown that the 11 kb segment actually interrupts the *nifD* reading frame in vegetative cell DNA (Golden, Robinson & Haselkorn, 1985; see Fig 5). However, during heterocyst development, this region of the genome undergoes several rearrangements. In the first of these the 11 kb segment is excised from the chromosome as a result of a site-specific recombination between directly repeated 11 bp sequences at each end of the element (Golden *et al.*, 1985). This has the effect of restoring the *nifHDK* operon which can then be transcribed from the *nifH* promoter as a polycistronic message (Fig. 5). It was later shown that the excision occurred at low frequency when the 11 kb element was cloned into a plasmid vector and propagated in *E. coli* (Lammers, Golden & Haselkorn, 1986). The excision requires the activity of a gene, *xisA* (for excisase A), located on the 11 kb element and thought to code for a site-specific recombinase (Lammers *et al.*, 1986; Fig. 5).

Recent work has shed some light on the control of *xisA* gene expression.

Fig. 2. Photomicrographs illustrating some of the differentiated cell types of cyanobacteria. (*a*) *Anabaena* sp. strain CA showing the regular spacing of heterocysts, which are the sites of dinitrogen fixation, and a developing proheterocyst (arrow). (*b*) *Anabaena* sp. strain CA grown in the presence of nitrate which completely suppresses heterocyst development. (*c*) *Anadaena cylindrica* showing large, granular akinetes developing immediately adjacent to a heterocyst. (*d*) An old culture of nitrate-grown *Anabaena* sp. strain CA in which all vegetative cells have transformed into spherical akinetes. Although the akinetes have become separated, the line of the original filaments can still be seen. Dilution of such a culture leads to germination of the akinetes. If the medium used for dilution does not contain a source of fixed nitrogen the short filaments which emerge from the akinete coats each contain a heterocyst (*e*). However, if the medium contains nitrate then the filaments remain undifferentiated (*f*). Bar markers represent 10 μm. Figure (*c*) reproduced with permission from Nichols & Adams (1982), Blackwell Scientific Publications.

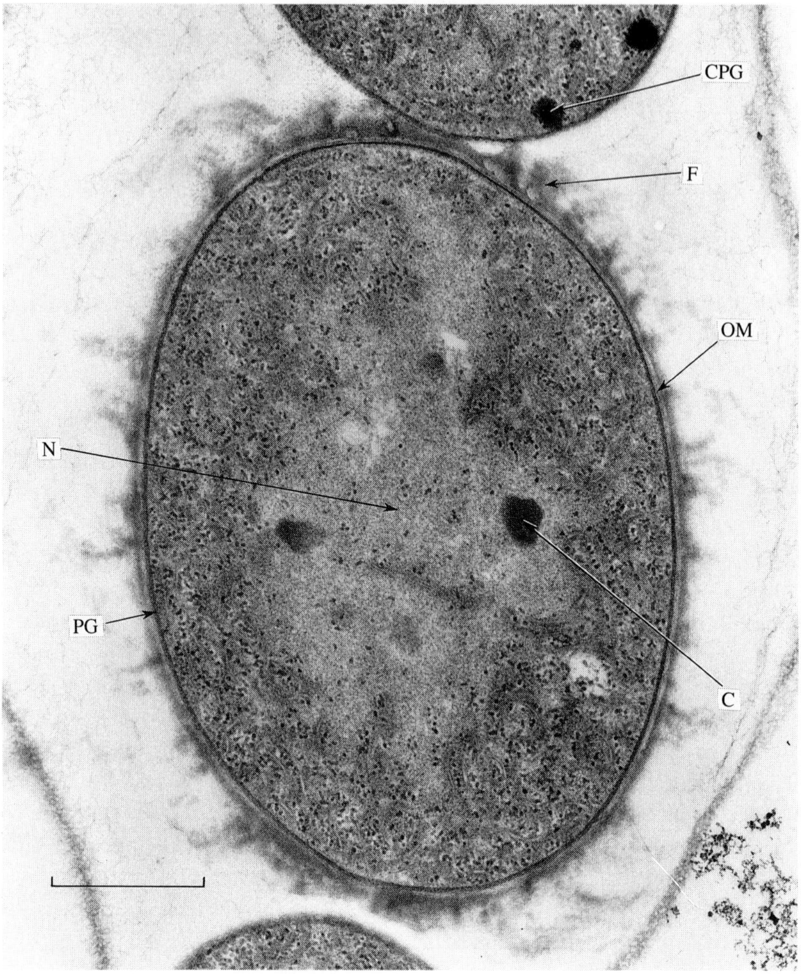

Fig. 3. Electron micrograph of a thin, longitudinal section of a filament of *Nostoc* sp. strain LBG1. The cell shown is at a very early stage of heterocyst development, showing little change from the neighbouring vegetative cells, other than the deposition of loose, fibrous material (F) external to the normal cell envelope.

N, nucleoplasmic region; C. carboxysome; CPG, cyanophycin granule; PG, peptidoglycan layer; OM, outer membrane. Sample fixed in osmium tetroxide and stained with uranyl acetate. Bar, 1 μm. Electron microscopy D. Ashworth.

Brusca, Chastain & Golden (1990) have shown that *xisA* can be expressed in vegetative cells of *Anabaena* PCC 7120 once a 127bp 5′ region has been deleted. Thus, when this 127 bp regulatory region is intact, expression of *xisA* is blocked in vegetative cells and hence the *nifD* rearrangement occurs only in heterocysts. To see if the *xisA* gene product alone was required for excision, the authors constructed a plasmid carrying only the borders

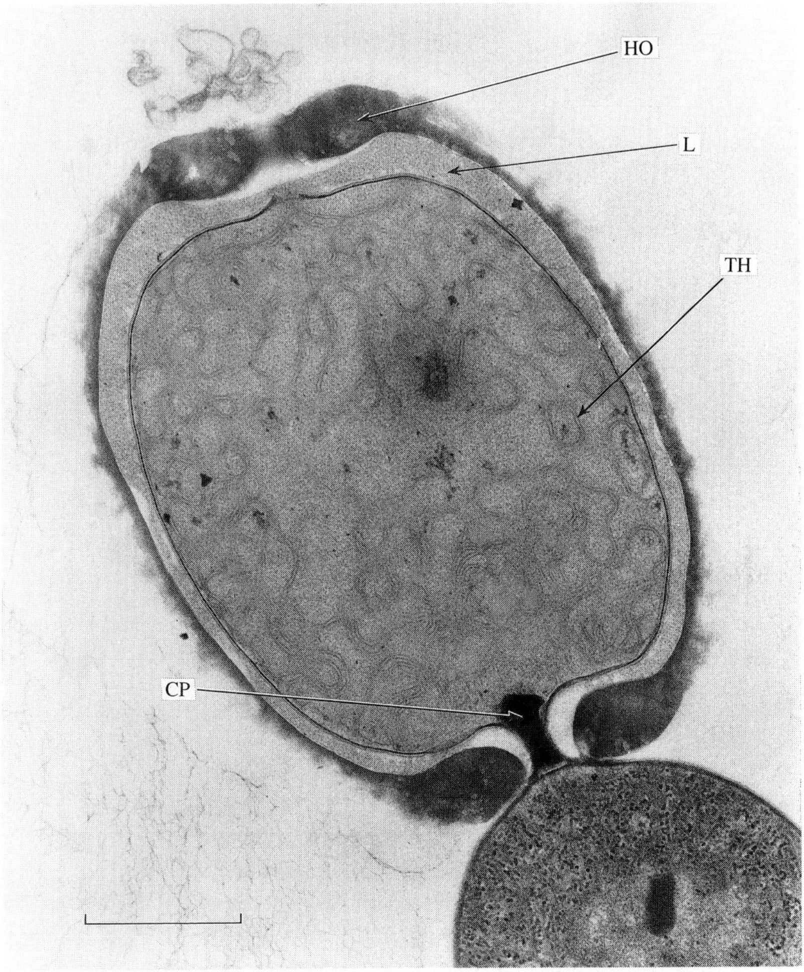

Fig. 4. Electron micrograph of a thin, longitudinal section of a filament of *Nostoc* sp. strain LBG1. The cell shown is a mature heterocyst showing the presence of the laminated (L) and homogeneous (HO) layers external to the normal cell wall. The junction between the heterocyst and neighbouring vegetative cell is very narrow and this polar region shows the characteristic deposition of a plug of cyanophycin (CP). The relatively agranular nature of the cytoplasm of the heterocyst can be seen when compared with the vegetative cell at the bottom of the picture. TH, thylakoids. Sample fixed in osmium tetroxide and stained with uranyl acetate. Bar 1 μm. Electron microscopy D. Ashworth.

of the *nifD* element with no other *Anabaena* open reading frame larger than 180 bp. When transferred to *E. coli* no rearrangement occurred unless a complementary plasmid, containing only the *xisA* gene, was also present. Thus, the *xisA* protein seems likely to be the only *Anabaena* gene product required for the rearrangement, although there remains the possibility that

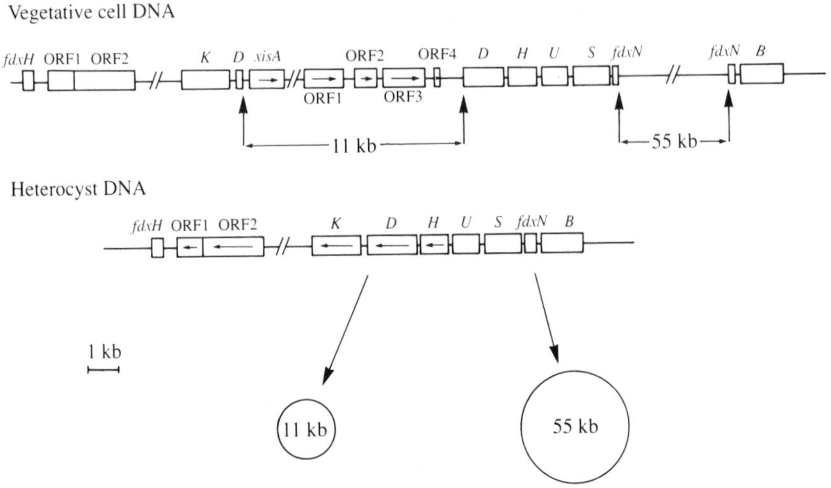

Fig. 5. Arrangement of genes in the *nif* region of the *Anabaena* sp. strain PCC 7120 chromosome. The *nif* genes *nifK*, *nifD*, *nifH*, *nifU*, *nifS* and *nifB* are indicated by their respective letters only. In vegetative cell DNA the *nifD* gene is interrupted by an 11 kb element containing the *xisA* gene and four open reading frames (ORF1–4), and the *fdxN* gene by a 55 kb element. During heterocyst development both of these elements are excised to restore the *nifD* and *fdxN* genes. Arrows within genes indicate the direction of transcription. See text for full details.

it works together with other *Anabaena* proteins whose function can also be provided by *E. coli* proteins.

A sequence-specific DNA-binding factor (VF1) has been isolated from vegetative cells of *Anabaena* sp. strain PCC 7120 (Chastain *et al.*, 1990). VF1 binds within the *xisA* regulatory region and may, therefore, prevent transcription of the *xisA* gene in vegetative cells. It also binds weakly to the *nifH* promoter which is expressed only in heterocysts. However, VF1 is unlikely to be involved solely in the regulation of heterocyst-specific genes since it also forms complexes with the promoter regions of *rbcL*, which is expressed only in vegetative cells, and *glnA*, which is expressed in both heterocysts and vegetative cells (Chastain *et al.*, 1990).

The second DNA rearrangement to occur during heterocyst development involves another deletion, this time near *nifS* and involving a 55 kb element (Golden *et al.*, 1988). It was subsequently shown that the 55 kb element interrupts a gene, *fdxN*, which codes for a bacterial type ferrodoxin (Mulligan, Buikema & Haselkorn, 1988; Mulligan & Haselkorn, 1989; see below and Fig. 5) This rearrangement, in common with that of the 11 kb element, involves site-specific recombination between 5 bp directly repeated sequences at either end of the element. However, the sequences of the recombination sites for the 11 kb and 55 kb deletions show no sequence similarity (Golden, Mulligan & Haselkorn, 1985), indicating that they are recognized by different site-specific recombinases. The excised 11 kb element remains as a stable, circular molecule in heterocysts (Golden *et al.*, 1985). The excised 55 kb element has not been detected in heterocyst DNA preparations (Golden *et al.*, 1988).

Both the *nifD* and *fdxN* DNA rearrangements occur exclusively in heterocysts during the late stages of development, vegetative cell DNA showing no sign of either rearrangement (Golden *et al.*, 1988). Despite this similarity there are a number of differences. The 55 kb deletion occurs during heterocyst development under aerobic conditions and also microaerobic conditions in which cultures are gassed with argon, but no nitrogen, oxygen or carbon dioxide is supplied. However, the *nifD* 11 kb deletion occurs under aerobic but not microaerobic conditions (Golden *et al.*, 1988). Site-directed inactivation of the *xisA* gene has no effect on the rearrangement of the 55 kb element, nor the differentiation and spacing of heterocysts, but it does block both the 11 kb excision and dinitrogen fixation (Golden & Wiest, 1988). Thus proper expression of the *nif* genes requires prior excision of the *nifD* 11 kb element. The presence of this element, however, seems to be restricted to heterocystous cyanobacteria, since many non-heterocystous, N_2-fixing strains have a contiguous *nifHDK* arrangement (Kallas *et al.*, 1983; Kallas, Coursin & Rippka, 1985). However, the *nifD* 11 kb insert is not present in the filamentous heterocystous cyanobacterium *Fischerella* sp. ATCC 27929 (Saville, Straus & Coleman, 1987). This organism belongs to group V in the classification of Rippka *et al.* (1979) which contains filamentous, heterocystous cyanobacteria which form true branches. These interesting cyanobacteria have complex growth morphologies and appear to show poor control over hetero-cyst spacing (Nierzwicki-Bauer, Balkwill & Stevens, Jr, 1984; Balkwill, Nierzwicki-Bauer & Stevens, Jr, 1984; Stevens, Jr, Nierzwicki-Bauer & Balkwill, 1985).

Less is known about the 55 kb element although it is not found in *Anabaena variabilis* ATCC 29413 which does, however, possess the 11 kb *nifD* element which is excised during heterocyst development and is flanked by 11 bp repeats identical to those in *Anabaena* PCC 7120 (Brusca *et al.*, 1989). Indeed, the 11 kb element in *Anabaena variabilis* ATCC 29413 contains a *xisA* gene capable of complementing a defective *xisA* gene in *Anabaena* PCC 7120 (Brusca *et al.*, 1989).

A second *nif* operon, located upstream from the *nifHDK* operon, has now been identified and sequenced (Mulligan & Haselkorn, 1989) and shown to contain four genes, *nifB, fdxN, nifS* and *nifU*(Fig. 5). The three *nif* genes are similar to their counterparts in other diazotrophs, but the *fdxN* gene codes for a bacterial type ferredoxin which has not been reported previously in cyanobacteria (Mulligan *et al.*, 1988). The 55 kb element is actually located within the *fdxN* gene and must be excised before the *nifB–fdxN–nifS–nifU* operon can be expressed (Mulligan & Haselkorn, 1989; Fig. 5).

A second ferredoxin gene *fdxH* is located approximately 7 kb downstream of *nifK* (Fig. 5). This codes for a heterocyst ferredoxin (Schrautemeier &

Böhme, 1985; Böhme & Schrautemeier, 1987a,b) and has been cloned and sequenced (Böhme & Haselkorn, 1988). During heterocyst development transcription of *fdxH* occurs at about the same time as the *nifD* rearrangement and the first appearance of dinitrogenase activity (Böhme & Haselkorn, 1988). Two further open reading frames, ORF1 and ORF2, have been identified in this region of the *Anabaena* PCC 7120 chromosome (Borthakur *et al.*, 1990; Fig. 5). These are located 4 kb downstream of *nifK* and have now been sequenced (Borthakur *et al.*, 1990). A mutant, generated by site-directed mutagenesis in ORF1, grows very slowly on medium lacking combined nitrogen and produces only 45% of wild-type acetylene reducing activity. This mutant can be complemented by a 2.8 kb fragment of wild-type *Anabaena* PCC 7120 DNA containing only ORF1 and ORF2, suggesting that one or both of these ORFs are required for efficient dinitrogen fixation (Borthakur *et al.*, 1990).

More than half of the *nifD* 11 kb element of *Anabaena* sp. strain PCC7120 has now been sequenced and shown to contain four open reading frames (Lammers *et al.*, 1990; Fig. 5). Three of these do not match any known protein sequences, while the fourth shows considerable similarity with the cytochrome P-450 family of monooxygenases (Lammers *et al.*, 1990). Although the excision of the 11 kb element is required for *nif* gene expression, it seems to have no significance for the control of heterocyst development and spacing. Although many filamentous, heterocystous cyanobacteria possess the 11 kb insert, others such as *Fischerella* sp. ATCC 27929 (Saville *et al.*, 1987) and the major cyanobacterial partners in several symbioses with the water-fern *Azolla* (Meeks, Joseph & Haselkorn, 1988) do not. Site-directed inactivation of the *xisA* gene blocks the 11 kb excision and dinitrogen fixation but has no effect on heterocyst differentiation and spacing (Golden & Wiest, 1988). In addition, strains completely lacking the 11 kb element display no obvious phenotypic changes in the presence or absence of fixed nitrogen showing normal heterocyst development and spacing in the latter case (Brusca *et al.*, 1990).

Development of a cloning system for heterocystous cyanobacteria

Wolk and co-workers have directed their efforts to developing shuttle vectors capable of replication and selection both in *E. coli* and in strains of filamentous, heterocystous cyanobacteria. Although transformation of unicellular cyanobacteria was well established by the early 1980s (see Tandeau de Marsac & Houmard, 1987), there was no such system for filamentous cyanobacteria. An alternative method for the transfer of cloned DNA is conjugation and Delaney & Reichelt (1982) had described a low frequency of transfer of the conjugative plasmid R68.45, a close relative of RP-4, from *E. coli* to the unicellular cyanobacterium *Synechococcus* sp. strain PCC 6301. Although the establishment of RP-4 in filamentous cyanobacteria had not been demonstrated, Wolk *et al.* (1984) reasoned that it should be possible to employ the conjugal properties of RP-4 to develop a system for the transfer of shuttle vectors between *E. coli* and dinitrogen-fixing filamentous cyanobacteria. It was known that several conjugative plasmids, including RP-4, could promote the transfer of derivatives of pBR322 from *E. coli* to other Gram-negative bacteria (Van Haute *et al.*, 1983; Taylor *et al.*, 1983). This process required an intact *bom* (basis of mobility) region in the plasmid being transferred and the presence of certain *trans*acting factors which could be provided by helper plasmids pDS4101 or pGJ28 (Finnegan & Sherratt, 1982). It did not matter that RP-4 was unable to replicate in the new host. Wolk *et al.* (1984) constructed hybrids between pBR322 and a cyanobacterial plasmid pDU1 which had been isolated from *Nostoc* (Reaston *et al.*, 1982). The hybrids were modified to remove the recognition sites for restriction enzymes present in several strains of *Anabaena*, since these were known to reduce the retention of DNA introduced into cyanobacteria (Currier & Wolk, 1979; Buzby, Porter & Stevens, Jr, 1983). A final modification was to include further antibiotic resistance determinants to produce four shuttle vectors (pRL1, pRL5, pRL6 and pRL8). These shuttle vectors could be readily transferred by conjugation from *E. coli* to *Anabaena*, providing RP-4 and a helper plasmid were also present (Wolk *et al.*, 1984). It was later shown that the pRL vectors could also be used for *Nostoc*, producing gene transfer frequencies as high as 1×10^{-3}–3×10^{-3} (Flores & Wolk, 1985). Unlike the *Anabaena* spp. used previously, which are obligate photoautotrophs, the *Nostoc* spp. are facultative heterotrophs and therefore potentially useful for the study of oxygenic photosynthesis (Flores & Wolk, 1985).

Towards an understanding of heterocyst genes and their expression

The establishment and maintenance of the regular spacing of heterocysts in filamentous cyanobacteria clearly requires differential gene expression, such that from the very earliest stages of the process new genes are expressed solely in those cells destined to become heterocysts. The vectors developed by Wolk and co-workers provided the opportunity to visualize the expression of such genes in individual cells of a cyanobacterial filament. To achieve this it was necessary to modify these vectors by incorporating bacterial luciferase genes to act as reporters of transcription.

Bacterial luciferase catalyses the reaction:

$$RCHO + O_2 + FMNH_2 \rightarrow R.COOH + FMN + H_2O + h\nu\ (490\,\text{nm})$$

in which R is a straight chain aliphatic moiety with seven or more carbon atoms in the absence of which no light is produced (Ziegler & Baldwin, 1981). The two subunits of luciferase are co-transcribed from the *luxA* and *luxB* genes of *Vibrio* spp. The genes of *V. fischeri* had already been cloned and used successfully as reporter genes in *E. coli* (Engebrecht, Simon & Silverman, 1985).

The first task was to see if bacterial *lux* genes would function in heterocystous cyanobacteria. The pRL shuttle vectors were used to clone the *lux* genes from both *Vibrio harveyi* and *Vibrio fischeri* (Schmetterer, Wolk & Elhai, 1986). Expression of these genes was monitored, in the presence of n-decanal as the required aliphatic moiety, using an ATP photometer. The *lux* genes from both *Vibrio* sp. were successfully expressed in several strains of *Anabaena* spp. and the level of expression was greatly enhanced if a strong promotor, such as that for the *Anabaena* ribulose bisphosphate carboxylase structural genes (Curtis & Haselkorn, 1983), was placed upstream of the genes and in the correct orientation (Schmetterer *et al.*, 1986). More importantly, perhaps, it was possible to observe microscopically the light produced by single cells of *Anabaena* PCC 7120, opening up the possibility of observing the expression of heterocyst genes in individual cells of a filament. The next requirement, however, was to isolate the heterocyst genes themselves.

The development of the conjugation-based cloning system for *Anabaena* and *Nostoc* (Wolk *et al.*, 1984) finally made possible the genetic analysis of mutants of such cyanobacteria, where previously only biochemical and physiological studies were possible. The simplest mutants to isolate are the Nif^- phenotype since they can be selected by their lack of growth on medium lacking a source of fixed nitrogen. Wolk *et al.* (1988) isolated such mutants of *Anabaena* PCC 7120 by mutagenesis with UV irradiation. To allow mutational repair of damaged DNA without photoreactivation (which requires wavelengths of <500 nm) cells were grown under yellow illumination (Flores & Schmetterer, 1986). Following penicillin enrichment a number of mutants were obtained which were unable to grow in the absence of fixed nitrogen. Two of these mutants were chosen for complementation analysis because they showed cytological differences in the heterocyst envelope layers. Mutant EF113 retained only a vestigal heterocyst envelope and in mutant EF116 the heterocyst envelope polysaccharide was less cohesive than in the wild-type (Wolk *et al.*, 1988). These mutants are important because the heterocyst envelope layers have been implicated as an oxygen barrier which helps maintain an anaerobic environment within the cell (Walsby, 1985; Murry & Wolk, 1989). In addition, the formation of the fibrous layer of the envelope is the first morphological manifestation of the development of a heterocyst. This layer consists of probably the same polysaccharide as that of the bulk of the outer layer of the envelope, which is altered in mutant EF116. It would seem, therefore, that the gene for this envelope polysaccharide is under developmental control and is activated very early during heterocyst differentiation (Wolk *et al.*, 1988).

For complementation studies Wolk *et al.* (1988) constructed a cosmid vector, pRL25C, from one of the previously developed pRL shuttle vectors (Wolk *et al.*, 1984). Both EF113 and EF116 could be complemented by clones from an *Anabaena* PCC 7120 gene library developed in pRL25C. In a later paper Holland and Wolk (1990) described the analysis of one of the clones, 3.5kb in length and designated pRL52, which complemented EF116. A series of deletion derivatives of pRL52 was tested for their ability to complement the mutant. When cells of EF116 were complemented with pRL52 green colonies developed within 7 days on selective medium lacking combined nitrogen. Removal of short sections from both the 3′ and 5′ ends of pRL52 resulted in what the authors described as delayed complementation, in which yellowish clumps of cells developed initially, followed by greenish colonies after 10–14 days. The authors concluded that delayed complementation resulted from recombination between the introduced plasmids and the mutated chromosomal DNA. It was clear then that pRL52 contained the gene, designated *hetA*, that was altered in mutant EF116. Plasmid pRL52 was later sequenced and shown to contain the whole of one open reading frame and the start of a second (Holland & Wolk, 1990). Transcription of *hetA* was induced by nitrogen starvation, yielding a monocistronic message which was most abundant 7 hours after nitrate removal. This was before proheterocysts could be distinguished by light microscopy and long before the peak of expression of the nitrogenase structural genes, *nifHD* (Holland & Wolk, 1990). A mutant with a phenotype resembling that of EF116 has since been produced by site-directed inactivation of *hetA* (Cai & Wolk, 1990). In medium free of combined nitrogen this mutant develops cells with the shape and spacing of heterocysts but showing no deposition of the envelope polysaccharide (Cai & Wolk, 1990) in contrast to EF 116 which shows irregular deposition (Wolk *et al.*, 1988). The genes which have been altered in mutants EF116 (*hetA*) and EF113 (*hetB*) have now been physically mapped on the *Anabaena* sp. strain PCC 7120 genome and shown to be well separated (Bancroft, Wolk & Oren, 1989).

Since heterocyst development is known to involve an ordered sequence of events, such as the appearance and disappearance of specific proteins at defined times (Fleming & Haselkorn, 1974), it is tempting to speculate that its regulation is transcriptionally controlled. The rearrangement which occurs within the *nifHDK* operon and the expression of the operon itself could then be seen as a part of the overall process of heterocyst development, which requires gene products specific to heterocyst differentiation. Hence, expression of the *nifHDK* operon would only occur in cells in which heterocyst genes were also being expressed. However, an alternative to such a developmental control would be an environmental one in which all that was required for the rearrangement and expression of the *nifHDK* operon was nitrogen starvation and anaerobiosis. In this case the expression of these genes would only occur in heterocysts at the final stages of maturation when the interior of the cells became anaerobic. However, expression of the operon could also occur in vegetative cells if they were starved of nitrogen under anaerobic conditions. Elhai and Wolk (1990) recently tested these two possibilities using two cyanobacteria, *Anabaena* PCC 7120 and a very closely related organism, *Anabaena* PCC 7118. The latter is a mutant unable to produce heterocysts, but which can fix nitrogen when incubated anaerobically in the absence of combined nitrogen (Rippka & Stanier, 1978). Under the same conditions *Anabaena* PCC 7120 develops heterocysts which lack the glycolipid layer needed to protect nitrogenase from oxygen (Rippka & Stanier, 1978). Having shown previously that bacterial *lux* genes could be used as transcriptional reporters (Schmetterer *et al.*, 1986) Elhai and Wolk (1990) produced a series of plasmids in which the promoters for *nifHDK, rbcLS* (encoding the large and small subunits of ribulose bisphosphate carboxylase/oxygenase) or *glnA* (encoding glutamine synthetase) were fused to either *lacZ* (encoding β-galactosidase) or *luxAB* (encoding bacterial luciferase). These plasmids were tranferred into *Anabaena* strains by conjugation from donors carrying the helper plasmid pRL528 (Elhai & Wolk, 1988). However, unlike the shuttle vectors constructed previously, they lacked the cyanobacterial replicon pDU1, derived from *Nostoc* PCC 7524, and were therefore unable to propagate when transferred to either *Anabaena* PCC 7120 or *Anabaena* PCC 7118. However, they were shown to integrate readily into the chromosome at a site directed by the cloned region of homology (Elhai & Wolk, 1990). This meant that the activity of *nifHDK*, *rbcLS* or *glnA* could be monitored, by measuring either β-galactosidase activity or light emission, by transfer of the plasmid containing the appropriate promoter-reporter fusion. In the case of the *luxAB* fusions, light emission from individual cells could be seen, allowing a precise localization of the expression of *nifHDK*, *rbcLS* or *glnA* under a variety of culture conditions (Elhai & Wolk, 1990).

When the plasmid containing the promoter for *nifHDK* (P_{nifHDK}) fused to *luxAB* was transfered to *Anabaena* PCC 7120 grown aerobically in the absence of fixed nitrogen, light emission was seen only in heterocysts, confirming that nitrogenase components are present only in heterocysts under these conditions. Under the same conditions *Anabaena* PCC 7120 carrying the P_{rbcLS}-*luxAB* fusion showed light emission only in vegetative cells, neatly confirming previous knowledge that the enzyme is absent from heterocysts (Winkenback & Wolk, 1973; Codd & Stewart, 1977) as is *rbcLS* mRNA (Golden *et al.*, 1988). When these experiments were repeated with anaerobically starved cultures, the same results were obtained, with P_{nifHDK} active only in heterocysts (confirming the immunochemical evidence of Murry, Hallenbeck & Benemann (1984) that nitrogenase in *A. cylindrica* is restricted to heterocysts even under anaerobic conditions) and P_{rbcLS} active only in vegetative cells. P_{glnA} was shown to be active in all cells, again confirming what was known about the activity of glutamine synthetase. Elhai & Wolk (1990) concluded that these results tended to support the developmental regulation of P_{nifHDK} since it showed no activity in vegetative cells under anaerobic nitrogen-starved conditions. The expression of P_{hetA} shows the same spatial pattern as P_{nifHDK}, occurring only in heterocysts under both aerobic or anaerobic conditions (Elhai & Wolk, 1990).

Elhai & Wolk (1990) then examined *nifHDK* expression in the mutant *Anabaena* PCC 7118 using the P_{nifHDK}-*lacZ* fusion. Induction of P_{nifHDK}-*lacZ* (measured as β-galactosidase activity) began between 13 and 16 hours after the removal of fixed nitrogen from the medium of cultures incubated anaerobically. This induction was accompanied by a decrease in average filament length which was not a result of random filament breakage since the resulting distribution of filament lengths was reminiscent of the spacing between heterocysts in wild-type strains (Wolk & Quine, 1975). No such fragmentation occurred in the presence of nitrate. Although no mature heterocysts were found, certain cells showed a loss of phycocyanin fluorescence and these were distributed in a well spaced pattern, not unlike that of heterocysts in wild-type strains. When the P_{nifHDK}-*luxAB* fusion was introduced into *Anabaena* PCC 7118 and the culture starved for fixed nitrogen under anaerobic conditions, light emission was only seen in single, well-spaced cells which showed certain morphological characteristics in common with the non-fluorescent cells. The authors concluded that the non-random filament fragmentation and the regular spacing of non-fluorescent cells were a result of aborted heterocyst differentiation and that the observed *nifHDK* expression was associated with these 'failed' heterocysts. Thus, in *Anabaena* PCC 7118, which is incapable of mature heterocyst development, under anaerobic conditions *nifHDK* is still expressed only in cells which in the wild-type would form heterocysts. This, Elhai & Wolk (1990) argued, strongly favoured a developmental regulation for *nif* gene expression.

Proteolysis

Growth of heterocystous cyanobacteria in the presence of NH_4^+ or NO_3^- inhibits heterocyst development such that filaments consist entirely or mostly of vegetative cells. The process of heterocyst differentiation can then be triggered by the removal of the fixed nitrogen from the medium. Since it is some time before mature heterocysts are produced and N_2 fixation can commence (typically 15–20 h for an organism with an 18 h generation time), the filaments must pass through a period of nitrogen starvation. Despite this they must also synthesize many new proteins to complete their morphological differentiation and since the nitrogen for this is not available externally, nor from nitrogen fixation, it must clearly be derived from within the cells of the filament. The potential sources of this nitrogen are the reserve materials cyanophycin and phycocyanin and those vegetative proteins which will not be required in heterocysts.

Cyanophycin

Cyanophycin is a copolymer of arginine and aspartic acid which is unique to cyanobacteria (Simon, 1987) and which most cyanobacteria accumulate at some stage of their growth cycle (Lawry & Simon, 1982; Allen, 1984). It serves as a nitrogen reserve which is degraded in nitrogen starved cells (Allen & Hutchinson, 1980). The enzymes for the synthesis and breakdown of cyanophycin are more active in *Anabaena* heterocysts than in vegetative cells (Gupta & Carr, 1981) and this has prompted the interesting hypothesis that the polymer serves as a dynamic reservoir of newly assimilated nitrogen which can then be used for biosynthesis (Carr, 1983, 1988; Carr & Wyman, 1986). Such a role for cyanophycin would seem not to occur in *A. cylindrica* growing on a 12 h light–12 h dark cycle since the level of the polymer varies little during the light and dark periods (Mackerras *et al.*, 1990b). Nitrogen fixation occurs only in the light period yet protein synthesis continues in the dark at 80% of the light rate, implying that the nitrogen required must come from reserves, but apparently not from cyanophycin. In similar experiments with the non-heterocystous cyanobacterium *Gloeothece*, which fixes N_2 only during the dark period, there is no decrease in cyanophycin levels during the light to support protein synthesis (Mackerras *et al.*, 1990*b*). However, there are circumstances when cyanophycin does show a dynamic metabolism. When *A. cylindrica* is grown on a 12 h light–12 h dark cycle with a low level of NH_4^+ (0.8 mM) in the medium a transient accumulation of cyanophycin occurs as the NH_4^+ level decreases (Mackerras, de Chazal & Smith, 1990*a*). The cyanophycin is degraded once more just before the NH_4^+ level reaches zero. If, once NH_4^+ is depleted, the culture is gassed with $Ar/O_2/CO_2$, to prevent N_2 fixation and therefore maintain the nitrogen starvation, and NH_4^+ is added back, another transient accumulation of cyanophycin occurs, followed by a decline as exponential growth resumes. However, if the culture is returned to nitrogen-fixing conditions (gassing with CO_2 in air) only a very small cyanophycin peak is detected. Mackerras *et al.* (1990*a*) concluded from these results that cyanophycin 'serves as a reservoir of newly assimilated nitrogen, which allows maximum acquisition of fixed nitrogen over relatively short time periods, and subsequent use of that nitrogen for growth or survival'.

These experiments give no indication of the possible involvement of cyanophycin in heterocyst development. However, work by the same group (Daday, Mackerras & Smith, 1988) does suggest that any disturbance of the normal metabolism of cyanophycin can affect the transition from non-nitrogen-fixing to nitrogen-fixing growth. They have shown that when a nickel-depleted culture of *A. cylindrica*, in which heterocysts and nitrogenase are repressed by growth on NH_4^+, is used to inoculate fresh nickel-depleted medium, there is a long lag before growth begins. There are equivalent delays in heterocyst development, nitrogenase activity and the re-synthesis of phycocyanin. During this lag period cyanophycin levels initially increase and then decline rapidly once more. This decline signals the end of the lag period. None of these changes occurs in nickel replete cultures and the authors concluded that nickel plays a role in maintaining rapid turnover of cyanophycin. In the absence of nickel the rate of synthesis exceeds that of breakdown, resulting in ammonia (released by turnover of proteins, especially phycocyanin) being diverted into cyanophycin production instead of *de novo* protein synthesis. The significance of this in terms of heterocyst development remains unclear, however, since the omission of nickel from the medium has no effect on the exponential growth rate of dinitrogen-fixing *Anabaena cylindrica* (Daday, Mackerras & Smith, 1985).

Phycocyanin

Although the involvement of cyanophycin metabolism in heterocyst development remains unclear, there are no doubts about the importance of phycocyanin. The phycobiliprotein phycocyanin is one of the major light harvesting pigments of cyanobacteria (see Glazer, 1987). During the process of heterocyst differentiation, induced by transfer of a culture from medium containing combined nitrogen to one lacking it, the culture becomes visibly yellow as the blue pigment phycocyanin is degraded. Since phycobiliproteins can constitute up to 50% of total cell protein in cyanobacteria (see Cohen-Bazire & Bryant, 1982) they are a major potential source of amino acids for the *de novo* protein synthesis required during heterocyst development.

Anabaena spp. have been shown to contain an enzyme (phycocyaninase) which specifically degrades phycocyanin (Foulds & Carr, 1977; Wood & Haselkorn, 1980). The activity of the enzyme increases six- to ten- fold in cells starved of nitrogen either by transfer to medium lacking in fixed nitrogen, or by the addition of the glutamine analogue, methionine sulphoximine, to block assimilation of fixed nitrogen (Wood & Haselkorn,1980). However, phycobiliprotein degradation is not induced when a *met* auxotroph is starved for methionine in the presence of ammonia and this led Wood & Haselkorn (1980) to propose that in *Anabaena* the limitation of glutamine is involved in the induction of both phycobiliprotein degradation and heterocyst development. Recently, however, Thiel (1990) has shown that in *Anabaena variabilis* ATCC 29413 phycobiliprotein degradation occurs to about the same extent in cultures transferred from ammonia to glutamine as nitrogen source as those transferred to medium lacking any source of fixed nitrogen.

As would be expected, phycocyanin (and allophycocyanin) gene expression is switched off in response to nitrogen starvation (Johnson *et al.*, 1988; Wealand, Myers & Hirschberg, 1989), probably as a result of transcriptional regulation (Wealand *et al.*, 1989). There is some evidence that in *Anabaena variabilis* ATCC 29413 proteolysis releases sufficent nitrogenous compounds to alleviate repression of the phycocyanin and allophycocyanin genes and permit their renewed expression several hours before nitrogen fixation commences (Wealand *et al.*, 1989).

Non-specific proteolysis

During heterocyst development a large number of proteins are synthesized, while others are degraded, in a defined temporal sequence (Fleming & Haselkorn, 1974). Following nitrogen starvation *Anabaena* PCC 7120 synthesizes *de novo* a soluble Ca^{2+} requiring protease which *in vitro* degrades many of the vegetative cell proteins that turn over during heterocyst development (Wood & Haselkorn, 1979). Recently Lockau, Massalsky & Dirmeier (1988) described the partial purification of a Ca^{2+} stimulated protease from *Anabaena variabilis* ATCC 29413 with a very similar activity to that described for *Anabaena* PCC 7120. A surprising difference, however, is that the enzyme in *A. variabilis* is just as active in cultures grown in the presence of ammonium nitrate as in dinitrogen-fixing cultures. The effect on enzyme activity of removal of the ammonium nitrate was not tested, however, and so it remains unclear whether or not the enzyme is induced during such a nitrogen stepdown when all cells of a filament experience nitrogen starvation at the same time. Indeed, Thiel (1990) has recently shown that proteolysis can be induced in the same cyanobacterium by nitrogen starvation, although this induction is inhibited by chloramphenicol. Suprisingly, the products of this early proteolysis, which begins within 2 h of nitrogen stepdown, are mostly excreted from the cells and are presumably, therefore, not immediately used for *de novo* protein synthesis. This early proteolysis is probably not due to phycocyaninase since yellowing of the culture does not occur until 16–24 h after nitrogen stepdown.

The precise involvement in heterocyst development of the different forms of proteolysis which occur as a result of nitrogen starvation is still unclear. It seems likely that the non-specific Ca^{2+} stimulated protease(s) induced during the early stages is involved more in the removal of vegetative proteins, rather than the provision of amino acids for *de novo* protein synthesis. These are most likely derived from the specific degradation of phycocyanin which occurs rather later during heterocyst development. To what extent the proteolysis of vegetative proteins by the Ca^{2+} protease is an essential part of heterocyst development is not known. It is interesting, however, that evidence is beginning to accumulate which demonstrates the potential importance of Ca^{2+} in cyanobacteria. *Anabaena* contains a Ca^{2+}-ATPase which is probably involved in calcium export (Lockau & Pfeffer, 1983); a number of cyanobacteria have been reported to contain a calmodulin-like protein (Kerson, Miernyk & Budd, 1984; Pettersson & Bergman, 1989; Lea, Onec & Smith, 1990; Onek, Lea & Smith, 1990) and the external calcium concentration has an influence on heterocyst frequency (Smith, Hobson & Ellis, 1987*a,b*; Smith 1988; Smith & Wilkins, 1988). Should the Ca^{2+} protease be stimulated *in vivo* by Ca^{2+} (Lockau *et al.*, 1988) then this might provide a link with the observed effects on heterocyst frequency. However, very recently Rodriquez *et al.* (1990) have described results which cast doubt on this explanation.

They have shown that in *Anabaena* ATCC 33047 and an unidentified *Anabaena* sp. only trace levels of Ca^{2+} are required for growth in the presence of combined nitrogen, or for microaerobic growth on N_2. However, much higher levels of Ca^{2+} (10^{-4} M) are required for aerobic growth on N_2. No significant differences in heterocyst frequencies are apparent in either aerobic or microaerobic cultures grown on N_2 in the presence or absence of Ca^{2+}. The authors concluded that the effect of Ca^{2+} was to protect nitrogenase from inactivation by O_2 and that a lack of calcium produced heterocysts with morphological alterations in their envelopes which might reduce their efficiency as a barrier against oxygen. Interestingly, a role for Ca^{2+} in the respiratory protection of nitrogenase in the non-heterocystous *Gloeothece* has been proposed by Hamadi and Gallon (1981) and Gallon and Hamadi (1984).

Thus, the true role, if any, of calcium in heterocyst development remains unclear and will remain so until much more conclusive data are available.

Regulation of heterocyst development and spacing

Two models have been proposed to explain the mechanism which enables the *de novo* establishment of a regular spacing of heterocysts (Wolk, 1989; Fig. 6). The first assumes that following the removal of fixed nitrogen from the medium heterocysts are initiated randomly in space and time. Each developing heterocyst produces a substance, inhibitory to heterocyst development, which diffuses away in both directions and is metabolized as it does so. This establishes a decreasing gradient of the inhibitor, being highest in the cell adjacent to the developing heterocyst (Fig. 6(*a*)). Where two cells begin to differentiate in the same region of the filament, the inhibitor will be at its lowest level midway between them (assuming they both produce inhibitor at the same rate). The first cells to respond to nitrogen starvation by commencing heterocyst development will thus prevent adjacent cells from beginning the process themselves (Fig. 6(*b*)).

The second model is derived from a suggestion made by Haselkorn (1978) that amino acids liberated by proteolysis may provide the diffusible inhibitors of differentiation. However, since these would initially be available in all vegetative cells, he proposed that the establishment of the necessary gradients would require amino acid pumps. In this model the first cells to respond to nitrogen starvation would activate such pumps, draining fixed nitrogen from adjacent cells (Wolk, 1989). These cells would respond to this depletion of fixed nitrogen by activating their own pumps, and so on along the filament (Fig. 6(*c*)). Wolk (1989) compared this propagation of pump activation to the one dimensional growth of a crystal from a point of nucleation. If crystal growth is assumed to occur at constant velocity, the predicted final distribution of crystals matches poorly with the observed distribution of heterocysts in growing cultures. However, the distributions of crystals and heterocysts

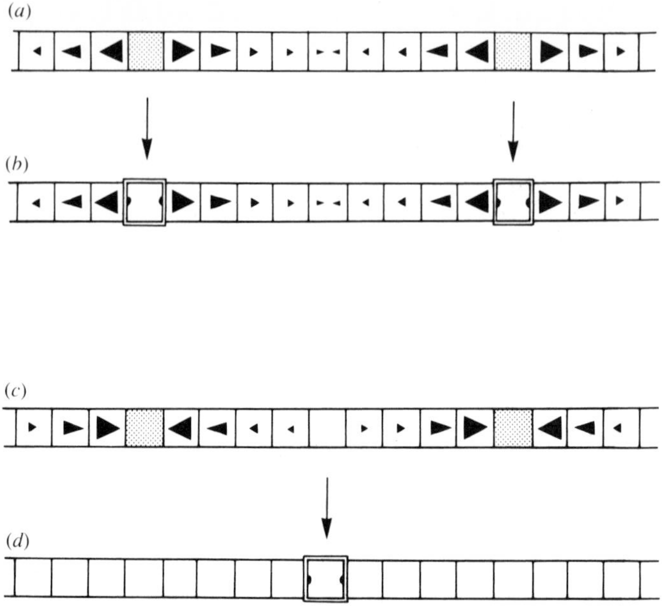

Fig. 6. Diagrammatic representation of two alternative models (Wolk, 1989) to explain the development of a spaced pattern of heterocysts in a filamentous cyanobacterium following the removal of fixed nitrogen from the medium. In the first model the first cells to respond to nitrogen deprivation (stippled cells) arise in random locations and produce an inhibitor of heterocyst development which diffuses along the filament in the direction indicated by the arrowheads (*a*). The inhibitor is degraded as it diffuses, establishing a decreasing concentration gradient (indicated by the size of the arrowheads). The production of the inhibitor prevents neighbouring cells from differentiating and the first cells to respond to nitrogen starvation therefore develop into mature heterocysts (*b*). The inhibitor gradients established in this inhibitor–diffusion model will help define the location of subsequent heterocysts which will only develop if the inhibitor concentration is low enough, i.e. if the cell is sufficiently far away from an existing or developing heterocyst. In the second model the first cells to respond to nitrogen starvation (stippled cells) activate amino acid pumps which draw the products of proteolysis from adjacent cells, which in turn activate pumps and drain their neighbours of fixed nitrogen (*c*). In this way pump activation is propagated along the filament but the magnitude of the pumps decreases with distance from the initial cell. The arrowheads in each cell indicate the direction of flow of fixed nitrogen and their size indicates the magnitude of the pump in each cell. In this model the cells which will differentiate are not the ones which first respond to nitrogen deprivation but are the ones which are being pumped from both directions and are therefore severely nitrogen starved (*d*).

match very well if it is assumed that the velocity of crystal growth decreases rapidly with distance from its point of nucleation. The growth of the crystal stops when it encounters another crystal and in a similar way propagation of pump activation will cease when one propagation 'wave' meets another travelling in the opposite direction. A vegetative cell at this point will be drained of fixed nitrogen from both directions, being unable to establish pumps of its own and it is this cell which is destined to become a heterocyst

(Fig. 6(*d*)). It is here that the predictions of the two models differ. In the inhibitor–diffusion model, the first cells to repond to nitrogen starvation are those which go on to differentiate into heterocysts (Fig., 6(*b*)). In the crystal growth model the first cells to respond do not differentiate themselves but establish amino acid pumps. It is the cells mid-way between these that eventually form heterocysts (Fig. 6(*d*)). This difference between the two models may permit them to be tested *in vivo* and Wolk (1989) has suggested that this might be achieved by using the promoter-luciferase fusions described earlier (Elhai & Wolk, 1990). If promoters for genes activated very soon after the removal of fixed nitrogen from the medium are put in control of the expression of luciferase, it might permit a comparison between the initial sites of activation of the genes (indicated by the luminescence of cells) and the subsequent positions of heterocysts. These sites should be the same if the first model is correct, but separate if the second model is correct.

Although such a test has not yet been performed, we can speculate on the merits of one of the two models in the light of results of experiments on proteolysis in *Anabaena*. Thiel (1990) has recently shown that proteolysis in *Anabaena variabilis* ATCC 29413 is evident 2–4 hours following ammonia removal. However, the products of proteolysis are not retained by cells, but are excreted into the medium. A similar observation had previously been made for *Anabaena* sp. strain PCC 7120 (Ownby, Shannahan & Hood, 1979). This calls into question the ability of the proposed amino acid pumps to establish the necessary gradients of the products of proteolysis. A second potential problem with the establishment of gradients by amino acid pumps is the speed with which the process can occur. Following the removal of fixed nitrogen the first cells to respond will draw the products of proteolysis from neighbouring cells which will in turn establish their own pumps. This will generate gradients of fixed nitrogen and it is at the low point of such gradients (specifically, cells which are being pumped from both directions) that a cell will begin to differentiate (Fig. 6(*c*) and (*d*)). All of this will clearly take time and yet in *Anabaena* sp. strain CA a regular pattern of proheterocysts is apparent by light microscopy within 2 hours of nitrate removal (Adams, unpublished observations). A final observation should be considered. The proteolysis which occurs during the early stages of nitrogen starvation and which provides the products for the putative amino acid pumps is presumably catalysed by the Ca^{2+} stimulated protease described by Wood & Haselkorn (1979) and Lockau *et al.* (1988). However, Rodriquez *et al.* (1990) have shown that the absence of Ca^{2+} has no effect on heterocyst frequencies of dinitrogen-fixing cultures of *Anabaena* ATCC 33047.

The *de novo* establishment of a regular heterocyst pattern clearly demands that only a selected number of appropriately placed cells are permitted to differentiate. It seems likely, therefore, that even though all cells have the capability of differentiating, only a limited number begin the process following nitrogen step-down. What it is that causes such cells to be more suitable for differentiation than others is not known, but, as discussed below, there is evidence to implicate both cell division and the cell cycle.

Cell division

Cell division in at least some *Anabaena* spp. is asymmetrical and heterocysts develop from only the smaller daughter cell (Wilcox, Mitchison & Smith, 1975*b*). Incubation of cultures of *A. cylindrica* for short periods at an elevated light intensity results in a 3- to 4-fold increase in the frequency of symmetrical cell divisions (Adams & Carr, 1981*a*). If this period of high light coincides with the induction of heterocyst development a high frequency of adjacent (double) heterocysts is produced. Adams & Carr (1981*a*) interpreted this as an indication that the daughter cells of a symmetrical cell division, being identical in size, were unable to compete effectively with one another, permitting both to differentiate. Thus, a change in the usual asymmetry of cell division results in a disturbance in the normal heterocyst pattern, implying a link, at least in *A. cylindrica*, between cell division and the control of heterocyst development.

In *Chlorogloeopsis fritschii* the first detectable response to nitrogen starvation is an alteration in the pattern of cell division (Foulds & Carr, 1981). In the presence of NH_4^+ most cell divisions are symmetrical, while in the absence of fixed nitrogen most are asymmetrical. This change in cell division precedes heterocyst development, although its significance for differentiation remains to be established in *C. fritschii* which is a complex organism growing frequently as clumps of cells with no readily discernable heterocyst pattern (see Castenholz, 1989*b*). Nevertheless, it is as important for *Chlorogloeopsis* to produce the correct frequency of heterocysts as it is for *Anabaena*, and asymmetrical cell division may play a role in this process.

Cell cycle

The requirement that, to be able to differentiate, a cell must be the smaller daughter of a division (Wilcox *et al.*, 1975*b*) effectively links heterocyst development and the cell cycle since the cell only remains a small daughter for a short period following its formation. This linkage can be investigated in another way. The use of inhibitors has shown that the differentiation of heterocysts requires DNA, RNA and protein synthesis (Adams & Carr, 1989). Indeed, it has been shown that, as they develop, heterocysts pass through a series of commitment times at which the addition of, say, chloramphenicol to inhibit protein synthesis cannot prevent the continued development of the cell. Following nitrogen step-down not all heterocysts begin development at the same time. For example, in *A. cylindrica* the maximum number of proheterocysts is not present until 12–15 h after ammonia removal (Fig. 7(*b*)), indicating that each proheterocyst commenced differentiation at a different time (assuming that they all take about the same length of time to develop once they have begun). Because of this slight asynchrony, the addition of an inhibitor at different times to a differentiating culture produces a commitment curve of the type shown in Fig. 7(*a*), from which commitment times can be derived (Adams & Carr, 1989). Commitment times can be obtained for many different treatments such as transfer to darkness, re-addition of NH_4Cl, and the inhibition of DNA synthesis (using mitomycin C), RNA synthesis (using rifampicin) and protein synthesis (using chloramphenicol). Each of these times is different, indicative of the ordered sequence of events occurring during differentiation. In addition, very different commitment times are obtained for proheterocysts and mature heterocysts, suggesting a wide temporal separation in the expression of the genes required for their respective formation (Adams & Carr, 1989).

These observations have been used to formulate a model for the control of heterocyst development in which DNA replication serves as a timer mechanism for the process (Adams & Carr, 1989). This model suggests that the genes, whose expression is required for the control of proheterocyst and mature heterocyst development (*phc* and *mhc* respectively), are widely separated on the *Anabaena* genome. These genes can only be expressed for a brief period immediately after their replication and hence their physical separation introduces an effective timer for proheterocyst and heterocyst development. *Phc* and *mhc* can be independently regulated and are repressed in the presence of high levels of fixed nitrogen. It is for this reason that the readdition of NH_4Cl gives rise to different commitment times for proheterocysts and heterocysts. The independent and temporally separated regulation of *phc* and *mhc* also endows these cyanobacteria with a very flexible response to nitrogen starvation since it provides a long period following the initiation of a proheterocyst's development during which the cell is not committed to full differentiation. A brief decrease in fixed nitrogen supply will therefore not lead to an over-production of mature heterocysts which are no longer needed when the supply increases once more. This is extremely important since the production of heterocysts is expensive in both energetic terms and in loss of division capacity (since the cells cannot divide).

Competition

The requirement that to become a 'candidate' for differentiation a cell must be a small daughter at the correct stage of its cell cycle will clearly limit the number of cells in a filament which can respond when it is transferred from medium containing fixed nitrogen to one lacking it. However, this cannot be sufficient to generate a regularly spaced pattern of single heterocysts, since the initial placement of candidate cells will be random and they will frequently occur in close proximity to one another. There must therefore be a mechanism for competition between candidates to establish the regular pattern which is apparent by light microscopy at even the earlier stages of proheterocyst development. Evidence for such a competitive mechanism has been presented in detail by Wilcox *et al.* (1975*b*; see also Adams &

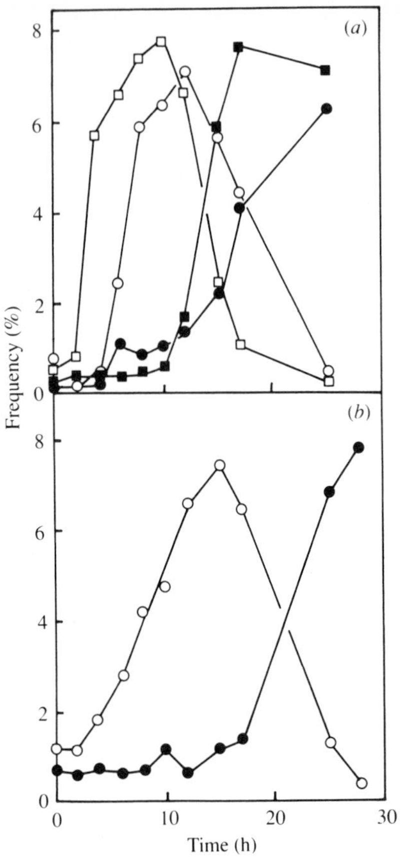

Fig. 7. Estimation of the commitment to heterocyst development in *Anabaena cylindrica*. Heterocyst development was initiated by transfer of a culture from medium containing NH_4^+ to one lacking NH_4^+ (0 h). (*b*) Proheterocyst (open circles) and heterocyst (closed circles) frequencies (expressed as a percentage of total cells) in the control culture at various times following NH_4^+ removal. (*a*) Proheterocyst (open symbols) and heterocyst (closed symbols) frequencies (counted 48 h after NH_4^+ removal) in samples of the control culture to which mitomycin C (to inhibit DNA synthesis; squares) or rifampicin (to inhibit RNA synthesis; circles) was added at the times indicated. Each time point therefore represents the ability of cells to differentiate into proheterocysts or mature heterocysts in the absence of any further RNA synthesis (circles) or DNA synthesis (squares). Note that the two inhibitors produce a separate commitment curve for the two cell types, implying a temporal separation in the requirement for RNA and DNA synthesis for proheterocysts and mature heterocysts. (Redrawn from Adams & Carr, 1989).

Carr, 1981*b*) and will not be discussed here. Such a competition probably exists between daughter cells, although under normal circumstances it is always the smaller daughter which wins. Treatments which disrupt the normal pattern of heterocysts, such as incubation with 7-azatryptophan (see below) or exposure to high light intensities, probably do so by interfering with competition.

Disruption of heterocyst pattern

There are a number of ways in which the normal spacing of single heterocysts can be disrupted. As described above incubation at high light intensity can result in the production of double heterocysts, although this has little effect on the overall spacing of heterocysts (i.e. the 'inhibitory zones' surrounding heterocysts remain unchanged) and is probably not a direct effect on the pattern control mechanism. However, mutants of *A. cylindrica* have been isolated which show dramatic changes in the normal spacing of heterocysts (Wilcox, Mitchison & Smith, 1975*a*). One of these mutants (mutant M7) produces a very high frequency of heterocysts in a very irregular arrangement and with many multiple heterocysts, consisting of up to 6 to 8 of the differentiated cells adjacent to one another. A remarkably similar alteration to heterocyst pattern is induced by incubation of *A. cylindrica* with the amino acid analogue 7-azatryptophan (AZAT; Mitchison & Wilcox, 1973; see also Adams & Carr, 1981*b*). Unlike the influence of high light intensity, this analogue has no effect on the asymmetry of cell division in *A. cylindrica* (Adams, unpublished observations) and so the effects must lie elsewhere. Indeed, AZAT seems to have a profound effect on the inhibitory zones of heterocysts since the number of vegetative cells between heterocysts decreases considerably as new heterocysts develop much closer to existing ones than in the absence of the analogue. However, the high frequency of multiple heterocysts induced by AZAT implies that it also reduces the ability of daughter or adjacent cells to compete effectively. The alterations to heterocyst pattern displayed in the mutant M7 and induced in wild-type *A. cylindrica* by incubation with AZAT are important to our understanding of heterocyst development since they represent highly specific modifications to the pattern control mechanism(s). This contrasts with the plethora of treatments which will induce small increases or decreases in heterocyst frequency due, presumably, to minor changes in the physiology of the cells. Therefore, clarification of the mechanisms involved in the influence of AZAT on heterocyst development, or an ability to complement mutants such as M7, would greatly improve our understanding of heterocyst pattern control.

What is the heterocyst repressor?

Ever since Fogg (1949) concluded that the formation of a heterocyst from a vegetative cell occurred when 'the concentration within it of a specific nitrogenous inhibitory substance probably ammonia or some simple derivative of ammonia falls below a critical level' there has been speculation as to the nature of the inhibitory substance. It has still not been identified although we can eliminate some of the contenders. That the initial candidate, ammonia, is not involved has been demonstrated by the use of the glutamine analogue, methionine sulphoximine (MSX). This inhibits glutamine synthetase (GS), which along with glutamate synthase (GOGAT) constitutes the major route for ammonia assimilation in cyanobacteria (see Guerrero & Lara, 1987). Addition of MSX to a culture of *A. cylindrica* growing on ammonium chloride leads to the development of heterocysts and nitrogenase activity, clearly indicating that ammonia *per se* does not inhibit either process (Stewart & Rowell, 1975). Since MSX results in a decrease in the glutamine pool this experiment raises the possibility that glutamine is the inhibitor of heterocyst development. Indeed, it has been a popular candidate for some time (see, for example, Neuer *et al.*, 1983; Wolk, 1989). However, recent evidence points against glutamine being involved in this way. Thiel & Leone (1986) have shown that *Anabaena variabilis* ATCC 29413 grown on glutamine has a high intracellular pool of glutamine but still produces heterocysts, although nitrogen fixation is largely repressed. The amino acid analogue DL-7-azatryptophan (AZAT) has been shown to induce the development of heterocysts in undifferentiated cultures of *Anabaena* sp. strain CA (ATCC 33047) grown in the presence of nitrate (Stacey *et al.*, 1979; Bottomley, Van Baalen & Tabita, 1980) and in other *Anabaena* spp. grown in the presence of ammonium chloride (see Adams & Carr, 1981*b*). Unlike MSX, azatryptophan does not inhibit GS activity in *Anabaena* sp. strain CA, although it does cause a slight inhibition of GOGAT activity (Chen, Van Baalen & Tabita, 1987). Addition of the analogue to nitrate-grown (and therefore undifferentiated) continuous cultures of *Anabaena* CA causes a significant decrease in the intracellular glutamate level and a slight transient increase in glutamine. These changes are accompanied by the development of heterocysts (Chen *et al.*, 1987), confirming that a high intracellular pool of glutamine does not inhibit heterocyst development. The decrease in the glutamate pool in this experiment might implicate this amino acid in heterocyst repression. However, in an equivalent experiment the addition of azaserine, an inhibitor of GOGAT, causes a similar decline in the intracellular glutamate pool, but does not induce heterocyst development (Chen *et al.*, 1987).

It seems then that all of the first three intermediates in the primary route for the assimilation of fixed nitrogen, namely ammonia, glutamine and glutamate, are not repressors of heterocyst development and its true identity remains to be established. The derepressive effects of AZAT on heterocyst development may imply an involvement of tryptophan or its metabolism in the control of differentiation (Bottomley *et al.*, 1980; Chen *et al.*, 1987). This possibility merits further investigation.

AKINETES

Akinetes are specialized cells which are resistant to both cold and desiccation, but not to elevated temperatures and are capable of germination under the appropriate conditions (see Adams & Carr, 1981*b*; Nichols & Adams, 1982;

Herdman, 1987). These properties have led to the generally accepted idea that they serve as perennating structures, enabling survival during cold winters or dry summers. The structure, properties and germination of akinetes have been extensively reviewed (see Herdman, 1987; Nichols & Adams, 1982; Adams & Carr, 1981*b*) and will not be discussed here.

Akinetes are produced by heterocystous cyanobacteria of sections IV and V of the classification of Rippka *et al.* (1979; see also Castenholz, 1989*a,b*). Although akinetes generally develop in filaments containing heterocysts, they can be produced in cultures grown in the presence of combined nitrogen and therefore lacking heterocysts. For example, in *Anabaena* sp. strain CA grown in the presence of $NaNO_3$, almost complete transformation of vegetative cells to akinetes can occur (Nichols & Adams, 1982 and see Fig. 2). Similarly, in *Nostoc* sp. PCC 7524, grown in the presence of NH_4Cl, akinetes develop at high frequency (Sutherland, Herdman & Stewart, 1979). Indeed, even in *A. cylindrica*, in which akinetes generally develop only in the presence of heterocysts, they can occasionally be seen in cultures grown in the presence of NH_4Cl containing a very low frequency of heterocysts (Nichols, Adams & Carr, 1980). Thus, it is clear that, at least in some cyanobacteria the presence of heterocysts is not essential for the development of akinetes. Nevertheless, it is equally apparent that, when present, heterocysts impose a rigid constraint on the positions at which akinetes develop within the filaments. For example, akinetes develop adjacent to heterocysts in *A. cylindrica* (Fig. 2(*c*)), midway between heterocysts in *Anabaena* sp. strain CA (see Nichols & Adams, 1982) and from the third vegetative cell away from heterocysts in *Anabaena circinalis* (Fay, Lynn & Majer, 1984). Subsequent differentiation can result in the formation of chains of akinetes which frequently display gradients of maturity (see Herdman, 1987).

Since heterocysts are the sites of dinitrogen fixation and are symmetrically placed within the filaments of vegetative cells it might be expected that gradients of fixed nitrogen would be produced, being at their highest adjacent to a heterocyst and lowest mid-way between heterocysts. It could then be argued that the development of akinetes occurs in positions dependent on this gradient. However, several pieces of evidence refute this suggestion. Akinetes develop in *A. cylindrica* next to the small number of proheterocysts and heterocysts present in cultures grown on ammonium chloride (Nichols *et al.*, 1980). Ammonium chloride represses dinitrogen fixation and can be assimilated by all vegetative cells thus destroying any gradients of fixed nitrogen. Similarly, in *Nostoc* PCC 7524 the possibility of such gradients can be removed by providing nitrate as a nitrogen source and incubating cultures under Ar/CO_2 to prevent dinitrogen fixation. Under these conditions heterocysts still develop and akinetes form in their usual position, 9 to 10 cells away from the heterocysts (Sutherland *et al.*, 1979).

Although the removal of possible gradients of fixed nitrogen has no effect on the normal placement of akinetes there are ways in which this can be altered. The amino acid analogue 7-azatryptophan, which causes increases in heterocyst frequency and the formation of multiple (adjacent) heterocysts in *Anabaena* (see earlier section), causes similar effects to both heterocyst and akinete development in *Nostoc* PCC 7524 (Sutherland *et al.*, 1979). Incubation of *A. cylindrica* with the arginine analogue, canavanine, results in the development of akinetes at random locations between heterocysts, in addition to their normal position adjacent to them (Nichols *et al.*, 1980). Another arginine analogue, cyanoalanine, and a serine analogue, aminobutyric acid, cause increases in akinete frequency but have no effect on their placement in relation to heterocysts (Nichols *et al.*, 1980).

Whatever the nature of the regulatory mechanism which controls the spatial arrangement of akinetes it seems highly unlikely that a single mechanism will function in all cyanobacteria. Sutherland *et al.* (1979) have suggested that in *Nostoc* PCC 7524 a common control mechanism exists for both heterocyst and akinete pattern because, at the end of exponential growth, akinetes develop in positions at which heterocysts would have developed had growth continued, and because 7-azatryptophan has similar effects on the placement of both heterocysts and akinetes (see above). Clearly the same argument cannot be applied to cyanobacteria such as *A. cylindrica* in which akinetes develop adjacent to heterocysts, a position which new heterocysts never normally occupy. Nichols *et al.* (1980) have proposed a model for *A. cylindrica* in which the heterocyst produces a diffusible compound which negates the inhibitory influence within vegetative cells which prevents them from becoming akinetes. Thus the cell closest to the heterocyst will be the first to form an akinete. A similar model had previously been proposed by Wolk (1965).

Factors which influence akinete development

A wide variety of nutrients has been implicated in the induction or inhibition of akinete development (see Nichols & Adams, 1982; Herdman, 1987, 1988), yet no common factor emerges from this plethora of data. Whether this is a true reflection of the diversity of cyanobacteria, misinterpretation of data or bad experimental technique is unclear (for an interesting discussion of this see Herdman, 1987).

It has been known for some time that in dinitrogen-fixing batch cultures akinete formation occurs towards the end of, or after cessation of exponential growth (Fay, 1969 and see Herdman, 1987). In cultures such as these, in which an excess of inorganic nutrients is provided in the medium, the factor which becomes limiting and results in the decline from exponential growth is light and hence energy. Indeed a direct correlation can be shown between the light intensity at which *A. cylindrica* is grown and the cell density at which akinete formation begins (Nichols *et al.*, 1980). The importance of light (energy) limitation in the induction of akinete formation was clearly demonstrated by Sutherland *et al.* (1979). They showed that in cultures of the facultative photoheterotroph *Nostoc* PCC 7524, the addition of a utilizable carbon source such as sucrose prolonged the exponential growth phase and resulted in akinetes developing at a higher cell density than in control cultures growing at the same light intensity without sucrose.

Fay *et al.* (1984) have confirmed the importance of light (energy) limitation in another cyanobacterium, *Anabaena circinalis*. They grew cultures at a fixed light intensity and then used centrifugation to concentrate samples to a range of different cell densities. Akinete development occurred at a different time in each sample, first in the sample concentrated to the highest cell density and last in the least concentrated sample. However, in all cases it occurred at the same optical density.

Thus, although factors such as phosphate limitation (Wolk, 1965) may play a role in triggering akinete development, the major factor in the restricted range of cyanobacteria so far examined is light and hence energy limitation.

HORMOGONIA

Hormogonia are produced by a wide variety of cyanobacteria and consist of short, non-heterocystous filaments of cells which are often smaller than the cells of the vegetative trichome from which they are derived (Rippka, Waterbury & Stanier, 1981; see Fig. 1). Within the family Nostocaceae hormogonia formation is restricted to the genus *Nostoc* and, unlike the vegetative trichomes, the hormogonia usually possess gliding motility (Rippka *et al.*, 1981; Castenholz, 1989*a*). Indeed, the formation of motile hormogonia seems to be mostly, if not exclusively, restricted to those cyanobacteria which possess immotile vegetative filaments. Many hormogonia also produce gas vacuoles giving them buoyancy (Tandeau de Marsac *et al.*, 1988) and this, together with their motility, makes them effective agents of dispersal (Dow, Whittenbury & Carr, 1983; Herdman & Rippka, 1988). Hormogonia are a relatively transient morphological form. Their development is often characterized by a round of rapid and relatively synchronous cell division and a period of motility (Tandeau de Marsac *et al.*, 1988; Campbell & Meeks, 1989). After a short time motility is lost and, under nitrogen deficient conditions, heterocysts develop at positions which are specific to the genus involved. In *Nostoc* spp. two heterocysts develop initially, both usually terminal, while in members of the genera *Scytonema* and *Calothrix* a single terminal heterocyst is produced (Rippka *et al.*, 1979, 1981). Growth and elongation of hormogonia leads to the typical vegetative morphology with many intercalary heterocysts in *Nostoc* and *Scytonema*, but a single terminal heterocyst in *Calothrix* (Rippka *et al.*, 1979, 1981). When gas vacuoles are produced during hormogonia development they are usually lost on return to vegetative growth (Armstrong, Hayes & Walsby, 1983; Tandeau de Marsac *et al.*, 1988).

Factors which trigger hormogonia formation

Formation of hormogonia seems to be triggered by environmental stimuli although the mechanisms remain obscure. Transfer of *Nostoc muscorum* from dark, heterotrophic growth conditions to red light (650 nm) induces hormogonia formation, but this can be reversed by exposure to green light (Lazaroff, 1973). A similar response has been reported for *Calothrix* PCC 7601 which produces 90 to 100% transformation to hormogonia when cells in early exponential phase are transferred to red light (Tandeau de Marsac *et al.*, 1988). Green light produces only 30% transformation. Hormogonia of *Calothrix* PCC 7601 contain gas vesicles, unlike vegetative trichomes, and the genes for gas vesicle formation are transcribed during hormogonia differentiation (Tandeau de Marsac *et al.*, 1988). Transfer of old cultures to fresh medium can also serve as a trigger, as with some *Nostoc* spp (Rippka *et al.*, 1979). Gas vacuole formation is also associated with hormogonia differentiation in *Nostoc muscorum* PCC 6719 and Armstrong *et al.* (1983) have shown that both are triggered by removal of $NaNO_3$ from the medium or by increasing the incident light intensity. Addition of equi-osmolar NaCl, KCl or KNO_3, but not glucose or sucrose, immediately following $NaNO_3$ removal, prevented the induction of gas vacuoles and hormogonia. Transfer of *Mastigocladus laminosus* from liquid medium to the surface of solid medium induces hormogonia formation (Hernandez-Muniz & Stevens, Jr, 1987). In this case the only external factor which appears to be necessary is Ca^{2+}, since the differentiation occurs in the dark on 1.5% agarose containing 0.2 mM Ca^{2+}, if the sample is taken from an actively growing culture. However, samples derived from resting cultures require light as well as Ca^{2+} (Hernandez-Muniz & Stevens, Jr, 1987).

The limited amount of information presently available concerning hormogonia provides no indication of a common trigger for their development. Whilst it is clear that environmental stimuli are involved, these are extremely varied and there seems to be no common element. It may indeed be a mistake to expect one since the formation of hormogonia may be employed by different cyanobacteria to fulfil a specific role dictated by their particular environmental niche. For example, the tendency of *Mastigocladus laminosus* to form hormogonia upon transfer from liquid to solid media (Hernandez-Muniz & Stevens, Jr, 1987) may reflect the use of hormogonia to allow rapid colonization of newly available areas of substratum, such as rock (Hernandez-Muniz, personal communication). Nevertheless, it would be surprising if control at the molecular level did not show many common elements, at least within, say, the genus *Nostoc*. Of this we presently know nothing, other than some aspects of the regulation of gas vesicle synthesis which sometimes accompanies hormogonia formation (Tandeau de Marsac *et al.*, 1988).

It is interesting that there have been several reports of cyanobacteria which produce extracellular metabolites which either stimulate or inhibit hormogonia formation. Cultures of *Nostoc muscorum* and *Nostoc commune* containing hormogonia excrete a heat labile substance which induces hormogonia formation when added to dark-grown cultures (Robinson & Miller, 1970). On the other hand, *Nostoc* PCC 7119 (originally designated as *Anabaena* PCC 7119) excretes, in late exponential to stationary phase, a compound which inhibits hormogonia formation even when diluted 10- to 20-fold (Herdman & Rippka, 1988). Although not identified, the inhibitor is destroyed by autoclaving and is dialysable. Such observations imply that extracellular communication might play a role in coordinating hormogonia formation and this may be particularly important where high population densities occur such as in large cyanobacterial colonies. It is interesting, though perhaps not surprising, that bryophytes, such as liverworts and hornworts, involved in symbioses with cyanobacteria have taken advantage of these possibilities by excreting compounds which stimulate hormogonia formation and presumably thereby enhance the chances of their becoming infected and maintaining the symbiosis (see next section).

CYANOBACTERIA–BRYOPHYTE SYMBIOSES

N_2-fixing, heterocystous cyanobacteria form symbiotic associations with a variety of plants including fungi (as lichens) and the water-fern *Azolla* (see Gallon & Chaplin, 1987; Smith & Douglas, 1987; Peters & Meeks, 1989). One of the most experimentally amenable systems is that formed between cyanobacteria of the genus *Nostoc* and bryophytes such as *Blasia* and *Anthoceros* (Rodgers & Stewart, 1977). In the liverwort *Blasia* the cyanobacteria occupy small, hemispherical structures, known as auricles or domatia, on the ventral side of the thallus (Schuster, 1966; Renzaglia, 1982; Kimura & Nakano, 1990) while in the hornwort *Anthoceros* they inhabit slime cavities (Rodgers & Stewart, 1977; Duckett *et al.*, 1977). With both of these associations it has proved possible to isolate the plant and cyanobacterial partners, grow them separately, and re-form the symbiosis at will (Rodgers and Stewart, 1977; Enderlin & Meeks,1983; Kimura & Nakano, 1990). This has greatly aided in developing our understanding of the interactions between the plant and cyanobacterial partners.

Bryophyte-induced changes in the cyanobacterial partner

These associations provide an interesting comparison with the intercellular communications (between heterocysts and vegetative cells, for example) which occur in free-living filamentous cyanobacteria, since inter-symbiont interactions are essential to the establishment and maintenance of the symbiosis. The cyanobacteria change both morphologically and physiologically once they become associated with the bryophytes and a number of these changes are discussed in the following sections.

Heterocyst frequency

The heterocyst frequency of most free-living *Nostoc* spp. is commonly 5 to 10%, yet in symbiotically associated *Nostoc* this can be as high as 80% (Silvester & McNamara, 1976; Duckett *et al.*, 1977; Hill, 1989). This may be seen as a means of producing greater amounts of fixed nitrogen for the partnership, although in the bryophtye-cyanobacteria symbioses there seems to be no increase in N_2 fixation rates (Rodgers & Stewart, 1977; Stewart & Rodgers, 1977) and it is not clear how many of the heterocysts are functional. Certainly in *Anthoceros* the degeneration of both vegetative cells and heterocysts has been observed by electron microscopy (Duckett *et al.*, 1977). Once isolated from the symbiosis, the cyanobacteria all appear to develop normal heterocyst patterns and frequencies which implies that the high frequencies in the symbiotically associated organism are a result of the direct influence of the plant host or the unusual nature of the tightly enclosed environment in the cavity. In this context the observations of Shi & Hall (1988) are of interest. These authors isolated the cyanobacterial partner from the symbiosis involving the water-fern *Azolla*, which is used as a green manure in rice cultivation (Shi & Hall, 1988). Grown in free culture this cyanobacterium produced a heterocyst frequency of 5 to 8%, but once it was immobilized in small cubes of polyurethane foam, the cells became larger, the heterocyst frequency increased to 10 to 18% and nitrogenase activity increased several times.

It has been suggested that the high heterocyst frequency in symbiotically associated cyanobacteria is a consequence of decreased glutamine synthetase activity (Stewart, Rowell & Rai, 1980). However, in at least two different plant-cyanobacteria symbioses, this seems not to be the case. Treatment of cultures of the *Nostoc-Anthoceros* association with NH_4^+ causes a decrease in both the heterocyst frequency and nitrogenase activity of the *Nostoc* (Enderlin & Meeks, 1983). However, the specific activity of glutamine synthetase and the amount of enzyme is very similar in *Nostoc* present in N_2-grown or NH_4^+-treated associations (Joseph & Meeks, 1987). In the plant-cyanobacterium association involving cycads there is no variation in the levels of glutamine synthetase protein in symbiotic *Nostoc* spp. present in different areas of the coralloid (coral-like) roots of the plant, despite a considerable variation in heterocyst frequency (Lindblad & Bergman, 1986).

Carbon dioxide fixation

Nostoc spp. freshly isolated from a reconstituted association with *Blasia pusilla* are incapable of light-dependent O_2 evolution or CO_2 fixation (Rodgers & Stewart, 1977). Similar results were recently obtained by Steinberg and Meeks (1989) who reconstituted the association between the hornwort *Anthoceros punctatus* and the cyanobacterium *Nostoc* strain UCD 7801 by mixing pure cultures of the two partners. *Nostoc* freshly isolated from the reconstituted association is capable of photosynthetic CO_2 fixation but at an eight-fold lower rate than the free-living isolate. The specific activity of ribulose bisphosphate carboxylase/oxygenase (RuBPC/O) is also eight-fold lower in cell extracts from freshly isolated *Nostoc* strain UCD 7801 than from the free-living form. However, there is little difference in the amount of RuBPC/O protein in the symbiotic and free-living *Nostocs* (Steinberg & Meeks, 1989; Rai *et al.*, 1989). Steinberg & Meeks (1989) concluded that the regulation of RuBPC/O activity in symbiosis was achieved by a post-translational rather than a transcriptional mechanism.

Glutamine synthetase

Assimilation of NH_4^+ by dinitrogen-fixing cyanobacteria occurs via the glutamine synthetase/glutamate synthase pathway, whether the NH_4^+ is derived exogenously or from N_2 (see Guerrero & Lara, 1987). The same is true of *Nostoc* symbiotically associated with *Anthoceros* (Meeks *et al.*, 1983, 1985). The role of dinitrogen-fixing cyanobacteria in symbioses such as those with bryophytes is to provide fixed nitrogen for the eukaryotic partner. It is no surprise to find therefore that the ability of such symbiotically associated cyanobacteria to assimilate ammonium is decreased, leaving a large proportion of the dinitrogen-derived NH_4^+ available for the plant. In the *Nostoc/Anthoceros* symbiosis this is achieved by a three- to four-fold decrease in the specific activity of glutamine synthetase in the symbiotically associated *Nostoc* (Joseph & Meeks, 1987). This involves primarily a post-translational mechanism, since the overall level of glutamine synthetase protein is similar in free-living and symbiotic *Nostoc*. However, part of the decrease in glutamine synthetase activity may result from a lower level of the protein specifically in the heterocysts of the symbiotically associated *Nostoc* compared with those in the free-living form (Rai *et al.*, 1989).

Hormogonia formation

Nostoc spp. which are capable of establishing a symbiotic association with *Anthoceros punctatus* can be induced to form hormogonia by co-culture with the hornwort (Campbell & Meeks, 1989). This hormogonia inducing activity is mediated by an extracellular compound(s) produced by *Anthoceros*. The compound(s) has not been identified but is unstable to autoclaving and is less than 12 to 14 kD in size.

There is little doubt that in establishing the symbiosis between *Nostoc* spp. and *Anthoceros* or *Blasia* the infective agents are hormogonia (Campbell & Meeks, 1989; Kimura & Nakano, 1990; Adams & Babic, unpublished observations). Their small size compared with vegetative filaments and their gliding motility enables them to gain entry through the small openings to the slime cavities of *Anthoceros* and auricles of *Blasia*. There is some indirect evidence that chemotaxis may also be involved in guiding the hormogonia to these openings (Adams & Babic, unpublished observations). It is essential for the continued growth of the bryophytes that newly formed cavities or auricles are rapidly colonized to provide the plant with a continued supply of combined nitrogen. It is therefore not surprising that a system exists which allows the plant to maximize the chances of successful infection by stimulating hormogonia formation in nearby *Nostoc* spp. Such a system is exemplified by the extracellular hormogonia-inducing factor(s) detected by Campbell & Meeks (1989). It is interesting that the production of this factor(s) occurs in response to the requirement for fixed nitrogen since none is detected when *Anthoceros* is incubated in the presence of NH_4NO_3 (Campbell & Meeks, 1989).

Hormogonia formation can be induced in another cyanobacterium, *Nostoc* LBG1, isolated from symbiosis with *Anthoceros* (Adams & Babic, 1988). This organism grows poorly under dinitrogen-fixing conditions unless supplied with glucose (30 mM). The removal of glucose elicits a series of highly synchronized changes culminating in the formation of motile hormogonia. Approximately two hours after glucose removal a round of synchronous cell division begins in which all vegetative cells divide within a 2 hour period (the generation time for *Nostoc* LBG1 is approximately 20 hours). These cell divisions occur without cell growth and the result is that cells become considerably smaller in both length and diameter. At this point the cells adjacent to heterocysts being to 'pinch off' at the point of attachment to the heterocyst, weakening the junction which subsequently breaks, liberating free heterocysts and short filaments of cells. These filaments become motile (hormogonia) for a short period and then lose motility and develop heterocysts, initially at either end of the filament, and subsequently at intercalary positions once cell division resumes and the filament increases in length.

The triggering of hormogonia formation by extracellular products of *Anthoceros* not only provides an interesting insight into the symbiotic association, but, because of the complete and highly synchronized transformation of vegetative filaments to hormogonia, it should allow a more controlled study of hormogonia formation than has been possible previously.

CONCLUSION

Studies of the regulation of heterocyst development have been hampered by the lack of a gene transfer system for heterocystous cyanobacteria. Happily, through the efforts of Wolk and others, this problem has now been overcome and we can expect a more rapid increase in our understanding of the molecular biology of heterocyst development. Indeed, we have already seen the isolation and mapping of the heterocyst-specific genes, *hetA* and *hetB*. The development of a conjugation-based cloning system for heterocystous cyanobacteria and the use of promoter-luciferase fusions has made possible the direct visualization of gene expression in individual cells. These techniques will allow us to examine the ways in which the spatial as well as temporal regulation of genes leads to the regular pattern of mature heterocysts.

Our present understanding of the regulation of other aspects of cyanobacterial development, such as akinete and hormogonia formation, is still rudimentary. However, it is now possible to induce synchronous and complete transformation of certain cyanobacterial cultures to hormogonia and this will provide an excellent model system for studying the process. The formation of hormogonia involves a form of signal transduction in which an environmental stimulus causes a change within the cell resulting in rapid cell division. Systems in which synchronous hormogonia formation can be induced should be well suited to the study of such signal transduction mechanisms in cyanobacteria. During hormogonia formation the period during which all cells divide is far too short to permit the completion of DNA replication, indicating that the process overrides the normal cell division controls. This system therefore offers the possibility of a greater understanding of cell cycle regulation in cyanobacteria.

ACKNOWLEDGEMENTS

I should like to thank Denise Ashworth for the electron microscopy and Margaret Oldfield, Gloria Rawle and Freddie Webster for their unending patience in the typing of the manuscript.

SUPPLEMENTARY READING

The following books cover a range of general aspects of cyanobacteria including structure, function, biochemistry, dinitrogen fixation and photosynthesis:

The Prokaryotes (1981). ed. M. P. Starr, H. Stolp, H. G. Trüper, A. Balows & H. G. Schlegel., (1981). Springer-Verlag.
The Biology of Cyanobacteria (1982). ed. N. G. Carr & B. A. Whitton. Oxford: Blackwell Scientific Publications.
The Cyanobacteria (1987). ed. P. Fay & C. Van Baalen. Amsterdam: Elsevier Science Publishers.
Methods in Enzymology (1988). ed. L. Packer & A. N. Glazer. London: Academic Press.
Bergey's Manual of Systematic Bacteriology (1989). ed. J. T. Staley, M. P. Bryant, N. Pfennig & J. G. Holt, vol. 3, Baltimore: Williams & Wilkins.

For a recent review of the molecular biology of cyanobacteria see Tandeau de Marsac & Houmard (1987).

REFERENCES

Adams, D. G. & Babic, S. (1988). Hormogonia formation in a cyanobacterium isolated from a symbiotic association with the bryophyte *Anthoceros*. In *Abstracts of the Sixth International Symposium on Photosynthetic Prokaryotes*, ed. L. R. Mur & T. Burger-Wiersma, p. 53.

Adams, D. G. & Carr, N. G. (1981*a*). Heterocyst differentiation and cell division in the cyanobacterium *Anabaena cylindrica*: effect of high light intensity. *Journal of Cell Science*, **49**, 341–52.

Adams, D. G. & Carr, N. G. (1981*b*). The developmental biology of heterocyst and akinete formation in cyanobacteria. *CRC Critical Reviews in Microbiology*, **9**, 45–100.

Adams, D. G. & Carr, N. G. (1989). Control of heterocyst development in the cyanobacterium *Anabaena cylindrica*. *Journal of General Microbiology*, **135**, 839–49.

Allen, M. M. (1984). Cyanobacterial cell inclusions. *Annual Review of Microbiology*, **38**, 1–25.

Allen, M. M. & Hutchinson, F. (1980). Nitrogen limitation and recovery in the cyanobacterium *Aphanocapsa* 6308. *Archives of Microbiology*, **128**, 1–7.

Armstrong, R. E., Hayes, P. K. & Walsby, A. E. (1983). Gas vacuole formation in hormogonia of *Nostoc muscorum*. *Journal of General Microbiology*, **128**, 263–70.

Balkwill, D. L., Nierzwicki-Bauer, S. A. & Stevens Jr, S. E. (1984). Modes of cell division and branch formation in the morphogenesis of the cyanobacterium *Mastigocladus laminosus*. *Journal of General Microbiology*, **130**, 2079–88.

Bancroft, I., Wolk, C. P. & Oren, E. V. (1989). Physical and genetic maps of the genome of the heterocyst-forming cyanobacterium *Anabaena* sp. strain PCC 7120. *Journal of Bacteriology*, **171**, 5940–8.

Böhme, H. & Haselkorn, R. (1988). Molecular cloning and nucleotide sequence analysis of the gene coding for heterocyst ferredoxin from the cyanobacterium *Anabaena* sp. strain PCC 7120. *Molecular and General Genetics*, **214**, 278–85.

Böhme, H. and Schrautemeier, B. (1987*a*). Electron donation to nitrogenase in a cell free system from heterocysts of *Anabaena variabilis*. *Biochimica et Biophysica Acta*, **891**, 115–20.

Böhme, H. and Schrautemeier, D. (1987*b*). Comparative characterization of ferredoxins from heterocysts and vegetative cells of *Anabaena variabilis*. *Biochimica et Biophysica Acta*, **891**, 1–7.

Borthakur, D., Basche, M., Buikema, W. J., Borthakur, P. B., & Haselkorn, R. (1990). Expression, nucleotide sequence and mutational analysis of two open reading frames in the *nif* gene region of *Anabaena* sp. strain PCC 7120. *Molecular and General Genetics*, **221**, 227–34.

Bottomley, P. J., Van Baalen, C. & Tabita, F. R. (1980). Heterocyst differentiation and tryptophan metabolism in the cyanobacterium *Anabaena* sp. CA. *Archives of Biochemistry and Biophysics*, **203**, 204–13.

Brusca, J. S., Chastain, C. J. & Golden, J. W. (1990). Expression of the *Anabaena* sp. strain PCC 7120 *xis* A gene from a heterologous promoter results in excision of the *nif* D element. *Journal of Bacteriology*, **172**, 3925–31.

Brusca, J. S., Hale, M. A., Carrasco, C. D. & Golden, J. W. (1989). Excision of an 11-kilobase-pair DNA element from within the *nifD* gene in *Anabaena variabilis* heterocysts. *Journal of Bacteriology*, **171**, 4138–45.

Buzby, J. S., Porter, R. D., & Stevens, S. E. Jr. (1983). Plasmid transformation in *Agmenellum quadruplicatum* PR-6: construction of biphasic plasmids and characterization of their transformation properties. *Journal of Bacteriology*, **154**, 1446–50.

Cai, Y. & Wolk, C. P. (1990). Use of a conditionally lethal gene in *Anabaena* sp. strain PCC 7120 to select for double recombinants and to entrap insertion sequences. *Journal of Bacteriology*, **172**, 3138–45.

Campbell, E. L., & Meeks, J. C. (1989). Characteristics of hormogonia formation by symbiotic *Nostoc* spp. in response to the presence of *Anthoceros punctatus* or its extracellular products. *Applied and Environmental Microbiology*, **55**, 125–31.

Carr, N. G. (1983). Biochemical aspects of heterocyst differentiation and function. In *Photosynthetic Prokaryotes: Cell Differentiation and Function*, ed. G. C. Papageorgiou & L. Packer, pp. 265–80. Amsterdam: Elsevier Science Publishing.

Carr, N. G. (1988). Nitrogen reserves and dynamic reservoirs in cyanobacteria. In *Biochemistry of the Algae and Cyanobacteria. Annual Proceedings of the Phytochemical Society of Europe*, ed. L. J. Rogers & J. R. Gallon, pp. 13–21. Oxford: Clarendon Press.

Carr, N. G., & Wyman, M. (1986). Cyanobacteria: their biology in relation to the oceanic picoplankton. Canadian Bulletin of Fisheries and Aquatic Sciences, **214**, 159–204.

Castenholz, R. W. (1989*a*). Subsection IV. Order *Nostocales*. In *Bergey's Manual of Systematic Bacteriology*, ed. J. T. Staley, M. P. Bryant, N. Pfennig & J. G. Holt, vol. 3, pp. 1780–99. Baltimore: Williams & Wilkins.

Castenholz, R. W. (1989*b*). Subsection V. Order *Stigonematales*. In *Bergey's Manual of Systematic Bacteriology*, ed. J. T. Staley, M. P. Bryant, N. Pfennig & J. G. Holt, vol. 3, pp. 1794–9. Baltimore: Williams & Wilkins.

Castenholz, R. W. & Waterbury, J. B. (1989). In *Bergey's Manual of Systematic Bacteriology*, ed. J. T. Staley, M. P. Bryant, N. Pfennig & J. G. Holt vol. 3, pp. 1710–27. Baltimore: Williams & Wilkins.

Chastain, C. J., Brusca, J. S., Ramasubramanian, T. S., Wei, T.-F. & Golden, J. W. (1990). A sequence-specific DNA-binding factor (VF1) from *Anabaena* sp. strain PCC 7120 vegetative cells binds to three adjacent sites in the *xisA* upstream region. *Journal of Bacteriology*, **172**, 5044–51.

Chen, C., Van Baalen, C. & Tabita, F. R. (1987). Nitrogen starvation mediated by DL-7-azatryptophan in the cyanobacterium *Anabaena* sp. strain CA. *Journal of Bacteriology*, **169**, 1107–13.

Codd, G. A., & Stewart, W. D. P. (1977). Ribulose-1,5-diphosphate carboxylase in heterocysts and vegetative cells in *Anabaena cylindrica. FEMS Microbiology Letters*, **2**, 247–9.

Cohen-Bazire, G., & Bryant, D. A. (1982). Phycobilisomes: composition and structure. In *The Biology of Cyanobacteria*, ed. N. G. Carr & B. A. Whitton, pp. 143–90. Oxford: Blackwell Scientific Publications.

Currier, T. C. & Wolk, C. P. (1979). Characteristics of *Anabaena variabilis* influencing plaque formation by cyanophage N-1. *Journal of Bacteriology*, **139**, 88–92.

Curtis, S. E. & Haselkorn, R. (1983). Isolation and sequence of the gene for the large subunit of ribulose-1,5-bisphosphate carboxylase from the cyanobacterium *Anabaena* 7120. *Proceedings of the National Academy of Sciences, USA*, **80**, 1835–39.

Daday, A., Mackerras, A. H. & Smith, G. D. (1985). The effect of nickel on hydrogen metabolism and nitrogen fixation in the cyanobacterium *Anabaena cylindrica. Journal of General Microbiology*, **131**, 231–8.

Daday, A., Mackerras, A. H. & Smith, G. D. (1988). A role for nickel in cyanobacterial nitrogen fixation and growth via cyanophycin metabolism. *Journal of General Microbiology*, **134**, 2659–63.

Delaney, S. F. & Reichelt, B. Y. (1982). Integration of the R plasmid, R68.45, into the genome of *Synechococcus* PCC 6301. *Abstracts of the Fourth International Symposium on Photosynthetic Prokaryotes*, pD5. Bombannes.

Dow, C. S., Whittenbury, R., & Carr, N. G. (1983). The 'shut down' or 'growth precursor' cell – an adaptation for survival in a potentially hostile environment.

In *Thirty-Fourth Symposium of the Society for General Microbiology: Microbes in their Natural Environment*, ed. J. H. Slater, R. Whittenbury & J. W. T. Wimpenny, pp. 187–247. Cambridge: Cambridge University Press.
Duckett, J. G., Prasad, A.K.S.K., Davies, D. A. & Walker, S. (1977). A cytological analysis of the *Nostoc* – bryophyte relationship. *New Phytologist*, **79**, 349–62.
Elhai, J. & Wolk, C. P. (1988). Conjugal transfer of DNA to cyanobacteria. In *Methods in Enzymology*, ed. L. Packer & A.N. Gazer, vol. 167, pp. 747–54. London: Academic Press.
Elhai, J. & Wolk, C. P. (1990). Developmental regulation and spatial pattern of expression of the structural genes for nitrogenase in the cyanobacterium *Anabaena. EMBO Journal*, **9**, 3379–88.
Enderlin, C. S. & Meeks, J. C. (1983). Pure culture and reconstitution of the *Anthoceros–Nostoc* symbiotic association. *Planta*, **158**, 157–65.
Engebrecht, J., Simon, M. & Silverman, M. (1985). Measuring gene expression with light. *Science*, **227**, 1345–47.
Fay, P., (1969). Cell differentiation and pigment composition in *Anabaena cylindrica. Archives of Microbiology*, **67**, 62–70.
Fay, P., Lynn, J. A. & Majer, S. C. (1984). Akinete development in the planktonic blue-green alga *Anabaena circinalis. British Phycological Journal*, **19**, 163–73.
Finnegan, J. S. & Sherratt, D. (1982). Plasmid Co1E1 conjugal mobility: the nature of *bom*, a region required in *cis* for transfer. *Molecular and General Genetics*, **185**, 344–51.
Fleming, H. & Haselkorn, R. (1974). Program of protein synthesis during heterocyst differentiation in nitrogen-fixing blue-green algae. *Cell*, **3**, 159–70.
Flores, E. & Schmetterer G. (1986). Interaction of fructose with the glucose permease of the cyanobacterium *Synechocystis* sp. strain PCC 6803. *Journal of Bacteriology*, **166**, 693–6.
Flores, E. & Wolk, C. P. (1985). Identification of facultatively heterotrophic, N_2fixing cyanobacteria able to receive plasmid vectors from *Escherichia coli* by conjugation. *Journal of Bacteriology*, **162**, 1339–41.
Fogg, G. E. (1949). Growth and heterocyst production in *Anabaena cylindrica* Lemm. II. In relation to carbon and nitrogen metabolism. *Annals of Botany*, **13**, 241–59.
Foulds, I. J. & Carr, N. G. (1977). A proteolytic enzyme degrading phycocyanin in the cyanobacterium *Anabaena cylindrica. FEMS Microbiology Letters* **2**, 117–19.
Foulds, I. J. & Carr, N. G. (1981). Unequal cell division preceding heterocyst development in *Chlorogloeopsis fritschii. FEMS Microbiology Letters*, **10**, 223–6.
Gallon, J. R. & Chaplin, A. E. (1987). *An Introduction to Nitrogen Fixation*. London: Cassell Education Limited.
Gallon, J. R. & Hamadi, A. F. (1984). Studies on the effects of oxygen on acetylene reduction (nitrogen fixation) in *Gloeothece* sp. ATCC 27152. *Journal of General Microbiology*, **130**, 495–503.
Glazer, A. N. (1987). Phycobilisomes: assembly and attachment. In *The Cyanobacteria*, ed. P. Fay & C. Van Baalen, pp. 69–94. Amsterdam: Elsevier Science Publishers.
Golden, J. W., Garrasco, C. D., Mulligan, M. E., Schneider, G. J. & Haselkorn, R. (1988). Deletion of a 55-kilobase-pair DNA element from the chromosome during heterocyst differentiation of *Anabaena* sp. strain PCC 7120. *Journal of Bacteriology*, **170**, 5034–41.
Golden, J. W., Mulligan, M. E. & Haselkorn, R. (1987). Different recombination site specificity of two developmentally regulated genome rearrangements. *Nature*, London, **327**, 5426–9.
Golden, J. W., Robinson, S. J. & Haselkorn, R. (1985). Rearrangement of nitrogen

fixation genes during heterocyst differentiation in the cyanobacterium *Anabaena. Nature*, London, **314**, 419–23.

Golden, J. W. & Wiest, D. R. (1988). Genome rearrangement and nitrogen fixation in *Anabaena* blocked by inactivation of *xisA* gene. *Science*, **242**, 1421–3.

Guerrero, M. G. & Lara, C. (1987). Assimilation of inorganic nitrogen. In *The Cyanobacteria*, ed. P. Fay & C. Van Baalen, pp. 163–86. Amsterdam: Elsevier Science Publishers.

Gupta, M. & Carr, N. G. (1981). Enzyme activities related to cyanophycin metabolism in heterocysts and vegetative cells of *Anabaena* spp. *Journal of General Microbiology*, **125**, 17–23.

Hamadi, A. F. & Gallon, J. R. (1981). Calcium ions, oxygen and acetylene reduction (nitrogen fixation) in the unicellular cyanobacterium *Gloeocapsa* sp. 1430/3. *Journal of General Microbiology*, **125**, 391–8.

Haselkorn, R. (1978). Heterocysts. *Annual Reviews of Plant Physiology*, **29**, 319–44.

Haselkorn, R. (1986). Organization of the genes for nitrogen fixation in photosynthetic bacteria and cyanobacteria. *Annual Reviews of Microbiology*, **40**, 525–47.

Herdman, M. (1987). Akinetes: structure and function. In *The Cyanobacteria*, ed. P. Fay & C. Van Baalen, pp. 227–50. Amsterdam: Elsevier Science Publishers.

Herdman, M. (1988). Cellular differentation: Akinetes. In *Methods in Enzymology*, ed. L. Packer & A. N. Glazer, vol. 167, pp. 222–32. London: Academic Press.

Herdman, M. & Rippka, R. (1988). Cellular differentiation: hormogonia and baeocytes. In *Methods in Enzymology*, ed. L. Packer, & A. N. Glazer, vol. 167, pp. 232–42. London: Academic Press.

Hernandez-Muniz, W. & Stevens, S. E., Jr (1987). Characterization of the motile hormogonia of *Mastigocladus laminosus. Journal of Bacteriology*, **169**, 218–23.

Hill, D. J. (1989). The control of the cell cycle in microbial symbionts. *New Phytologist*, **112**, 175–84.

Holland, D. & Wolk, C. P. (1990). Identification and characterization of *hetA*, a gene that acts early in the process of morphological differentiation of heterocysts. *Journal of Bacteriology*, **172**, 3131–7.

Johnson, T. R., Haynes, II, J. I., Wealand, J. L., Yarbrough, L. R. & Hirschburg, R. (1988). Structure and regulation of genes encoding phycocyanin and allophycocyanin from *Anabaena variabilis* ATCC 29413. *Journal of Bacteriology*, **170**, 1858–65.

Joseph, C. M. & Meeks, J. C. (1987). Regulation of expression of glutamine synthetase in a symbiotic *Nostoc* strain associated with *Anthoceros punctatus. Journal of Bacteriology*, **169**, 2471–5.

Kallas, T., Coursin, T. & Rippka, R. (1985). Different organization of *nif* genes in nonheterocystous and heterocystous cyanobacteria. *Plant Molecular Biology*, **5**, 321–9.

Kallas, T., Rebiere, M-C., Rippka, R. & Tandeau de Marsac, N. (1983). The structural *nif* genes of the cyanobacteria *Gloeothece* sp. and *Calothrix* sp. share homology with those of *Anabaena* sp., but the *Gloeothece* genes have a different arrangement. *Journal of Bacteriology*, **155**, 427–31.

Kerson, G. W., Miernyk, J. A. & Budd, K. (1984). Evidence for the occurrence of, and possible physiological role for, cyanobacterial calmodulin. *Plant Physiology*, **75**, 222–4.

Kimura, J. & Nakano, T. (1990). Reconstitution of a *Blasia–Nostoc* symbiotic association under axenic conditions. *Nova Hedwigia*, **50**, 191–200.

Lammers, P. J., Golden, J. W. & Haselkorn, R. (1986). Identification and sequence of a gene required for a developmentally regulated DNA excision in *Anabaena. Cell*, **44**, 905–11.

Lammers, P. J., McLaughlin, S., Papin, S., Trujillo-Provencio, C. & Ryncarz II, A. J. (1990). Developmental rearrangement of cyanobacterial *nif* genes: nucleotide sequence, open reading frames, and cytochrome P-450 homology of the *Anabaena* sp. strain PCC 7120 *nifD* element. *Journal of Bacteriology*, **172**, 6981–90.

Lawry, N. H. & Simon, R. D. (1982). The normal and induced occurrence of cyanophycin inclusion bodies in several blue-green algae. *Journal of Phycology*, **18**, 391–9.

Lazaroff, N. (1973). Photomorphogenesis and nostocacean development. In *The Biology of Blue-Green Algae*, ed. N. G. Carr & B. A. Whitton, pp. 279–319. Oxford: Blackwell Scientific Publications.

Lea, P. J., Onec, L. A. & Smith, J. (1990). Calmodulin inhibitors enhance dinitrogen fixation in a heterocystous cyanobacterium. *Plant Physiology*, **79**, A75.

Lindblad, P. & Bergman, B. (1986). Glutamine synthetase: activity and localization in cyanobacteria of the cycads *Cycas revoluta* and *Zamia skinneri*. *Planta*, **169**, 1–7.

Lockau, W., Massalsky, B. & Dirmeier, A. (1988). Purification and partial characterization of a calcium-stimulated protease from the cyanobacterium, *Anabaena variabilis*. *European Journal of Biochemistry*. **172**, 433–8.

Lockau, W. & Pfeffer, S. (1983). ATP-dependent calcium transport in membrane vesicles of the cyanobacterium, *Anabaena variabilis*. *Biochimica et Biophysica Acta*, **733**, 124–32.

Mackerras, A. H., de Chazal, N. M. & Smith, G. D. (1990*a*). Transient accumulations of cyanophycin in *Anabaena cylindrica* and *Synechocystis* 6308. *Journal of General Microbiology*, **136**, 2057–65.

Mackerras, A. H., Youens, B. N., Weir, R. C. & Smith, G. D. (1990*b*). Is cyanophycin involved in the integration of nitrogen and carbon metabolism in the cyanobacteria *Anabaena cylindrica* and *Gloeothece* grown on light/dark cycles? *Journal of General Microbiology*, **136**, 2049–56.

Mazur, B. J., Rice, D. & Haselkorn, R. (1980). Identification of blue-green algal nitrogen fixation genes by using heterologous DNA hybridization probes. *Proceedings of the National Academy of Sciences, USA*, **77**, 186–90.

Meeks, J. C., Joseph, C. M. & Haselkorn, R. (1988). Organization of the *nif* genes in cyanobacteria in symbiotic association with *Azolla* and *Anthoceros*. *Archives of Microbiology*, **150**, 61–71.

Meeks, J. C., Enderlin, C. S., Joseph, C. M., Chapman, J. S. & Lollar, M. W. L. (1985). Fixation of $[^{13}N]N_2$ and transfer of fixed nitrogen in the *Anthoceros–Nostoc* symbiotic association. *Planta*, **164**, 406–14.

Meeks, J. C., Enderlin, C. S., Wycoff, K. L., Chapman, J. S. & Joseph, C. M. (1983). Assimilation of $^{13}NH_4^+$ by *Anthoceros* grown with and without symbiotic *Nostoc*. *Planta*, **158**, 384–91.

Mitchison, G. J. & Wilcox, M. (1973). Alteration in heterocyst pattern of *Anabaena* produced by 7-azatryptophan. *Nature*, London, **246**, 229–33.

Mulligan, M. E., Buikema, W. J. & Haselkorn, R. (1988). Bacterial-type ferredoxin genes in the nitrogen fixation regions of the cyanobacterium *Anabaena* sp. strain 7120 and *Rhizobium meliloti*. *Journal of Bacteriology*, **170**, 4406–10.

Mulligan, M. E. & Haselkorn, R. (1989). Nitrogen fixation (*nif*) genes of the cyanobacterium *Anabaena* species strain PCC 7120. The *nifB–fdxN–nifS–nifU* operon. *Journal of Biological Chemistry*, **264**, 19200–7.

Murry, M. A., Hallenbeck, P. C. & Benemann, J. R. (1984). Immunochemical evidence that nitrogenase is restricted to the heterocysts in *Anabaena cylindrica*. *Archives of Microbiology*, **137**, 194–9.

Murry, M. A. & Wolk, C. P. (1989). Evidence that the barrier to the penetration of oxygen into heterocysts depends upon two layers of the cell envelope. *Archives of Microbiology*, **151**, 469–74.

Neuer, G., Papen, H., & Bothe, H. (1983). Heterocyst biochemistry and differentiation. In *Photosynthetic Prokaryotes: Cell Differentiation and Function*. ed. G. C. Papageorgiou & L. Packer, pp. 219–42. Amsterdam: Elsevier Science Publishing.

Nichols, J. M. & Adams, D. G. (1982). Akinetes. In *The Biology of Cyanobacteria*, ed. N. G. Carr & B. A. Whitton, pp. 387–412. Oxford: Blackwell Scientific Publications.

Nichols, J. M., Adams, D. G. & Carr, N. G. (1980). Effect of canavanine and other amino acid analogues on akinete formation in the cyanobacterium *Anabaena cylindrica*. *Archives of Microbiology*, **127**, 67–75.

Nierzwicki-Bauer, S. A., Balkwill, D. L. & Stevens, Jr., S. E., (1984). Heterocyst differentiation in the cyanobacterium *Mastigocladus laminosus*. *Journal of Bacteriology*, **157**, 514–25.

Onek, L., Lea, P. & Smith, R. J. (1990). Cyanobacterial calmodulin. *Plant Physiology*, **93**, supplement, 147.

Ownby, J. D., Shannahan, M. & Hood, E. (1979). Protein synthesis and degradation in *Anabaena* during nitrogen starvation. *Journal of General Microbiology*, **110**, 255–61.

Peters, G. A. & Meeks, J. C. (1989). The *Azolla–Anabaena* symbiosis: Basic biology. *Annual Reviews of Plant Physiology and Plant Molecular Biology*, **40**, 193–210.

Pettersson, A. & Bergman, B. (1989). Calmodulin in heterocystous cyanobacteria: biochemical and immunological evidence. *FEMS Microbiology Letters*, **60**, 95–100.

Rai, A. N., Borthakur, M., Singh, S. & Bergman, B. (1989). *Anthoceros–Nostoc* symbiosis: immunoelectronmicroscopic localization of nitrogenase, glutamine synthetase, phycoerythrin and ribulose-1, 5-bisphosphate carboxylase/oxygenase in the cyanobiont and the cultured (free-living) isolate *Nostoc* 7801. *Journal of General Microbiology*, **135**, 385–95.

Reaston, J., van den Hondel, C. A. M. J. J., van Arkel, G. A. & Stewart, W. D. P. (1982). A physical map of plasmid pDU1from the cyanobacterium *Nostoc* PCC 7524. *Plasmid*, **7**, 101–4.

Renzaglia, K. S. (1982). A Comparative Developmental Investigation of the Gametophyte Generation in the Metzgeriales (Hepatophyta). In *Bryophytorum Bibliotheca*, vol. 24, Vaduz: J. Cramer.

Rice, D., Mazur, B. J. & Haselkorn, R. (1982). Isolation and physical mapping of nitrogen fixation genes from the cyanobacterium *Anabaena* 7120. *Journal of Biological Chemistry*, **257**, 13157–63.

Rippka, R., Deruelles, J., Waterbury, J. B., Herdman, M. & Stanier, R. Y. (1979). Generic assignments, strain histories and properties of pure cultures of cyanobacteria. *Journal of General Microbiology*, **111**, 1–61.

Rippka, R. & Stanier, R. Y. (1978). The effects of anaerobiosis on nitrogenase synthesis and heterocyst development by Nostocacean cyanobacteria. *Journal of General Microbiology*, **105**, 83–94.

Rippka, R., Waterbury, J. B. & Stanier, R. Y. (1981). Provisional generic assignments for cyanobacteria in pure culture. In *The Prokaryotes*, ed. M. P. Starr, H. Stolp, H. G. Trüper, A. Balows & H. G. Schlegel, vol. 1, pp. 247–56. Berlin: Springer-Verlag.

Robinson, B. L. & Miller, J. H. (1970). Photomorphogenesis in the blue-green alga *Nostoc commune* 584. *Physiologia Plantarum*, **23**, 461–72.

Rodgers, G. A. & Stewart, W. D. P. (1977). The cyanophyte-hepatic symbiosis I. Morphology and physiology. *New Phytologist*, **78**, 441–58.

Rodriquez, H., Rivas, J., Guerrero, M. G. & Losada, M. (1990). Ca^{2+} requirement for aerobic nitrogen fixation by heterocystous blue-green algae. *Plant Physiology*, **92**, 886–90.

Saville, B., Straus, N. & Coleman, J. R. (1987). Contiguous organization of nitrogenase genes in a heterocystous cyanobacterium. *Plant Physiology*, **85**, 26–9.

Schmetterer, G., Wolk, C. P. & Elhai, J. (1986). Expression of luciferases from *Vibrio harveyi* and *Vibrio fischeri* in filamentous cyanobacteria. *Journal of Bacteriology*, **167**, 411–14.

Schrautemeier, B. & Böhme, H. (1985). A distinct ferredoxin for nitrogen fixation isolated from heterocysts of the cyanobacterium *Anabaena variabilis*. *FEBS Letters*, **184**, 304–8.

Schuster, R. M. (1966). *The Hepaticae and Anthocerotae of North America*, vol. 1. pp. 249–54. New York: Columbia University Press.

Shi, D-J. & Hall, D. O. (1988). *Azolla* and immobilized cyanobacteria (blue-green algae): from traditional agriculture to biotechnology. *Plants Today*, **1**, 5–12.

Silvester, W. B. & McNamara, P. J. (1976). The infection process and ultrastructure of the *Gunnera–Nostoc* symbiosis. *New Phytologist*, **77**, 135–41.

Simon, R. D. (1987). Inclusion bodies in the cyanobacteria: cyanophycin, polyphosphate, polyhedral bodies. In *The Cyanobacteria*, ed. P. Fay & C. van Baalen, pp. 199–225. Amsterdam: Elsevier Science Publishers.

Smith, D. C. & Douglas, A. E. (1987). *The Biology of Symbiosis*. London: Edward Arnold.

Smith, R. J. (1988). Calcium-mediated regulation in the cyanobacteria? In: *Proceedings of the Phytochemical Society of Europe: Biochemistry of the Algae and Cyanobacteria*, ed. L. J. Rogers & J. R. Gallon, pp. 185–99. Oxford: Clarendon Press.

Smith, R. J., Hobson, S. & Ellis, I. R. (1987*a*). Evidence for calcium-mediated regulation of heterocyst frequency and nitrogenase activity in *Nostoc* 6720. *New Phytologist*, **105**, 531–41.

Smith, R. J., Hobson, S. & Ellis, I. R. (1987*b*). The effect of abscisic acid on calcium-mediated regulation of heterocyst frequency and nitrogenase activity in *Nostoc* 6720. *New Phytologist*, **105**, 543–9.

Smith, R. J. & Wilkins, A. (1988). A correlation between intracellular calcium and incident irradiance in *Nostoc* 6720. *New Phytologist*, **109**, 157–61.

Stacey, G., Bottomley, P. J., Van Baalen, C, & Tabita, F. R. (1979). Control of heterocyst and nitrogenase synthesis in cyanobacteria. *Journal of Bacteriology*, **137**, 321–6.

Steinberg, N. A. & Meeks, J. C. (1989). Photosynthetic CO_2 fixation and ribulose bisphosphate carboxylase/oxygenase activity of *Nostoc* sp. strain UCD 7801 in symbiotic association with *Anthoceros puctatus*. *Journal of Bacteriology*, **171**, 6227–33.

Stevens, S. E., Jr., Nierzwicki-Bauer, S. A. & Balkwill, D. L. (1985). Effect of nitrogen starvation on the morphology and ultrastructure of the cyanobacterium *Mastigocladus laminosus*. *Journal of Bacteriology*, **161**, 1215–18.

Stewart, W. D. P. (1980). Some aspects of structure and function in N_2-fixing cyanobacteria. *Annual Reviews of Microbiology*, **34**, 497–536.

Stewart, W. D. P. & Rowell, P. (1975). Effects of L-methionine-DL-sulphoximine on the assimilation of newly fixed NH_3, acetylene reduction and heterocyst production in *Anabaena cylindrica*. *Biochemical and Biophysical Research Communications*, **65**, 846–56.

Stewart, W. D. P. & Rowell, P. (1986). Biochemistry and physiology of nitrogen fixation with particular emphasis on nitrogen-fixing phototrophs. *Plant and Soil*, **90**, 167–91.

Stewart, W. D. P., Rowell, P. & Rai, A. N. (1980). Nitrogen metabolism in symbiotic cyanobacteria. In *Nitrogen Fixation*, ed. W. D. P. Stewart & J. R. Gallon, pp. 239–77. London: Academic Press.

Stewart, W. D. P. & Rodgers, G. A. (1977). The cyanophyte–hepatic symbiosis. II. Nitrogen fixation and the interchange of nitrogen and carbon. *New Phytologist*, **78**, 459–71.

Sutherland, J. M., Herdman, M. & Stewart, W. D. P. (1979). Akinetes of the cyanobacterium *Nostoc* PCC 7524: macromolecular composition, structure and control of differentiation. *Journal of General Microbiology*, **115**, 273–87.

Tandeau de Marsac, N. & Houmard, J. (1987). Advances in cyanobacterial molecular genetics. In *The Cyanobacteria*, ed. P. Fay & C. Van Baalen, pp. 251–301. Amsterdam: Elsevier Science Publishers.

Tandeau de Marsac, N., Mazel, D., Damerval, T., Guglielmi, G., Capuano, V. & Houmard, J. (1988). Photoregulation of gene expression in the filamentous cyanobacterium *Calothrix* sp. PCC 7601: light-harvesting complexes and cell differentiation. *Photosynthesis Research*, **18**, 99–132.

Taylor, D. P., Cohen, S. N., Clark, W. G. & Marrs, B. L. (1983). Alignment of genetic and restriction maps of the photosynthesis region of the *Rhodopseudomonas capsulata* chromosome by a conjugation-mediated marker rescue technique. *Journal of Bacteriology*, **154**, 580–90.

Thiel, T. (1990). Protein turnover and heterocyst differentiation in the cyanobacterium *Anabaena variabilis*. *Journal of Phycology*, **26**, 50–4.

Thiel, T. & Leone, M. (1986). Effect of glutamine on growth and heterocyst differentiation in the cyanobacterium *Anabaena variabilis*. *Journal of Bacteriology*, **168**, 769–74.

Van Haute, E., Joos, H., Maes, M., Warren, G., Van Montagu, M. & Schell, J. (1983). Intergeneric transfer and exchange recombination of restriction fragments cloned in pBR322: a novel strategy for the reversed genetics of the Ti plasmids of *Agrobacterium tumefaciens*. *EMBO Journal*, **2**, 411–17.

Walsby, A. E, (1985). The permeability of heterocysts to the gases nitrogen and oxygen. *Proceedings of the Royal Society of London, Series B*, **226**, 345–66.

Wealand, J. L., Myers, J. A. & Hirschberg, R. (1989). Changes in gene expression during nitrogen starvation in *Anabaena variabilis* ATCC 29413. *Journal of Bacteriology*, **171**, 1309–13.

Wilcox, M., Mitchison, G. J. & Smith, R. J. (1975*a*). Mutants of *Anabaena cylindrica* altered in heterocyst spacing. *Archives of Microbiology* **103**, 219–23.

Wilcox, M., Mitchison, G. J. & Smith, R. J. (1975*b*). Spatial control of differentiation in the blue-green alga *Anabaena*. In *Microbiology 1975*, ed. D. Schlessinger, pp. 453–63. Washington DC: American Society for Microbiology.

Winkenback, F. & Wolk, C. P. (1973). Activities of enzymes of the oxidative and the reductive pentose phosphate pathways in heterocysts of a blue-green alga. *Plant Physiology*, **52**, 480–3.

Wolk, C. P. (1965). Control of sporulation in a blue-green alga. *Developmental Biology*, **12**, 15–35.

Wolk, C. P. (1982). Heterocysts. In *The Biology of Cyanobacteria*, ed. N. G. Carr & B. A. Whitton, pp. 359–86. Oxford: Blackwell Scientific Publications.

Wolk, C. P. (1989). Alternative models for the development of the pattern of spaced heterocysts in *Anabaena* (Cyanophyta). *Plant Systematics and Evolution*, **164**, 27–31.

Wolk, C. P., Cai, Y., Cardemil, L., Flores, E., Hohn, B., Murry, M., Schmetterer, G., Schrautemeier, B. & Wilson, R. (1988). Isolation and complementation of mutants of *Anabaena* sp. strain PCC 7120 unable to grow aerobically on dinitrogen. *Journal of Bacteriology*, **170**, 1239–44.

Wolk, C. P. & Quine, M. P. (1975). Formation of one-dimensional patterns by stochastic processes and by filamentous blue-green algae. *Developmental Biology*, **46**, 370–82.

Wolk, C. P., Vonshak, A., Kehoe, P. & Elhai, J. (1984). Construction of shuttle vectors capable of conjugative transfer from *Escherichia coli* to nitrogen-fixing filamentous cyanobacteria. *Proceedings of the National Academy of Sciences, USA*, **81**, 1561–5.

Wood, N. B. & Haselkorn, R. (1979). Proteinase activity during heterocyst differentiation in nitrogen-fixing cyanobacteria. In *Limited Proteolysis in Microorganisms*, ed. G. N. Cohen & H. Holzer, pp. 159–66. Washington, DC: U.S. Department of Health, Education and Welfare Publications.

Wood, N. B. & Haselkorn, R. (1980). Control of phycobiliprotein proteolysis and heterocyst differentiation in *Anabaena*. *Journal of Bacteriology*, **141**, 1375–85.

Ziegler, M. M. & Baldwin, T. O. (1981). Biochemistry of bacterial bioluminescence. *Current Topics in Bioenergenics*, **12**, 65–113.

CELL–CELL INTERACTIONS CONTROLLING FRUITING BODY DEVELOPMENT OF *MYXOCOCCUS XANTHUS*

LAWRENCE J. SHIMKETS

Department of Microbiology, University of Georgia, Athens, Georgia, 30602, USA

INTRODUCTION

The myxobacteria are found in soil and fresh water habitats worldwide (for reviews, see Kaiser, 1989; Shimkets, 1989, 1990; Zusman *et al.*, 1990). They can be divided into two physiological groups based on their preference of carbon and energy sources: most species utilize protein but some prefer cellulose. The myxobacteria are microbial predators that can locate and consume other bacterial species. They secrete a variety of extracellular enzymes that help solubilize the prey by hydrolyzing peptidoglycans, lipids, nucleic acids, polysaccharides, and proteins (Rosenberg & Varon, 1984). While the ecological role(s) of the myxobacteria has never been properly studied, it is likely that they control the population sizes of certain prey microorganisms.

More than half of the myxobacterial isolates produce antibiotics with unusual chemical structures that differ from those of other antibiotic-producing bacteria (Reichenbach *et al.*, 1988). It is likely that some of these antibiotics, or their derivatives, will play a prominent role in medicine in the coming century. The clinical usefulness of several of the antibiotics is under investigation. Sorangicin is a broad-spectrum antibacterial compound that inhibits transcription initiation in eubacteria (Irschick *et al.*, 1987). Ambruticin seems particularly promising in the treatment of systemic mycoses (Levine, Ringel & Cobb 1978). The primary reason that studies of the myxobacterial antibiotics lag so far behind that of the Streptomycetes is that myxobacteria are notoriously difficult to cultivate from environmental samples. They grow slowly or not at all on most conventional media, and selective media have not yet been developed (Reichenbach & Dworkin, 1981; McCurdy, 1989). Myxobacteria are typically isolated by enrichments in which soil or water samples are incubated with dung pellets or on agar seeded with prey bacteria. If myxobacteria are present in the sample, fruiting bodies are generally observed in about a week. Once they are isolated in pure culture, many myxobacterial species are relatively easy to cultivate in liquid growth media.

The myxobacteria belong to the delta group of the purple eubacteria along with *Bdellovibrio* and the sulfur- and sulfate-reducing bacteria (Oyaizu &

Woese, 1985; Woese, 1987). Unlike their close relatives, myxobacteria have evolved the most sophisticated social behaviors within the prokaryotes. They appear to have separated from their closest relatives 0.9 to 1.1 billion years ago based on the divergence of the 16S rRNA molecules of the members of the delta group (Shimkets, 1990), well before the appearance of the first eukaryotic cell (Ochman & Wilson, 1987). The myxobacteria appear, therefore, to be the most ancient group of organisms to evolve complex social behaviors.

GENOME SIZE AND GENETIC MAP

If social and developmental complexity is due to genetic complexity, myxobacteria would be expected to have a genome substantially larger than that of most other prokaryotes. The genome size is currently available for only one myxobacterial species, *Myxococcus xanthus* FB, which is 9454 kbp long (Chen, Keseler & Shimkets, 1990). The genome of this species is about twice the size of the *Escherichia coli* chromosome and is the largest prokaryotic genome analyzed by pulsed field gel electrophoresis (Kraweic & Riley, 1990). It is, nevertheless, smaller than the smallest known eukaryotic genome, that of *Saccharomyces cerevisiae*. The *M. xanthus* genome has been examined by physical mapping techniques and consists of a single, circular chromosome (Chen *et al.*, 1991). Many genes involved in the social and developmental behaviors of *M. xanthus* have been mapped and are distributed with relative uniformity around the chromosome (Fig. 1 & Table 1). Large deletions of the chromosome would eliminate genes that are essential for the social behaviors of the organism, providing a distinct selective disadvantage for the organisms in which these deletions arose. Therefore, having evolved a large genome size, it appears unlikely that the size will diminish without seriously jeopardizing the ability to perform these social behaviors.

Myxobacterial social behavior is observed during feeding, movement, and development. *M. xanthus* is the most carefully studied species and has become a model system for the analysis of these social behaviors because of its well-developed genetic system. Four different social behaviors have been described in this species. In order of their perceived genetic and physiological complexity they are cooperative growth, social motility, rippling, and fruiting body formation. Each of these behaviors will be discussed in the following sections.

COOPERATIVE GROWTH

Cooperative growth refers to the fact that the rate of growth on insoluble carbon sources increases with the cell density (Rosenberg, Keller & Dworkin, 1977). Cooperative growth appears to be due to the cooperative hydrolysis of extracellular polymers. In the case of growth on casein, the concentration of extracellular proteases and the amount of casein hydrolyzed are directly

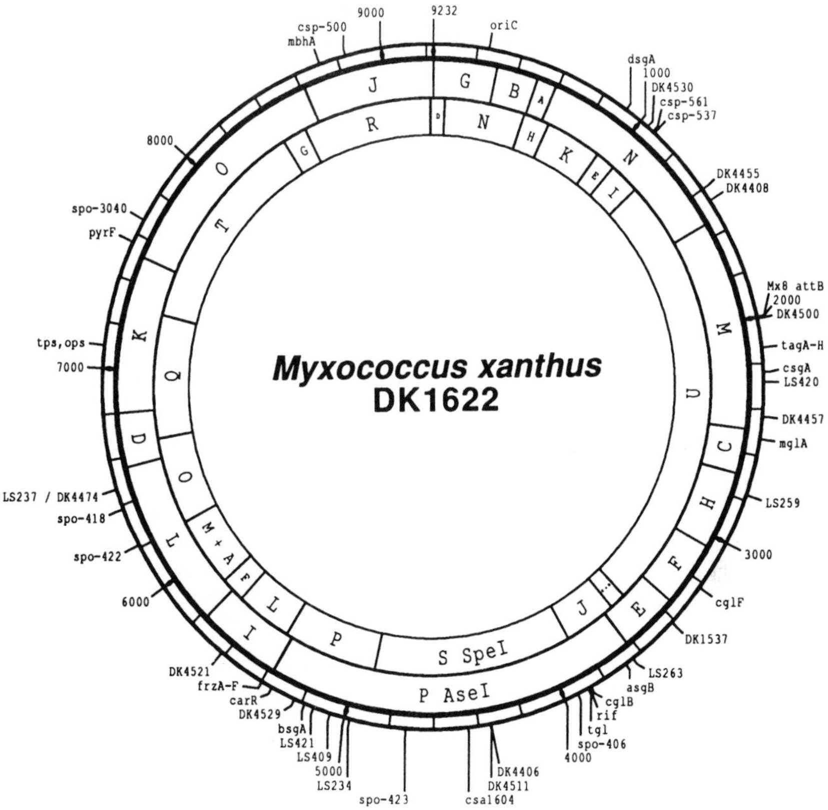

Fig. 1. Genetic and physical map of the *Myxococcus xanthus* chromosome.

proportional to the cell density (Rosenberg *et al.*, 1977). These results suggest that more efficient feeding occurs at higher cell densities due to communal use of the extracellular enzymes. Cooperative growth is likely to provide a direct selection for communal interactions and may have led to the evolution of the other myxobacterial social behaviors.

SOCIAL MOTILITY

Myxobacteria are rod-shaped cells that move by gliding, a form of motility that requires contact with a surface (for review, see Burchard, 1984). Movement occurs in the direction of the long axis of the cell with occasional stops and reversals of direction. Little is known about the mechanism of gliding since the organelle(s) providing propulsion are unknown. Social motility refers to the tendency of gliding cells to travel in large groups known as swarms. Cells within a swarm maintain contact with one another. A selective advantage for the evolution of social motility appears to exist in

Table 1. *Location and function of* Myxococcus xanthus *genetic loci for use in conjunction with Figure 1. Not all of these loci have been placed on the circular map yet.*

Locus	Location (*AseI* restriction fragment)	Function
ade	P	AMP biosynthesis
aglB	J	A (adventurous) motility
aglR	L	A (adventurous) motility
asgA	K	A signal
asgB	P	A signal
asgC	K	A signal
bsgA	P	B signal
carR	P	caroteinoid biosynthesis
cglB	P	A (adventurous) motility
cglC	F	A (adventurous) motility
cglE	L	A (adventurous) motility
cglF	F,L & P	A (adventurous) motility
csa	P	cell surface antigen
csgA	M	C signal
csp	K & N	suppressor of csgA
dsgA	N	D signal synthesis
dsp	O	S (social) motility
fprA	M	flavoprotein
frzA-F	I	directional cell movement
igl	P	directional cell movement
lps	K, L & P	lipopolysaccharide biosynthesis
mbhA	J	hemagglutinin
mglA	C	A and S motility
Mxl^r	K	bacteriophage Mx1 resistance
Mx8 attB	M	bacteriophage Mx8 integration site
ops	K	cytoplasmic spore protein
oriC	G	origin of replication
pyrF	O	pyrimidine biosynthesis
rif	P	rifampicin resistance
sgl	K & O	S (social) motility
spo	many	spore formation
stk	P	cell adhesion
tagA-H	M	34 °C aggregation system
tgl	P	S (social) motility
tps	K	spore coat protein S

light of accelerated growth rates due to cooperative growth. Nevertheless, there are occasions when individual cells venture away from the swarm. These two opposing aspects of gliding cell behavior, adventurous and social, are controlled by two nearly separate multigene systems known as A and S (Hodgkin & Kaiser, 1977, 1979*a*,*b*). The A system allows cells to glide away from the swarm as individuals and is regulated by at least 22 genetic loci. The S system controls the social aspect of cell behavior and allows gliding only when cells are in direct contact (Kaiser & Crosby, 1983). Figure 2 shows the effect of A and S mutations on the behavior of cells. To be nonmotile, cells must lose both adventurous and social motility. This usually

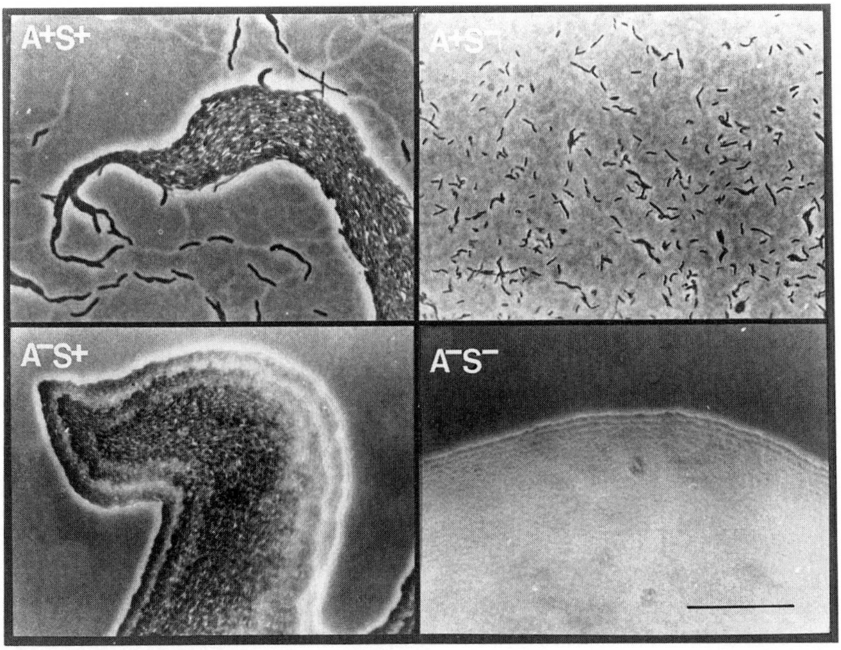

Fig. 2. Effect of A (adventurous) and S (social) motility mutations on *M. xanthus* cell behaviour (Shimkets, 1986*a*). Bar, 50 μm.

takes two mutations, one in each system (Hodgkin & Kaiser, 1979*a,b*). However, single mutations in the *mglA* gene can make fully motile cells nonmotile in a single step (Hodgkin & Kaiser, 1979*a,b*; Stephens & Kaiser, 1987; Stephens, Hartzell & Kaiser, 1989). It would appear that each behavioral system contains its own method of propulsion since only one gene is shared by both systems. In addition, each system would have those features that accentuate the type of behavior it regulates.

One of the most prominent features of the social motility system is the production of a glycocalyx that enables cells to adhere to one another. Inhibition of glycocalyx synthesis through genetic or biochemical means abolishes the ability to move socially and to participate in fruiting body formation (Shimkets, 1986*a,b*; Arnold & Shimkets, 1988*a,b*). Similarly, treatment of cells with Congo red, a dye that prevents the formation of the fibrils, inhibits glycocalyx production, social motility, and fruiting body formation (Arnold & Shimkets, 1988*a,b*). The glycocalyx is formed from 50 nm thick fibrils that coat the surface of the cell and extend outward, often attaching to the surfaces of adjacent cells. Figure 3 shows fibrils produced by wild-type cells. Gliding cells also deposit this material in trails which appear to be the preferred route for travel by other cells. In older colonies, the secreted material forms a thick matrix in which the cells are embedded. The extracellu-

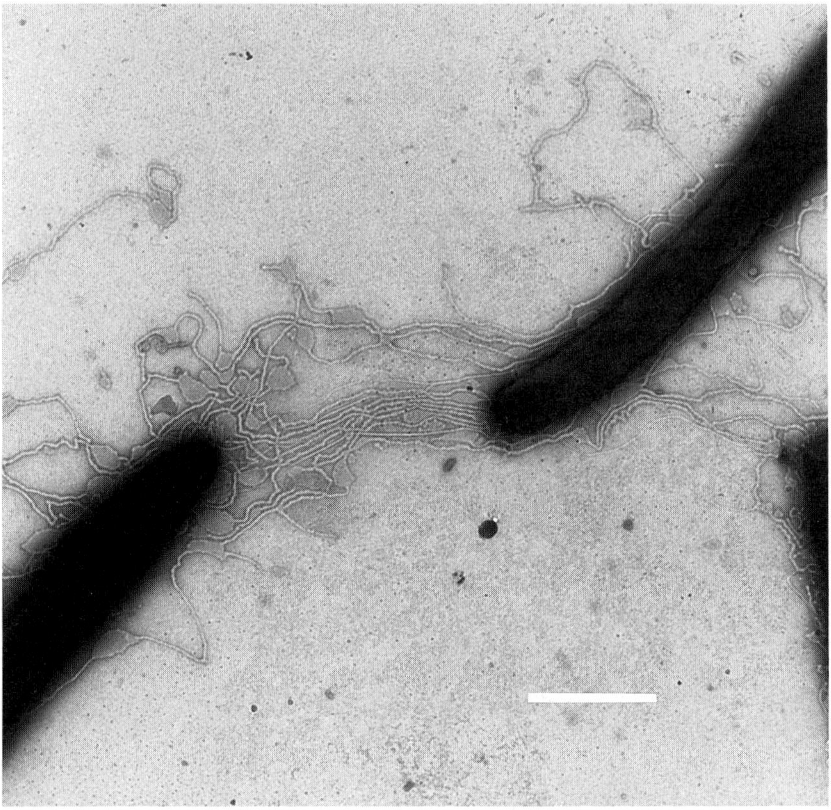

Fig. 3. Transmission electron micrograph of negatively stained wild-type cells showing the 50 nm thick fibrils. Courtesy of Judy W. Arnold & Lawrence J. Shimkets. Bar, 1 μm.

lar matrix of *Myxococcus virescens* is composed of polysaccharide, protein and lipid (14 : 8 : 16) and appears to be a passive means of acquiring nutrients by binding, denaturing, and hydrolyzing proteins (Gnosspelius, 1978*a,b*). At least three extracellular enzymes are observed in this complex as well as bacteriolytic enzyme activity. This matrix would, therefore, accentuate cooperative growth.

The glycocalyx also has an essential role in social motility which has been investigated using the social motility mutants. The social motility gene system contains at least 11 loci, *mglA*, *tglA*, *sglA-H* (Hodgkin & Kaiser, 1979*b*), and *dsp* (Shimkets, 1986*b*). While all mutations in the social motility genes inhibit glycocalyx formation, transfer of the *stk* mutation to each of these mutants restores glycocalyx production to all but the *dsp* mutants (J. Dana, MS Thesis, 1991, University of Georgia). These results suggest that the *mgl*, *tgl* and *sgl* genes control the movement process while the *dsp* genes are the structural genes for glycocalyx synthesis (Shimkets, 1986*b*;

Arnold & Shimkets, 1988*a,b*; J. Dana, MS Thesis, 1991, University of Georgia). Since the phenotypes of the *sgl*, *tgl*, and *dsp* mutants are similar in the absence of the *stk* mutation, these results also suggest that social motility and glycocalyx synthesis are mutually dependent. A model has been formulated in which social motility and glycocalyx synthesis regulate each other in a positive feedback loop (J. Dana, MS Thesis, 1991, University of Georgia). In its simplest form, the model proposes that cell–cell contact that is mediated by the glycocalyx is necessary for social movement. This aspect of the model is based on the observation that cells must be within one cell length of each other in order to move socially (Kaiser & Crosby, 1983). To complete the loop, social movement then stimulates glycocalyx synthesis (J. Dana, 1991, MS Thesis, University of Georgia).

Social motility is essential for the remaining two types of social behavior. Rippling and fruiting body formation require carefully orchestrated movement of tens of thousands of cells.

RIPPLING

Rippling is a rhythmic behavior in which cells arrange themselves in a series of equidistant ridges that move processively to give the appearance of pulsating waves (Reichenbach, 1965; 1966; Reichenbach, Heunert & Kuczka, 1965*a,b*). The waves emanate from several points in the interior of the colony and appear to travel outward to the periphery. Rippling is also observed during the early stages of development but does not appear to play an essential role in the developmental aggregation process. It is observed in those situations where there is substantial hydrolysis of eubacterial cell walls, for example, when *M. xanthus* preys on *Micrococcus luteus* cells (Shimkets & Kaiser, 1982). The natural inducer of rippling appears to be peptidoglycan and its components *N*-acetylglucosamine, *N*-acetylmuramic acid, diaminopimelate, and D-alanine. Figure 4 shows ripples that were induced by peptidoglycan components. Rippling requires the A and S motility system as well as the CsgA protein, which is an intercellular signal also required for fruiting body formation and sporulation. Myxobacteria appear to be the only prokaryotes that perform this remarkable multicellular behavior (Reichenbach, 1986).

FRUITING BODY FORMATION

Fruiting body formation is an alternative to the vegetative growth cycle and is induced by nutritional downshift. Tens of thousands of cells glide to a central location where they arrange themselves in a structure known as a fruiting body that is over 100 μm tall. The shape of the fruiting body varies with the species and may be a simple raised mound, as in the case of *M. xanthus* (Shimkets & Seale, 1974; Kuner & Kaiser, 1982; O'Conner & Zusman, 1989), or an elaborately branched tree-like structure as in the

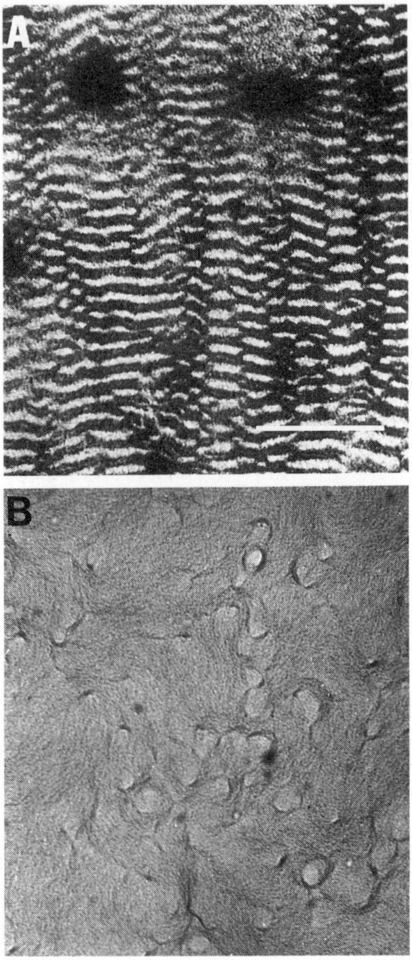

Fig. 4. Induction of ripples by peptidoglycan components (Shimkets & Kaiser, 1982*a*). (A) Culture with 2.5 mM each of *N*-acetylglucosamine, *D*-alanine, and diaminopimelate. (B) Culture with no additions. Bar in panel A, 400 μm.

case of *Chondromyces crocatus* and *Stigmatella aurantiaca* (Grilione & Pangborn, 1975; Qualls, Stephens & White, 1978). Figure 5 shows scanning electron micrographs of several myxobacterial fruiting bodies which emphasize the morphological diversity. Inside the fruiting body the vegetative cells undergo a series of structural changes that lead to the formation of dormant, resistant spores. An *M. xanthus* vegetative cell is rod-shaped, about 5 μm in length and 0.5 μm in diameter, and rounds up during sporulation to produce a single spherical myxospore with a diameter of 4–5 μm.

The *M. xanthus* developmental cycle appears to involve two distinct types

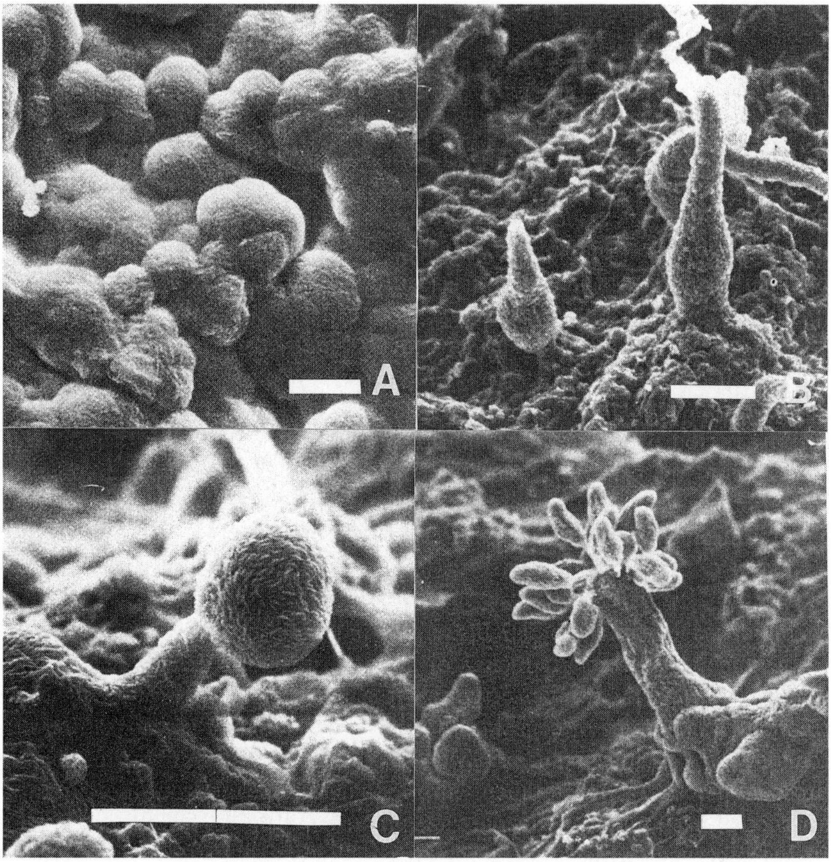

Fig. 5. Scanning electron micrographs of myxobacterial fruiting bodies indicating the morphological diversity (Brockman & Todd, 1974). (A) *Polyangium cellulosum.* (B) *Myxococcus (Corallococcus) coralloides.* (C) *Melittangium lichenicola.* (D) *Stigmatella aurantiaca.* Bar, 20 μm.

of communication processes: those involved with inducing and regulating the pace of development and those involved with bringing the cells together to form a fruiting body. These interactions were initially identified using a phenotypic complementation assay in which two mutant cell types were mixed and examined for fruiting body formation. The germinated spores retain the mutant phenotype of the parental strains indicating that the original mutation has not been repaired (McVittie, Messik & Zahler, 1962; Hagen, Bretscher & Kaiser, 1978; Shimkets 1986*a,b*). This experiment suggests the existence of extracellular molecules controlling fruiting body development. These phenotypic complementation experiments can be divided into two groups depending on whether fruiting body formation alone (Shimkets, 1986*a,b*), or fruiting body formation and myxosporulation (Hagen *et al.*,

1978) are restored. In the former case, the mutations appear to affect those signaling processes that are necessary for directed movement, leading to the aggregation of cells and the construction of a fruiting body. In the latter case, the signaling steps appear to affect some early developmental processes on which both aggregation and sporulation depend. These two elements of developmental cell–cell interactions will be discussed below.

Directed movement during development

The recruitment of cells into aggregation centers occurs over relatively long distances. It is not unusual to see aggregation fields with diameters of over 1 mm. Cells at the far reaches of the aggregation field would have to move about 100 cell lengths to reach the center of the fruiting body. The movement of cells during fruiting body formation has been recorded by time-lapse photography, and *M. xanthus* cells move directly to the aggregation center in large streams (Reichenbach, 1966; Reichenbach *et al.*, 1965*a,b*). Many smaller streams feed into the main stream, making the overall pattern resemble a river with many tributaries. The mechanism of directed movement during fruiting body formation is unknown. The two most intensely studied examples of tactic behavior that could lead to aggregation include chemotaxis of the enteric bacteria and developmental aggregation of *Dictyostelium discoideum.* A brief summary of these types of directed movement is useful in pointing out the obvious differences with aggregating *M. xanthus* cells.

The enteric bacteria also belong to the purple eubacteria (Woese, 1987). However, enteric bacteria have flagella and move by swimming in liquids. Two types of swimming motions are observed: runs, movements along relatively straight paths, and tumbles, rapid somersaults. After a tumble, cells choose a new course randomly in any of the three-dimensional directions. Chemotaxis of the enteric bacteria is thus a random-biased walk in which net movement up the gradient occurs because cells swim up the gradient for longer intervals than in the many other directions they are forced to travel (McNab, 1987). Myxobacterial gliding movement is confined to a solid surface and does not involve such pronounced directional changes, primarily just forward and reverse. Movement of myxobacteria toward a fruiting body is not a random-biased walk, because a course is followed with directional precision.

The eukaryotic cellular slime molds also have a developmental cycle involving aggregation and fruiting body formation. While *Dictyostelium* amoebae move into the aggregate in large streams like myxobacteria, the rhythmic cadence of *Dictyostelium* is distinctly different from the continuous motion of myxobacteria. Developmental chemotaxis of *D. discoideum* occurs in a rhythmic manner that corresponds to the release of and response to the chemoattractant 3′–5′ adenosine monophosphate (Tomchick & Devreotes, 1981). Based on the films of *M. xanthus* aggregation, it appears that the method of directed movement used by myxobacteria is distinctly different from that of either the enteric bacteria or *Dictyostelium* (Reichenbach, 1966; Reichenbach *et al.*, 1965*a,b*).

Both theoretical and experimental evidence suggests that chemotaxis towards a small diffusible substance is unlikely to be involved in myxobacterial fruiting body formation. Myxobacteria move slowly, about 2 μm/min, which means that a small molecule moving by diffusion would move faster than the cells. Given the short duration of time during which such a gradient would be useful and the slow response of cells to it, chemotaxis to a diffusible molecule seems inefficient for myxobacteria. Perhaps the most compelling evidence against a model involving a diffusible chemoattractant is that directed movement occurs even when cells are submerged under a layer of buffer that is vigorously shaken (Shimkets & Kaiser, 1982). In addition, *M. xanthus* is not attracted to a variety of diffusible substances including amino acids, nucleotides, cell wall components, and cell extracts (Dworkin & Eide, 1983).

A clue to the mechanism of myxobacterial-directed movement might be found in the numerous reports that myxobacteria produce and follow trails on surfaces. Trail-following as a means of developmental aggregation is suggested at low cell densities where the cell layer is discontinuous, or late in development when the stragglers at the far reaches of the aggregation field make their way to the aggregation center. In both cases, movement in the direction of the fruiting body is usually unerring, indicating the presence of a defined signaling mechanism. *M. xanthus* secretes a tremendous amount of extracellular polysaccharides, about 10% of the dry weight of cells (Sutherland & Thomson, 1975). This material covers the surface of the cells (Arnold & Shimkets, 1988*b*) and is deposited in trails on the solid surface by gliding cells. Other cells encountering a trail often follow it.

Two types of models can be invoked to account for the trail-following ability of myxobacteria. The first type of model is based on elasticotaxis of myxobacteria in which cells move parallel to stress lines in the agar rather than crossing over the ridges generated by these stress lines (Stanier, 1942). If the polysaccharide chains of the trail on which the cells move assume an orientation that physically resembles the stress lines in agar, the number of directions in which a cell can easily move is reduced to just two, forward and backward along these stress lines. Having oriented themselves on such a matrix, however, it is not clear how the cells know to move toward the aggregation focus rather than away from it by elasticotaxis alone. While elasticotaxis might help reduce the probability of cells venturing in many directions, it would not be capable of supplying all the information necessary to point the cells unambiguously in the direction of the aggregation center. A second type of model involves taxis toward directional signals that are attached to the extracellular matrix. Cells would respond to such signals in a manner similar to the trail-following behavior of ants. Such a mechanism could account for trail-following over protracted intervals and for directed movement during development. While this model is consistent with the observations, it is presently just a speculation that remains to be tested by genetic and biochemical approaches.

Several genetic loci are required for aggregation: *dsp, tag*, and *frz*. The *dsp* locus, which appears to encode the structural genes for the synthesis of the glycocalyx and extracellular matrix material (J. Dana, MS Thesis, 1991, University of Georgia), is essential for fruiting body formation (Shimkets, 1986*b*). The ability to form fruiting bodies may be restored to *dsp* mutants by providing the extracellular matrix from other cells. The *dsp* mutants form fruiting bodies when mixed with wild type cells or other developmental mutants that produce the matrix (Shimkets, 1986*b*). If production of the extracellular matrix is inhibited by treatment of wild type cells with Congo red, fruiting body formation is also inhibited (Arnold & Shimkets, 1988*a,b*). Suppression of mutations that inhibit matrix production restores fruiting body formation. Two groups of mutants that normally do not produce the extracellular matrix, *sgl* and *tgl*, are also defective in development. Unlike *dsp* mutants, *sgl* and *tgl* mutants may be stimulated to produce the extracellular matrix material with a second mutation known as *stk* (J. Dana, 1991, MS Thesis, University of Georgia). The *stk sgl* or *stk tgl* double mutants also regain the ability to form fruiting bodies although their primary defect, movement via the social motility system, remains unchanged. Taken together, these results demonstrate that the extracellular matrix is essential for fruiting body formation. The particular role it plays is an interesting topic for future investigation.

The *tag* locus is 8.5 kbp long and contains eight genotypic complementation groups (Torti & Zusman, 1981; O'Connor & Zusman, 1990). All *tag* mutants, even those formed by transposon insertion, are unable to form fruiting bodies at 34 °C but display normal development at 28 °C. These results imply the existence of at least two aggregation systems: the *tag* system mediates aggregation at 34 °C while the other unidentified system mediates aggregation at 28 °C. The discovery of two aggregation systems remains an intriguing aspect of myxobacterial development.

The *frz* locus contains six genes and controls the frequency with which cells reverse their direction of movement (Blackhart & Zusman, 1985*a,b*). Gliding motility is limited to forward and backward because of the association of the cells with a solid surface. Wild-type cells reverse their direction of movement every 6.8 min on the average, but net movement in one direction occurs because of large variations in the interval between switching directions. The majority of the *frz* mutants reverse their direction of movement every 2 hours, and, as a result, tend to glide in long, thin streams. The *frzCD* mutants reverse their direction of movement every 2.2 min, but there is little variation in the interval between reversals so the cells show little net movement. Instead of establishing an aggregation focus and accumulating there, *frz* mutants eventually sporulate in the thin stream in which they travel (Zusman, 1982). The Frz proteins have substantial amino acid similarity with proteins that control the directional movement of enteric bacteria in response to chemoattractants. FrzCD has 40% amino acid homology with the *Salmonella typhimurium* Tar protein over about one-third the length of the protein (McBride, Weinberg & Zusman, 1989). The Tar protein, a methyl-accepting chemotaxis protein, is involved in chemoreception and adaptation to chemoattractants in enteric bacteria, and controls the frequency of changes in the direction of flagellar rotation. Like Tar, FrzCD is modified by methylation at glutamate residues (McCleary, McBride & Zusman, 1990). FrzF probably encodes the methyltransferase since methylation of FrzCD does not occur in FrzF cells. The amino acid sequence of FrzG is similar to CheB from *Escherichia coli* which encodes the methylesterase. FrzA contains 28% amino acid identity with CheW from *S. typhimurium*, which is a protein that controls the clockwise rotation of the flagellum (McBride *et al.*, 1989). FrzE has amino acid similarities to both CheA and CheY of *S. typhimurium*, which are members of a family of two component response regulators (McCleary & Zusman, 1990). FrzE may be a second messenger that relays information between FrzCD and the gliding motor in response to chemoattractants. It appears, then, that *M. xanthus* regulates directional movement during development with a mechanism that is evolutionarily related to the system controlling flagellar rotation during chemotaxis of enteric bacteria. If the methylation state of FrzCD is influenced by the chemical signal used in aggregation, analysis of the methylation state of this protein might provide an assay for the identification of the signal(s) controlling aggregation.

In summary, many of the genes controlling directed movement during fruiting body formation have been identified. The *dsp* locus controls the production of the extracellular matrix in which the cells move. The *tag* locus is involved with the production of the signal that controls aggregation at 34 °C. Another, as yet unidentified locus appears to control production of the aggregation signal for 28 °C aggregation. The *frz* locus controls the response of cells to the aggregation stimulus by coupling perception of the signal with operation of the transmission that controls the direction of cell movement. A thorough analysis of the gene products of these loci should allow the formulation of a clearer model of the regulation of directed movement.

Regulation of development by intercellular signaling

In addition to the intercellular signals controlling directed movement, other signaling systems are also produced during the early stages of development. Signaling mutants of the latter type show obvious defects in both aggregation and sporulation and appear to be defective in steps on which both processes depend. The mutants are conditional mutants in that they may be stimulated to develop by mixture with other nonsporulating mutants. Pairwise mixture of different mutants allows them to be divided into four phenotypic complementation groups which have been designated Asg, Bsg, Csg, and Dsg for A signal, B signal, C signal, and D signal (Hagen *et al.*, 1978). All of the *asg* mutants map to one of three genetic loci *asgA*, *asgB* and *asgC* (Mayo & Kaiser 1989; Kuspa & Kaiser, 1989). Developing asg^+ cells produce a heat-labile substance that restores development to *asg* mutants, which produce less than 5% of the wild-type activity (Kuspa, Kroos & Kaiser, 1986; Kuspa & Kaiser, 1989). The addition of certain amino acids to the *asg* mutants also restores development. The possibility that A factor is an extracellular protease is under examination.

Several of the *bsg* mutations map to a gene known as *bsgA*, which encodes an 89 kD protein located primarily in the cytoplasm (Gill, Cull & Fly, 1988; Gill & Bornemann, 1988; Gill & Cull, 1986). The mechanism of extracellular complementation in this group is particularly puzzling since the BsgA gene product is not extracellular during development. DNA sequencing of this gene is currently in progress in an effort to identify the developmental role of this gene.

Half of the *dsg* mutations map to a gene known as *dsgA* (Cheng & Kaiser, 1989*a,b*). Although some point mutations in the gene affect only development, transposon insertions in the gene are lethal, indicating that the gene is essential for growth. Treatment of *dsgA* mutants with the AMI restores fruiting body formation and sporulation (Rosenbluh & Rosenberg, 1989). AMI is a mixture of saturated and unsaturated fatty acids produced from the phospholipase-mediated hydrolysis of the major membrane phospholipid phosphotidylethanolamine (Gelvan, Varon & Rosenberg, 1987). AMI accumulates in the culture medium during vegetative growth and kills the producing cells (Varon, Cohen & Rosenberg, 1984; Varon, Tietz & Rosenberg, 1986). Whether or not AMI is the actual D signal is unclear. In lysing a small portion of the population of developing *dsgA* cells, AMI may release the true D signal or other substances that can bypass this developmental step.

The Csg group is the least complex and most extensively characterized group. All *csg* mutations are located in a single gene known as *csgA* (Shimkets, Gill & Kaiser, 1983; Arnold & Shimkets, 1988*a,b*; Hagen & Shimkets, 1990). The CsgA gene product was overproduced in *E. coli* to obtain enough protein to raise antibodies in rabbits (Shimkets & Rafiee, 1990). Immunopurified polyclonal antibodies raised against the CsgA gene product were used to determine the cellular location of the protein in developing cells by colloidal-gold labeling and transmission electron microscopy. CsgA is associated with the cell surface and extracellular matrix at a concentration of about 1000 to 2000 molecules per cell. It apparently binds tightly to the cell surface and matrix since it remains attached even after extensive washing of the cells. The antibody preparation inhibits the development of wild-type cells, suggesting that CsgA performs an essential extracellular role. Similar conclusions were drawn from a complementary approach. A protein known as C-factor was extracted from wild-type cells using detergent and purified using a bioassay based on its ability to restore development to *csgA* mutants (Kim & Kaiser, 1990*a,b*). Since addition of C-factor to developing cells restores development, C-factor probably acts from the outside in, as would be expected of an intercellular signal. C-factor is present at about 8000 molecules per cell at the peak of its production during development. C-factor is almost certainly the protein product of the *csgA* gene. Antibodies prepared against the CsgA gene product cross-react with C-factor (Kim & Kaiser, 1990*b*; Shimkets & Rafiee, 1990). In addition, a partial amino acid sequence of C-factor is identical to that predicted from the DNA sequence of the *csgA* gene (Kim & Kaiser, 1990*a*; Hagen & Shimkets, 1990). Taken together, these data demonstrate that the CsgA gene product is the actual C signal. The CsgA gene product does not appear to have substantial amino acid similarity with any known developmental hormones (Hagen & Shimkets 1990).

Function of the C-signal

If CsgA is not the attractant involved in aggregation, then what is its function? The developmental role of *csgA* was probed by modifying the regulatory region upstream from the *csgA* gene (S. Li, PhD Dissertation, University of Georgia, 1991). A series of 5′-deletions was constructed across the upstream regulatory region and the strains were assayed for developmental proficiency and for *csgA* expression using a *lacZ* reporter gene coupled to the *csgA* upstream regulatory region. The results were remarkable in two respects. First, the upstream regulatory region was at least as long as the gene itself and appeared to be regulated in an unusual if not unique manner: expression of *csgA* increased incrementally with increasing lengths of the upstream regulatory region. Other developmental genes with complex expression patterns have previously been reported for *M. xanthus*, but little is known at present about the mechanism of developmental gene expression (Downard, 1987; Downard, Kim & Kil, 1988; Kil, Brown & Downard, 1990). Continued investigation of expression of these genes is a worthy project since it appears that they are distinctly different from traditional enteric bacterial genes. Second, the stage at which cells ceased development increased with the length of the upstream regulatory sequences. CsgA null mutants are defective in rippling, fruiting body construction, and sporulation, three partially overlapping developmental stages that occur in a precise temporal framework (Shimkets & Asher, 1988). With short upstream sequences one could observe rippling, the first developmental event, and the initial stages of aggregation, the second developmental event. With intermediate length upstream sequences one could observe rippling and fruiting body formation but little sporulation. Only with full length sequences could one observe all aspects of development. It appears that when CsgA is limiting, development ceases at a stage that is determined by the concentration of CsgA that is achieved. These results suggest a simple model for development involving a molecular clock, an activator molecule and a set of target genes that are differentially activated by the effector molecule. The role of the developmental clock is to increase the concentration of the activator molecule incrementally according to a defined temporal program. The role of the activator molecule, in this case CsgA, is to couple the movement of the clock with the regulation of gene expression. The target genes are those that mediate different stages of development. Target genes for rippling and aggregation are expected to be induced at lower CsgA concentrations than sporulation genes. Analysis of the upstream regulatory region of CsgA is expected to provide some insight into the nature of the developmental clock.

Mechanism of action of CsgA

The mechanism of perception of CsgA is unknown but presumably involves a signal transduction pathway involving a receptor, a secondary messenger, and a means of regulating developmental gene expression. Nonmotile mutants abort the developmental pathway at about the same time as *csgA* mutants (Kroos *et al.*, 1988). The nonmotile *mgl* mutants are unable to serve as either CsgA donors or recipients even though they produce normal amounts of a biologically active CsgA protein (Kim & Kaiser, 1990*c*). Motility is obviously essential for transmission of the C signal but the precise role of motility remains unknown. One role of motility might be the achievement of an optimal cell–cell alignment for signal exchange (Kim & Kaiser, 1990*d*). At present, however, it is not clear that cell–cell contact is essential for transmission of the signal, and many other possibilities exist. Motility might be required to bring the CsgA receptor into contact with extracellular matrix-associated CsgA or to process information about the relative concentration of CsgA over specific movement or time intervals.

In order to identify genes that might be involved in the signal transduction pathway, sporulating pseudorevertants were isolated from *csgA* mutants (Rhie & Shimkets, 1989). The suppressor mutations exhibited a variety of phenotypes and mapped to at least six different loci. The analysis of these suppressor groups is still in its initial stages. However, two groups show particular promise as members of the signal transduction pathway. One suppressor group, represented by a single allele referred to as *soc-500*, restored fruiting body formation, but not rippling (Rhie & Shimkets, 1991). This mutation is *trans*-dominant and has properties one might expect of a class of signal transduction mutations that are constantly on, even in the absence of the appropriate stimulus.

Another group exhibited normal rippling and fruiting body formation in spite of the absence of CsgA. This group contains two alleles, *soc-559* and *soc-560*, that were formed by transposon insertions which are separated by about 500 base pairs (K. Lee, MS Thesis, University of Georgia, 1991). The *soc-559/soc-560* merodiploid suppresses CsgA mutations, suggesting that both mutations lie in the same transcriptional unit and most likely affect the same gene, which has been designated *socA*. This gene is a strong negative autoregulator of its own expression and possibly a negative regulator of rippling, aggregation, and sporulation genes. The gene also appears to be regulated by CsgA. Continued analysis of these suppressor mutations is likely to help identify the elements of the signal transduction pathway.

CONCLUSIONS

Of the procaryotes, myxobacteria have the most complex social behaviors, which provide an evolutionarily ancient solution to many developmental and behavioral problems faced by eukaryotic cells. The manner in which the genetic code specifies the formation of fruiting bodies and the synchronized movement of cells in ripples are fundamental biological problems. Four different behavioral systems have been described for the myxobacterium *M. xanthus*. The results obtained so far indicate that the complex behaviors of the myxobacteria can be understood with the genetic and biochemical approaches currently in place. Therefore, the myxobacteria provide an ideal system for the analysis of these social behaviors.

ACKNOWLEDGEMENTS

Research in my laboratory has been supported by grant DCB-9001755 from the National Science Foundation and by the Georgia Power Company.

REFERENCES

Arnold, J. W. & Shimkets, L. J. (1988*a*). Inhibition of cell–cell interactions in *Myxococcus xanthus* by Congo red. *Journal of Bacteriology*, **170**, 5765–70.

Arnold, J. W. & Shimkets, L. J. (1988*b*). Cell surface properties correlated with cohesion in *Myxococcus xanthus. Journal of Bacteriology*, **170**, 5771–7.

Blackhart, B. D. & Zusman, D. R. (1985*a*). Cloning and complementation analysis of the *Frizzy* genes of *Myxococcus xanthus. Molecular and General Genetics*, **198**, 243–54.

Blackhart, B. D. & Zusman, D. R. (1985*b*). *Frizzy* genes of *Myxococcus xanthus* are involved in control of the frequency of reversal of gliding motility. *Proceedings of the National Academy of Sciences, USA*, **82**, 8767–71.

Brockman, E. R. & Todd, R. L. (1974). Fruiting myxobacters as viewed with a scanning electron microscope. *International Journal of Systematic Bacteriology*, **24**, 118–24.

Burchard, R. P. (1984). Gliding motility and taxes. In *Myxobacteria: Development and Cell Interactions*, Rosenberg, E., ed., pp. 139–61. Springer-Verlag, New York.

Chen, H-W., Keseler, I. M. & Shimkets, L. J. (1990). The genome size of *Myxococcus xanthus* determined by pulsed field gel electrophoresis. *Journal of Bacteriology*, **172**, 4206–13.

Chen, H.-W., Kuspa, A., Keseler, I. M. & Shimkets, L. J. (1991). A physical map of the *Myxococcus xanthus* chromosome. *Journal of Bacteriology*, **173**, 2109–15.

Cheng, Y. & Kaiser, D. (1989*a*). *dsg*, a gene required for cell–cell interaction early in *Myxococcus* development. *Journal of Bacteriology*, **171**, 3719–26.

Cheng, Y. & Kaiser, D. (1989*b*). *dsg*, a gene required for *Myxococcus* development, is necessary for cell viability. *Journal of Bacteriology*, **171**, 3727–31.

Downard, J. 1987. Identification of the RNA products of the *ops* gene of *Myxococcus xanthus* and mapping of *ops* and *tps* RNAs. *Journal of Bacteriology*, **169**, 1522–8.

Downard, J. S., Kim, S.-H. & Kil, K-S. (1988). Localization of the *cis*-acting regulatory DNA sequences of the *Myxococcus xanthus tps* and *ops* genes. *Journal of Bacteriology*, **170**, 4931–8.

Dworkin, M. & Eide, D. (1983). *Myxococcus xanthus* does not respond chemotactically to moderate concentration gradients. *Journal of Bacteriology*, **154**, 437–42.

Gelvan, I., Varon, M. & Rosenberg, E. (1987). Cell-density-dependent killing of *Myxococcus xanthus* by autocide AMV. *Journal of Bacteriology*, **169**, 844-8.

Gill, R. E. & Bornemann, M. C. (1988). Identification and characterization of the *Myxococcus xanthus bsgA* gene product. *Journal of Bacteriology*, **170**, 5289–97.

Gill, R. E. & Cull, M. G. (1986). Control of developmental gene expression by cell-to-cell interaction in *Myxococcus xanthus. Journal of Bacteriology*, **168**, 341–7.

Gill, R. E., Cull, M. G. & Fly, S. (1988). Genetic identification and cloning of a gene required for developmental cell interactions in *Myxococcus xanthus. Journal of Bacteriology*, **170**, 5279–88.

Gnosspelius, G. (1978*a*). Purification and properties of an extracellular protease from *Myxococcus virescens. Journal of Bacteriology*, **133**, 17–25.

Gnosspelius, G. (1978*b*). Myxobacterial slime and proteolytic activity. *Archives of Microbiology*, **116**, 51–9.

Grilione, P. L. & Pangborn, J. (1975). Scanning electron microscopy of fruiting body formation by myxobacteria. *Journal of Bacteriology*, **124**, 1558–65.

Hagen, D. C., Bretscher, A. P. & Kaiser, D. (1978). Synergism between morphogenetic mutants of *Myxococcus xanthus. Developmental Biology*, **64**, 284–96.

Hagen, T. J. & Shimkets, L. J. (1990). Nucleotide sequence and transcriptional products of the *csg* locus of *Myxococcus xanthus. Journal of Bacteriology*, **172**, 15–23.

Hodgkin, J. & Kaiser D. (1977). Cell-cell stimulation of movement in nonmotile mutants of *Myxococcus. Proceedings of the National Academy of Sciences, USA*, **74**, 2938–42.

Hodgkin, J. & Kaiser, D. (1979*a*). Genetics of gliding motility in *Myxococcus xanthus* (Myxobacterales): genes controlling movements of single cells. *Molecular and General Genetics*, **171**, 167–76.

Hodgkin, J. & Kaiser, D. (1979*b*). Genetics of gliding motility in *Myxococcus xanthus* (Myxobacterales): two genes systems control movement. *Molecular and General Genetics*, **172**, 177–91.

Irschick, H., Jansen, R. Gerthe, K., Holfe, G. & Reichenbach, H. 91987). The sorangicins, novel and powerful inhibitors of eubacterial RNA polymerase isolated from myxobacteria. *Journal of Antibiotics* **40**, 7–13.

Kaiser, D. (1989) Multicellular development in myxobacteria. In *Genetics of Bacterial Diversity*, Hopwood, D. A. & Chater, K. F. eds, pp. 243–66, Academic Press Inc., San Diego CA.

Kaiser, D. & Crosby, C. 1983. Cell movement and its coordination in swarms of *Myxococcus xanthus. Cell Motility*, **3**, 227–45.

Kil, K.-S., Brown, G. & Downard, J. (1990). A segment of *Myxococcus xanthus ops* DNA functions as an upstream activation site for *tps* gene transcription. *Journal of Bacteriology*, **172**, 3081–8.

Kim, S. K. & Kaiser, D. (1990*a*). C-factor: a cell–cell signaling protein required for fruiting body morphogenesis of *M. xanthus, Cell* **61**, 19–26.

Kim, S. K. & Kaiser, D. (1990*b*). Purification and properties of *Myxococus xanthus* C-factor, an intercellular signaling protein. *Proceedings of the National Academy of Sciences, USA*, **87**, 3635–9.

Kim, S. K. & Kaiser, D. (1990*c*). Cell motility is required for the transmission of C-factor, an intercelular signal that coordinates fruiting body morphogenesis of *Myxococcus xanthus, Genes and Development*. **4**, 896–905.

Kim, S. K. & Kaiser, D. (1990*d*). Cell alignment required in differentiation of *Myxococcus xanthus. Science*, **249**, 926–8.

Kraweic, S. & Riley, M. (1990). Organization of the bacterial chromosome. *Microbiological Reviews*, **54**, 502–39.

Kroos, L., Hartzell, P., Stephens, K. & Kaiser, D. (1988). A link between cell movement and gene expression argues that motility is required for cell–cell signaling during fruiting body development. *Genes and Development*, **2**, 1677–85.

Kuner, J. M. & Kaiser, D. (1982). Fruiting body morphogenesis in submerged cultures of *Myxococcus xanthus. Journal of Bacteriology*, **151**, 458–61.

Kuspa, A. & Kaiser, D. (1989). Genes required for developmental signalling in *Myxococcus xanthus*: three *asg* loci. *Journal of Bacteriology*, **171**, 1762–72.

Kuspa, A., Kroos, L. & Kaiser, D. (1986). Intercellular signalling is required for developmental gene expression in *Myxococcus xanthus. Developmental Biology*, **117**, 267–76.

Levine, H. B., Ringel, S. M. & Cobb, J. M. (1978). Therapeutic properties or oral ambruticin (W7783) in experimental pulmonary coccidiomycosis of mice. *Chest*, **73**, 202–6.

Mayo, K. A. & Kaiser, D. (1989). *asgB*, a gene required early for developmental signalling, aggregation, and sporulation of *Myxococcus xanthus. Molecular and General Genetics*, **218**, 409–18.

McBride, M. J., Weinberg, R. A. & Zusman, D. R. (1989). *Frizzy* aggregation genes of the gliding bacterium *Myxococcus xanthus* show sequence similarities to the chemotaxis genes of enteric bacteria. *Proceedings of the National Academy of Sciences, USA*, **86**, 424–8.

McCleary, W. R., McBride, M. J. & Zusman, D. R. (1990). Developmental sensory transduction in *Myxococcus xanthus* involves methylation and demethylation of FrzCD. *Journal of Bacteriology*, **172**, 4877–87.

McCleary, W. R. & Zusman, D. R. (1990). FrzE of *Myxococcus xanthus* is homologous to both CheA and CheY of *Salmonella typhimurium. Proceedings of the National Academy of Sciences, USA*, **87**, 5898–902.

McCurdy, H. D. (1989). Fruiting gliding bacteria: the myxobacteria. In *Bergey's manual of systematic bacteriology* volume 3, Staley, J. T., Bryant, M. P., Pfennig, N. & Holt J. G. eds, pp. 2139–68, Williams and Wilkens, Baltimore.

McNab, R. M. (1987). Motility and chemotaxis. In Escherichia coli *and* Salmonella typhimurium: *Cellular and Molecular Biology*. F. C. Neidhardt, ed., pp. 732–59. American Society for Microbiology, Washington DC.

McVittie, A., Messik, F. & Zahler, S. A. (1962). Developmental biology of *Myxococcus. Journal of Bacteriology*, **84**, 546–51.

Ochman, H. & Wilson, A. C. (1987). Evolution in bacteria: evidence for a universal substitution rate in cellular genomes. *Journal of Molecular Evolution*, **26**, 74–86.

O'Connor, K. A. & Zusman, D. R. (1989). Patterns of cellular interactions during

fruiting body formation in *Myxococcus xanthus. Journal of Bacteriology*, **171**, 6013–24.
O'Connor, K. A. & Zusman, D. R. (1990). Genetic analysis of the *tag* mutants of *Myxococcus xanthus* provides evidence for two developmental aggregation systems. *Journal of Bacteriology*, **172**, 3868–78.
Oyaizu, H. & Woese, C. R. (1985). Phylogenetic relationships among the sulfate respiring bacteria, myxobacteria and purple bacteria. *Systematics and Applied Microbiology*, **6**, 257–63.
Qualls, G. T., Stephens, K. & White, D. (1978). Light stimulated morphogenesis in the fruiting myxobacterium *Stigmatella aurantiaca. Science* **201**, 444–5.
Reichenbach, H. (1965). Rhythmische Vorange bei der Schwarmentfaltung von Myxobacterien. *Ber. Deutsch. Bot. Ges.* **78**: 102–5.
Reichenbach, H. (1966). *Myxococcus* spp. (Myxobacterales) Schwarmentwicklung und Bildung von Protocysten, pp. 557–8. In *Encylop. Cinematogr. Film E778/1965.* Wolf, G., ed., Inst. Wiss. Film, Gottingen.
Reichenbach, H. (1986). The myxobacteria: common organisms with an uncommon behavior. *Microbiological Sciences*, **3**, 268–74.
Reichenbach, H. & Dworkin, M. (1981). The order Myxobacterales. In *The prokaryotes: A Handbook on Habitats, Isolation, and Identification of Bacteria.* Starr, M. P., Stolp, H., Truper, H. G., Balows, A. & Schlegel, H. G. eds, pp. 328–55, Springer-Verlag, Berlin.
Reichenbach, H., Gerth, K., Irschik, H., Kunze, B. & Holfe, G. (1988). Myxobacteria: a new source of antibiotics. *Biotechnology*, **6**, 115–21.
Reichenbach, H., Heunert, H. H. & Kuczka, H. (1965*a*). Schwarmentwicklung und Morphogenese bei Myxobakterien-*Archangium, Myxococcus, Chondrococcus*, and *Chondromyces*. Film C893. Inst. Wiss. Film, Gottingen, Germany.
Reichenbach, H., Heunert, H. H. & Kuczka, H. (1965*b*). *Myxococcus* spp. (Myxobacterales)–Schwarmentwicklung und Bildung von Protocysten. Film E778. Inst. Wiss. Film, Gottingen, Germany.
Rhie, H.-G. & Shimkets, L. J. (1989). Developmental bypass suppression of *Myxococcus xanthus csgA* mutations. *Journal of Bacteriology*, **71**, 3268–76.
Rhie, H.-G. & Shimkets, L. J. (1991). Low temperature induction of *Myxococcus xanthus* developmental gene expression in wild type and *csgA* suppressor cells. *Journal of Bacteriology*, in press.
Rosenberg, E., Keller, K. H. & Dworkin, M. (1977). Cell density-dependent growth of *Myxococcus xanthus* on casein. *Journal of Bacteriology*, **129**, 770–7.
Rosenberg, E. & Varon, M. (1984). In *Myxobacteria: Development and Cell Interactions*, Rosenberg, E., ed., pp. 109–25, Springer-Verlag, New York.
Rosenbluh, A. & Rosenberg, E. (1989). Autocide AMI rescues development in *dsg* mutants of *Myxococcus xanthus. Journal of Bacteriology*. **171**, 1513–18.
Shimkets, L. J. (1986*a*). Correlation of energy-dependent cell cohesion with social motility in *Myxococcus xanthus. Journal of Bacteriology*, **166**, 837–41.
Shimkets, L. J. (1986*b*). Role of cell cohesion in *Myxococcus xanthus* fruiting body formation. *Journal of Bacteriology*, **166**, 842–8.
Shimkets, L. J. (1989). The role of the cell surface in the social and adventurous behavior of the myxobacteria. *Molecular Microbiology*, **3**, 1295–8.
Shimkets, L. J. (1990). Social and developmental biology of the myxobacteria. *Microbiology Reviews*, **54**, 473–501.
Shimkets, L. J. & Asher, S. J. (1988). Use of recombination techniques to examine the structure of the *csg* locus of *Myxococcus xanthus. Molecular and General Genetics*, **211**, 63–71.
Shimkets, L. J., Gill, R. E. & Kaiser, D. (1983). Developmental cell interactions

in *Myxococcus xanthus* and the *spoC* locus. *Proceedings of the National Academy of Sciences*, USA, **80**, 1406–10.

Shimkets, L. J. & Kaiser, D. (1982). Induction of coordinated cell movement in *Myxococcus xanthus. Journal of Bacteriology*, **152**, 451–61.

Shimkets, L. J. & Rafiee, H. (1990). CsgA, an extracellular protein that is essential for *Myxococcus xanthus* development. *Journal of Bacteriology*, **172**, 5299–306.

Shimkets, L. J. & Seale, T. W. (1974). Fruiting-body formation and myxospore differentiation and germination in *Myxococcus xanthus* viewed by scanning electron microscopy. *Journal of Bacteriology*, **121**, 711–20.

Stanier, R. Y. (1942). A note on elasticotaxis in myxobacteria. *Journal of Bacteriology*, **44**, 405–12.

Stephens, K., Hartzell, P. & Kaiser, D. (1989). Gliding motility in *Myxococcus xanthus: mgl* locus, RNA, and predicted protein products. *Journal of Bacteriology*, **171**, 819–30.

Stephens, K. & Kaiser, D. (1987). Genetics of gliding motility in *Myxococcus xanthus*: molecular cloning of the *mgl* locus. *Molecular and General Genetics*, **207**, 256–66.

Sutherland, I. W. & Thomson, S. (1975). Comparison of the polysaccharides produced by *Myxococcus* strains. *Journal of General Microbiology*, **89**, 124–32.

Tomchik, K. J. & Devreotes, P. N. (1981). Adenosine 3', 5'- monophosphate waves in *Dictyostelium discoideum*: a demonstration by isotope dilution-fluorography. *Science*, **212**, 443–6.

Torti, S. & Zusman, D. (1981). Genetic characterization of aggregation-defective developmental mutants of *Myxococcus xanthus. Journal of Bacteriology*, **147**, 768–75.

Varon, M., Cohen, S. & Rosenberg, E. (1984). Autocides produced by *Myxococcus xanthus. Journal of Bacteriology*, **167**, 356–61.

Varon, M., Tietz, A. & Rosenberg, E. (1986). *Myxococcus xanthus* autocide AMI. *Journal of Bacteriology*, **167**, 356–61.

Woese, C. R. (1987). Bacterial evolution. *Microbiological Reviews*, **51**, 221–71.

Zusman, D. R. 1982. *Frizzy* mutants: a new class of aggregation-defective developmental mutants of *Myxococcus xanthus. Journal of Bacteriology*, **150**, 1430–7.

Zusman, D. R., McBride, M. J., McCleary, W. R. & O'Conner, K. A. (1990). Control of directed motility in *Myxococcus xanthus. Symposium of the Society for General Microbiology* **46**, 199–218.

DIFFERENTIATION IN ACTINOMYCETES

DAVID A. HODGSON

Department of Biological Sciences, University of Warwick, Coventry, CV4 7AL, UK

INTRODUCTION

Actinomycetes are soil dwelling, Gram positive bacteria with DNA with a high G + C content. They include some genera capable of unicellular growth, e.g. *Arthrobacter, Corynebacterium* and *Mycobacterium*; and others only capable of mycelial growth, e.g. *Streptomyces, Actinoplanes* and *Thermomonospora* (Goodfellow, 1989). The latter group include exospore forming bacteria (sporoactinomycetes). Studies of the sequence of 16S rRNA from a wide range of bacteria have demonstrated that the high G + C content, Gram positive bacteria are a phylogenetically coherent group (Fig. 1). They also revealed that thermoactinomycetes, which had been included with the actinomycetes because of their mycelial growth habit despite the fact they produced heat-resistant endospores and had a G + C content closer to 50% (Stackebrandt & Woese, 1981), are more closely related to *Bacillus* and *Clostridium* than to any actinomycete.

The sporoactinomycetes include genera with life cycles surprisingly complex for prokaryotes. They can all form non-motile, branched, filamentous cell networks (mycelia) which grow over and into the substrate supporting the colony. This substrate mycelium is believed to promote efficient solubilization of high molecular weight biopolymers (proteins, polysaccharides, etc) by secretion of extracellular enzymes and also the efficient uptake of the resulting products. As this mycelium is fixed to the substrate, some form of dispersal mechanism is necessary if the bacteria are to seek out and colonize new environments. In this respect, the role of the dispersal cells, usually referred to as 'spores', is analogous to the role of the flagellated swarmer cell of *Caulobacter* spp. and the gliding hormogonia of some cyanobacteria (Hodgson, 1989). The morphological appearance, development and properties of these 'spores' are characteristic of the different genera.

The actinoplanetes can produce two types of dispersal cell: motile 'zoospores' and non-motile spores. In some cases both types are present in the same organism, e.g. *Dactylosporangium* in which the non-motile spores are sessile. It is perhaps misleading to use the term 'spore' with respect to the motile cells as this implies that the cells are metabolically quiescent. *Actinoplanes* zoospores synthesize and use flagella following hydration. This process is distinct from germination and hyphal outgrowth. Clearly, therefore, they cannot be metabolically inactive. Ensign (1982) has reported that *Dactylos-*

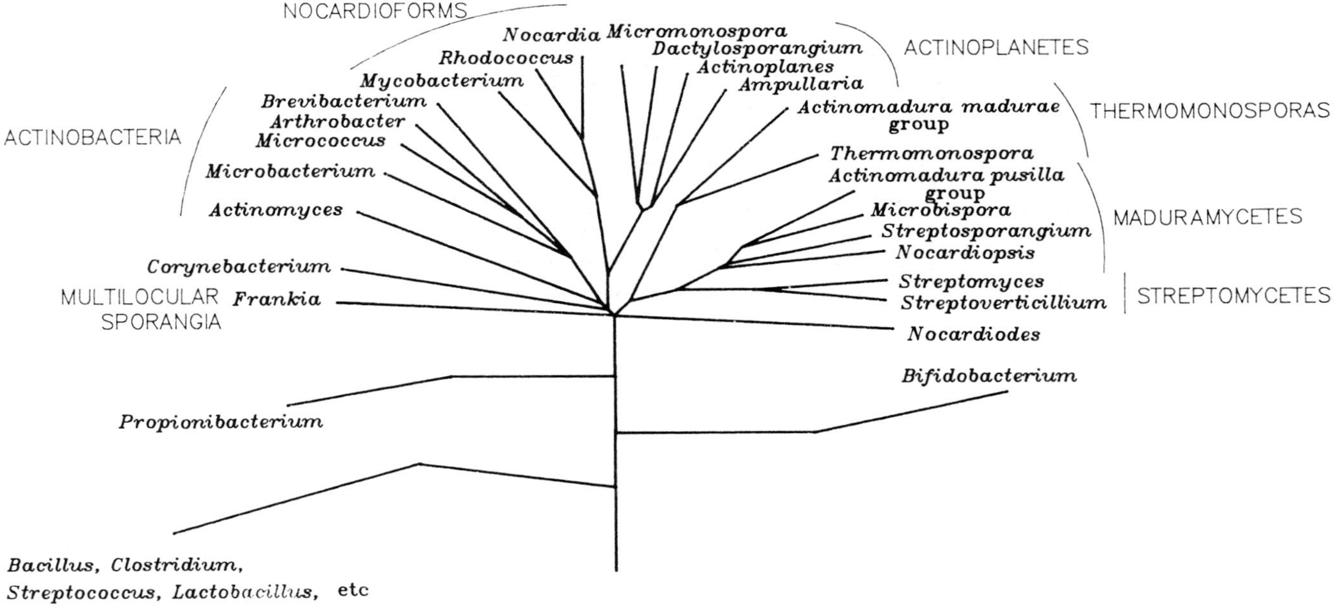

Fig. 1. Taxonomic relationships of the actinomycetes as revealed by analysis of 16S rRNA sequences. The figure illustrates the suprageneric relationships between a selected sample of actinomycetes. The length of the lines represent the S_{AD} (relatedness) values between the individual genera. Data taken from Goodfellow (1989). The suprageneric group names (e.g. actinoplanetes) are also indicated.

porangium zoospores have a 20-fold higher respiratory quotient than do the sessile spores. The zoospores of actinoplanetes are formed within sporangia and are released following hydration. They can remain motile for several hours. Some work has been reported on the morphological aspects of the developmental pathway of*Pilimelia* spp. (Vobis, 1984), and some of the physiological characteristics of the sessile spores and the motile zoospores of *Dactylosporangium* (Ensign, 1982). Unfortunately, little has been reported on the molecular genetic characterization of the developmental cycles of these fascinating bacteria.

The best characterized developmental cycle of the actinomycetes is that of the streptomycetes and the rest of this review is focussed on them. It is hoped that the structural aspects and control of development will be covered in this chapter and equal emphasis will be placed on physiological, morphological and genetic approaches to these questions. The reader is recommended to the excellent review of Chater (1989) for a more molecular genetic emphasis. Further information on the developmental biology of other actinomycetes is to be found in the extensive review of Kalakoutskii and Agre (1976) and the more recent review of Locci (1984).

Interest in the streptomycetes centres, on the whole, on their important biotechnological role. Some two-thirds of the antibiotics/secondary metabolites so far described are produced by this one genus and they include the great majority of antibiotics and growth promotants used in medicine and veterinary practice. The development of a wide range of techniques for genetic analysis in *Streptomyces coelicolor* A3(2) and its close relative *S. lividans* (Hopwood *et al.*, 1985) has allowed rapid advances in the study of a number of aspects of streptomycete biology, including secondary metabolism and morphological differentiation. However, it is true to say that the latter is still an interest of a minority of researchers and, for this reason, far less is known about streptomycete sporulation than about the sporulation processes of bacilli and myxobacteria.

THE STRUCTURE OF THE STREPTOMYCETE COLONY

A previous symposium volume of this Society which addressed microbial differentiation included a review of streptomycete life cycle and the structure of the streptomycete colony (Chater & Hopwood, 1973). The reader is recommended to this review for references to the early morphological studies of streptomycete colony formation.

Macroscopic aspects

Figure 2(*a*) is a picture of a colony of *Streptomyces coelicolor* A3(2) growing on a solid medium. The colony has a powdery or hairy appearance because of the hydrophobic chains of spores. The desiccation-resistant, but heat- and UV- sensitive, spores are derived from specialized hyphae (aerial hyphae) which originate from substrate hyphae but grow perpendicularly into the air. Figure 2(*b*) is a diagrammatic representation of a vertical section through such a colony. This structure appears to be common to most streptomycete colonies when growing on solid media. The two forms of hyphae, substrate and aerial, and the helical chains of ellipsoidal spores are illustrated.

Within the closely packed aerial mycelium above the colony there is a mixture of mature spores, developing spore chains, redundant aerial hyphae and germinating or germinated spores. It is perhaps surprising to observe that spores that have been formed during one round of sporulation may germinate and form new hyphae. The formation of the fully mature spore mass seems to be the result of successive waves of aerial hyphae formation and sporulation. The primary aerial hyphae grow and develop into spore chains. New aerial hyphae grow through these spore chains and in turn sporulate. Any redundant portion of an aerial hypha which does not metamorphose into spores is destined to lyse. Lysis is also the fate of the substrate hyphae which gave rise to the aerial hyphae in the first place (Wildermuth, 1970).

Whether or not the streptomycete colony idealized in Figure 2(*b*) is representative of the structure of streptomycete colonies in their natural habitat, soil, is a reasonable question. Scanning electron microscope analysis of soil particles reveals structures similar to substrate hyphae, aerial hyphae and the coiled spore chains. The extent of streptomycete colony development appears to depend on the nutritional richness of the microenvironment. In conditions of localized high concentration of nutrients, microcolonies of densely packed substrate hyphae surmounted by aerial hyphae and spore chains can be seen when *S. scabies*, the potato pathogen, grows on its host (Wellington, personal communication). Under other conditions microcycle development occurs, i.e. spores germinate and short hyphae are formed which then immediately develop into spore chains. It is assumed that this is due to the colonization of nutrient-limited micro-niches (Locci, 1988).

Microscopic aspects – aerial hypha and spore coats

Substrate hyphae usually have a mean diameter of 0.7 μm (Wolf & Schoppmann, 1989) and are bounded by a mucopeptide cell wall 0.01–0.02 μm wide (Glauert & Hopwood, 1960). Capsules have not been reported. One identifying characteristic shared by aerial hyphae and spore chains but not substrate hyphae is the 'spore sheath'. In some streptomycetes this sheath is highly ornamented. The ornaments include spines, knobs, warts, wrinkles and hairs and have been used in species classification (Locci, 1989). Szabo *et al.* (1977) reported that the formation of spines in some 'hairy' spore streptomycetes appeared to be the product of a morphogenic organizing centre situated at the tip of the developing aerial hypha. The unidirectional propagation of the morphogenic signal did not always appear to be complete

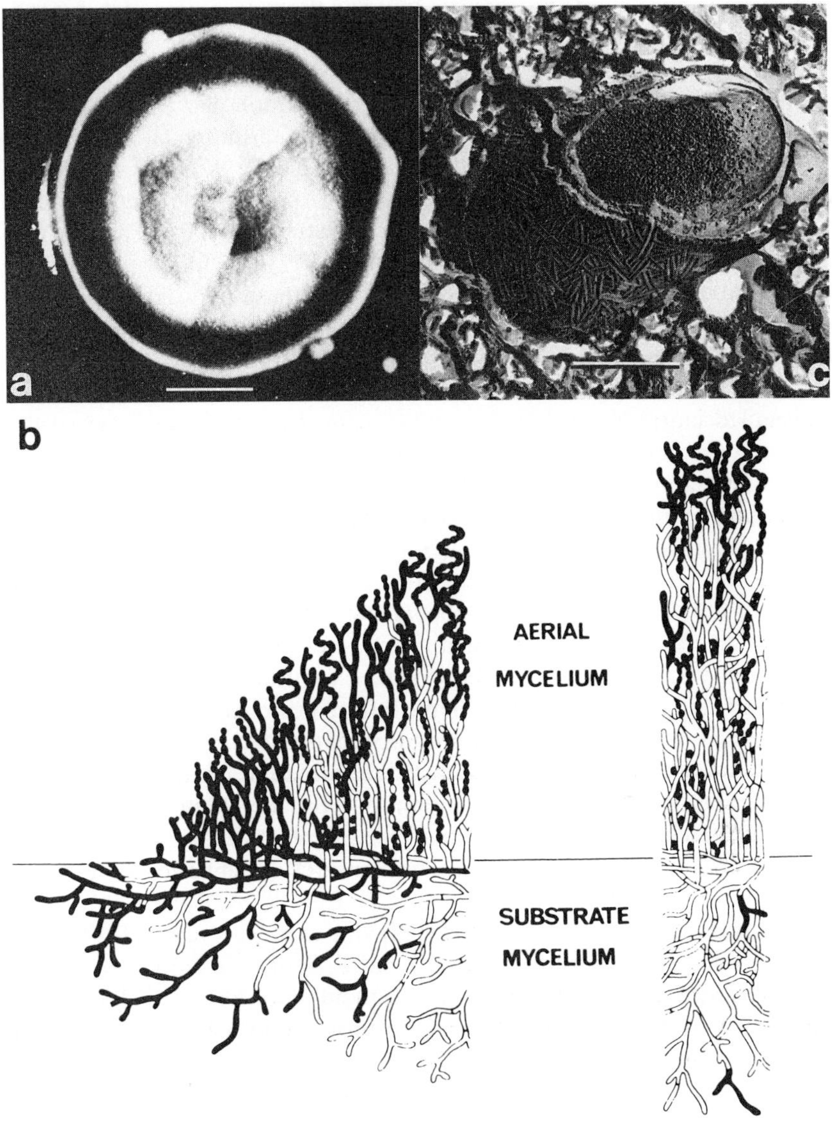

Fig. 2. Structural aspects of the streptomycete colony. a. A streptomycete colony growing on a solid medium. The bar represents 1 mm. b. A diagrammatic representation of a cross-section through a streptomycete colony (Wildermuth, 1970). c. A freeze-fractured, freeze-etched spore demonstrating the double rodlet pattern and the multiple wall structure (Chater & Merrick, 1979). The bar represents 0.5 μm.

and the spores at the bottom of the spore chain were often found to be devoid of spines, although disorientated spine precursors were present. Hence, hairy and smooth spores were found in these streptomycetes, depending on the position of the spore in the spore chain.

The role of the decorations on streptomycete spores is not clear. The spores of streptomycetes appear to be very hydrophobic and it has been suggested that this hydrophobicity is important in the dispersal of the spores by water on the surface of water droplets (Ruddick & Williams, 1972). The observation that removal of the sheath from spores made them wettable implies that the sheath may, therefore, be important in spore dispersal (Williams *et al.*, 1972). Ruddick and Williams (1972) also proposed that the spore coat ornaments, in those species that bear them, may aid the dispersal of the spores in soil. Altenatively, or additionally, the sheath may have a role in the observed desiccation resistance of streptomycete spores (Chater *et al.*, 1991).

Detailed morphological studies of *S. coelicolor* A3(2) spores revealed a number of structural layers around the spores. Smucker and Pfister (1978) reported: 'Mature spore envelope layers from the inner surface to the external surface are plasma membrane, inner spore wall, outer spore wall, rodlet mosaic, an undefined granular matrix and the sheath'. In a number of early reports this rodlet mosaic (Figure 2(*c*)) was thought to be a component of the sheath. The rodlet mosaic appears to be built up of individual fibrils that, rather surprisingly, are made up of chitin, a polysaccharide not usually associated with prokaryotes. The evidence for the involvement of chitin was: purification and comparison of the infrared spectrum and X-ray diffraction pattern with that of crab chitin; dissociation of the mosaic *in situ* by chitinase treatment; and disruption of the formation of the rodlet mosaic in a number of streptomycetes by chitin synthase inhibitors (Smucker, 1984*a*,*b*; Smucker & Pfister, 1978; Smucker & Simon, 1986).

Guijarro *et al.* (1988) reported the presence of a number of Spore-Associated Proteins (Saps) that could be released from *S. coelicolor* A3(2) spores following treatment with detergent (SDS) and a reducing agent (DTT). Five such proteins were identified (SapA-E). SapC, SapD and SapE have proved to be encoded by plasmid SCP1 (Fig. 3) and since loss of SCP1 did not lead to loss of ability to sporulate these proteins must be inessential. The remaining two proteins, SapA and SapB, of 13 and 3 kDa molecular mass, respectively, appear to be induced during aerial mycelium formation and to be specific to aerial hyphae and spores. SapB isolated from spores reacts as a glycoprotein to Schiff's reagent (Willey *et al.*, 1991).

Microscopic aspects – hyphal septation

The septa of substrate hyphae are oblique and irregular, enclosing compartments that usually contain many nucleoids. In contrast, the septa of aerial

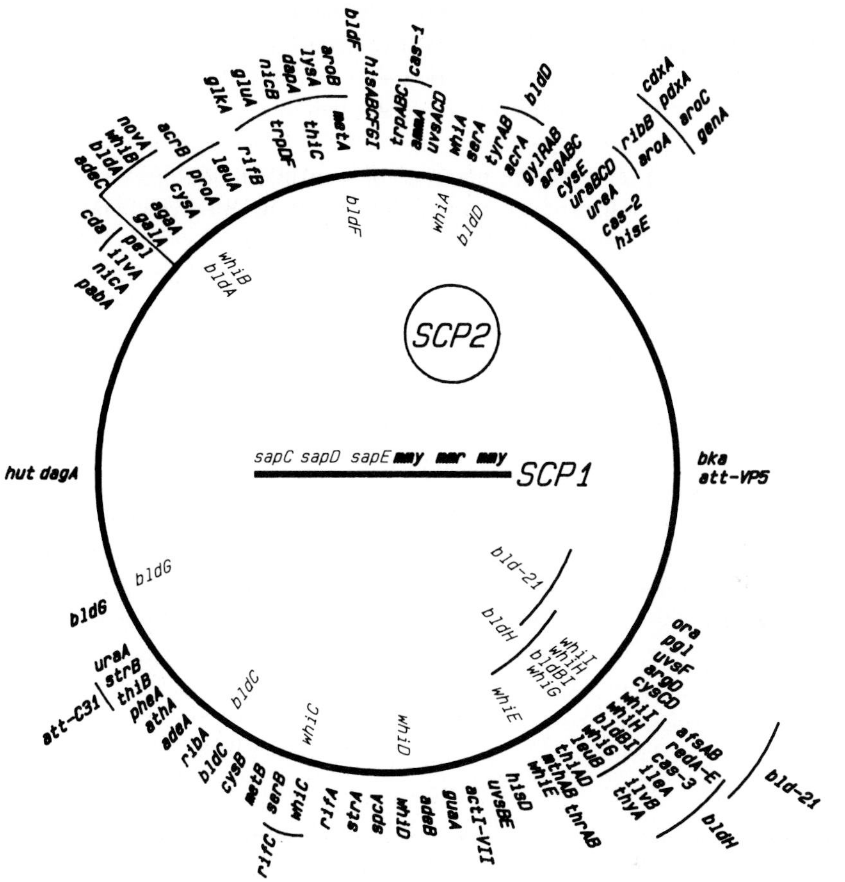

Fig. 3. The genetic map of *Streptomyces coelicolor* A3(2). Those genes concerned with development are repeated within the circle. The bars represent ambiguous gene positions. SCP1 is a large (410–560 kb) linear plasmid which encodes methylenomycin production (*mmy*) and resistance (*mmr*) and three of the five spore associated proteins. SCP2 is a smaller (31 kb) circular plasmid. See Hopwood *et al.* (1985) for explanation of other symbols.

hyphae formed during sporulation are perpendicular to the cell wall and include one nucleoid per cell compartment (Chater & Hopwood, 1973). These differences reflect the fact that aerial hyphal compartments are destined to separate whilst the substrate hyphal compartments are not. However, the cross walls in the latter would limit loss of cellular components from other compartments during lysis. Manzanal and Hardisson (1978) surveyed streptomycete aerial hyphae septation processes. They proposed three distinct septation patterns which seemed to have application to phylogenetic studies. Brana, Manzanal & Hardisson (1981) reported that sporulation septa, which formed initially as annuli around the cell wall and then extended inwards, appear to be made of a modified peptidoglycan that was resistant to lysozyme treatment. Ephemeral, lysozyme-sensitive material was deposited within the cell at the point of annulus formation, but this was degraded during spore maturation and was presumed to represent temporary organizing or storage structures.

Strunk (1978) reported the presence of microplasmodesmata connecting the cytoplasm of adjacent cell compartments of substrate hyphae in *S. melanochromogenes*. Microplasmodesmata also perforated the septa of immature spore chains but were sealed before the spore chains broke up. The structure and size of the microplasmodesmata were the same as those observed between vegetative cells and heterocysts of filamentous cyanobacteria and it was postulated that they allowed intercellular transport.

THE STREPTOMYCETE LIFE CYCLE

Germination

Simply wetting streptomycete spores with distilled water induces a low level of respiration, i.e. QO_2s of 3.9 and 10.8 μl 0_2 h^{-1} mg dry wt^{-1}, in *S. viridochromogenes* and *S. antibioticus*, respectively (Hirsch & Ensign, 1978; Salas & Hardisson, 1981). The wetted spores did not germinate but the ease of inducing respiration implied that dormancy was shallow. This suggestion appeared to be confirmed by the presence of high levels of AMP, ADP and ATP, when compared with *Bacillus* endospores (Hirsch & Ensign, 1978; Hardisson *et al.*, 1984). This probably contributes to the poor resistance qualities of streptomycete spores. Germination has been characterized as involving three distinct stages: the change from a phase bright to a phase dark appearance; swelling of the spore; and outgrowth of the germ tube. During the swelling process the spore coat has been seen to resolve into two distinct layers. The inner of these spore coat layers forms the germ tube wall, the outer layer being ruptured as outgrowth proceeds (Sharples & Williams, 1976; Hardisson *et al.*, 1978).

Germination of all streptomycetes can be initiated in complex media which include amino acids, nucleosides and vitamins. However, when attempts were made to obtain defined germination media, the requirements proved species specific. In *S. viridochromogenes* germination occurs most efficiently after a mild heat shock (Hirsch & Ensign, 1976*a,b*), whilst *S. antibioticus* spores do not require such treatment (Hardisson *et al.*, 1978). Spore darkening was shown to require the presence of divalent cations, usually Ca^{2+} (Eaton & Ensign, 1980), but Mg^{2+} and Fe^{2+} can substitute in some species (Hardisson *et al.*, 1978).

Grund and Ensign (1978) reported that CO_2 was an absolute requirement for spore darkening in minimal media. It was incorporated into oxaloacetate, presumably *via* pyruvate carboxylase. This implies that the role of CO_2 was to stimulate the Krebs cycle. This was confirmed by the observations that it can be replaced by Krebs cycle intermediates and that germination is inhibited by Krebs cycle inhibitors. Spore darkening can be inhibited by respiratory inhibitors but not by RNA, protein or DNA synthesis inhibitors (Hardisson *et al.*, 1978), which implies that endogenous carbon reserves, i.e. trehalose, are being mobilized (McBride & Ensign, 1987*b*).

Swelling of *S. antibioticus* spores leads to a 1.5-fold increase of volume without any concomitant increase in weight. This stage depends on the supply of an exogenous carbon source and RNA and protein but not DNA synthesis. The respiratory quotient increases 110-fold above the wetted dormant spore value to 1210 $\mu l\ 0_2\ h^{-1}$ mg dry wt^{-1} (Hardisson *et al.*, 1978). Later work has shown that a glucose transport system is present in the spores, whilst fructose and galactose transport systems are induced during outgrowth (Salas & Hardisson, 1981).

Outgrowth of *S. antibioticus* spores requires DNA, RNA and protein synthesis, during which exogenous carbon and nitrogen sources are required. There is a concomitant increase in cell mass, but a reduction of respiratory quotient to that of vegetative cells (680 $\mu l\ 0_2\ h^{-1}$ mg dry wt^{-1}) (Hardisson *et al.*, 1978).

Studies on the germination process of *S. viridochromogenes* revealed a potential biological role for some streptomycete antibiotics (Grund & Ensign, 1985). These authors reported that the spores contain a germination inhibitor that appears to inhibit calcium-dependent ATPase activity reversibly in spores. Initial respiratory activity of germinating spores appears to be uncoupled from ATP synthesis, coupling beginning when the inhibitor is released from the spore. It has been suggested that the inhibitor might function to delay germination of spores during the dispersal process.

Studies on macromolecular synthesis and turnover in *S. antibioticus* spores have revealed some surprising results. Using radioactive tracer experiments to study the RNA population as a whole in streptomycete spores has revealed that a number of species contain stable mRNA ($t_{1/2}$ approximately 20 minutes). During the first hour following germination this mRNA is increasingly turned over. When the spores are wetted, even in the absence of spore darkening, translation of these mRNAs begins. The proteins encoded by these mRNAs appear to be similar in the different strains studied (Quiros, Hardisson & Salas, 1985). Surprisingly, it appears that these proteins are degraded soon after synthesis (Guijarro *et al.*, 1983). The biological significance, if any, of these events is difficult to assess. It is possible that this mRNA represents a pool of undegraded molecules left over from the spore formation process.

Substrate hyphae growth and branching

The primary germ tube which emerges from a spore is followed by the emergence of a second germ tube from the same spore some time later. Branching of these primary cell filaments leads to the formation of the substrate mycelium. Recently Wolf and Schoppmann (1989) reported that streptomycete hyphae could grow through filter pores less than one-fifth the width of the normal cell, although often the emerging hypha produced morphological variations such as helical filaments.

Light microscope observations show that *S. coelicolor* A3(2) hyphae grow by apical extension. Hyphal growth is linear but exponential growth of the colony can be achieved by increasing the number of hyphal branches exponentially (Allan & Prosser,1983, 1985). Electron microscopic autoradiography with a tritiated cell wall precursor indicated that cell wall synthesis occurred exclusively in a localized area behind the hyphal tip. The hyphal tip was also shown to be semi-ellipsoid rather than hemispherical as had previously been presumed (Gray, Gooday & Prosser, 1990). Miguelez *et al.* (1988) reported that *S. antibioticus* hyphae increase in length at a linear rate but that, in addition to the cell wall synthesis at the apex of the hypha, there was a more diffuse incorporation of cell wall material to its rear.

Exclusion of crystal violet stained cytoplasm from discrete regions across the hyphae was taken to indicate septa. These regions were not present at recent hyphal branch points, but were present to the rear of the older branches rather than in front of them (Prosser, Gray & Gooday, 1988; Gray *et al.*, 1990). This observation has recently been confirmed in *S. tendae* following computer enhancement of epifluorescence microscope images taken of acridine orange stained hyphae (Reichl *et al.*, 1990).

DNA staining of *S. coelicolor* A3(2) revealed small, rounded, discrete structures that were interpreted as individual nucleoids. Paired, dumb-bell shapes were thought to be nucleoids caught in the act of division. Near the hyphal branch points the discrete nucleoids appeared to aggregate. In total, 50% of the cytoplasm appeared to be nucleoid material. Using light microscope autoradiography Gray *et al.* (1990) demonstrated that RNA synthesis occurred uniformly throughout the mycelium. The presence of DNA synthesis in old and new hyphae was also observed. However, shorter labelling periods revealed that DNA synthesis was more localized than RNA synthesis and the majority occurred in the hyphal tip proximal regions. The observation that hyphal extension is linear but DNA replication is exponential implies that there must be a mechanism for segregation of nucleoids into hyphal branches.

Following microscopic analysis of a large number of colonies growing under a variety of nutrient conditions, Allan and Prosser (1985) concluded that growth of *S. coelicolor* A3(2) colonies was limited, not by nutrient limitation, but rather by the production of 'staling' compounds. This conclusion arose from the observation that radial extension of the colony decreased as the colony size increased. These staling compounds could have been related to the secondary metabolites observed to be produced during growth under these conditions. To test this hypothesis Allan and Prosser (1987) grew *S. coelicolor* A3(2) on a cellophane membrane under which a continuous supply of nutrients was pumped. Therefore, nutrient supply was constant but staling compounds would be washed away. Under these conditions, the colony diameter increased at a constant rate, independent of the diameter of the colony. Cellophane-growth did lead to a decrease in the rate of hyphal extension and hence the rate of radial growth of the colony, compared to colonies growing on agar; however, the hyphal branching rate increased so that they were more compact than agar-grown colonies. The authors proposed that colonies secrete staling compounds that normally suppress hyphal branching so allowing concentration of growth into primary hyphal extension.

The role of the substrate mycelium

The observations that substrate hyphal lysis and secondary metabolite formation occur at the onset of aerial mycelium formation led to the suggestions that the substrate mycelium acts as a source of nutrient during aerial hyphae formation and that secondary metabolites with antibiotic activities might serve to stop other microorganisms from exploiting this food source (Chater & Merrick, 1979). Mendez *et al.* (1985) reported that macromolecules present in the substrate mycelium of *S. antibioticus* were mobilized into the aerial mycelium. The activities of two serine proteases and a cell wall-associated DNAase were correlated with the onset of sporulation in *S. lactamdurans* and *S. antibioticus*, respectively (Ginther, 1979; de los Reyes-Gavilan *et al.*, 1991). It is tempting to speculate that such enzymes are involved in the mobilization of macromolecules in substrate mycelium (Table 1).

The formation of the aerial mycelium

During growth on solid minimal medium with glucose as sole carbon source, aerial hyphal initials of *S. coelicolor* A3(2) were formed approximately 350 μm from the margin of the colony (Allan & Prosser, 1985). However, these authors did not specify the characteristics they used to identify them as aerial hyphae. Granozzi *et al.* (1990) reported that, when colonies of *S. coelicolor* A3(2) growing on solid media were about to initiate aerial mycelium formation, a transitory cessation of growth was observed. During this growth pause DNA, RNA and protein synthesis also ceased. Macromolecular synthesis and growth were then reinitiated and the formation of the aerial mycelium then became apparent. Mendez *et al.* (1985) had observed

Table 1. *Developmentally regulated biochemical markers of streptomycetes.*

Marker	Streptomycete	Comment	Reference
Glycogen metabolism	*S. antibioticus*	Glycogen accumulated in substrate hyphae during aerial mycelium initiation and was then degraded. A further round of synthesis and degradation occurred in septated aerial hyphae.	Brana *et al.*, 1986
Trehalose synthesis	*S. antibioticus*	Trehalose synthesis occurred in substrate hyphae, aerial hyphae and spores but the disaccharide accumulated in the latter two.	Brana *et al.*, 1986
Spore protesin (SapA, SapB, SapC, SapD and SapE)	*S. coelicolor* A3(2)	Spore associated proteins which can be released by detergent treatment. SapC, SapD and SapE are encoded by SCPI and are dispensable. The SapA gene has been cloned. SapB acts as a stimulator of aerial mycelium formation in a *bld* mutant (Table 2).	Guijarro *et al.*, 1988 Willey *et al.*, 1991
Chitin synthesis	*S. coelicolor* A3(2)	Chitin forms the rodlet assembly on the spore coat in a number of streptomyces species.	Smucker & Pfister, 1978
Spore pigment	*S. coelicolor* A3(2)	A brown polyketide pigment associated with the spore. Antibiotic function not known.	Davis & Chater, 1990
Sporulation pigment	*S. venezuelae*	A pH indicating pigment which induced submerged sporulation in media not normally capable of it.	Scribner *et al.*, 1973
Serine proteases	*S. lactamdurans*	Coordinate expression with antibiotic induced at the end of exponential phase. A *bld* mutant was unable to produce proteases or antibiotic.	Ginther, 1979
Cell wall-associated DNAse	*S. antibioticus*	Induction preceded aerial mycelium formation Nutritional repression of the latter repressed DNAase function also.	de los Reyes-Gavilan *et al.*, 1991
Penicillin binding proteins	*S. griseus*	A 27 kDa PBP was present in spore and substrate hyphae membranes. A 58 kDa PBP was unique to the former and a 38 kDa PBP was unique to the latter.	Barabas *et al.*, 1988

a similar but less dramatic pause in growth rate before commencement of aerial mycelium formation in *S. antibioticus*.

A number of carbon storage compounds have been implicated in streptomycete development. Glycogen granules accumulate in aging substrate hyphae but disappear during aerial mycelium development, only to reappear within recently septated aerial hyphae and finally disappear during spore maturation (Brana *et al.*, 1986; Chater *et al.*, 1991). Trehalose is a glucose disaccharide which acts as the spore carbon storage compound of streptomycetes and is also believed to have a role in spore resistance (McBride & Ensign, 1987*a,b*). Synthesis is found in all cell types but accumulation is greatest in aerial hyphae and spores (Table 1).

ANALYSIS OF STREPTOMYCETE SPORULATION

Sporulation of streptomycetes in liquid culture

Because streptomycetes are multicellular organisms with a number of differentiated cell types physically attached to one another, it is difficult to carry out biochemical and physiological analysis of them. If synchronous sporulation on a large scale was achievable, then such analyses would be more meaningful. Although aerial mycelium formation, and hence sporulation, is usually suppressed in liquid media, there were early reports that some streptomycete strains were capable of forming spores in liquid culture (Carvajal, 1947); this was exploited in the case of *S. griseus* (see below: *Factor C*).

Kendrick and Ensign (1983) reported that nutrient shiftdown, involving either nitrogen or phosphate, greatly stimulated sporulation of *S. griseus* in submerged culture, whilst nutrient-rich media suppressed it. However, the spores that formed (submerged spores) had thinner walls and were more lysozyme-sensitive than those produced on solid media (aerial spores). The criteria used to classify them as spores were: their microscopic appearance; the presence of trehalose; the spore storage catabolite; their low endogenous respiration rate; the stimulatory effect of mild heat treatment on germination and outgrowth; and the observation that a non-spore-forming mutant did not produce them. Two-dimensional SDS PAGE comparison of the proteins present in aerial spores and submerged spores revealed the presence of a number common to both but also some unique to aerial spores.

Koepsel and Ensign (1984) reported that, when arginine was the sole carbon and nitrogen source and was present in growth-limiting amounts, *S. viridochromogenes* and *S. coelicolor* A3(2) spores germinated and formed a primary hypha. This hypha then developed into a spore chain. This microcycle sporulation is very similar to that seen in many streptomycetes during growth within some soil particles (see above: *Macroscopic aspects*). The microcycle spores formed by *S. viridochromogenes* had the species-specific spore coat decorations, the heat, lysozyme and sonication resistance, the endogenous respiration levels, and germination characteristics of aerial spores. They differed in trehalose content but this could be remedied by incubation in buffer containing glucose.

Phosphate downshift, achieved by addition of Ca^{2+} to the medium, induced submerged sporulation by a number of *Streptomyces* species. (Daza *et al.*, 1989). Glazebrook *et al.* (1990) and Huber, Piper & Mertz (1987) reported conditions that promoted submerged sporulation in other streptomycetes. This area of research is one that holds promise for the future. However, there is still much to be done to demonstrate that submerged spores not only appear to be structurally and biochemically similar to aerial spores, but also that the pathways of development are the same or similar. Babcock and Kendrick (1988) recently reported that some classes of *S. griseus* mutants were blocked in aerial and submerged sporulation. However, other mutant classes were blocked only in aerial spore formation, although it should be noted that these mutants could be induced to sporulate on solid medium if the carbon source was changed.

Genetic analysis of sporulation

Various approaches have been developed to identify the genes that play a role in development and differentiation. One way is to isolate mutations that block the life cycle at specific points (forward genetics). Another approach is to identify a protein or enzyme which is developmentally regulated and then clone the gene that encodes it (reverse genetics). A more recent approach is to identify promoters that are activated only during sporulation and/or only in specific cell types and then identify the gene transcribed from the promoter (lateral genetics).

Forward genetics has been used extensively to study streptomycete development. The use of *S. coelicolor* A3(2) as a model of streptomycete development has proved very powerful because two classes of mutant that were affected in their ability to form spores could easily be discerned by simple visual inspection of mutant colonies. Mutant colonies that have lost the hairy appearance of the wildtype and appear flat and shiny are referred to as bald (*bld*) or aerial mycelium minus (Amy^- or Am^-). I will use the former term. White (*whi*) mutants of *S. coelicolor* A3(2) produce an aerial mycelium but it is not the characteristic grey/brown colour of the wildtype.

It is tempting to look upon *bld* mutants as those that are blocked in the initiation of aerial mycelium formation, whilst *whi* mutants are blocked at different stages in the transformation of aerial hyphae into spore chains. However, it should be noted that some *bld* mutants, e.g. *bldA* and *bldD*, produce prostrate, sheathed and perpendicularly septated hyphae in addition to substrate hyphae (Chater & Merrick, 1979). Therefore, the block appears to affect erection and maturation of aerial hyphae rather than their initiation.

Bald mutants

The first systematic study of *bld* mutants of *S. coelicolor* A3(2) was made by Merrick (1976) who identified three different phenotypic classes spread between four different genetic loci (*bldA*, *bldB*, *bldC* and *bldD*) (Table 2, Fig. 3). The mutants were identified on complex media and all retained their phenotype on minimal medium with glucose as sole carbon source. However, three classes, *bldA*, *bldC* and *bldD*, could be induced to sporulate on minimal media in which glucose had been replaced by an alternative carbon source such as mannitol. If both mannitol and glucose were present then the Bld$^-$ phenotype was seen, which led to the conclusion that sporulation in these mutants was glucose-repressible. Other classes of *bld* mutants have since been isolated and three of these (*bldE*, *bldG* and *bldH*) were capable of glucose-repressible sporulation. Of the other three loci that have been tested, *bldB* appears to be a gene that is not present in streptomycetes other than *S. coelicolor* A3(2) and *bldI* maps closely to *bldB* (Table 2 and Fig. 3).

The possibility arises, therefore, that there are two pathways to aerial hyphae formation and hence sporulation in *S. coelicolor* A3(2), one of which is glucose-repressible. As far as is known by the author no one has attempted to isolate mutations in a *bldA* background that abolish sporulation on mannitol. Would an outcross of any new mutations (excluding new *bldBI* alleles) into a wildtype background give rise to a Bld$^-$ phenotype? Glucose repression of primary metabolism has been demonstrated in *S. coelicolor* A3(2) and mutants that had lost this system proved to have lost glucose kinase (*glkA*) (Ikeda *et al.*, 1984). Would a *bldA glkA* double mutant be Bld$^+$ on glucose media?

Two *bld* genes, *bldA* and *bldB*, have been cloned. When sequenced, the *bldA* gene appeared to encode a tRNA which recognised the rare leucine codon, TTA (Lawlor, Baylis & Chater, 1987). This codon has only been found in streptomycete genes involved in antibiotic production and resistance (Chater, 1989) and an *S. griseus* development gene (Babcock & Kendrick, 1990). Promoter analysis demonstrated that initiation of *bldA* gene transcription occurred in late growth phase, about the time of initiation of aerial hyphae formation (Lawlor *et al.*, 1987).

The *bldB* genetic locus proved to be complex. A DNA clone was isolated that complemented a number of *bld* alleles that had been mapped to the *bldB* region. However, two further alleles, *bld-28* (Harasym *et al.*, 1990) and *bld-249* (Champness, 1988), were not complemented by the clone. It was proposed that these two alleles affected a new *bld* gene, *bldI*. Subcloning of the *bldB* clone revealed that different mutations did not have the same complementation pattern, implying that there were two closely linked genes which gave rise to mutants with the *bldB* phenotype (Harasym *et al.*, 1990).

A number of Bld$^-$ mutants of *S. griseus* have been isolated. One class

Table 2. *Genes of* Streptomyces coelicolor *A3(2) identified by bald phenotype mutations*

Gene	Phenotype and Comments*	Reference
bldA	Wrinkled, soft, fragmenting colonies. Prostrate, sheathed, aerial hyphae present. Sporulation on mannitol MM. Undecylprodigiosin production only on low phosphate media. *bldA* has been cloned and identified as a tRNA with a TAA anti-codon (TTA codon).	Merrick, 1976; Lawlor *et al.*, 1987; Guthrie & Chater, 1990
bldB	Smooth, hard, non-fragmenting colonies. No evidence of aerial hyphae. No sporulation on any media tested, but some immature aerial hyphae formed on mannitol MM. One gene cloned, some evidence of genetic heterogenicity. Gene cloned present only in *S. coelicolor* A3(2).	Merrick, 1976; Harasym *et al.*, 1990
bldC	Smooth, non-fragmenting colonies. No evidence of aerial hyphae. Sporulation on mannitol MM. Antibiotic production apparently unaffected. Only one allele identified.	Merrick, 1976
bldD	Wrinkled, soft, fragmented colonies. Prostrate, sheathed, aerial hyphae present. Sporulation on mannitol MM. No antibiotic production. Only one *bldD* mutant isolated.	Merrick, 1976
bldE	Smooth but sculpted colonies. Undecylprodigiosin production. Overproduction of agarase. Isolated as able to grow on agar as sole carbon source in the presence of homoserine (Hmr^r). Sporulation on mannitol MM. Same as *bldF*?	Hodgson, 1980
bldF	Smooth but sculpted colonies. Undecylprodigiosin production. Overproduction of agarase. No sporulation on any media tested. Same as *bldE*?	Puglia & Cappelletti, 1984; Chater, 1989
bldG	Smooth, soft, fragmenting colonies. Some evidence of aerial hyphae but no spores. Sporulation on mannitol MM. No antibiotic production.	Champness, 1988
bldH	Smooth, hard, fragmenting colonies. No antibiotic production. Sporulation and antibiotic production on mannitol medium.	Champness, 1988
bldI	Colony morphology and properties similar to *bldB*. Maps close to the *bldB* locus but not complemented by DNA clones that complement true *bldB* mutants.	Champness, 1988; Harasym *et al.*, 1990
bld-830	Smooth, soft colonies. No antibiotic production. Placement next to wild type colonies led to the development of a zone of sporulation and antibiotic production. 'Sporulation Factor' could pass through dialysis membrane. Originally thought to derive from *bldC*. Fertility status ruled out this possibility.	Hodgson, 1980; Wood & Hodgson, unpublished.
bld-21	Aerial hyphae form in close proximity to wildtype and *whi*, but not *bld*, colonies. SapB protein induced aerial hyphae formation.	Willey *et al.*, 1991
bld-17	Aerial hyphae form when in close proximity to wildtype, *whi* and some *bld* strains. Related to *bld-830*?	Willey *et al.*, 1991

*Phenotype on complex media and minimal medium with glucose as sole carbon source.

(*afsA*) regained the ability to form aerial mycelium when they were placed close to wildtype cells (Hara & Beppu, 1982). These mutants have lost the ability to synthesize but not to respond to A-factor, a butyrolactone (see below). Babcock and Kendrick (1988, 1990) isolated three further classes of *S. griseus* Bld$^-$ mutants and characterized them phenotypically and genetically. Two classes could be induced to sporulate by growth on solid minimal medium and also to form submerged spores in liquid media, though less efficiently than wildtype.

A DNA clone was obtained that complemented one of these classes and homologous DNA fragments could be identified in other streptomycetes. Sequencing and S1 nuclease analysis revealed a promoter followed by an open reading frame (ORF) for a 55.5 kDa protein that contained a rare TTA codon (see above). There was evidence of a second ORF encoding a 49.5 kDa protein transcribed by an independent promotor within the first ORF. Analysis of transcription revealed that both ORFs were transcribed during vegetative growth but following initiation of submerged sporulation the 49.5 kDa ORF promoter was inactivated. Analysis of the predicted amino acid sequence of the 55.5 kDa protein revealed a possible DNA binding site and similarity to the *E. coli nusA* gene product, which is involved in transcription anti-termination.

A number of auxotrophic mutants have been isolated that are Bld$^-$. The first of these mutants described in detail were those affecting the *argG* (argininosuccinate synthetase) genes of *S. alboniger, S. scabies* and *S. violaceusruber* (Redshaw *et al.*, 1979). In many cases the loss of the *argG* gene was accomplished by the deletion of a large segment of the streptomycete chromosome, thus raising the possibility that the Bld$^-$ phenotype was not due to loss of the *argG* function *per se*, but rather to the loss of some linked gene. This possibility was excluded by Meade (1985), who demonstrated the concomitant restoration of the Arg$^+$ nd the Bld$^+$ phenotypes of an *argG* mutant of *S. cattleya* with a short DNA fragment bearing the *argG*$^+$ gene. The discovery that *S. fradiae* mutants lacking ornithine carbamoyltransferase (*argG*) were also Bld$^-$ and that aerial mycelium, but not spore formation, could be restored by citrulline, implied that the ornithine cycle was important in aerial mycelium initiation/formation (Vargha, Karsai & Szabo, 1983).

White mutants

Fifty *whi* mutations were subjected to phenotypic and genetic analysis and shown to map to nine genetic loci (*whiA–whiI*) on the *S. coelicolor* A3(2) genetic map (Fig. 3). Following microscopic examination of the development of aerial hyphae in wildtype, single *whi* and double *whi* strains, it was proposed that the order of expression and role of the *whi* genes was: *whiG*; *whiH*; *whiA*, *whiB* and *whiI*; *whiD*; *whiF*; and *whiE* (Fig. 4). The proposers of this scheme (Chater & Merrick, 1979) warned that caution should be taken to avoid 'naive interpretation of epistasis data in terms of dependent

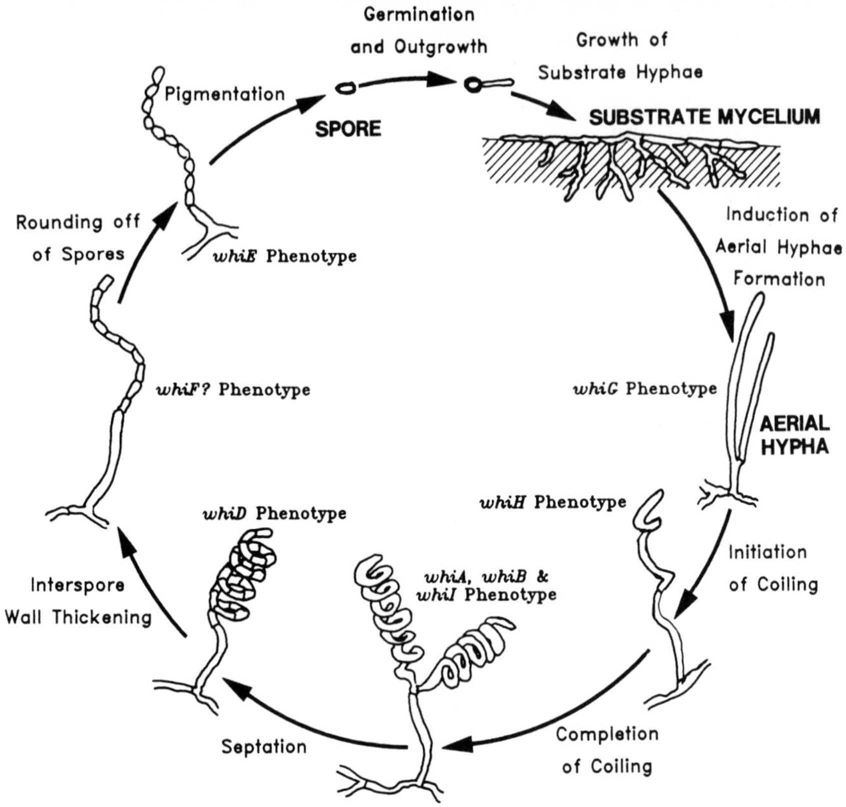

Fig. 4. The pathway of aerial hyphae development and sporulation in *Streptomyces coelicolor* A3(2). Modified from Chater and Merrick (1979).

sequences of gene expression when only morphological criteria are used'.

Several of the *whi* genes have now been cloned and sequenced. The *whiG* gene was cloned on a phage vector and complemented a number of alleles. When the complete wildtype gene was transferred to a multicopy plasmid it caused hyper-sporulation, i.e. formation of branched spore chains; formation of spore chains below the surface on solid media; and formation of spores in liquid culture. This implied that *whiG* was a positive regulator of sporulation (Mendez & Chater, 1987). Sequencing of the gene revealed an ORF homologous with RNA polymerase sigma factors, in particular the *Bacillus subtilis* motility sigma factor, σ^D (Chater *et al.*, 1989). It was noted that the *B. subtilis* sigma factor with the lowest level of similarity to *whiG* was σ^H. This might appear surprising as σ^H is the earliest sporulation sigma factor and so might be expected to be the functional equivalent of *whiG*. However, as discussed above, spores act as the dispersal phase of the streptomycete life cycle and in this sense streptomycete sporulation and bacillus motility can be looked upon as functionally equivalent. Probing the DNA from a wide range of *Streptomyces* species revealed the presence of a *whiG* homologue in all but one of them.

The *whiB* gene has also been cloned and sequenced (N. K. Davis, quoted in Chater, 1989) and shown to encode a small protein rich in charged amino acids. It is tempting to speculate that this protein is a structural molecule involved in sporulation septa formation. The gene locus thought to act latest in aerial hyphae maturation, *whiE*, has also recently been cloned and sequenced (Davis & Chater, 1990). Mutations in *whiE* produce spores that appear to be morphologically identical to wildtype spores. The only difference is that the spore mass lacks the brown/grey coloration. The DNA sequence revealed a complex locus of seven ORFs, of which six had high homology to genes encoding polyketide antibiotic synthesis. In fact, this gene cluster had already been cloned and shown to confer the ability to form a brown pigment. The implication is that the brown pigment is developmentally regulated. This discovery raises an interesting question: Is the brown sporulation pigment an antibiotic and, if so, is it functionally equivalent to the *S. viridochromogenes* germination inhibitor discussed above?

Reverse genetics

Guijarro *et al.* (1988) reported the identification and cloning of the gene for SapA, the 13 kDa spore-associated protein. They presented evidence that the protein was synthesized with a 38 amino acid sequence which was absent from the protein isolated from the spore. Promoter probe studies confirmed earlier S1 nuclease studies that had revealed that the gene was transcribed poorly during vegetative (substrate mycelium) growth, but was activated at the start of aerial mycelium formation (Schauer *et al.*, 1988). The identification of a number of developmentally regulated biochemical markers (Table 1) suggests a number of genes to which reverse genetics could be applied. Attempts to isolate the gene which encodes SapB led to the identification of two new *bld* genes, one of which responds to purified SapB by forming an immature aerial mycelium (Table 2).

Lateral genetics

The promoters from streptomycete genes that had previously been shown to be expressed at different times and in different cell types were coupled to a promoterless *lux* gene cassette and their temporal and spatial expression patterns monitored by the release of light. These studies confirmed previous studies. Once the reliability of the system has been demonstrated a DNA library was constructed in the *lux* promoter probe plasmid and a number of promoters were identified that showed interesting temporal and spatial expression patterns (Schauer *et al.*, 1988).

Now that a number of sporulation-specific promoters are available, they can be sub-cloned into suitable promoter probe vectors and introduced into different mutant backgrounds. The level and timing of promoter expression can then be compared to the wildtype situation. Promoter fusion can also potentially be used in the isolation of new regulatory genes by selecting up-regulatory and down-regulatory mutants. It should then be possible to construct gene dependency pathways, as has been done so successfully in *B. subtilis* (Losick *et al.*, 1989) and *Myxococcus xanthus* (Kaiser, 1989).

SPORULATION FACTORS

A special class of Bld$^-$ mutants have been identified in a number of *Streptomyces* species that can sporulate when placed in close proximity to the wildtype (see above and Table 2). One interpretation of this phenotype is that the mutant has lost the ability to produce a small molecular mass compound that is essential for aerial mycelium formation and that this molecule is supplied in *trans* by the wildtype. This raises the possibility that the compound has a regulatory role, i.e. it is a 'sporulation factor' that serves as a signal for the induction of sporulation by the colony as a whole. Alternatively, the compound may have a structural role, i.e. it is required for the assembly of an essential part of the aerial mycelium.

A number of researchers have noted that some streptomycetes produce extracts that specifically stimulate the formation of their own aerial mycelium. This phenomenon could be due to the presence of a sporulation factor or a structural component in the extract that is of limiting concentration within a colony. There is also the possibility that the compound differentially inhibits or stimulates substrate or aerial mycelium formation.

A-factor

This butyrolactone was the first streptomycete sporulation factor to be identified. It was originally characterized as a stimulator of streptomycin production and spore formation in *S. griseus*. The structure of the compound was determined and confirmed by synthesis (Fig. 5). As discussed above Bld$^-$ *afsA* mutants incapable of producing streptomycin or A-factor could produce spores and antibiotic if treated with natural or synthetic A-factor (Hara & Beppu, 1982). A number of structurally related butyrolactones, capable of substituting for A-factor in *S. griseus afsA* strains, were isolated from different streptomycetes (Fig. 5). A related factor was also found to be produced by *S. coelicolor* A3(2) (Anisova *et al.*, 1984). Two mutant classes of *S. coelicolor* A3(2) were isolated that were unable to produce the factor. One class (*afsA*) was unaffected in any other discernible way. The second class (*afsB*) had lost the ability to produce a number of secondary metabolites, i.e. actinorhodin and undecylprodigiosin (Hara *et al.*, 1983). The fact that neither of the mutants were Bld$^-$ demonstrates that the factor is not involved in the control of development in this strain.

A DNA clone from *S. bikiniensis* – renamed *S. griseus* (Horinouchi *et*

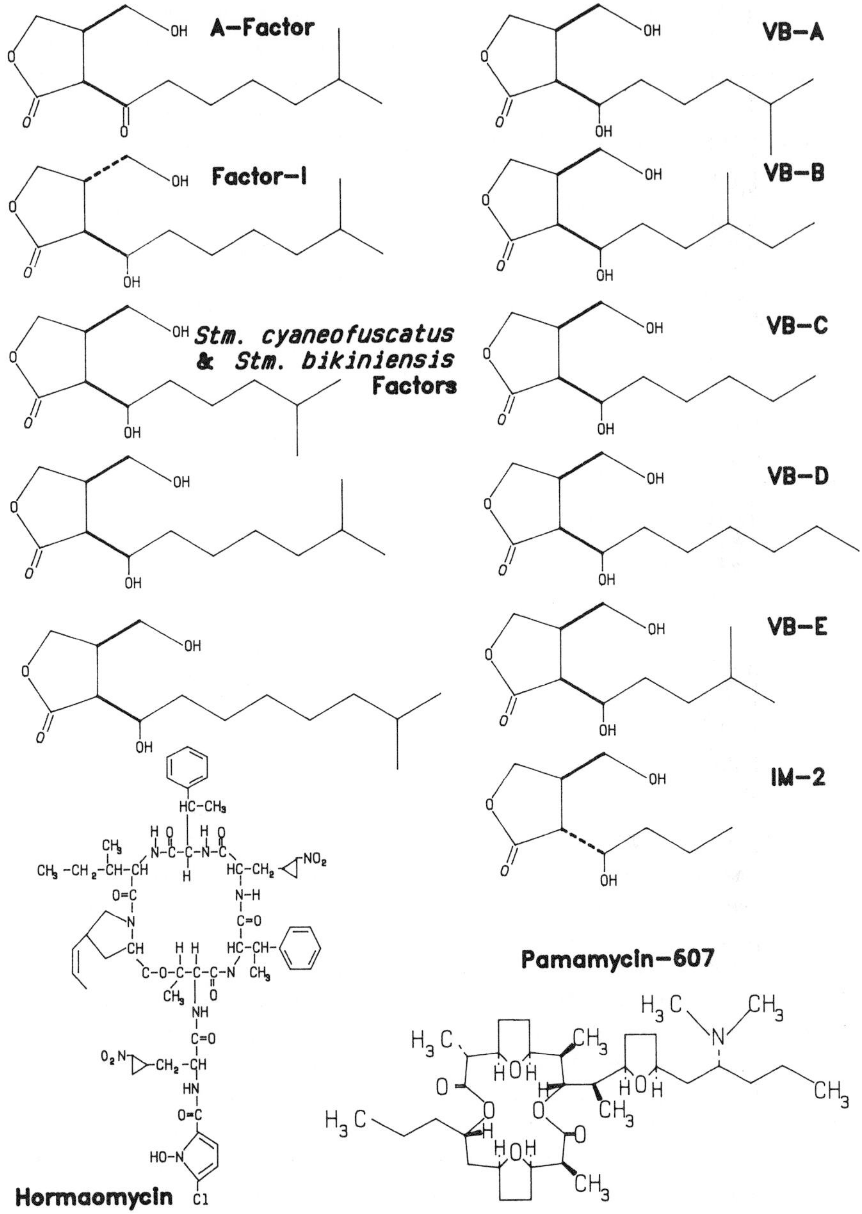

Fig. 5. Aerial mycelium and antibiotic inducing factors from streptomycetes. See: Sato *et al.* (1989) for references to A-factor, factor-I, *S. bikiniensis* and *S. cyaneofuscatus* factors and VB-A–VB-E factors and structure of IM-2; Kondo *et al.* (1987) for structure of pamamycin; and Rossner *et al.* (1990) for structure of hormaomycin.

al., 1989) – that complemented *S. bikiniensis, S. griseus* and *S. coelicolor* A3(2) *afsA* mutants was isolated and a 1.1 kb fragment was shown to encode the complementing function. Multiple copies of the fragment greatly stimulated A-factor production and endowed production on a streptomycete strain previously shown not to contain the gene nor to produce A-factor. This implied that the fragment encoded all the information necessary to synthesize A-factor from normal intermediary metabolites (Horinouchi, Kumada & Beppu, 1984; Horinouchi *et al.*, 1985). Sequencing of the gene revealed the presence of a 301 amino acid potential ORF which had no homology with any proteins in the data banks analysed (Horinouchi *et al.*, 1989).

A cytoplasmic 26 kDa protein that bound radiolabelled A-factor was identified in *S. griseus*. The protein bound A-factor very tightly ($K_d = 0.7$ nM) with 1 : 1 stoichiometry and did not bind related butyrolactones. The protein was in low abundance, i.e. about 37 copies per genome. No evidence of the protein could be found in *S. coelicolor* A3(2) (Miyake *et al.*, 1989). *S. griseus afsA*$^-$ pseudo-revertants were isolated which still did not produce A-factor, yet could sporulate and produce streptomycin. In all cases they had lost the A-factor binding activity. From this it was concluded that the binding protein acted as a repressor of sporulation and antibiotic production unless inactivated by A-factor. All of the mutants initiated sporulation sooner and produced antibiotic earlier and to a greater extent than the wild-type (Miyake *et al.*, 1990).

A-factor is the best example of a sporulation factor in streptomycetes. Production of the factor is temporally regulated (Hara & Beppu, 1982) but nothing about the regulation of the A-factor receptor protein has been reported. The implication is that A-factor synthesis is developmentally regulated and this then acts as a trigger for activation of the developmental gene cascade. Presumably this pheromone-based system permits the coordinated induction of development within the colony and, possibly, neighbouring colonies. It is perhaps surprising that a bacterium with a mycelial growth habit has evolved such a system when cells are already in such intimate contact. Such cell–cell signalling systems are to be expected in motile microorganisms with life cycles dependent on cellular cooperation, e.g. myxobacteria (Kaiser, 1989).

Virginiamycin butanolides

A family of five related compounds (VB-A–E) were identified as inducers of virginiamycin production by *S. virginiae* (Kondo *et al.*, 1989). Compounds that could substitute for the VB compounds were identified from a number of streptomycetes (Ohashi *et al.*, 1989). It should be noted that A-factor and related compounds could not substitute for the VB factors in bioassays and *vice-versa*. A cytoplasmic VB-C factor binding protein was identified in *S. virginiae* using radiolabelled factor. The protein was purified and had a molecular weight of 35.8 kDa and a high affinity ($K_d = 1.1$ nM) and specificity for VB-A, VB-B, VB-C, and VB-D. A-factor did not compete with VB-C binding. Again the protein seemed to be in low abundance, i.e. 30–40 binding sites per genome. Indirect evidence was presented that the protein had some DNA binding activity (Kim *et al.*, 1989, 1990). It should be noted that, whilst the virginiamycin inducing factor system has a large number of similarities to the A-factor system, there is no reported evidence that they play any role in morphological differentiation. To my knowledge no mutant unable to produce the VB factors has been isolated.

Recently a new factor, IM-2, which stimulated blue pigment production in *Streptomyces* sp. FRI-5, has been identified (Fig. 5). This factor has structural similarities to both the A-factor families and the VB families (Sato *et al.*, 1989).

Factor C

An *S. griseus* mutant, 52-1, was isolated that was unable to make submerged spores. However, when it was transferred into liquid media that had been conditioned by growth of the wildtype strain, sporulation was restored (Szabo *et al.*, 1984; 1988). Fractionation of the conditioned medium revealed a 34.5 kDa hydrophobic protein, factor C, with a high specific activity of sporulation induction, i.e. 8 ng per ml of purified protein was sufficient to repair fully the sporulation deficiency. Glucosamine made up 2% of the molecular mass of the protein (Biro *et al.*, 1980). Recently, polyclonal and monoclonal antibodies have been raised against factor C and these revealed that a wide range of streptomycetes produced a protein of similar weight and immunological properties. There were two exceptions, including *S. coelicolor* A3(2), in which the cross-reacting protein had a molecular weight of 70 kDa. The factor C homologues were usually secreted into the medium. The *S. griseus* wildtype produced 64 times more of the factor than did mutant 52-1 (Szeszak *et al.*, 1990).

The mode of action of factor C has also been investigated. Addition of the 0.7 ng per ml of factor to a number of streptomycetes, including 52-1, caused the release of K^+ after an initial lag time of 3 to 10 minutes. This release did not appear to be dependent on RNA or protein synthesis (Szeszak, Vitalis & Szabo, 1989). The role of factor C in the induction or repression of expression of specific developmentally regulated genes has as yet to be demonstrated. The observation that exogenously supplied proteins like factor C and SapB (Table 2) can repair developmental defects might reflect a structural role rather than a signalling one (Willey *et al.*, 1991). A protein with a signalling role has been identified in *Myxococcus xanthus*; this is the coincidentally named C-factor (Kim & Kaiser, 1990).

Pamamycin and hormaomycin

These two structurally unrelated compounds are antibiotics which have a stimulatory effect on the morphological differentiation of the producer. The pamamycin complex of at least eight homologous compounds is produced by *S. alboniger* and the unusual structure of one of these, pamamycin-607, has been solved (Fig. 5). The compound had anti-fungal and anti-bacterial (Gram positive) activity and at high concentrations (10 μg per disc) inhibited growth of the producer. At concentrations of 0.1 μg per disc significant stimulation of aerial mycelium formation by a spontaneous *bld* mutant was observed. Evidence was presented that the molecule could act as an anion transporter, as MnO_4^- could be transferred from water into benzene (Kondo *et al.*, 1988).

Hormaomycin is a complex cyclic peptide (Fig. 5) produced by *S. griseoflavus* that potently inhibits growth of *Arthrobacter* spp. and corynebacteria. In addition, at low concentration, the compound stimulated aerial mycelium formation in 5 out of 56 streptomycetes in which sporulation had been suppressed by addition of amino acids. It was also shown that it stimulated antibiotic production. It is always difficult to assess the physiological significance of stimulation of aerial mycelium formation by compounds with antibiotic function, especially on media that suppress development. Many antibiotics cause differential inhibition or stimulation of aerial and substrate mycelium, i.e. if the compound is an inhibitor of substrate mycelium formation but not aerial mycelium formation it would appear to stimulate the latter. The ultimate proof of relevance is to isolate a specific receptor protein(s), loss of which has a specific effect on development.

Calcium

A role for calcium in streptomycetes development was suggested following the observation that high concentrations (0.1 to 1.5 mM) of calcium acetate could induce aerial mycelium formation by *S. ambofaciens* on media that normally suppressed it. The specific calcium chelating agent EGTA was found to inhibit development. A range of streptomycete species were tested for inhibition of aerial mycelium formation by EGTA and stimulation by calcium. Four classes were found: two strains were induced by calcium; five strains were stimulated by calcium; 14 strains were inhibited by EGTA but were not stimulated/induced by calcium; and 15 strains were unaffected by calcium and EGTA (Natsume, Yasui & Marumo, 1989). It was not clear whether the effects, where present, were due to: phosphate sequestration; a structural requirement for calcium in aerial mycelium formation; or calcium acting as a cellular regulator.

IS AERIAL MYCELIUM FORMATION INDUCED BY STARVATION?

In many of the descriptions of the life cycles of actinomycetes the tacit assumption is made that starvation is the inducer of aerial mycelium formation. This assumption derives from the clear demonstration of starvation induction of resting stage formation by microorganisms more amenable to physiological study, e.g. bacilli, myxobacteria, azotobacter, cyanobacteria, etc. It is suggested here that there is little evidence for environmental induction of aerial mycelium formation by streptomycetes, although there is evidence that nutritional status can modify the timing, pathway and extent of spore formation (see below). It is hoped to show here that there is evidence that initiation of aerial hyphae formation is determined by the age and/or the position of the parental substrate hypha within the colony rather than its nutritional status. If this hypothesis is correct then a *Caulobacter* may prove a better model for streptomycete development and differentiation than *Bacillus*.

The formation of spores by *Bacillus* spp. can be looked upon as insurance against deterioration of the local environment. The swarmer cell of *Caulobacter*, on the other hand, is both an essential stage in the reproduction of the cell and the active dispersal form. As discussed above, the formation of spores is the dispersal strategy used by streptomycetes. It is perhaps relevant to remember in this context that *whiG*, the sporulation sigma factor of RNA polymerase, showed the greatest similarity to the motility sigma factor of all the sigma factors characterized in *Bacillus*. Like the streptomycete developmental cycle, the timing and morphological aspects of the *Caulobacter* developmental cycle are subject to modification by nutritional status (Poindexter, 1987). However, the analogy with *Caulobacter* can be taken too far. *Caulobacter* differentiation is an integral part of the cell cycle, which is clearly not the case for streptomycetes.

One observation that argues against the importance of nutritional status is that streptomycetes appear to be able to sporulate on rich media and on simple salts media without much alteration of timing. In addition it has been noted that if *S. coelicolor* A3(2) spores are spread onto cellophane discs and these are then placed on top of solidified minimal media the colonies go through the same growth cycle whether or not the medium is fresh or has been used to support several previous rounds of growth. Granozzi *et al.* (1990) reported that transfer of colonies on cellophane to fresh media did not alter developmental timing, when compared with colonies that were not transferred. In this context it is worth remembering the observation of Wildermuth (1970), that spores often germinated in the aerial mycelium, implying that nutritional limitation was not a factor.

Allan and Prosser (1987) examined colony formation on top of a semipermeable membrane that was continuously washed with nutrient medium on the underside. Under these conditions aerial mycelium and spore formation were substantially normal. Aerial hyphae appeared to be initiated closer to the edge of the colony. This 'earlier' initiation was correlated with the greater branching of substrate hyphae which implied that the extent of branching rather than nutrient limitation was the controlling element in aerial mycelium formation. This indicates that the age and position of the parental hyphae are the key to aerial hyphae initiation. It is difficult to assess the importance or otherwise of localized starvation within the colony.

Huber *et al.* (1987) reported that sporulation of *S. roseosporus* in submerged culture 'was not purposefully initiated by critical manipulation of either nutritional or environmental conditions'. Ensign (1988) re-examined the timing of submerged spore formation in *S. griseus* and came to the conclusion that starvation was not responsible for the initiation of microcycle sporulation, but it could affect a predetermined developmental program which had been initiated using defined nutritional conditions. Once spores had germinated and formed hyphae, sporulation occurred 26 hours later. The cells became committed to sporulation after 10 to 12 hours. Before this execution point, starvation inhibited growth and the cells became destined to lyse – the substrate mycelium programme? – but after that time starvation did not affect the timing of sporulation, but it did cause growth to cease.

The role of catabolite repression

Redshaw *et al.* (1976) reported that *S. alboniger* aerial mycelium formation was sensitive to glucose repression and that this repression could be reversed by the inclusion of adenine in the medium. They also discovered that accumulation of organic acids might also be involved. The fact that acid accumulation appears to be the cause of the 'glucose repression' might suggest that really this was a case of differential glucose toxicity. Surowitz and Pfister (1985) revealed that the acid which accumulated was pyruvate and that the reason for accumulation was that glucose stimulated glycolysis but not the citric acid cycle. Addition of adenine readjusted the balance between these two pathways which suppressed pyruvate production and so aerial hyphae were formed, although they did not mature into spores.

Coleman and Ensign (1982) reported that *S. viridochromogenes* aerial mycelium formation on minimal medium could be prevented by addition of casein hydrolysate, but only in the presence of high concentrations of phosphate (15 mM). Aerial hyphae formation, but not sporulation, could be recovered by addition of adenine. It is tempting to speculate that some form of metabolic imbalance was responsible for this phenomenon. In both these cases it seems that 'glucose repression' and 'casein hydrolysate repression' were examples of nutritional inhibition of development rather than examples of starvation induction. It is not yet clear how the observed glucose-repressible sporulation seen in the majority of *S. coelicolor* A3(2) *bld* mutants relates to this point (see above, Table 2). Does this streptomycete have two pathways, one of which is nutritionally controlled and one which is not?

The role of the stringent response

Amino acid starvation in streptomycetes leads to the formation of ppGpp and pppGpp, as in other bacteria. Ochi (1986*a,b*) reported a correlation between antibiotic synthesis with ppGpp accumulation and aerial mycelium formation with a fall in GTP levels in an uncharacterized streptomycete. He also reported that suppression of aerial mycelium formation by excess nutrient could be reversed by the inhibitor of GMP synthesis, decoyinine. A *rel* mutant still formed an aerial mycelium, but it was delayed in onset and extent. The same correlations were reported with *S. griseus* with the additional information that decoyinine induced submerged spore formation. Addition of decoyinine to *S. griseus* and *S. antibioticus rel* mutants restored their ability to produce submerged spores (Ochi, 1987*a,b,c*). A bicozamycin-overproducing mutant of *S. griseoflavus* was obtained which had an increased ability to form aerial hyphae, produced higher levels of ppGpp earlier than the wildtype, and had an accentuated decrease in GTP after nutritional shiftdown in liquid medium (Ochi *et al.*, 1988).

Glazebrook *et al.* (1990) reported that both submerged and surface sporulation by *S. venezuelae* could be inhibited by addition of peptone but that neither was reversed by decoyinine. However, a fall in GTP concentration was observed before submerged sporulation. More detailed studies of *relC* mutants of *S. griseus* and *S. coelicolor* A3(2) (Ochi, 1990*a,b*) confirmed that the delay in aerial mycelium formation and the Rel$^-$ phenotypes co-segregated. However, it was revealed that in *S. coelicolor* A3(2) the fall in GTP concentration observed in the wildtype still occurred in the *relC* mutant. This implies that fall in GTP concentration and hence nutritional control was not enough to initiate aerial mycelium formation. Recently, Strauch *et al.* (1991) reported that accumulation of ppGpp alone was not sufficient for induction of secondary metabolism in *S. coelicolor* A3(2).

When we consider that the *bldA* gene encodes a tRNA, it is perhaps not surprising that changes in ppGpp metabolism affect development. However, that effect may reflect the requirement for ppGpp in the expression of an essential gene that is otherwise developmentally regulated rather than a regulatory role for ppGpp. As discussed above, there is a danger of over- interpreting the effect of antibiotics, such as decoyinine, on strains that are developmentally blocked by nutritional excess (see *Pamamycin* and *Hormaomycin*).

It is potentially misleading to assume that inhibition of sporulation by a metabolite implies that starvation for that metabolite is what initiates sporulation. As discussed above that assumption can be invalidated by the phenomenon of differential nutrient sensitivity of different cell types. The observation of correlations between cellular parameters, such as nucleotide pool changes and differentiation events, can also be misleading, as the changes in concentration could be an *effect* rather than *a cause* of development initiation. For example, changes in cell membrane structure (Barabas *et al.*, 1988) could lead to changes in transport of nucleotide precursors, etc. In conclusion, I submit that the case for nutritional control of initiation of streptomycete differentiation is not proven.

ACKNOWLEDGEMENTS

I thank Dr Elizabeth Wellington, Professor Richard Losick and Professer Keith Chater for helpful discussions and permission to use unpublished information. I also thank Professor Chater for a critical reading of, and useful comments on, an early draft of this chapter.

REFERENCES

Allan, E. J. & Prosser, J. I. (1983). Mycelial growth and branching of *Streptomyces coelicolor* A3(2) on solid medium. *Journal of General Microbiology*, **129**, 2029–36.

Allan, E. J. & Prosser, J. I. (1985). A kinetic study of the colony growth of *Streptomyces coelicolor* A3(2) and J802 on solid medium. *Journal of General Microbiology*, **131**, 2521–32.

Allan, E. J. & Prosser, J. I. (1987). Colony growth of *Streptomyces coelicolor* A3(2) under conditions of continuous nutrient supply. *FEMS Microbiology Letters*, **43**, 139–142.

Anisova, L. N., Blinova, I. N., Efremenkova, O. V., Kozmin, Y. P., Onoprienko, V. V., Smirnova, G. M. & Khokhlov, A. S. (1984). Development regulators in *Streptomyces coelicolor* A3(2). *Biological Bulletin of the Academy of Sciences, USSR*, **11**, 75–84. (English translation)

Babcock, M. J. & Kendrick, K. E. (1988). Cloning of DNA involved in sporulation of *Streptomyces griseus. Journal of Bacteriology*, **170**, 2802–8.

Babcock, M. J. & Kendrick, K. E. (1990). Transcriptional and translational features of a sporulation gene of *Streptomyces griseus. Gene*, **95**, 57–63.

Barabas, J., Barabas, G., Szabo, I., Veenhuis, M. & Harder, W. (1988). Penicillin-binding proteins of protoplast and sporoplast membranes of *Streptomyces griseus* strains. *Archives of Microbiology*, **150**, 105–8.

Biro, S., Bekesi, I., Vitalis, S. & Szabo, G. (1980). A substance affecting differentiation in *Streptomyces griseus. European Journal of Biochemistry*, **103**, 359–63.

Brana, A. F., Manzanal, M. B. & Hardisson, C. (1981). Cytochemical and enzymatic characterization of the sporulation septum of *Streptomyces antibioticus. Canadian Journal of Microbiology*, **27**, 1060–5.

Brana, A. F., Mendez, C., Diaz, L. A., Manzanal, M. B. & Hardisson, C, (1986). Glycogen and trehalose accumulation during colony development in *Streptomyces antibioticus. Journal of General Microbiology*, **132**, 1319–26.

Carvajal, F. (1947). The production of spores in submerged cultures by some *Streptomyces. Mycologia*, **39**, 425–40.

Champness, W. C. (1988). New loci required for *Streptomyces coelicolor* morphological and physiological differentiation. *Journal of Bacteriology*, **170**, 1168–74.

Chater, K. F., (1989). Sporulation in *Streptomyces*. In *Regulation of Procaryotic*

Development, Smith, I., Slepecky, R. A. & Setlow, P. eds, pp. 277–99. American Society for Microbiology, Washington.

Chater, K. F., Bruton, C. J., Davis, N. K., Plaskitt, K., Soliveri, J. & Huarong, T. (1991). Gene expression during sporulation in *Streptomyces coelicolor* A3(2). In *Genetics and Product Formation in Streptomyces*, Noack, D., Krugel, H., & Baumberg, S. eds. Plenum, London (in press).

Chater, K. F., Bruton, C. J., Plaskitt, K. A., Buttner, M. J., Mendez, C. & Helmann, J. D. (1989). The developmental fate of *S. coelicolor* hyphae depends upon a gene product homologous with the motility σfactor of *B. subtilis*. *Cell*, **59**, 133–43.

Chater, K. F. & Hopwood, D. A. (1973). Differentiation in actinomycetes. In *Microbial Differentiation*, Ashworth, J. M. & Smith, J. E. eds, pp. 143–160. Society for General Microbiology Symposium 23. Cambridge University Press, Cambridge.

Chater, K. F. & Merrick, M. J. (1979). Streptomycetes. In *Developmental Biology of Prokaryotes*, Parish, J. H. ed., pp. 93–114, Blackwell Scientific Publications, London.

Coleman, R. H. & Ensign, J. C. (1982). Regulation of formation of aerial mycelia and spores of *Streptomyces viridochromogenes*. *Journal of Bacteriology*, **149**, 1102–11.

Davis, N. K. & Chater, K. F. (1990). Spore colour in *Streptomyces coelicolor* A3(2) involves the developmentally regulated synthesis of a compound biosynthetically related to polyketide antibiotics. *Molecular Microbiology*, **4**, 1679–91.

Daza, A., Martin, J. F., Dominguez, A. & Gil, J. (1989). Sporulation of several species of *Streptomyces* in submerged culture after nutritional downshift. *Journal of General Microbiology*, **135**, 2483–91.

Eaton D. & Ensign, J. C. (1980). *Streptomyces viridochromogenes* spore germination initiated by calcium ions. *Journal of Bacteriology*, **143**, 377–82.

Ensign, J. C. (1982). Developmental biology of actinomycetes. In *Overproduction of Microbial Products*, Krumphanzl, V., Sikyta, B. & Vanek, Z. eds, pp. 127–40, Academic Press, London.

Ensign, J. C. (1988). Physiological regulation of sporulation of *Streptomyces griseus*. In *Biology of Actinomycetes '88*, Okami, Y., Beppu, T. & Ogawara, H. eds, pp. 305–19. Japan Scientific Societies Press, Tokyo.

Ginther, C. L. (1979). Sporulation and the production of serine protease and cephamycin C by *Streptomyces lactamdurans*. *Antimicrobial Agents and Chemotherapy*, **15**, 522–6.

Glauert, A. M. & Hopwood, D. A. (1960). The fine structure of *Streptomyces coelicolor*. I. The cytoplasmic membrane system. *Journal of Biophysical and Biochemical Cytology*, **7**, 479–88.

Glazebrook, M. A., Doull, J. L., Stuttard, C. & Vining, L. C. (1990). Sporulation of *Streptomyces venezuelae* in submerged cultures. *Journal of General Microbiology*, **136**, 581–88.

Goodfellow, M. (1989). The actinomycetes I: Suprageneric classification of actinomycetes. In *Bergey's Manual of Systematic Bacteriology*, volume 4, Williams, S. T., Sharpe, M. E., & Holt, J. G. eds, pp. 2333–9. Williams & Wilkins, Baltimore.

Granozzi, C., Billetta, R., Passantino, R., Sollazzo., M. & Puglia, A. M. (1990). A breakdown in macromolecular synthesis preceding differentiation in *Streptomyces coelicolor* A3(2). *Journal of General Microbiology*, **136**, 713–16.

Gray, D. I., Gooday, G. W. & Prosser, J. I. (1990). Apical hyphal extension in *Streptomyces coelicolor* A3(2). *Journal of General Microbiology*, **136**, 1077–84.

Grund, A. D. & Ensign, J. C. (1978). Role of carbon dioxide in germination of spores of *Streptomyces viridochomogenes*. *Archives of Microbiology*, **118**, 279–88.

Grund, A. D. & Ensign, J. C. (1985). Properties of the germination inhibitor of

Streptomyces viridochromogenes spores. *Journal of General Microbiology*, **131**, 833–47.

Guijarro, J. A., Suarez, J. E., Salas, J. A. & Hardisson, C. (1983). Pattern of protein degradation during germination of *Streptomyces antibioticus* spores. *Canadian Journal of Microbiology*, **29**, 637–43.

Guijarro, J., Santamaria, R., Schauer, A. & Losick, R. (1988). Promoter determining the timing and spatial localization of transcription of a cloned *Streptomyces coelicolor* gene encoding a spore-associated polypeptide. *Journal of Bacteriology*, **170**, 1895–901.

Guthrie, E. P. & Chater, K. F. (1990). The level of transcript required for production of a *Streptomyces coelicolor* antibiotic is conditionally dependent on a tRNA gene, *Journal of Bacteriology*, **172**, 6189–93.

Hara, O. & Beppu, T. (1982). Mutants blocked in streptomycin production in *Streptomyces griseus* – the role of A-factor. *Journal of Antibiotics*, **35**, 349–58.

Hara, O., Horinouchi, S., Uozumi, T. & Beppu, T. (1983). Genetic analysis of A-factor synthesis in *Streptomyces coelicolor* A3(2) and *Streptomyces griseus*. *Journal of General Microbiology*, **129**, 2939–44.

Harasym, M., Zhang, L.-H., Chater, K. & Piret, J., (1990). The *Streptomyces coelicolor* A3(2) *bldB* region contains at least two genes involved in morphological development. *Journal of General Microbiology*, **136**, 1543–50.

Hardisson, C., Manzanal, M., B., Salas, J. A. Suarez, J. E. (1978). Fine structure, physiology, and biochemistry of arthrospore germination in *Streptomyces antibioticus*. *Journal of General Microbiology*, **105**, 203–14.

Hardisson, C., Guijarro, J. A., Suarez, J. E. & Salas, J. A. (1984). Early biochemical events during the germination of *Streptomyces* spores. In *Biological, Biochemical, and Biomedical Aspects of Actinomycetes*, Ortiz-Ortiz, L., Bojalil, L. F. & Yakoleff, V. eds, pp. 179–95, Academic Press, Orlando.

Hirsch, C. F. & Ensign, J. C. (1976*a*). Nutritionally defined conditions for germination of *Streptomyces viridochromogenes* spores. *Journal of Bacteriology*, **126**, 13–23.

Hirsch, C. F. & Ensign, J. C. (1976*b*). Heat activation of *Streptomyces viridochromogenes* spores. *Journal of Bacteriology*, **126**, 24–30.

Hirsch, C. G. & Ensign, J. C. (1978). Some properties of *Streptomyces viridochromogenes* spores. *Journal of Bacteriology*, **134**, 1056–63.

Hodgson, D. A. (1980). Carbohydrate utilization in *Streptomyces coelicolor* A3(2). Ph D Thesis, University of East Anglia, Norwich.

Hodgson, D. A. (1989). Bacterial diversity: the range of interesting things that bacteria do. In *Genetics of Bacterial Diversity*, Hopwood, D. A. & Chater, K. F. eds, pp. 3–22, Academic Press, London.

Hopwood, D. A., Bibb, M. J., Chater, K. F., Keiser, T., Bruton, C. J., Keiser, H. M., Lydiate, D. J., Smith, C. P., Ward, J. M. & Schrempf, H. (1985). *Genetic Manipulation of Streptomyces: A Laboratory Manual*, The John Innes Foundation, Norwich.

Horinouchi, S., Kumada, Y. & Beppu, T. (1984). Unstable genetic determinant of A-factor biosynthesis in streptomycin-producing organisms: cloning and characterization. *Journal of Bacteriology*, **158**, 481–7.

Horinouchi, S., Nishiyama, M., Suzuki, H., Kumada, Y. & Beppu, T. (1985). The cloned *Streptomyces bikiniensis* A-factor determinant. *Journal of Antibiotics*, **38**, 636–41.

Horinouchi, S., Suzuki, H., Nishiyama, M. & Beppu, T. (1989). Nucleotide sequence and transcription analysis of the *Streptomyces griseus* gene (*afsA*) responsible for A-factor biosynthesis. *Journal of Bacteriology*, **171**, 1206–10.

Huber, F. M., Piper, R. L. & Mertz, F.P. (1987). Sporulation of *Streptomyces roseosporus* in submerged culture. *Journal of Industrial Microbiology*, **2**, 235–41.

Ikeda, H., Seno, E. T., Bruton, C. J. & Chater, K. F. (1984). Genetic mapping, cloning and physiology aspects of the glucose kinase gene of *Streptomyces coelicolor. Molecular and General Genetics*, **196**, 501-7.

Kaiser, D. (1989). Multicellular development in myxobacteria. In *Genetics of Bacterial Diversity*, Hopwood, D. A. & Chater, K. F., eds, pp. 243–63, Academic Press, London.

Kalakoutskii, L. V. & Agre, N. S. (1976). Comparative aspects of development and differentiation in actinomycetes. *Bacteriological Reviews*, **40**, 469–524.

Kendrick, H. E. & Ensign, J. C. (1983). Sporulation of *Streptomyces griseus* in submerged culture. *Journal of Bacteriology*, **155**, 357–66.

Kim, S. K. & Kaiser, D. (1990). C-factor: A cell-cell signalling protein required for fruiting body morphogenesis of *M. xanthus. Cell*, **61**, 19–26.

Kim, H. S., Nihira, T., Tada, H., Yanagimoto, M. & Yamada, Y. (1989). Identification of binding protein of virginiae butanolide C, an autoregulator in virginiamycin production, from *Streptomyces virginiae, Journal of Antibiotics*, **42**, 769–78.

Kim, H. S., Tada, H., Nihira, T. & Yamada, Y. (1990). Purification and characterization of virginiae butanolide C-binding protein, a possible pleiotropic signal-transducer in *Streptomyces virginiae. Journal of Antibiotics*, **43**, 692–706.

Koepsel, R. & Ensign, J. C. (1984). Microcycle sporulation of *Streptomyces viridochromogenes. Archives of Microbiology*, **140**, 9–14.

Kondo, S., Yasui, K., Katayama, M., Marumo, S., Kondo, T. & Hattori, H. (1987). Structure of pamamycin-607, an aerial mycelium-inducing substance of *Streptomyces alboniger. Tetrahedron Letters*, **28**, 5861–4.

Kondo, S., Yasui, K., Natsume, M., Katayama, M. & Marumo, S. (1988). Isolation, physico-chemical properties and biological activity of pamamycin-607, an aerial mycelium-inducing substance from *Streptomyces alboniger. Journal of Antibiotics*, **41**, 1196–204.

Kondo, K., Higuchi, Y., Sakuda, S., Nihira, T. & Yamada, Y. (1989). New virginiae butanolides from *Streptomyces virginiae. Journal of Antibiotics*, **42**, 1873–6.

Lawlor, E. J., Baylis, H. A. & Chater, K. F. (1987). Pleiotropic morphological and antibiotic deficiences result from mutations in a gene encoding a tRNA-like product in *Streptomyces coelicolor* A3(2). *Genes and Development*, **1**, 1305–10.

Locci, R. (1984). Actinomycetes as models of bacterial morphogenesis. In *Biological, Biochemical and Biomedical Aspects of Actinomycetes*, Ortiz-Ortiz, L., Bojalil, L. F. & Yakoleff, V. eds, pp. 395–408, Academic Press, Orlando.

Locci, R. (1988). Comparative morphology of actinomycetes in natural and artificial habitats. In *Biology of Actinomycetes '88*, Okami, Y., Beppu, T. & Ogawara, H. eds, pp. 482–9, Japan Scientific Societies Press, Tokyo.

Locci, R. (1989). Streptomycetes and related genera. In *Bergey's Manual of Systematic Bacteriology* volume 4, Williams, S. T., Sharpe, M. E. & Holt, J. G., eds, pp. 2451–8. Williams & Wilkins, Baltimore.

Losick, R., Kroos, L., Errington, J., & Youngman, P. (1989). Pathways of developmentally regulated gene expression in *Bacillus subtilis*. In *Genetics of Bacterial Diversity*, Hopwood, D. A. & Chater, K. F., eds, pp. 221–42, Academic Press, London.

Manzanal, M. B. & Hardisson, C. (1978). Early stages of arthrospore maturation in *Streptomyces. Journal of Bacteriology*, **133**, 293–7.

McBride, M. J. & Ensign, J. C. (1987*a*). Effects of intracellular trehalose content on *Streptomyces griseus* spores. *Journal of Bacteriology*, **169**, 4995–5001.

McBride, M. J. & Ensign, J. C. (1987*b*). Metabolism of endogenous trehalose by

Streptomyces griseus spores and by spores of other actinomycetes. *Journal of Bacteriology*, **169**, 5002–7.

Meade, H. (1985). Cloning of *argG* from *Streptomyces*: loss of gene in Arg mutants of *S. cattleya*. *Bio/Technology*, **3**, 917–18.

Mendez, C. & Chater, K. F. (1987). Cloning of *whiG*, a gene critical for sporulation of *Streptomyces coelicolor* A3(2). *Journal of Bacteriology*, **169**, 5715–20.

Mendez, C., Brana, A. F., Manzanal, M. B. & Hardisson, C. (1985). Role for substrate mycelium in colony development in *Streptomyces*. *Canadian Journal of Microbiology*, **31**, 446–50.

Merrick, M. J. (1976). A morphological and genetic mapping study of bald colony mutants of *Streptomyces coelicolor*. *Journal of General Microbiology*, **96**, 299–315.

Miguelez, E. M., Martin, M. C., Manzanal, M. B. & Hardisson, C. (1988). Hyphal growth in *Streptomyces*. In *Biology of Actinomycetes '88*, Okami, Y., Beppu, T. & Ogawara, H., eds, pp. 490–5, Japan Scientific Societies Press, Tokyo.

Miyake, K., Horinouchi, S., Yoshida, M., Chiba, N., Mori, K., Nogawa, N., Morikawa, N. & Beppu, T. (1989). Detection and properties of A-factor binding protein from *Streptomyces griseus*. *Journal of Bacteriology*, **171**, 4298–302.

Miyake, K., Kuzuyama, T., Horinouchi, S. & Beppu, T. (1990). The A-factor binding protein of *Streptomyces griseus* negatively controls streptomycin production and sporulation. *Journal of Bacteriology*, **172**, 3003–8.

Natsume, M., Yasui, K. & Marumo, S. (1989). Calcium ion regulates aerial mycelium formation in actinomycetes. *Journal of Antibiotics*, **42**, 440–7.

Ochi, K. (1986*a*). A decrease in GTP content is associated with aerial mycelium formation in *Streptomyces* MA406-A-1. *Journal of General Microbiology*, **132**, 299–305.

Ochi, K. (1986*b*). Occurrence of the stringent response in *Streptomyces* sp. and its significance for the initiation of morphological and physiological differentiation. *Journal of General Microbiology*, **132**, 2621–31.

Ochi, K. (1987*a*). Metabolic initiation of differentiation and secondary metabolism by *Streptomyces griseus*: significance of the stringent response (ppGpp) and GTP content in relation to A-factor. *Journal of Bacteriology*, **169**, 3608–16.

Ochi, K. (1987*b*). Changes in nucleotide pools during sporulation of *Streptomyces griseus* in submerged culture. *Journal of General Microbiology*, **133**, 2787–95.

Ochi, K. (1987*c*). A *rel* mutation abolishes the enzyme induction needed for actinomycin synthesis by *Streptomyces antibioticus*. *Agricultural and Biological Chemistry*, **51**, 829–35.

Ochi, K. (1990*a*). *Streptomyces relC* mutants with an altered ribosomal protein ST-L11 and genetic analysis of a *Streptomyces griseus relC* mutant. *Journal of Bacteriology*, **172**, 4008–16.

Ochi, K. (1990*b*). A relaxed (*rel*) mutant of *Streptomyces coelicolor* A3(2) with a missing ribosomal protein lacks the ability to accumulate ppGpp, A-factor and prodigiosin. *Journal of General Microbiology*, **136**, 2405–12.

Ochi, K., Tsurumi, Y., Shigematsu, N., Iwami, M., Umehara, K. & Okuhara, M. (1988). Physiological analysis of bicozamycin high-producing *Streptomyces griseoflavus* used at industrial level. *Journal of Antibiotics*, **41**, 1106–15.

Ohashi, H., Zheng, Y.-H., Nihira, T. & Yamada, Y. (1989). Distribution of virginiae butanolides in antibiotic-producing actinomycetes, and identification of the inducing factor from *Streptomyces antibioticus* as virginiae butanolide A. *Journal of Antibiotics*, **42**, 1191–5.

Poindexter, J. (1987). Bacterial responses to nutrient limitation. In *Ecology of Microbial Communities*, Fletcher, M., Gray, T. R. G. & Jones, J. G., eds, pp.

283–317, Society for General Microbiology Symposium 41, Cambridge University Press, Cambridge.

Prosser, J. I., Gray, D. I. & Gooday, G. W. (1988). Cellular mechanisms for growth and branch formation in streptomycetes. In *Biology of Actinomycetes '88*, Okami, Y., Beppu, T. & Ogawara, H., eds, pp. 316–20. Japan Scientific Societies Press, Tokyo.

Puglia, A. M. & Cappelletti, E. (1984). A bald ultrafertile U.V.-resistant strain in *Streptomyces coelicolor* A3(2). *Microbiologica*, **7**, 263–6.

Quiros, L. M., Hardisson, C. & Salas, J. A. (1985). Stable mRNA is a common feature of some *Streptomyces* spores. *FEMS Microbiology Letters*, **26**, 323–7.

Redshaw, P. A., McCann, P. A., Sankaran, L. & Pogell, B. M. (1976). Control of differentiation in streptomycetes: involvement of extrachromosomal deoxyribonucleic acid and glucose repression in aerial mycelium development. *Journal of Bacteriology*, **125**, 698–705.

Redshaw, P. A., McCann, P. A., Pentella, M. A. & Pogell, B. M. (1979). Simultaneous loss of multiple differentiated functions in aerial mycelium-negative isolates of streptomycetes. *Journal of Bacteriology*, **137**, 891–9.

Reichl, U., Yang, H., Gilles, E.-D. & Wolf, H. (1990). An improved method for measuring the interseptal spacing in hyphae of *Streptomyces tendae* by fluorescence microscopy coupled with image processing. *FEMS Microbiology Letters*, **67**, 207–10.

de los Reyes-Gavilan, C. G., Cal, S., Barbes, C., Hardisson, C. & Sanchez, J. (1991). Nutritional regulation of differentiation and synthesis of an exocytoplasmic deoxyriboendonuclease in *Streptomyces antibioticus. Journal of General Microbiology*, **136**, 299–305.

Rossner, E., Zeeck, A. & Konig, W. A. (1990). Elucidation of the structure of hormaomycin. *Angewandte Chemie, International Edition (English)*, **29**, 64–5.

Ruddick, S. M. & Williams, S. T. (1972). Studies on the ecology of actinomycetes in soil. V. Some factors influencing the dispersal and absorption of spores in soil. *Soil Biology and Biochemistry*, **4**, 93–103.

Salas, J. A. & Hardisson, C. (1981). Sugar uptake during germination of *Streptomyces antibioticus* spores. *Journal of General Microbiology*, **125**, 25–31.

Sato, K., Nihira, T., Sakuda, S., Yanagimoto, M. & Yamada, Y. (1989). Isolation and structure of a new butyrolactone autoregulator from *Streptomyces* sp. FRI-5. *Journal of Fermentation and Bioengineering*, **68**, 170–3.

Schauer, A., Ranes, M., Santamaria, R., Guijarro, J., Lawlor, E., Mendez, C., Chater, K. & Losick, R. (1988). Vizualizing gene expression in time and space in the filamentous bacterium *Streptomyces coelicolor. Science*, **240**, 768–72.

Scribner III, H. E., Tang, T. & Bradley, S. G. (1973). Production of a sporulation pigment by *Streptomyces venezuelae. Applied Microbiology*, **25**, 873–9.

Sharples, G. P. & Williams, S. T. (1976). Fine structure of spore germination in actinomycetes. *Journal of General Microbiology*, **96**, 323–32.

Smucker, R. A. (1984*a*). Biochemistry of the *Streptomyces* spore sheath. In *Biological, Biochemical, and Biomedical Aspects of Actinomycetes*, Ortiz-Ortiz, L., Bojalil, L. F. & Yakoleff, V., eds, pp. 171–7, Academic Press, Orlando.

Smucker, R. A. (1984*b*). *Streptomyces bambergiensis* spore envelope ultrastructure. In *Biological, Biochemical, and Biomedical Aspects of Actinomycetes*, Ortiz-Ortiz, L., Bojalil L. F. & Yakoleff, V., eds, pp. 409–21, Academic Press, Orlando.

Smucker, R. A. & Pfister, R. M. (1978). Characteristics of *Streptomyces coelicolor* A3(2) aerial spore rodlet mosaic. *Canadian Journal of Microbiology*, **24**, 397–408.

Smucker, R. A. & Simon, S. L. (1986). Some effects of diflubenzuron on growth and sporogenesis in *Streptomyces* spp. *Applied and Environmental Microbiology*, **51**, 25–31.

Stackebrandt, E. & Woese, C. R. (1981). The evolution of prokaryotes. In *Molecular and Cellular Aspects of Microbial Evolution*, Carlile, M. J., Collins, J. F. & Moseley, B. E. B. eds, pp. 1–31, Society for General Microbiology Symposium 32, Cambridge University Press, Cambridge.

Strauch, E., Takano, E., Baylis, H. A. & Bibb, M. J. (1991). The stringent response in *Streptomyces coelicolor* A3(2). *Molecular Microbiology*, **5**, 289–98.

Strunk, C. (1978). Sporogenesis in *Streptomyces melanochromogenes. Archives of Microbiology*, **118**, 309–316.

Surowitz, K. G. & Pfister, R. M. (1985). Glucose metabolism and pyruvate excretion by *Streptomyces alboniger. Canadian Journal of Microbiology*, **31**, 702–6.

Szabo, I. M., Kondics, L., Marton, M. & Buti, I. (1977). Changes in the cell surface layer (sheath) of the cell wall of streptomycetes during sporulation. *Acta Microbiologica (Academiae Scientiarum Hungaricae)*, **24**, 237–46.

Szabo, G., Biro, S., Tron, L., Valu, G. & Vitalis, S. (1984). Mode of action of factor C upon the differentiation process of *Streptomyces griseus*. In *Biological, Biochemical, and Biomedical Aspects of Actinomycetes*, Ortiz-Ortiz, L., Bojalil, L. F. & Yakoleff, V., eds, pp. !97–214, Academic Press, Orlando.

Szabo, G., Szeszak, F., Vitalis, S. & Toth, F. (1988). New data on the formation and mode of action of factor C. In *Biology of Actinomycetes '88*, Okami, Y. Beppu, T. & Ogawara, H., eds, pp. 324–9. Japan Scientific Societies Press, Tokyo.

Szeszak, F., Vitalis, S. & Szabo, G., (1989). Factor C a regulatory protein of *Streptomyces griseus* induces release of potassium from the mycelium. *Journal of Basic Microbiology*, **29**, 233–40.

Szeszak, F., Vitalis, S., Toth, F., Valu, G., Fachet, J. & Szabo, G., (1990). Detection and determination of factor C – a regulatory protein – in *Streptomyces* strains by antiserum and monoclonal antibody. *Archives of Microbiology*, **154**, 82–4.

Vargha, G., Karsai, T. & Szabo, G., (1983). A conditional aerial mycelium-negative mutant of *Streptomyces fradiae* with deficient ornithine carbamoyltransferase activity. *Journal of General Microbiology*, **129**, 539–42.

Vobis, G. (1984). Sporogenesis in the *Pilimelia* species. In *Biological, Biochemical, and Biomedical Aspects of Actinomycetes*, Ortiz-Ortiz, L., Bojalil, L. F. & Yakoleff, V., eds, pp. 423–39. Academic Press, Orlando.

Wildermuth, H. (1970). Development and organization of the aerial mycelium in *Streptomyces coelicolor. Journal of General Microbiology*, **60**, 43–50.

Willey, J., Santamaria, R., Guijarro, J., Geistlich, M. & Losick, R. (1991). Extracellular complementation of a developmental mutation implicates a small sporulation protein in aerial mycelium formation by *Streptomyces coelicolor. Cell*, **65**, 641–50.

Williams, S. T., Bradshaw, R. M., Costerton, J. W. & Forge, A. (1972). Fine structure of the spore sheath of some Streptomyces species. *Journal of General Microbiology*, **72**, 249–58.

Wolf, H. & Schoppmann, H. (1989). Streptomycetes can grow through small filters. *FEMS Microbiology Letters*, **57**, 259–64.